Proceedings of the 5th Ritsumeikan International Symposium

STOCHASTIC PROCESSES AND APPLICATIONS TO MATHEMATICAL FINANCE

Proceedings of the 5th Ritsumeikan International Symposium

STOCHASTIC PROCESSES AND APPLICATIONS TO MATHEMATICAL FINANCE

Ritsumeikan University, Japan 3 – 6 March 2005

Editors

Jiro Akahori

Shigeyoshi Ogawa

Shinzo Watanabe

Ritsumeikan University, Japan

World Scientific

NEW JERSEY • LONDON • SINGAPORE • BEIJING • SHANGHAI • HONG KONG • TAIPEI • CHENNAI

Published by

World Scientific Publishing Co. Pte. Ltd.

5 Toh Tuck Link, Singapore 596224

USA office: 27 Warren Street, Suite 401-402, Hackensack, NJ 07601

UK office: 57 Shelton Street, Covent Garden, London WC2H 9HE

British Library Cataloguing-in-Publication Data
A catalogue record for this book is available from the British Library.

STOCHASTIC PROCESSES AND APPLICATIONS TO MATHEMATICAL FINANCE
Proceedings of the 5th Ritsumeikan International Symposium

ISBN 981-256-519-1

Printed in Singapore by World Scientific Printers (S) Pte Ltd

PREFACE

The international colloquium on *Stochastic Processes and Applications to Mathematical Finance* was held at Biwako-Kusatsu Campus (BKC) of Ritsumeikan University, March 3–6, 2005. If counted from the first symposium on the same title held in the year 2001, this colloquium was the fifth one of that series of symposia.

The colloquium has been organized under the joint auspices of Research Center for Finance and Department of Mathematical Sciences of Ritsumeikan University, and financially supported by MEXT (Ministry of Education, Culture, Sports, Science and Technology) of Japan, the Research Organization of Social Sciences (BKC), Ritsumeikan University, and Department of Mathematical Sciences, Ritsumeikan University. The aim of this research project has been to hold assemblies of those interested in the applications of theory of stochastic processes and stochastic analysis to financial problems, in which several eminent specialists as well as active young researchers have been jointly invited to give their lectures.

In the organization of this colloquium, the committee chaired by Mr. Shigeyoshi Ogawa aimed to organize it as a Winter School especially for those younger researchers who intend to join in or just begin the research activity on the relevant subjects. For this reason we asked some of the invited speakers to give introductory talks composing of two or three unified lectures on the same themes (cf. the program cited below). As a whole we had about eighty participants with nine invited lecturers. The present volume is the proceedings of this colloquium based on those invited lectures.

We, members of the editorial committee of this proceedings listed below, would express our deep gratitude to those who contributed their works to this proceedings and to those who kindly helped us in refereeing them. We would express our cordial thanks to Professors Toshio Yamada and Keisuke Hara at the Department of Mathematical Sciences, of Ritsumeikan University, for their kind assistance in our editing this volume. We would thank also Mr. Satoshi Kanai for his works in editing TeX files and Ms. Chelsea Chin of World Scientific Publishing Co. for her kind and generous assistance in publishing this proceedings.

December, 2005, Ritsumeikan University (BKC)

Jiro Akahori
Shigeyoshi Ogawa
Shinzo Watanabe

PROGRAM

March, 3 (Thursday)

9:50–10:00 Opening Speech, by Shigeyoshi Ogawa (Ritsumeikan University)

10:00–10:50 Shinzo Watanabe (Ritsumeikan University, Kusatsu)
Martingale representation and chaos expansion I

11:10–11:50 Monique Jeanblanc (Université d'Evry, Val d'Essonne)
Hedging defaultable claims I (joint work with T. Bielecki and A. M. Rutkowski)

12:00–13:30 Lunch time

13:30–14:20 Paul Malliavin (Académie des Sciences, Paris)
Stochastic calculus of variations in mathematical finance I

14:30–15:20 Yoshio Miyahara (Nagoya City University)
Geometric Lévy process models in finance I

16:00–16:50 Arturo Kohatsu-Higa (Universitat Pompeu Fabra, Barcelona)
Insider modelling in financial market I

17:30– Welcome party

March, 4 (Friday)

10:00–10:50 Paul Malliavin (Académie des Sciences, Paris)
Stochastic calculus of variations in mathematical finance II

11:10–11:50 Shinzo Watanabe (Ritsumeikan University, Kusatsu)
Martingale representation and chaos expansion II

12:00–13:30 Lunch time

13:30–14:20 Monique Jeanblanc (Université d'Evry, Val d'Essonne)
Hedging defaultable claims II (joint work with T. Bielecki and A. M. Rutkowski)

14:30–15:20 Arturo Kohatsu-Higa (Universitat Pompeu Fabra, Barcelona)
Insider modelling in financial market II

16:00–16:50 Hideo Nagai (Osaka University)
A family of stopping problems of certain multiplicative functionals and utility maximization with transaction costs.

March, 5 (Saturday)

10:00–10:50 Monique Jeanblanc (Université d'Evry, Val d'Essonne)
Hedging defaultable claims III (joint work with T. Bielecki and A. M. Rutkowski)

11:10–11:50 Paul Malliavin (Académie des Sciences, Paris)
Stochastic calculus of variations in mathematical finance III

12:00–13:30 Lunch time

13:30–14:20 Shinzo Watanabe (Ritsumeikan University, Kusatsu)
Martingale representation and chaos expansion III

14:30–15:20 Makoto Yamazato (University of Ryukyus)
Lévy processes in mathematical finance

15:30–16:00 Break

16:00–16:50 Yoshio Miyahara (Nagoya City University)
Geometric Lévy process models in finance II

18:30– Reception (at Kusatsu Estopia Hotel)

March, 6 (Sunday)

10:00–10:50 Toshio Yamada (Ritsumeikan University, Kusatsu)
On stochastic differential equations driven by symmetric stable processes (joint work with H. Hashimoto and T. Tsuchiya)

11:00–12:20 Short Communications;
1. Romuald Elie (Centre de Recherche en Économie et Statistique, France)
Optimal Greek weights by kernel estimation
2. Kiyoshi Kawazu (Yamaguchi University)
The recurrence of product stochastic processes in random environment

CONTENTS

Harmonic Analysis Methods for Nonparametric Estimation of Volatility: Theory and Applications

Emilio Barucci[1], Paul Malliavin[2], and Maria Elvira Mancino[3]

[1] Dipartimento di Matematica, Politecnico di Milano, Italy
email: barucci@mate.polimi.it
[2]10 rue Saint Louis en l'Isle, 75004 Paris, France
email: sli@ccr.jussieu.fr
[3]DIMAD, Università di Firenze, Italy
email: mariaelvira.mancino@dmd.unifi.it

Key words: Volatility, Fourier analysis, time series, hedging.

1. Introduction

We have proposed in [41] a method to compute the volatility of a semimartingale based on Fourier series (*Fourier method*). The method allows to compute both the instantaneous volatility and the volatility in a time interval (*integrated* volatility). The method is well suited to employ high frequency data and therefore to compute volatility of financial time series. Since the method has been proposed, it has been extended/applied in several directions. This paper aims to review these contributions.

The benchmark to compute the volatility of a financial time series in a time interval with high frequency data (e.g. daily volatility) is provided by the cumulative squared intraday returns (*realized volatility*), see [2, 11]. In the limit, as the time interval between two consecutive observations converges to zero the realized volatility converges to the quadratic variation of the process and its derivative provides the instantaneous volatility of the process, e.g. the coefficient in front of the Brownian motion in case of a semimartingale. The major novelty of the Fourier method is that it allows to reconstruct the instantaneous volatility as a series expansion with coefficients gathered from the Fourier coefficients of the price variation, by using an integration procedure, instead of a differentiation one. The key point of Fourier approach is the realization of the volatility as a function of time; this fact makes possible to iterate the volatility functor and, for instance, to compute the volatility of the volatility function. As we will show below, this feature is useful in several applications: a double application of the volatility functor will lead to an effective computation of the *leverage effect*, which is a first order effect; by iterating three times the volatility functor we obtain the *feedback volatility rate* effect.

The contribution is organized as follows. Section 2 presents the Fourier methodology and the main theoretical results. We present the method to compute the volatility both in an univariate and in a multivariate setting and we show consistency of the estimator and a central limit theorem. In Section 3 we show how the method has been implemented to compute the volatility of financial time series and we compare its performance to the realized volatility. In Section 4 some applications of the Fourier methodology are presented in order to illustrate the potentiality of the method. In Section 5 we obtain an estimator of the instantaneous volatility using Laplace transform. In Section 6 we generalize the Fourier methodology to obtain a non-parametric estimation of the Heath-Jarrow-Morton generator for the interest rate curve and of the Lie bracket of the diffusion driving vectors.

A key fact is that all the infinitesimal generators of risk free measures have a drift which is completely determined by their second order terms: it vanishes for assets in Black-Scholes type model and for the interest rate curve the drift is fully determined by the HJM model. The second order terms can be computed in real time, and model free, by a volatility analysis made on a single trajectory of the market. By consequence pathwise volatility analysis gives access to a pathwise computation and model free of the infinitesimal generator of the risk free measure; by iterating this procedure the Greek Delta can be computed pathwise and model free; in the same spirit the hypoelliptic structure underlying the HJM infinitesimal generator can be computed pathwise and model free.

2. Fourier Methods for Volatility Computation

Let p be the asset price process, we will make the following assumption: $p(t)$ is a continuous Brownian semi-martingale satisfying the stochastic differential equation

$$\textbf{(H)} \qquad dp(t) = \sigma(t)\, dW(t) + b(t)\, dt$$

where W is a Brownian motion on a filtered probability space $(\Omega, (\mathcal{F}_t)_{t\in[0,T]}, P)$, σ and b are stochastic processes such that

$$E[\int_0^T \sigma^4(t)dt] < \infty \ , \ E[\int_0^T b^2(t)dt] < \infty.$$

We suppose that σ is adapted and b is not necessarily adapted.

The semi-martingale satisfying hypothesis **(H)** is the most familiar semi-martingale in econometrics and in finance. Note that this class includes stochastic volatility models.

The Fourier method for estimating the *instantaneous* (and *integrated) volatility* are based on an exact mathematical formula relating the Fourier transform of the price process $p(t)$ to the Fourier transform of the volatility process $\sigma^2(t)$. This identity is obtained in [43] and is proposed in Theorem 2.1.

Different methods which address the problem of estimating the instantaneous volatility have been proposed in [25, 20, 46]. They are based on the quadratic variation formula and use a double asymptotic in order to perform both the numerical derivative and the approximating procedure.

Before presenting the result, we recall some definitions from harmonic analysis theory (see for instance [38]). By change of the origin of time and rescaling the unit of time we can always reduce ourselves to the case where the time window $[0, T]$ becomes $[0, 2\pi]$. Then given a function ϕ on the circle S^1 its *Fourier transform* is defined on the group of integers $\mathbf{Z}$ by the formula

$$\mathcal{F}(\phi)(k) := \frac{1}{2\pi}\int_0^{2\pi} \phi(\vartheta)\, \exp(-ik\vartheta)\, d\vartheta, \quad \text{for } k \in \mathbf{Z}.$$

We define

$$\mathcal{F}(d\phi)(k) := \frac{1}{2\pi}\int_{]0,2\pi[} \exp(-ik\vartheta)\, d\phi(\vartheta)\,.$$

We emphasize that in the above formula we considered the open interval $]0, 2\pi[$, which means that the Dirac mass appearing in 2π by the jump $p(2\pi) - p(0)$ is not taken into account. Using integration by parts, we have

$$\mathcal{F}(\phi)(k) = \frac{i}{k}\,[\frac{1}{2\pi}(\phi(2\pi) - \phi(0)) - \mathcal{F}(d\phi)(k)].$$

Given two functions Φ, Ψ on the integers $\mathbf{Z}$, we say that the *Bohr convolution product* exists if the following limit exists for all integers k

$$(\Phi *_B \Psi)(k) := \lim_{N\to\infty} \frac{1}{2N+1} \sum_{s=-N}^{N} \Phi(s)\Psi(k-s). \tag{1}$$

The subindex B is the initial of Bohr, who developed similar ideas in the context of the theory of almost periodic functions.

Theorem 2.1. *Consider a process p satisfying assumption* **(H)**. *Then we have:*

$$\frac{1}{2\pi}\mathcal{F}(\sigma^2) = \Phi *_B \Phi, \text{ where } \Phi(k) = \mathcal{F}(dp)(k) \tag{2}$$

for any integer k. Convergence of the convolution product (2) is attained in probability.

The proof is based on the Itô energy identity for stochastic integrals (see for instance [39]), after having observed that the drift component $b(t)$ in the semimartingale $p(t)$ does not give contribution to the formula (2).

The reconstruction of the function $\sigma^2(t)$ from its Fourier coefficients computed by Theorem 2.1, can be obtained as follows. Define

$$\Phi_N(k) = \mathcal{F}(dp)(k) \text{ for } |k| \le 2N \text{ and } 0 \text{ otherwise}$$

$$\Psi_N(k) = \frac{1}{2N+1} \sum_{s=-\infty}^{+\infty} \Phi_N(s)\Phi_N(k-s).$$

Then the Fourier-Féjer summation gives the instantaneous volatility function

$$(3) \qquad \sigma^2(t) = \lim_{N\to\infty} \sum_{|k|\le N} (1 - \frac{|k|}{N})\Psi_N(k)\exp(ikt) \text{ for almost all } t \in (0, 2\pi).$$

Remark 2.2. We note that the approximating trigonometric functions appearing in the sequence of polynomials (3) are positive (see [43]).

It is possible to implement Theorem 2.1 in real terms, by expanding the Fourier transform of the volatility function as a linear combination of sines and cosines (see [41]). This procedure allows us to determine a formula for the volatility $\sigma^2(t)$ analogous to (3). The formalism is a little cumbersome and less conceptual than working with complex exponentials; nevertheless the numerical implementation of the real Fourier series is easier. Therefore we recall this procedure.

Suppose that the function $p(t)$ satisfies **(H)**. Define

$$a_k(p) = \frac{1}{\pi}\int_{]0,2\pi[} \cos(kt)p(t)dt, \quad b_k(p) = \frac{1}{\pi}\int_{]0,2\pi[} \sin(kt)p(t)dt, \quad k > 0,$$

then

$$a_0(dp) = 0, \quad a_k(dp) = kb_k(p) + \frac{1}{\pi}(p(2\pi) - p(0)), \quad b_k(dp) = -ka_k(p), \quad k > 0.$$

We define the prolongation to all integers k by parity for a_k and by imparity for b_k; more precisely let

$$(4)\ a^*_0 = b^*_0 = 0,\ a^*_k = \begin{cases} a_k(dp) & \text{for } k > 0 \\ a_{-k}(dp) & \text{for } k < 0 \end{cases} \text{ and } b^*_k = \begin{cases} b_k(dp) & \text{for } k > 0 \\ -b_{-k}(dp) & \text{for } k < 0. \end{cases}$$

The following Theorem provides the formula of the Fourier coefficients of $\sigma^2(t)$ in terms of a series of combination of the Fourier coefficients of dp, previously defined.

Theorem 2.3. *Consider a process p satisfying hypothesis* **(H)**. *Define, for $0 \le q \le 2N$,*

$$\alpha_q(N) = \frac{1}{2N+1} \sum_{s=-N}^{N-q} (a^*_{q+s} a^*_s + b^*_{q+s} b^*_s);$$

$$\beta_q(N) = \frac{1}{2N+1} \sum_{s=-N}^{N-q} (-a^*_{q+s} b^*_s + b^*_{q+s} a^*_s).$$

Denote by $a_q(\sigma^2)$, $b_q(\sigma^2)$ the Fourier coefficients of $\sigma^2(t)$. Then, for all fixed $q \ge 0$, the following convergence in probability holds:

$$\lim_{N\to+\infty} \alpha_q(N) = \frac{1}{\pi} a_q(\sigma^2), \quad \lim_{N\to+\infty} \beta_q(N) = \frac{1}{\pi} b_q(\sigma^2). \tag{5}$$

The above result provides an alternative formula to compute the so called *instantaneous volatility*. In fact the reconstruction of the function $\sigma^2(t)$ from its Fourier coefficients, derived by (5), can be obtained for instance by the classical Féjer inversion formula:

$$\sigma^2(t) = \lim_{N\to\infty} \sum_{k=0}^{N} (1 - \frac{k}{N})(a_k(\sigma^2)\cos(kt) + b_k\ (\sigma^2)\sin(kt)), \tag{6}$$

for almost all $t \in [0, 2\pi]$.

2.1 Integrated volatility

As a byproduct of formula (2), we can also compute the *integrated volatility* in the time interval $[0, 2\pi]$ through the following identity

$$2\pi\mathcal{F}(\sigma^2)(0) = \int_0^{2\pi} \sigma^2(t)dt.$$

Using (2) we have

$$\begin{aligned} \frac{1}{2\pi}\mathcal{F}(\sigma^2)(0) &= (\mathcal{F}(dp) * \mathcal{F}(dp))(0) \\ &= \lim_{N\to\infty} \frac{1}{2N+1} \sum_{s=-N}^{N} \mathcal{F}(dp)(s)\mathcal{F}(dp)(-s). \end{aligned} \tag{7}$$

This can be expressed using the Fourier coefficients, because (7) is equal to

$$(8)\qquad \lim_{N\to\infty}\frac{1}{2N+1}\sum_{s=-N}^{N}|\mathcal{F}(dp)(s)|^2=\lim_{N\to\infty}\frac{1}{2N+1}\sum_{s=-N}^{N}\frac{1}{4}((a_s^*)^2+(b_s^*)^2),$$

where a_s^* and b_s^* are defined in (4). Finally

$$(9)\qquad \int_0^{2\pi}\sigma^2(t)dt=\pi^2\lim_{N\to\infty}\frac{1}{2N+1}\sum_{s=-N}^{N}(a_s^*)^2+(b_s^*)^2.$$

Therefore (9) provides a measure of the integrated volatility, see [41].

2.2 Asymptotic analysis

The asymptotic properties of the Fourier estimator, i.e., consistency and central limit theorem, have been obtained in [43]. Apart from the statistical importance of these results, they are necessary to understand how the estimator behaves for very small time intervals, due to market frictions. In fact, the presence of market frictions makes this limiting argument not really accurate for very small time intervals. Such difficulties with limiting arguments are present in almost all area of econometrics and the development of a central limit theorem helps in understanding the behavior of the Fourier estimator for relatively small time intervals.

We consider the following discrete unevenly spaced sampling of the price process p. We fix a sequence $\mathcal{S}_n$ of finite subsets of $[0,2\pi]$, let $\mathcal{S}_n := \{0=t_{0,n}\le t_{1,n}\le\cdots\le t_{k_n,n}=2\pi\}$ for any $n\ge 1$ and denote

$$\rho(n):=\max_{0\le j\le k_n-1}|t_{j+1,n}-t_{j,n}|.$$

Remark that the Fourier estimation method allows to consider non-synchronous and random observations. Therefore we could choose the observations randomly in an independent way of W. Nevertheless by splitting the probability space, this kind of sampling reduces to a deterministic sampling. Therefore all our study will be made in the deterministic context.

We use the following interpolation formula

$$p_n(t):=\sum_{j=0}^{k_n-1}p(t_{j,n})1_{[t_{j,n},t_{j+1,n}[}(t),$$

where $1_{[t_{j,n},t_{j+1,n}[}$ denotes the indicator function of the interval $[t_{j,n},t_{j+1,n}[$. Then for any integer k, with $|k|\le 2N$ compute

$$c_k(dp_n)=\frac{1}{2\pi}\sum_{j=0}^{k_n-1}\exp(-ikt_{j,n})\delta_j(p)\quad\text{where }\delta_j(p):=p(t_{j+1,n})-p(t_{j,n}).$$

Consider now the estimator of the volatility function based on (3) defined by

$$\widehat{\sigma}^2_{n,N}(t) := \sum_{|k|\le N}(1-\frac{|k|}{N})\alpha_k(N,p_n)\exp(ikt) \tag{10}$$

where for any $|k| \le N$

$$\alpha_k(N,p_n) := \frac{2\pi}{2N+1}\sum_{s=-N}^{N} c_{k-s}(dp_n)c_s(dp_n). \tag{11}$$

The following theorem shows consistency in probability uniformly in t of the estimator (10).

Theorem 2.4. *Let $\rho(n) \to 0$ as $n \to \infty$, and consider the estimator of the volatility function defined by (10). Then the following convergence in probability holds:*

$$\lim_{n,N\to\infty}\sup_{t\in[0,2\pi]}|\widehat{\sigma}^2_{n,N}(t) - \sigma^2(t)| = 0.$$

We consider now the distribution of the error. The Central Limit Theorem is obtained as $\rho(n) \to 0$ and the interval $[0, 2\pi]$ remains fixed. For the existence of a limiting distribution it is necessary that the number of the Fourier coefficients N and the number of observations n increase with a suited order. Therefore let $\vartheta(n)$ be a function such that $\vartheta(n) \to \infty$ as $n \to \infty$. Consider the estimator

$$\widehat{\sigma}^2_{n}(t) := \sum_{|k|\le\vartheta(n)}(1-\frac{|k|}{\vartheta(n)})\alpha_k(\vartheta(n),p_n)\exp(ikt) \tag{12}$$

of $\sigma^2(t)$ as defined in (10), but in (12) we have highlighted dependence between the number of Fourier coefficients $\vartheta(n)$ and the number of observations n. Then:

Theorem 2.5. *Assume that as $n \to \infty$ then $\rho(n) \to 0$, $\vartheta(n) \to \infty$, $\rho(n)\vartheta(n) \to 0$ and $\rho(n)(\vartheta(n))^2 \to \infty$. Assume furthermore that*

$$\lim_{n\to\infty}\frac{\sup_j(t_{j+1,n}-t_{j,n})}{\inf_j(t_{j+1,n}-t_{j,n})} = 1. \tag{13}$$

Then, for any function $h \in L^2(0, 2\pi)$ satisfying the condition

$$\frac{1}{\rho(n)}\sum_{|k|>\vartheta(n)}|c_k(h)|^2 \to 0 \text{ as } n \to \infty, \tag{14}$$

the following result holds for $\widehat{\sigma_n^2}(t)$ defined in (12):

$$\frac{1}{\sqrt{\rho(n)}}\int_0^{2\pi} h(t)(\widehat{\sigma_n^2}(t)-\sigma^2(t))dt \tag{15}$$

converges in law to a mixture of Gaussian distributions with random variance equal to $2\int_0^{2\pi} h^2(t)\sigma^4(t)dt$.

Remark 2.6. Condition (13) is a measure of the regularity of the partition. A different condition called ε-balance, with $\varepsilon \in (0,1)$, is considered in [9]. Our condition implies ε-balance condition for any $\varepsilon \in (0,1)$. In [46] a general study of central limit result is done in the case where prices are recorded at irregular time intervals.

2.3 Cross-volatility computation

The Fourier method was proposed in [41] having in mind difficulties arising in the multivariate setting when applying the quadratic variation methodology because of non-synchronicity of prices observed for different assets.

Assume that $p(t) = (p^1(t), \ldots, p^n(t))$ are Brownian semi-martingales satisfying the following Itô stochastic differential equations

$$dp^j(t) = \sum_{i=1}^{d} \sigma_i^j(t)\, dW^i + b^j(t)\, dt \quad j = 1, \ldots, n \tag{16}$$

where $W = (W^1, \ldots, W^d)$ are independent Brownian motions, and σ_*^* and b^* are random processes satisfying hypothesis **H**.

There is a large literature in which cross-volatilities are estimated through the quadratic covariation formula. However this formula is not well suited to provide a good estimate of cross-volatilities. Difficulties arise from the absence of synchronous observations. The non-synchronicity trading problem has been studied for quite a long time in empirical finance, e.g. see [52, 37]. The bias caused by non-synchronicity and random sampling for the cross-correlations estimation has been recently highlighted in [29]. The Fourier methodology proposed in [41] is immune from these difficulties due to its own definition, being based on the integration of "all" data.

We recall now the Fourier method for computing multivariate volatilities. From the representation (16) we define the *volatility matrix*, which in our hypothesis *depends upon time*:

$$\Sigma^{j,k}(t) = \sum_{i=1}^{d} \sigma_i^j(t)\sigma_i^k(t). \tag{17}$$

The Fourier method reconstructs $\Sigma^{*,*}(t)$ on a fixed time window (which we can reduce always to $[0,2\pi]$ by change of origin and rescaling) using the Fourier coefficients of $dp^*(t)$. First we compute the Fourier coefficients of dp^j for $j = 1,\dots,n$ defined by

$$(18)\quad a_k(dp^j) = \frac{1}{\pi}\int_{]0,2\pi[} \cos(kt)dp^j(t),\ \ b_k(dp^j) = \frac{1}{\pi}\int_{]0,2\pi[} \sin(kt)dp^j(t).$$

We then consider the Fourier coefficients of the cross-volatilities

$$(19)\quad a_0(\Sigma^{i,j}) = \frac{1}{2\pi}\int_{]0,2\pi[} \Sigma^{i,j}(t)\,dt,\ \ a_k(\Sigma^{i,j}) = \frac{1}{\pi}\int_{]0,2\pi[} cos(kt)\Sigma^{i,j}(t)\,dt,$$

$$(20)\quad b_k(\Sigma^{i,j}) = \frac{1}{\pi}\int_{]0,2\pi[} \sin(kt)\Sigma^{i,j}(t)\,dt.$$

By a polarization argument in [41] it is proved the following Theorem.

Theorem 2.7. *Fix an integer $n_0 > 0$, the functions $\Sigma^{i,j}(t)$ of the volatility matrix have the following Fourier coefficients*

$$(21)\quad a_0(\Sigma^{i,j}) = \lim_{N\to\infty} \frac{\pi}{N+1-n_0}\sum_{s=n_0}^{N} \frac{1}{2}(a_s(dp^i)a_s(dp^j) + b_s(dp^i)b_s(dp^j)).$$

$$(22)\quad a_k(\Sigma^{i,j}) = \lim_{N\to\infty} \frac{\pi}{N+1-n_0}\sum_{s=n_0}^{N} (a_s(dp^i)a_{s+k}(dp^j) + a_s(dp^j)a_{s+k}(dp^i)).$$

$$(23)\quad b_k(\Sigma^{i,j}) = \lim_{N\to\infty} \frac{\pi}{N+1-n_0}\sum_{s=n_0}^{N} (a_s(dp^i)b_{s+k}(dp^j) + a_s(dp^j)b_{s+k}(dp^i)).$$

Finally using the Fourier-Féjer inversion formula we can reconstruct $\Sigma^{i,j}(t)$ from its Fourier coefficients:

$$(24)\quad \Sigma^{i,j}(t) = \lim_{N\to\infty} \Sigma^{i,j}_N(t)$$

where for any $t \in (0,2\pi)$

$$\Sigma^{i,j}_N(t) := \sum_{k=0}^{N}(1-\frac{k}{N})(a_k(\Sigma^{i,j})\cos(kt) + b_k(\Sigma^{i,j})\sin(kt)).$$

3. Time Series Analysis

This Section has two main goals: to show how the Fourier method can be implemented to compute both volatility and cross volatility; to compare the method with other methods proposed in the literature to compute the integrated volatility.

3.1 Volatility computation

Efficiency of the Fourier method to compute the integrated volatility of a stochastic process representing the asset price has been analyzed in several papers, see [15, 16, 33, 50, 48].

To implement the method, [15, 16] proceed as follows. Given a time series of N observations $(t_i, p(t_i))$, $i = 1, \dots N$, data are compressed in the interval $[0, 2\pi]$ and integrals are computed through *integration by parts*:

$$a_k(dp) = \frac{p(2\pi) - p(0)}{\pi} + k\, b_k(p). \tag{25}$$

To compute the integrals, we need an assumption on how data are connected. As a matter of fact, high frequency data are not equispaced and therefore there is no constant time length between two consecutive observations. To handle this problem, a time grid with a fixed time interval is chosen (e.g., 30 seconds, 1 minute, 5 minute): t_k, $k = 1, 2, \dots$. We may not have an observation on a point of the grid. To cope with this problem, two methods have been proposed in the high frequency data literature: *interpolation* and *imputation* of data. In the first case $p(t_i)$ and $p(t_{i+1})$ are connected through a straight line: if $t_k \in [t_i,\ t_{i+1}[$, then $p(t_k) = (t_k - t_i)\frac{p(t_{i-1}) - p(t_i)}{t_{i+1} - t_i}$. Instead according to the interpolation method $p(t_k) = p(t_i)$ (piecewise constant).

In [15, 16] the imputation method has been employed, then the integral in (25) in the interval $[t_i, t_{i+1}]$ becomes

$$\begin{aligned} \frac{k}{\pi}\int_{t_i}^{t_{i+1}} \sin(kt)p(t)dt &= p(t_i)\frac{k}{\pi}\int_{t_i}^{t_{i-1}} \sin(kt)dt \\ &= p(t_i)\frac{1}{\pi}\Big(\cos(kt_i) - \cos(kt_{i+1})\Big) \end{aligned} \tag{26}$$

thus avoiding the multiplication by k which amplifies cancellation errors when k becomes large.

The methodology has been applied to compute volatility in a standard GARCH setting. Let $p(t) = \log S(t)$, where $S(t)$ is a generic asset price, and $r_t = p(t) - p(t-1)$ is the logarithmic return. It is assumed that the asset price follows the continuous time GARCH(1,1) model proposed in [47]:

$$\begin{aligned} dp(t) &= \sigma(t)dW_1(t) \\ d\sigma^2(t) &= \theta(\omega - \sigma^2(t))dt + \sqrt{2\lambda\theta}\sigma^2(t)dW_2(t), \end{aligned} \tag{27}$$

where W_1, W_2 are independent Brownian motions. This model is closed under temporal aggregation in a weak sense, see [22], and its discrete time analogous is given by:

$$\begin{aligned} r_t &= \sigma_t\epsilon_t \\ \sigma_{t+1}^2 &= \psi + \alpha \cdot r_t^2 + \beta \cdot \sigma_t^2 \end{aligned} \tag{28}$$

where ϵ_t are i.i.d. Normal distributed random variables. The exact relation between (ψ, α, β) and $(\theta, \omega, \lambda)$ is derived in [22].

The task is to assess the capability of the Fourier method to reproduce the theoretical volatility of the GARCH(1,1) model. The analysis is based on Monte Carlo simulations. High frequency unevenly sampled observations have been generated as follows: one day of trading [0,1] has been simulated by discretizing (27) with a time step of one second, for a total of 86.400 observations a day. Then observation times have been extracted in such a way that time differences are drawn from an exponential distribution with mean equal to $\tau = 45$ seconds, which corresponds to the average value observed for many financial time series. As a result, we have a dataset $(t_k, p(t_k), k = 1, \ldots, N)$ with t_k unevenly sampled.

The most used way to compute the integrated volatility is to exploit the quadratic variation and therefore to compute $\int_0^1 \sigma^2(\tau)d\tau$ as the sum of squared intra-day returns (*realized volatility*), see [36] for the relation with the Fourier method. Provided a grid with m points $(\frac{1}{m}, \frac{2}{m}, \ldots 1)$, volatility in a day is computed as follows:

$$\hat{\sigma}^2(m) = \sum_{i=2}^{m} \left[p\left(\frac{i}{m}\right) - p\left(\frac{i-1}{m}\right) \right]^2 . \tag{29}$$

Theoretically, thanks to the Wiener theorem, by increasing the frequency of observations, an arbitrary precision in the estimate of the integrated volatility can be reached. Note that the Fourier method uses all observations, instead the sum of squared intraday returns uses only a fraction of observations, i.e., for low m some observations are lost.

In most of the papers estimating volatility with high frequency data, e.g. see [1], (29) is computed with $m = 288$ corresponding to five-minute returns. In the simulation setting it is also considered $m = 144$, corresponding to ten-minute returns, and $m = 720$ corresponding to two-minutes returns. In the literature the interpolation technique is employed.

The performance of the Fourier method is compared to that of (29) with $m = 144, 288, 720$ by the statistics:

$$\mu = \frac{\int_0^1 \sigma^2(s)ds - \hat{\sigma}^2}{\int_0^1 \sigma^2(s)ds}, \quad std = \left[\left(\frac{\int_0^1 \sigma^2(s)ds - \hat{\sigma}^2}{\int_0^1 \sigma^2(s)ds} \right)^2 \right]^{\frac{1}{2}},$$

where $\hat{\sigma}^2$ is the estimate and $\int_0^1 \sigma^2(s)ds$ is the integrated volatility generated in a simulation, whose value is known in the simulation setting.

We also evaluate the forecasting performance of the model (28), when ex-post volatility is measured by $\hat{\sigma}^2$. This is done by means of the R^2 of the linear regression

$$\hat{\sigma}^2_{t+1} = a + b \cdot \sigma^2_{t+1} + \varepsilon_t. \tag{30}$$

We recall that without manipulating the data, we should observe smaller μ and *std* when increasing the frequency. Figure 1 shows the results on simulated time series with $\alpha = 0.25, \beta = 0.7, \psi(1 - \alpha - \beta) = 1$.

First let us consider the realized volatility. The ten-minute estimator provides a downward biased estimate of the integrated volatility with a standard deviation larger than the bias. The five-minute estimator is also downward biased with a standard deviation of the same order of the bias in mean. Increasing further the frequency, the estimator is characterized by a smaller variance but a larger bias is observed. This effect can only be due to the interpolation scheme described above and therefore it can be linked to non-synchronous trading, see also [37]. The Fourier estimator is characterized by the smallest bias in mean and by a variance smaller than that of the 5-10 minutes estimate and slightly larger than that of the 2 minutes estimate. To check the robustness of these results, we repeated the Monte Carlo experiments on a grid of values $(\alpha, \beta, \psi = (1 - \alpha - \beta)^{-1})$ with 2 and 5 minutes returns. Results, reported in Table 1, can be summarized as follows: the estimator (29) turns out to be downward biased ($\mu > 0$), with a bias increasing in m, while the bias of the Fourier estimator is almost null. If m is chosen in such a way that the bias of (29) is less than its standard deviation, then the Fourier estimate provides a smaller standard deviation. Analyzing the forecasting capability of the discrete time GARCH model (28) we have that the Fourier estimate renders a better performance than the classical estimator computed with 2 and 5 minute returns.

The Fourier methods to compute volatility has been applied in several directions. Volatility is not constant over time; empirically it has been shown that days with high (low) volatility are followed by days with high (low) volatility, i.e., there is persistence in the volatility process. In the '90s, a large set of volatility models capturing a persistence component has been proposed in the financial time series literature. GARCH models provided the mile stone. The main problem with these models was that while their forecasting performance in sample was good, the forecasting performance out of sample using rough volatility proxies, e.g., square of the closing price minus the opening price, was very poor. [1] have shown that the problem was given by a poor ex post computation of volatility. Using high frequency data and the cumulative squared intraday returns to compute volatility they show that the forecasting performance of the standard GARCH(1,1) model in predicting the exchange rate volatility is quite good. [16] apply the Fourier method to compute volatility of the Deutsch Mark-Us dollar and of Yien-Us dollar exchange rate considering a one year of high frequency

Table 1 μ, std, R^2 (multiplied by 100) for the three estimators: (29) with $m = 720$ denoted by $2'$, (29) with $m = 188$ denoted by $5'$ and the Fourier estimator denoted by F, on a grid of values for (α, β) in (28), and $\psi \cdot (1 - \alpha - \beta) = 1$. All the values are computed with 10000 "daily" replications.

$\beta \downarrow$	$\alpha \rightarrow$	0.05			0.1			0.15			0.2			0.25		
		$2'$	$5'$	F	$2'$	$5'$	F	$2'$	$5'$	F	$2'$	$5'$	F	$2'$	$5'$	F
0.5	μ	23.9	10.0	0.1	23.9	10.0	0.1	23.9	10.0	0.1	23.9	10.0	0.1	23.9	10.0	0.1
	std	4.4	7.7	4.9	4.5	7.8	5.0	4.5	7.9	5.0	4.6	7.9	5.1	4.6	8.0	5.1
	R^2	4.92	4.79	4.98	9.58	9.44	9.70	14.0	13.9	14.2	18.6	18.5	18.8	23.8	23.8	24.0
0.6	μ	23.9	10.0	0.1	23.9	10.0	0.1	23.9	10.0	0.1	23.9	10.0	0.1	23.9	10.0	0.1
	std	4.4	7.7	4.9	4.5	7.8	4.9	4.5	7.8	5.0	4.5	7.8	5.0	4.5	7.8	5.0
	R^2	6.79	6.62	6.88	13.8	13.7	14.0	21.9	21.8	22.1	32.5	32.5	33.0	44.6	44.5	45.6
0.7	μ	23.9	10.0	0.1	23.9	10.0	0.1	23.9	10.0	0.1	23.9	10.0	0.1	23.9	10.0	0.1
	std	4.4	7.7	4.9	4.4	7.7	4.9	4.4	7.7	4.9	4.5	7.8	4.9	4.5	7.8	4.9
	R^2	9.67	9.43	9.79	21.5	21.3	21.7	38.2	37.9	38.7	51.3	51.1	52.2	48.2	48.1	48.8
0.8	μ	23.9	10.0	0.1	23.9	10.0	0.1	23.9	10.0	0.1						
	std	4.4	7.6	4.9	4.4	7.7	4.9	4.4	7.7	4.9						
	R^2	15.2	14.7	15.3	36.9	36.5	37.2	54.3	53.8	54.5						
0.9	μ	23.9	10.0	0.1												
	std	4.4	7.6	4.8												
	R^2	30.8	29.9	30.9												

Table 2 R^2 for the two time series.

Estimator	R^2	
	DM-\$	Y-\$
$\sum_{i=2}^{288}(p(\frac{i}{288}) - p(\frac{i-1}{288}))^2$	0.400	0.128
Fourier	0.470	0.143

observations (from October, 1st 1992 to September 30th 1993), i.e., the dataset analyzed in [1]. The dataset consists of 1.466.944 quotes for the Deutsch Mark-Us dollar and 567.758 quotes for the Yien-Us dollar exchange rate.

Performance of the GARCH(1,1) model has been evaluated when the integrated volatility is computed according to the Fourier method. The parameters of the model are those estimated in [1]. Table 2 provides the corresponding R^2.

We observe that the GARCH model performs well in forecasting when the Fourier method is employed to compute the integrated volatility. Its performance is better than that associated with the sum of squared intraday returns as an integrated volatility measure. On performance of volatility models and volatility computation see also [31].

As the Fourier method provides an accurate estimate of the volatility, we can handle the volatility as an observable process. In the literature, time varying volatility has been handled as a latent process, e.g., the GARCH process. In [13], this idea has been applied to compute the volatility of the overnight interest rate in the Italian money market and to test for the martingale hypothesis correcting for heteroschedasticity with a good proxy of the volatility and thus avoiding estimation problems connected with the use of a GARCH model. In [17] an autoregressive process for the volatility estimated according to the Fourier method is estimated to forecast one-day return volatility and to define the *value at risk* threshold. The method performs well in forecasting, better than the GARCH(1,1) model and the exponential smoothing proposed by RiskMetrics.

In recent years other volatility models have been proposed to cope with empirical regularities observed for volatility of financial time series. Among the models, we have models with long memory in the volatility process considering a fractional Brownian motion W_2 of order d independent of W_1 in (27), i.e., $W_2(t) = \int_0^t \frac{(t-s)^2}{\Gamma(d+1)} dW(s)$. To capture sharp increase in volatility, jumps in the volatility process have been introduced adding a Poisson process in (27). In these more general models, the sum of squared intraday returns and the Fourier method do not provide consistent estimators of the volatility. [48] compare through Monte Carlo simulations the bias and the root mean square error of the Fourier method, of the cumulative squared intraday returns and of a wavelet estimator. They show that the Fourier method provides the lowest bias and root mean squared error. They also compare the three methods when a bid-ask bounce effect is inserted, i.e., as a random buy/sell order arrives in the market there is a liquidity effect on the price creating spurious serial correlation in returns and volatility. In this case the realized volatility and the Fourier method are no longer consistent for the integrated volatility. As far as the bias and the root mean square error is concerned, the Fourier method performs better than the wavelet and the cumulative squared intraday returns method. On the comparison between Fourier method and the wavelet method see also [33].

3.2 Computation of cross volatility

In this section we analyze the performance of the method in the bivariate case using Monte Carlo simulations of high frequency asset prices as studied in [44].

As in [51], we simulate two correlated asset price diffusions with the

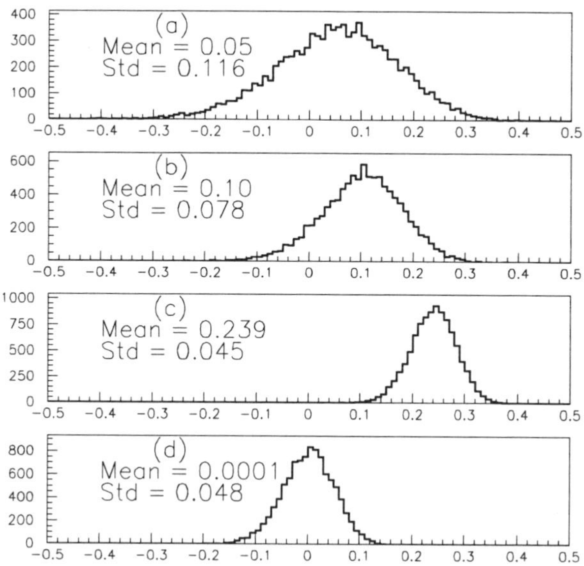

Figure 1. Distribution of $\frac{\int_0^1 \sigma^2(t)dt - \hat{\sigma}^2}{\int_0^1 \sigma^2(t)dt}$, where $\hat{\sigma}^2$ are three different estimators of the integrated volatility: (a) estimator (29) with $m = 144$; (b) estimator (29) with $m = 288$; (c) estimator (29) with $m = 720$; (d) Fourier estimator. The distribution is computed with 10,000 "daily" replications.

bi-variate continuous GARCH(1,1) model:

$$
\begin{aligned}
dp_1(t) &= \sigma_1(t)dW_1(t), \\
dp_2(t) &= \sigma_2(t)dW_2(t), \\
d\sigma_1^2(t) &= \lambda_1[\omega_1 - \sigma_1^2(t)]dt + \sqrt{2\lambda_1\theta_1}\sigma_1^2(t)dW_3(t), \\
d\sigma_2^2(t) &= \lambda_2[\omega_2 - \sigma_1^2(t)]dt + \sqrt{2\lambda_2\theta_2}\sigma_1^2(t)dW_4(t), \\
corr(dW_1, dW_2) &= \rho,
\end{aligned} \tag{31}
$$

and all the other correlations between the Brownian motions set to zero. The choice of this particular model comes from the fact that it is the continuous time limit of the very popular GARCH(1,1) model, and it has been studied extensively in the literature, e.g. see [28].

We will use the parameter values estimated by [1] on foreign exchange rates, i.e. $\theta_1 = 0.035, \omega_1 = 0.636, \lambda_1 = 0.296, \theta_2 = 0.054, \omega_2 = 0.476, \lambda_2 = 0.480$. We will instead analyze two mirror cases for the correlation coefficient: $\rho = 0.35$ and $\rho = -0.35$.

To get a representation of high-frequency tick-by-tick data, after discretizing (31) by a first-order Euler discretization scheme with a time step

Table 3 Average correlation measurement on 10,000 Monte Carlo replications of the model (31). Two generated values of the correlation are considered, $\rho = 0.35$ and $\rho = -0.35$. We compute the variance-covariance matrix via the Fourier estimator and via the realized volatility estimator (32). L.I. means Linear Interpolation, while P.T. means Previous Tick interpolation. Standard deviations of in-sample measurements are reported in the columns named Std.

Estimator	Generated correlation $\rho = 0.35$		Generated correlation ρ =-0.35	
	Measured	Std	Measured	Std
Fourier	0.350	0.039	−0.349	0.039
Realized 5′, L.I.	0.204	0.058	−0.203	0.055
Realized 5′, P.T.	0.181	0.060	−0.180	0.058
Realized 15′, L.I.	0.338	0.090	−0.337	0.090
Realized 15′, P.T.	0.329	0.091	−0.328	0.092
Realized 30′, L.I.	0.345	0.127	−0.344	0.126
Realized 30′, P.T.	0.342	0.127	−0.341	0.126

of one second, we extract observation times drawing the durations from an exponential distribution with mean 30 seconds and 60 seconds respectively. Observation times are drawn independently for the two time series. After simulating the process (31) we compute on it daily (86400 seconds, corresponding to 24 hours of trading, as for currencies) variance-covariance matrix according to the Fourier theory and according to the realized volatility measure of [3], given by:

$$\sigma_{ij}^2 = \sum_{j=1}^{m-1} \left[p_i\left(t + \frac{k+1}{m}\right) - p_i\left(t + \frac{k}{m}\right) \right] \cdot \left[p_j\left(t + \frac{k+1}{m}\right) - p_j\left(t + \frac{k}{m}\right) \right]. \tag{32}$$

The choice of m in (32) comes from a tradeoff between increasing precision and cutting out microstructure distortions. A typical choice is $m = 288$, corresponding to five minute returns. As pointed out above, to obtain an equally spaced time series we can rely upon a linear interpolation (L.I.) or a previous tick interpolation (P.T.). Both methods have been applied for $m = 288, 96, 48$ (corresponding respectively to 5,15,30 minute returns), when measuring correlations on Monte Carlo experiments. Table 3 shows the results. First of all we notice that the Fourier estimator performs considerably better than the realized volatility, which is biased toward zero. The bias in the correlation measurement of realized volatility is more and more severe when the sampling frequency is increased. For the five-minute estimator with the previous-tick interpolation, we get a mean value of 0.181 (−0.180), which is quite far from the true value of 0.35 (−0.35); this bias is completely due to the non synchronicity of quotes.

Realized volatility with linearly interpolated returns is closer to the right value, but this is because of the downward bias in the volatility measurement due to the linear interpolation documented in [15, 16]. In these papers, it is shown that the spurious positive serial correlation induced by the linear interpolation technique lowers volatility estimates. Since variances are spuriously measured to be lower, correlations turn out to be spuriously higher, thus compensating in some way the bias due to non-synchronicity. This is also true, but to a much lower extent, for the 15-minute and 30-minute realized volatility estimator.

The precision of the Fourier estimator, as measured by the standard deviation of measurement errors across Monte Carlo replications, is always better than the realized volatility estimator. We implemented the Fourier estimator with $N = 500$ coefficients for the first (30 seconds average spacing) time series, $N = 160$ coefficients for the second (60 seconds average spacing), and $N = 160$ coefficients for the computation of covariance. Increasing the number of coefficients would increase the variance measurement precision but not the covariance measurement precision, because of the Epps effect, see [51]. On the other hand, even the gain in precision of the realized covariance measurement obtained when increasing the sampling frequently is cancelled out by the bias.

4. Applications

The potentiality of our method relies in the fact that it reconstructs instantaneous volatility and cross-volatility as functions of time. This feature of the Fourier method is essential when a stochastic derivation of volatility along time evolution is performed as in contingent claim pricing-hedging. In this Section we present some developments of the multivariate Fourier estimation procedure which illustrate this point.

4.1 Leverage effect

There is a vast empirical literature showing that the asset price and its volatility are related. Among other phenomena, it has been observed the so called leverage effect, i.e., a negative return sequence is associated with a volatility increase. In some recent papers, see [12, 10], the integrated volatility estimation methodology based on the so called realized volatility has been extended to allow for a leverage effect. From the mathematical point of view the no leverage hypothesis means that $\sigma(t)$ is independent from the Brownian motion W. The no leverage hypothesis simplifies the study of the properties of the volatility estimator, but is not realistic for the analysis of equity returns.

The asset price model **H** is very general and includes stochastic volatility as well as level dependent volatility models, i.e. volatility is a time independent function of the asset price, it allows for feedback effects of asset price on volatility and in particular the leverage effect. In this section it is proved that these effects can be non-parametrically estimated using the Fourier methodology without knowledge of the exact form of the evolution equation for the volatility process. A first order indicator of the stochastic dependence between the asset price process and the volatility process is obtained and we refer to it as *leverage effect*, we also consider a second order indicator which is called *feedback effect rate*.

Consider the asset price model which satisfies hypothesis **H**. Moreover assume that the process $\sigma(t)$ is a random process satisfying

$$(\widetilde{\mathbf{H}}) \qquad d\sigma(t) = \alpha(t)dW(t) + \beta(t)dW_1(t) + \gamma(t)dt$$

where α, β and γ are predictable functions (for simplicity suppose they are almost sure bounded) and W_1 is another Brownian motion independent of W. Then the *leverage effect* is defined as

$$B(t) := \; dp(t) \star d\sigma^2(t),$$

where $\star$ denotes the Itô stochastic contraction divided by dt. Therefore

$$B(t) = 2\alpha(t)\sigma^2(t).$$

Nevertheless we have no knowledge of the random function $\alpha(t)$. Therefore it is interesting to find a formula for estimating $B(t)$ starting from the price observations.

By Itô calculus the following result holds:

Theorem 4.1. *Let p and σ satisfy* **H** *and* $\widetilde{\mathbf{H}}$, *then we have*

$$B(t) = \frac{1}{4}(Vol(p + \sigma^2)(t) - Vol(p - \sigma^2)(t)),$$

where $Vol(p \pm \sigma^2)$ means the volatility function of the stochastic process $p \pm \sigma^2$.

In order to estimate $B(t)$ from the asset prices we see that the Fourier transform of p can be computed from price observations; the Fourier transform of σ^2 is computed by Theorem 2.1; therefore the Fourier transform of $p + \sigma^2$ and $p - \sigma^2$ are known; a second application of Theorem 2.1 gives the Fourier transform of the volatility $Vol(p \pm \sigma^2)$ and finally the usual inversion formula produces the computation of $B(t)$.

4.2 Delta hedging (estimation of the gradient)

Suppose that the price process $p(t)$ in logarithmic scale satisfies

$$dp(t) = \sigma(p(t))dW(t) - \frac{1}{2}\sigma^2(p(t))dt \tag{33}$$

where W is a Brownian motion and σ is an unknown C^1-function . No hypothesis on the shape of this function are done.

In the following it is shown that is possible to estimate the pathwise gradient using Fourier transform methodology, therefore also the Greek Delta can be estimated non-parametrically. In financial applications the Greek Delta is the sensitive which allows to perform the so called Delta-hedging, that is to make the portfolio neutral with respect to small modifications of the initial value of the price. In model (33), the Greek Delta is defined as

$$\Delta(t) = \frac{\partial}{\partial \varepsilon} p^{\varepsilon}(t)_{|\varepsilon=0}$$

where $p^{\varepsilon}(t)$ corresponds to the solution of (36) with $p^{\varepsilon}(t_0) = p(t_0) + \varepsilon$. Then

$$d\Delta(t) = \Delta(t)[\sigma'(p(t))\, dW(t) - \sigma(p(t))\sigma'(p(t))dt]. \tag{34}$$

It follows that the computation of $\Delta(t)$ depends on the estimation of the volatility $\sigma(p(t))$ and the derivative of the volatility $\sigma'(p(t))$. Denote by $B(t)$ the function obtained in Theorem 4.1, then it holds

$$\sigma'(p(t)) = \frac{B(t)}{\sigma^2(p(t))}. \tag{35}$$

This first result expresses the derivative of the volatility function as the ratio of terms which can be estimated using prices data. By substitution of (35) in (34) and using (33), it follows

$$d\Delta(t) = \Delta(t)[\, \frac{1}{\sigma(p(t))}(\frac{B(t)}{\sigma^2(p(t))}dp(t) + (\frac{\sigma^2(p(t))}{2} - B(t))dt)\,].$$

Note that all terms in the above equation are given or estimated by Fourier method from market asset prices. A more general Delta hedging result has been obtained in [14] by using the Fourier method.

4.3 Feedback volatility rate

In [14] we have constructed a second order indicator of the feedback effects of asset price on volatility and we have used the Fourier estimation method in order to compute this effect from asset price data. Working under this general assumption, we produced a time-dependent price-volatility *feedback effect rate function* and in the multivariate case a time-dependent matrix, which will be called *elasticity matrix,* implementable in real time from asset price observations.

The mathematical theory suggests that eigenfunctions associated with positive eigenvalues of the elasticity matrix correspond to instability directions of the market; eigenfunctions associated with negative eigenvalues correspond to stability directions of the market.

The key mathematical tool to construct the elasticity matrix is a new methodology of transferring price perturbations through time, using the *inertial frame transport*; this transport is designed so that through time the variation is governed by a first order linear *ordinary differential equation*. The rate of variation through time of the initial perturbation is given by the elasticity matrix. Computation of the elasticity matrix is a three-steps outcome; at each steps it is necessary to compute volatilities of observed quantities or computed in the previous step. Therefore the Fourier methodology works out. We can write explicit expressions in terms of Fourier coefficients of asset prices leading to real-time determination of feedback rates.

For simplicity we present the univariate case, but the method can be developed for any finite number of assets (see [14]). Let $p(t)$ be the risky asset price at time t. Suppose it satisfies the stochastic differential equation in logarithmic scale:

$$dp(t) = \sigma(p(t))\, dW(t) - \frac{1}{2}\sigma^2(p(t))dt \tag{36}$$

where W is a Brownian motion, σ is a fixed but unknown $C^2(\mathbf{R})$ function. Let $\zeta(t)$ be the variation process which is solution of the linearized stochastic differential equation

$$d\zeta(t) = \zeta(t)\, [\sigma'(p(t))\, dW(t) - \sigma'(p(t))\sigma(p(t))dt].$$

We associate to $\zeta(t)$ the *rescaled variation* defined as

$$z(t) = \frac{\zeta(t)}{\sigma(p(t))}.$$

In Section 4.2 we have obtained the SDE driving the Delta propagation. It is a remarkable fact that this SDE can be reduced to an ODE at the price of a renormalization of the Greek Delta, as it is showed in the following result.

Theorem 4.2. *The rescaled variation is a differentiable function with respect to t; denote by $\lambda(t)$ its logarithmic derivative, then*

$$z(t) = \exp(\int_s^t \lambda(\tau)\, d\tau)\, z(s),$$

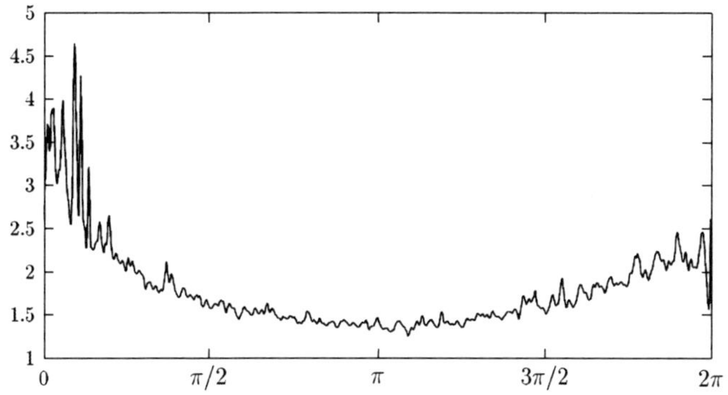

Figure 2. Average estimate of daily volatility $A(t)$ on IBM data. On the x axis, the time window $[0, 2\pi]$ corresponds to one trading day (6.5 hours). $A(t)$ displays the typical U-shape of volatility in stock markets.

where

$$\lambda(t) = -\frac{1}{2}(\sigma(p(t))\sigma'(p(t)) + \sigma(p(t))\sigma''(p(t))).$$

Definition 4.3. We will call $\lambda(t)$ the price-volatility *feedback rate function*.

Note that in the standard Black-Scholes framework $\lambda = 0$.

We derive a nonparametric estimation of the feedback volatility rate by using Fourier methodology. We suppose that we do not know the explicit expression of the function σ; we want to obtain a non parametric estimation in real time of $\lambda(t)$ from the observation of a single market evolution.

Theorem 4.4. *Denoting by $\star$ the Itô contraction divided by dt, and define the following cross-volatilities:*

$$dp \star dp =: A(t), \;\; dA \star dp =: B(t), \;\; dB \star dp =: C(t).$$

Then the price-volatility feedback effect rate function λ can be expressed as

$$\lambda(t) = \frac{3}{8}\frac{B^2}{A^3} - \frac{1}{4}\frac{B}{A} - \frac{1}{2}\frac{C}{A^2}.$$

We stress the point that all terms A, B, C can be obtained from the asset price data through the Fourier cross-volatility estimation. Therefore the feedback rate $\lambda(t)$ can be empirically estimated. The theory suggests that the sign of λ is associated with stability of the market: a negative λ would witness a period of stability, a positive λ would signal instability.

Precise estimation of quadratic and higher order variation asks for huge quantities of data; high frequency data are a natural candidate for this purpose. In [14] we have used two data sets: a data set containing quotes of the DEM-USD foreign exchange rates from October 1992 to September 1993 and a data set of IBM quotes from January to December 1999.

In the case of the IBM stock price, estimates of $A(t), B(t), C(t)$ have been computed for a one day time window (6.5 hours); the larger number of Fourier coefficients produces a higher resolution of the plots. Figure 2 gives the volatility $A(t)$ averaged over the full year; here we recognize the U-shape pattern which is typical of stock market intra-day volatility.

Short-horizon estimates of λ are the most important for traders. We present in Figure 3 as typical sample of daily (non-averaged) estimates the values for $A(t), B(t), C(t)$ and $\lambda(t)$ for two days in 1999.

It is noteworthy that taking the logarithm of the stock price mainly changes the scales of $A(t), B(t)$ and $C(t)$, but lets the shape of the curves more or less invariant. For this reason, in Figures 2, 3 the scales have been chosen according to the stock price (without taking logarithms).

On January 4th, 1999, the beginning of the trading day reveals positivity of λ which detects instability of the market and which is revealed by subsequent large picks of A. Over the whole day the positive values of λ dominate and indicate an unstable trading day. On April 9th, 1999, the beginning of the day reveals negativity of λ, which indicates stability of the market and which is revealed by a subsequent progressive damping of A. In contrast to January 4th, over the whole trading day, small and mainly negative values of λ dominate and indicate stability.

As argued above, precise estimation of λ could, in principle, result in important consequences for trading strategies. In this context, it can be of some help to analyze the sign of λ week-by-week or even day-by-day; our results show that, by using high frequency data, an estimate of this effects can be readily accomplished.

4.4 Dynamic principal component analysis

The Fourier method of cross-correlation estimation has been applied in [44] to develop a *dynamic principal component analysis*. The theory differs from the statistical principal component analysis because we do not assume any linear functional dependence; moreover we describe a dynamic situation, while usually principal component analysis describes a static situation. Classical PCA provides a linear sub-manifold of smaller dimension which carries the essential information coming from the data. Dynamic PCA will produce an abstract curve, which we call the *Core*, which allows us to determine the eigenvectors of multivariate volatility in continuous time. The analysis is model free.

Making mild assumptions on the multivariate prices behavior (Brownian semi-martingale hypothesis), we get by the Fourier method multivariate volatilities as a time dependent quadratic form, let $\Sigma(t)$. We suggest a

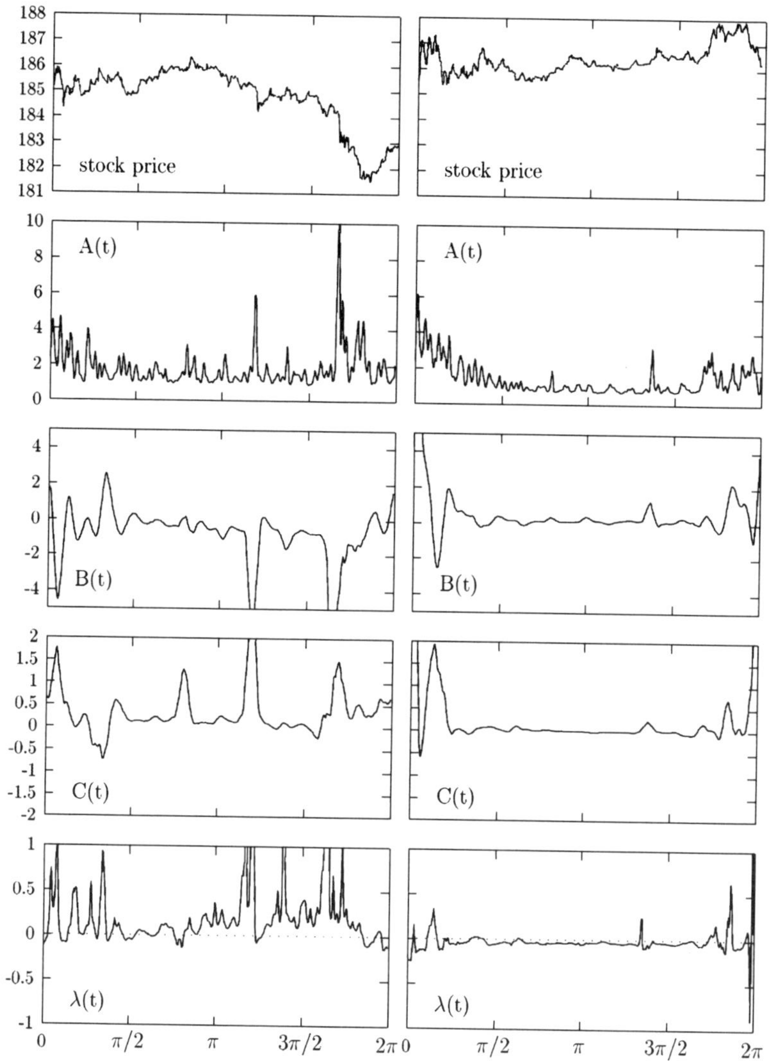

Figure 3. Daily values of $A(t), B(t), C(t), \lambda(t)$ on IBM data. The time windows $[0, 2\pi]$ correspond to two typical trading days in 1999 (6.5 hours each). Jan 4 (left-hand side) displays positivity of λ with large picks of A (instable market); April 9 (right-hand side) displays small and mainly negative values of λ along with a progressive damping of A (stable market). During the first two hours of trading, A has about the same shape on both days, but λ develops dramatically different shapes; computing λ in real time could give an indicator forecasting market instability.

method to construct an abstract curve which contains the essential information coming from multivariate volatilities. The analysis of the evolution in time of this curve can be used to decipher the stability of the market about an asset, or the degree of market integration of a given asset. We then introduce the concept of *reference assets*, via a geometric definition: a reference asset is one whose volatility is mainly due to the market volatility instead of its idiosyncratic noise.

We omit mathematical details of the construction, while we illustrate the ideas via the analysis of two months, April and May 2001, of high-frequency data for 98 stocks, selected into the *S&P* 100 index, among the most liquid ones. The results point out the need of a time-dependent analysis versus a static one, since the variance-covariance eigenvalues structure turns out to be deeply time-varying. In particular, there are some days in the market in which correlations are widely distributed, thus less factors are needed to explain the variance-covariance structure.

On the whole, we analyze 42 days; for each day, we compute the variance-covariance matrix using the Fourier method. We start by performing PCA for each day, after normalizing the variance-covariance matrix in order to have the variance of every stock equal to 1. The first factor explains, on average, 25.79% of the movements. However in some days the first factor's weight can be as large as 56.09% (April, 18th) and 73.26% (May, 16th). Moreover, and more interestingly for any financial application, this phenomenon seems to present some degree of persistence.

In the second step of our analysis, we define the Core of the market as the vector subspace spanned by the first 30 eigenvectors, and we divide our 42 market days into 6 periods of 7 market days each. In each subperiod, we perform principal component analysis on the aggregate variance-covariance matrix, and we obtain the coordinates of the Core. In each subperiod, we define 15 reference assets as those who have the largest projection onto the Core. We interpret these assets as those who are more correlated with the market itself, or alternatively as the basic constituents of the market. The list of the reference assets in any subperiod is shown in Table 4. Given the low value of the average correlation, we expect that the basket of reference assets is quite variable, given the unalienable noise in the correlation measurements. This is indeed the case. The month of April shows more persistence: 4 reference assets in the first period, out of 15, are in the second period too; and 6 of the second are still reference assets in the third. The month of May shows much more variability, or less "market integration"; only two stocks are reference assets in the third and fourth period, only one in the fourth and fifth period and none out of 15 in the fifth and sixth period. Looking at individual stocks, AES Corporation is a reference asset in the whole sample, with the exception of the sixth period,

Table 4 Lists the reference assets by ticker name in the six subperiods considered. They are ranked according to their projection on the core, which is reported in brackets. Also the five assets with the smallest projection on the Core are listed. Each subperiod is composed of 7 market days. In bold face, we indicate those stocks who remain reference assets in the subsequent period.

Periods					
1	2	3	4	5	6
F (74.4%)	**AES** (93.3%)	**AES** (84.0%)	WY (72.2%)	AES (88.9%)	HON (86.4%)
CCU (72.2%)	AOL (86.1%)	**WFC** (77.1%)	TOY (72.1%)	HIG (78.7%)	GM (74.7%)
AOL (71.1%)	INTC(84.1%)	ATI (73.9%)	**AES** (72.0%)	HAL (76.7%)	DOW (72.1%)
INTC(70.0%)	AA (82.0%)	LU (73.4%)	WMB (68.2%)	WMT (71.8%)	IBM (70.4%)
AES (69.5%)	JPM (80.0%)	AEP (72.5%)	CSC (67.7%)	HWP (71.7%)	GE (70.4%)
PHA (68.8%)	IBM (79.9%)	AVP (72.2%)	XOM (67.2%)	TYC (70.9%)	LTD (70.4%)
IP (64.6%)	**JNJ** (78.5%)	TXN (70.3%)	MSFT(66.3%)	ONE (68.8%)	BUD (69.7%)
UTX (64.1%)	**AEP** (77.5%)	BUD (69.8%)	PFE (65.8%)	HNZ (68.3%)	FDX (68.4%)
CPB (64.0%)	**ATI** (74.8%)	AIG (67.2%)	PEP (65.1%)	F (68.1%)	JNJ (67.6%)
GE (63.7%)	**MRK** (70.5%)	MRK (64.9%)	CSCO(65.1%)	UTX (67.3%)	MAY (67.1%)
TYC (63.6%)	IP (70.0%)	LTD (64.8%)	BAX (64.4%)	HCA (66.8%)	MSFT(66.4%)
EK (63.2%)	T (69.6%)	AMGN(64.7%)	BMY (64.3%)	ORCL(66.1%)	AIG (65.5%)
VZ (62.0%)	MCD (67.7%)	MDT (63.5%)	SLB (63.1%)	HET (65.5%)	ETR (65.3%)
BNI (62.0%)	**LU** (67.5%)	JNJ (61.7%)	WFC (63.1%)	DIS (64.9%)	MMM (64.6%)
UIS (61.6%)	MEDI(66.6%)	EMC (61.5%)	ETR (62.8%)	VIAB(64.1%)	AXP (63.3%)
BMY (43.0%)	USB (35.9%)	BNI (41.4%)	IBM (40.4%)	MER (37.4%)	HCA (37.4%)
AXP (42.8%)	HWP (35.2%)	HON (39.7%)	IP (40.3%)	LU (36.3%)	PFE (36.9%)
MMM (42.0%)	S (35.2%)	DIS (38.1%)	LEH (38.3%)	NT (34.4%)	SLE (36.7%)
MSFT(39.1%)	KO (35.1%)	IBM (37.9%)	AIG (37.6%)	AA (34.3%)	VIAB(35.4%)
MAY (37.4%)	GD (33.1%)	DD (37.5%)	HD (31.0%)	WFC (30.9%)	CL (34.6%)
BAX (37.0%)	VZ (28.5%)	T (29.1%)	T (30.3%)	HON (28.0%)	AES (30.2%)

and in three periods it is the asset with the largest absolute projection on the Core. Table 4 also shows the percentage projection on the Core, and the five assets, in any period, with the lowest projection onto it. For example, in period 2, 93.3% of the variance of AES can be considered to be driven by the market, and 6.7% is explained by idiosyncratic fluctuation; thus AES essentially lies on the Core, which is the subspace which explains most of the variance of the whole market; on the other hand, in the same period for Coca-Cola (KO), only 35.1% of its variance is driven by the market, while 64.9% is independent fluctuation.

Summarizing, out of the 98 assets, 32 are never reference assets; 46 are only once; 18 are twice, Johnson & Johnson (JNJ) is thrice and AES is five times. Then our analysis identifies nearly 20 assets which had a major role in market integration. In the set of this 20 assets, 8 are among the 20 most capitalized; thus capitalization plays an important role in defining leading assets, but it is not the only factor to be taken in account. For example, in our analysis AES turned out to be the most important stock, but its capitalization (measured as market value) is only about 0.5 of the capitalization of Microsoft, the most capitalized stock in our sample.

5. Laplace Transform Method for Volatility Computation

The Fourier theory has been extended in [42] where it is shown that the Laplace transform is an appropriate tool to build estimators of the instantaneous volatility based on a long time series of prices by smoothing past data and retaining recent price observations. The Laplace transform estimator has the same advantages of the Fourier estimation procedure in comparison with the quadratic variation methods, being based on the integration procedure of all the data as the Fourier estimation method. Moreover it has further good features: in [41] it was shown that the estimation efficiency is better in the center of the considered time window, while the use of Laplace transform estimator allows to obtain an estimator of instantaneous volatility which becomes less sensitive to boundary effects when approaching to the present time (say going from $(t = -\infty)$ to $(t = 0)$). From a conceptual point of view the introduction of Laplace transform has two advantages: firstly to avoid the artificial modification into periodic functions subjacent to Fourier series; secondly to lead to formula (41) which constitutes a bridge between the two different methods of computation of volatility, the method based on quadratic variation and our approach by Fourier series. On the other side the Laplace method is based on the use of integrals instead of series; the discretization of these integrals will introduce a drawback from the numerical point of view.

For simplicity consider

$$dp(t) = \sigma(t)dW(t) \tag{37}$$

the log-price process, where $W(t)$ is a Brownian motion on $(-\infty, 0]$ and $\sigma(t)$ is a stochastic process adapted to the filtration generated by W. Consider the Laplace transform of dp on $(-\infty, 0]$:

$$\Phi_W(z) := \int_{-\infty}^{0} dp(t)\exp(zt). \tag{38}$$

Integrating by parts we have that $\Phi_W(z)$ can be expressed as

$$\Phi_W(z) = \int_{-\infty}^{0} \exp(zt)\sigma(t)\, dW(t). \tag{39}$$

We take $z = a + is$.

In [42] it is proved that the Laplace transform of the volatility process $\sigma^2(t)$ (which is expressed by the right hand side of (40)) can be exactly computed through the *Bohr convolution product* of the Laplace transform of $dp(t)$, which is expressed by the function Φ_W defined in (39). Then the Laplace inversion formula allows to reconstruct the process $\sigma^2(t)$ for any $t \in (-\infty, 0]$.

Later on we denote by $\overline{\psi}$ the conjugate of any complex function ψ.

Theorem 5.1. *Let $\Phi_W(z)$ be the Laplace transform of dp as given by (39). Then the following convergence in probability holds*

$$(40)\quad \lim_{R\to+\infty} \frac{1}{2R}\int_{-R}^{R} \Phi_W(a+is)\,\overline{\Phi_W}(a+is+iv)\,ds = \int_{-\infty}^{0} \sigma^2(t)\exp\Big((2a-iv)t\Big)\,dt.$$

Remark 5.2. In formula (40) we have the choice of an arbitrary parameter a; we must consider always $a > 0$. If we want to damp quickly the effect of the past time we have to take $|a|$ large.

Using the exact formula (40) it is possible to derive an estimator of volatility given a discrete unevenly spaced sampling of the price process p, containing only an averaged sum of the jumps of the prices combined with universal kernels. Denote by t_i the times at which prices are observed, let $-\infty < \ldots < t_{i+1} < t_i < \ldots < t_0 \leq 0$ and let $\delta_i(p) := \; -p(t_1) + p(t_{i-1})$, for any $i \geq 1$.

Let the integration interval in (40) be fixed equal to $[-R, R]$, then the following approximation formula for the volatility function can be obtained:

$$(41)\qquad Vol(p)(t) \approx \sum_{i,j>0} \delta_i(p)\delta_j(p)u_\delta(t-t_j)\psi(R(t_i-t_j))$$

where u_δ is defined for $\delta > 0$ by

$$u_\delta(s) = \frac{1}{\delta}u_1(\frac{s}{\delta}),\quad u_1(s) = 1_{|s|<1}(1-|s|),\quad \text{and}\quad \psi(\lambda) = \frac{\sin\lambda}{\lambda}.$$

6. HJM Model Looked as an Hypoelliptic Operator

Consider the Heath-Jarrow-Morton model [32] for the interest rates in the Musiela parametrization, see [45]. In this parametrization a market state at time t_0 is described by the instantaneous interest rate curve $r_{t_0}(\xi)$ to a maturity ξ, defined for $\xi \in \mathbf{R}_+$. The price $P(t_0, T)$ of a default-free zero coupon bond traded at t_0 with maturity $t_0 + T$, has the value

$$P(t_0, T) = \exp(\int_0^T r_{t_0}(\xi)\,d\xi).$$

The interest rate curve takes its value in the *infinite dimensional* space C of continuous functions on $[0, \infty[$. A remarkable experimental fact is that the rank n of its volatility matrix is very low $n \simeq 4$, see [18] . Then elliptic models are ruled out and hypoelliptic models are the most regular models still available.

By restricting the HJM model to the case of a finite number of scalar-valued driving Brownian motions $W_1, \dots, W_n$, then the HJM model is expressed by the following SDE:

$$dr_t(\xi) = \sum_{k=1}^{n} B_k(t,\xi)\, dW_k(t) + \left\{ \frac{\partial r_t}{\partial \xi}(\xi) + \sum_{k=1}^{n} \int_0^{\xi} B_k(t,\xi) B_k(t,\eta)\, d\eta \right\} dt, \ \forall\, \xi \in \mathbf{R}_+,$$

where the drift of the risk-free process is completely determined under the no arbitrage condition by the volatility matrix $B_*(t, \cdot)$. In fact it is known from [21] that the risk-free generator associated to the risk free measure has its infinitesimal generator fully determined by its second order terms. For the stock price the drift associated vanishes under the risk free measure. For the interest rate model the drift is determined by the no-arbitrage condition implemented in the HJM model. This key property implies that the infinitesimal generator of the HJM process can be pathwise and model free computed through the computation of the volatility matrix.

Under the hypothesis of *Markovian completeness* of the market, the dependence of the volatility matrix $B_*(t, \cdot)$ is factorizable through the final value $r_t(\cdot)$, therefore it is possible to write

$$(42) \qquad dr_t(\xi) = \sum_{k=1}^{n} (A_k(r_t))(\xi)\, dW_k(t) + \left\{ \frac{\partial r_t}{\partial \xi}(\xi) + \sum_{k=1}^{n} \int_0^{\xi} (A_k(r_t))(\xi)(A_k(r_t))(\eta)\, d\eta \right\} dt,$$

where A_k are "driving vector fields" defined on C. An appropriate notion of "smoothness" of vector fields is a necessary hypothesis in order to prove existence and uniqueness of solutions for (42), see [24].

In [40] a non-parametric estimation of HJM generator is obtained and a method similar to the methodology of the price-volatility feedback rate is developed for the interest rate curve. The question of the efficiency of the mathematical theory proposed in this section to decipher the state of the market has not yet confirmed by numerical computations.

The first point is the possibility of measuring, in real time by high frequency market data, the full historical volatility matrix. We remark that, while the correlations between stocks prices at high frequency has no clear economic meaning, on the contrary the high frequency cross-correlations are clearly significant in the HJM model.

Assume that the HJM hypothesis for the infinitesimal generator is satisfied; no assumption on the explicit expression of the vector field A_k in (42) are done. It is proved in [40] that an estimation of the maps $t \mapsto A_k(r_t)$ is achieved by the observation of an unique evolution of the market.

Fix N different maturities (N can be extremely large and in particular it can be $n << N$), then define the time-dependent $N \times N$ covariance matrix

$$[\sigma^2(t)]_{\xi,\eta} := [dr_t(\xi) \; \star \; dr_t(\eta)]_{\xi,\eta} \tag{43}$$

where ξ, η vary between all maturities and $\star$ denotes the Itô contraction divided by dt.

For simplicity of notation consider two maturities ξ, η and the bi-dimensional process: $t \mapsto (r_t(\xi), r_t(\eta))$. Then it holds:

Theorem 6.1. *Given two different maturities ξ, η we consider the 2-dimensional process: $t \mapsto (r_t(\xi), r_t(\eta))$; then*

$$[\sigma^2(t)]_{\xi,\eta} = \frac{1}{4}\Big[\; Vol(r_t(\xi) + r_t(\eta)) - \; Vol(r_t(\xi) - r_t(\eta))\Big]. \tag{44}$$

The vector $A_j(r_t)$ is defined as the jth column of the matrix $\sqrt{\sigma^2(t)}$

Note that the r.h.s. of (44) can be computed using the Fourier methodology for the estimation of volatility.

6.1 Non parametric estimation of Hörmander bracket

Definition 6.2. Given two vector fields A_1, A_2 denote by A_k^ξ the ξ-th component of A_k. The Lie bracket of A_1 and A_2 is the vector field having the components:

$$[A_1, A_2]^\xi := \frac{\partial A_2^\xi}{\partial A_1} - \frac{\partial A_1^\xi}{\partial A_2} = \sum_\eta \left(A_1^\eta \frac{\partial A_2^\xi}{\partial r^\eta} - A_2^\eta \frac{\partial A_1^\xi}{\partial r^\eta} \right). \tag{45}$$

The Lie algebra $\mathcal{A}$ generated by n vector fields $A_1, \ldots, A_n$ is defined as the vector space of all fields obtained as linear combinations with constant coefficients of

$$A_k, [A_k, A_l], [[A_k, A_l], A_s], [[[A_k, A_l], A_s], A_u], \ldots \tag{46}$$

Given $r \in \mathbf{R}^N$, let $\mathcal{A} := \{\zeta \in \mathbf{R}^N | \; \zeta = Z(r) \text{ for some } Z \in \mathcal{A}\}$.

Definition 6.3. The vector fields $A_1, \ldots A_n$ satisfy the Hörmander criterion for hypoellipticity if $A_1, \ldots, A_n$ are infinitely often differentiable and if $\mathcal{A}(r) = \mathbf{R}^N$ for any $r \in \mathbf{R}^N$.

Due to the non-ellipticity of the market, the multivariate price-volatility feedback rate constructed in [14] cannot be applied, so [40] construct a pathwise econometric computation of the bracket of the driving vector of the diffusion.

Without any hypothesis on the explicit form of the vector fields $A_k(r_t)$, we estimate the bracket $[A_1, A_2](r_t)$ from a single trajectory. For simplicity we omit writing explicitly the dependence of A_k on r_t. In order to get this result, by the definition (45), the main point is the estimation of

$$\frac{\partial A_k}{\partial r^\eta}, \quad k = 1, 2.$$

This estimation is obtained by the relation

$$dA_2 \star dr_t^\lambda = \sum_\eta \frac{\partial A_2}{\partial r^\eta} \times [\sigma^2(t)]_{\eta,\lambda} \tag{47}$$

together with the fact that the l.h.s. of (47) can be computed by Fourier methodology. Therefore we can state the result:

Theorem 6.4. *The brackets of the driving vector fields of the HJM diffusion can be numerically computed from a single time series of market data, without any assumption on the model.*

6.2 Global and pathwise compartmentage by maturities

Let C the space of continuous functions on $[0, \infty[$. Define the *logarithm derivative of a measure* μ_λ depending on a parameter λ as the function satisfying

$$\partial_\lambda \int_C f(r)\ \mu_\lambda(dr) = \int_C f(r)\ (\partial_\lambda \log(\mu_\lambda))(r)\ \mu_\lambda(dr),$$

for every test function f. Denote $\pi_t(r_0, dr)$ the fundamental solution of the heat equation associated to the HJM model; define an Hilbert norm on tangent vectors at r_0 by

$$||z||_t^2 = \int_C [\frac{d}{d\varepsilon}_{|\varepsilon=0} (\log(\pi_t(r_0 + \varepsilon z, *))(r)]^2\ \pi_t(r_0, dr).$$

Definition 6.5. We say that the *global compartmentage* holds if the following condition holds:

the norms $||z||_s$ and $||z||_{s'}$, $s \neq s'$ are inequivalent.

In [40] it is proved that the global *compartmentage* condition holds generically in the HJM setting. As a consequence, it is showed that it is impossible to hedge an European option at maturity T by trading options on maturities larger than T.

The operator $Q^{\leftarrow}_{W}$ is defined in [40] (5.3) by solving the linearized HJM model:

$$d\zeta_W(t) = \sum_{k=1}^{n} \mathbf{A}_k(\zeta(t))dW_k(t) + \mathbf{A}_0(\zeta(t))dt, \tag{48}$$

where the function $\mathbf{A}_k$, taking values in the space of $N \times N$ matrices, is defined by

$$(\mathbf{A}_k)^{\xi}_{\eta} := \frac{\partial A^{\xi}_k}{\partial r^{\eta}}.$$

Choose on C an Hilbert metric $\| * \|_C$; by using the canonic Hilbert structure of $L^2([0,s]; R^n)$, the adjoint $(Q^{\leftarrow}_{W_s})^*$ is defined and it is possible to define the Malliavin covariance matrix:

$$\sigma^{\leftarrow}_{W_s} = Q^{\leftarrow}_{W_s} \circ (Q^{\leftarrow}_{W_s})^*.$$

Consider a fixed trajectory W of the Brownian motion. Denote $\sigma^{\leftarrow}_{W}(s)$ the covariance matrix computed on $[0,s]$.

Theorem 6.6. *The pathwise compartmentage property holds if the range of $\sigma^{\leftarrow}_{W}(s)$ is a strictly increasing function of s.*

Therefore it is possible to deduce the property of pathwise compartmentage from the behavior of Hörmander brackets, which we showed can be estimated non-parametrically using Fourier methodology.

References

1. Andersen, T. and Bollerslev, T. (1998) Answering the Skeptics: Yes, Standard Volatility Models do Provide Accurate Forecasts. *International Economic Review*, **39**, 885–905.
2. Andersen, T., Bollerslev, T. and Diebold, F. (2002) Parametric and Nonparametric Volatility Measurement. *Handbook of Financial Econometrics*.
3. Andersen, T., Bollerslev, T., Diebold, F. and Labys, P. (2003) Modeling and forecasting realized volatility. *Econometrica*, **71**, 579–625.
4. Andersen, T., Bollerslev, T., Diebold, F. and Labys, P. (2001) The distribution of exchange rate volatility. *Journal of the American Statistical Association*, **96**, 42–55.
5. Andersen, T., Bollerslev, T. and Lange, S. (1999) Forecasting Financial Market Volatility: Sample Frequancy vis-à-vis Forecast Horizon. *Journal of Empirical Finance*, **6**, 457–477.
6. Bandi, F. M. and Phillips, P. C. B. (2003) Fully nonparametric estimation of scalar diffusion models. *Ecometrica*, **71**(1), 241–283.
7. Bandi, F. M. and Russel, J. R. (2003) Microstructure noise, realized volatility and optimal sampling. *Preprint*.

8. Barndorff-Nielsen, O. E. and Shephard, N. (2002) Econometric analysis of realised volatility and its use in estimating stochastic volatility models. *Journal of the Royal Society*, Series B, **64**, 253–280.
9. Barndorff-Nielsen, O. E. and Shephard, N. (2003) Realised power variation and stochastic volatility. *Bernoulli*, **9**, 243–265.
10. Barndorff-Nielsen, O. E. and Shephard, N. (2005) Power Variation and Time Change. *Probability Theory and its Applications* (forthcoming).
11. Barndorff-Nielsen, O. E. and Shephard, N. (2005) Variation, jumps, market frictions and high frequency data in financial econometrics. Preprint.
12. Barndorff-Nielsen, O. E., Graversen, S. E., Jacod, J. and Shephard, N. (2005) Limit theorems for bipower variation in financial econometrics. *Preprint*.
13. Barucci, E., Impenna, C. and Renò, R. (2003). The Italian overnight market: microstructure effects, the martingale hyphotesis and the payment system. *Temi di Discussione, Bank of Italy*. N. 475.
14. Barucci, E., Malliavin, P., Mancino, M. E., Renò, R. and Thalmaier A. (2003) The price-volatility feedback rate: an implementable mathematical indicator of market stability. *Mathematical Finance*, **13**, 17–35.
15. Barucci, E. and Renò, R. (2001) On measuring volatility of diffusion processes with high frequency data. *Economics Letters*, **74**, 371–378.
16. Barucci, E. and Renò, R. (2002a) On measuring volatility and the GARCH forecasting performance. *Journal of International Financial Markets, Institutions and Money*, **12**, 183–200.
17. Barucci, E. and Renò, R. (2002b). Value at risk with high frequency data. In C. Zopounidis (Ed.), *New Trends in Banking Management*, 223–232. Physica-Verlag.
18. Bouchaud, J. P., Cont, R., El-Karoui, N., Potters, M. and Sagna, N. (1999). Phenomenology of the interest rate curve. *Appl. Math. Finance*, **6**(3), 209–232.
19. Campbell, J., Lo, A. and MacKinlay, C. (1997) *The Econometrics of Financial Markets*. Princeton Univesrity Press.
20. Comte, F. and Renault, E. (1998) Long memory in continuous time stochastic volatility models. *Math. Finance*, **8**, 291–323.
21. Delbaen, F. and Schachermayer, W.: The No-Arbitrage Property under a Change of Numéraire, Stochastics Stochastics Rep. 53 (1995), no. 3–4, 213–226.
22. Drost, F. and Werker, B. (1996). Closing the GARCH Gap: Continuous time GARCH Models. *Journal of Econometrics*, **74**, 31–57.
23. Engle, R. F. and Russel, J. R. (1998) Analysis of high frequency data. In Y. Ait-Sahalia and L. P. Hansen *Handbook of Financial Econometrics*, North-Holland.
24. Filipovic, D. and Teichmann, J. (2003) Existence of invariant manifolds for stochastic equations in infinite dimension. *J. of Funct. Anal.*, no. 2, 398–432.
25. Foster, D. P. and Nelson, D. B. (1996) Continuous record asymptotics for rolling sample variance estimators. *Econometrica*, **64**, 139–174.
26. Genon-Catalot, V., Laredo, C. and Picard, D. (1992) Nonparametric estimation of the diffusion coefficient by wavelet methods. *Scandinavian Journal of Statistics*, **19**, 317–335.
27. Ghysels, E., Harvey, A., and Renault, E. (1996) Stochastic volatility. *Handbook of Statistics*, vol.14 Statistical methods in finance, 119–191. North-Holland.
28. Kroner, K. and Ng, V. (1998) Modelling asymmetric comovements of asset

returns. *Review of Financial Studies*, **11**, 817–844.
29. Hayashi, T. and Yoshida, N. (2004) On covariance estimation of nonsynchronously observed diffusion processes. *Bernoulli*, **11**, 359–379.
30. Hansen, P. R. and Lunde, A. (2003) An optimal and unbiased measure of realized variance based on intermittent high-frequency data. *Preprint*.
31. Hansen, P. R., Lunde, A. and Nason, J. (2003) Choosing the best volatility models: the model confidence set approach. *Preprint*.
32. Heath, D., Jarrow, M. and Morton A. (1992) Bond pricing and the term structure of interest rates: a new methodology for contingent claims valuation. *Econometrica*, **60**(1), 77–105.
33. Hog, E. and Lunde, A. (2003) Wavelet estimation of integrated volatility. *Preprint*.
34. Hsieh, D. (1991) Chaos and nonlinear dynamics: application to financial markets. *Journal of Finance*, **46**, 1839–1877.
35. Jacod, J. and Protter, P. (1998) Asymptotic error distributions for the Euler method for stochastic differential equations. *Annals of Probability*, **26**, 267–307.
36. Kanatani, T. (2004) Integrated volatility measuring from unevenly sampled observations. *Economics Bullettin*, **3**, 1–8.
37. Lo, A. and MacKinlay, C. (1990) The econometric analysis of nonsynchronous trading. *Journal of Econometrics*, **45**, 181–211.
38. Malliavin, P. (1995) *Integration and Probability*. Springer-Verlag.
39. Malliavin, P. (1997) *Stochastic analysis*. A series of comprehensive studies in mathematics, vol. 313. Springer-Verlag, 1997.
40. Malliavin, P. and Thalmaier A. (2005) *Stochastic Calculus of Variations in Mathematical Finance*. Springer Finance, 2005.
41. Malliavin, P. and Mancino, M. E. (2002). Fourier series method for measurement of multivariate volatilities. *Finance and Stochastics*, **4**, 49–61.
42. Malliavin, P. and Mancino, M. E. (2004). Nonparametric Volatility Estimation using Laplace Transform. *Preprint Dept. Math. for Decisions, University of Firenze*.
43. Malliavin, P. and Mancino, M. E. (2005). A Fourier transform method for nonparametric estimation of volatility. *Preprint Dept. Math. for Decisions, University of Firenze*.
44. Mancino, M. E. and Renò, R. (2005). Dynamic Principal Component Analysis of multivariate volatilities via Fourier analysis. *Applied Mathematical Finance*, **12**(2), 187–199.
45. Musiela, M. (1993). Stochastic PDEs and term structure models. *Journées Internationales de Finance*, IGR-AFFI, La Baule, 1993.
46. Mykland, P. and Zhang, L. (2005). Anova for diffusions. *Annals of Statistics*, **33**.
47. Nelson, D. (1990). Arch Models as Diffusion Approximations. *Journal of Econometrics*, **45**, 7–38.
48. Nielsen, M. O. and Frederiksen, P. H. (2005). Finite sample accuracy of integrated volatility estimators. *Preprint Department of Economics, Cornell University*.
49. Ocone, D. and Karatzas, I. (1991). A generalized Clark representation formula, with application to optimal portfolios. *Stochastics and Stochastic Reports*, **34**, 187–220.
50. Precup, O. V. and Iori, G. (2003) A comparison of high-frequency cross-

correlation measures. *Preprint*.
51. Renó, R. (2003) A closer look at the Epps effect. *International Journal of Theoretical and Applied Finance*, **6**, 87–102.
52. Scholes, M. and Williams, J. (1977) Estimating betas from nonsinchronous trading. *Journal of Financial Economics*, **5**, 309–327.
53. Schwert, G. (1990) Stock market volatility. *Financial Analyst Journal*, **46**, 23–34.
54. Zamansky, M. (1950) Classes de saturation des procédés de sommation des séries de Fourier et applications aux séries trigonométriques. *Annales scientifiques de l'É.N.S.* 3e série, **t.67**, 161–198.
55. Zhang, L., Mykland, P. and Ait-Sahalia, Y. (2003) A tale of two time scales: determining integrated volatility with noisy high frequency data. Forthcoming *Journal of the American Statistical Association*.

Hedging of Credit Derivatives in Models with Totally Unexpected Default

Tomasz R. BIELECKI[1], Monique JEANBLANC[2], and Marek RUTKOWSKI[3]

[1]Department of Applied Mathematics, Illinois Institute of Technology, Chicago, IL 60616, USA
[2]Département de Mathématiques, Université d'Évry Val d'Essonne, 91025 Évry Cedex, France
[3]School of Mathematics, University of New South Wales, Sydney, NSW 2052, Australia & Faculty of Mathematics and Information Science, Warsaw University of Technology, 00-661 Warszawa, Poland

The paper analyzes alternative mathematical techniques, which can be used to derive hedging strategies for credit derivatives in models with totally unexpected default. The stochastic calculus approach is used to establish abstract characterization results for hedgeable contingent claims in a fairly general set-up. In the Markovian framework, we use the PDE approach to show that the arbitrage price and the hedging strategy for an attainable contingent claim can be described in terms of solutions of a pair of coupled PDEs.
Key words: arbitrage pricing, hedging, credit risk, default intensity.

Introduction

This paper presents some methods to hedge defaultable derivatives under the assumption that there exist tradeable assets with dynamics allowing for elimination of default risk of derivative securities. We investigate hedging strategies in alternative frameworks with different degrees of generality, an abstract semimartingale framework and a more specific Markovian set-up, and we use two alternative approaches.

On the one hand, we use the stochastic calculus approach in order to establish rather abstract characterization results for hedgeable contingent claims in a fairly general set-up. We subsequently apply these results to derive closed-form solutions for prices and replicating strategies in particular models.

On the other hand, we examine the PDE approach in a Markovian setting. In this method, the arbitrage price and the hedging strategy for an attainable contingent claim are described in terms of solutions of a pair of coupled PDEs. Again, for some standard examples of defaultable claims, we provide explicit formulae for prices and hedging strategies (for further examples of trading strategies involving tradeable credit derivatives, we refer to Laurent [31] or Bielecki *et al.* [8]). As expected, both methods yield identical results for some special cases considered in this work.

For the sake of simplicity, we only deal with financial models with no more than three primary assets (models with an arbitrary number of primary assets were studied in Bielecki *et al.* [6]). Also, it is postulated throughout that the default time is the same for all defaultable securities. An extension of our results to the case of several (possibly dependent) default times is crucial if someone wishes to cover the so-called basket credit derivatives (in this regard, see Section 6 in Bielecki *et al.* [7]).

Let us comment briefly on the terminology used in this work. Traditionally, credit risk models are classified either as *structural models* (also known as *value-of-the-firm models*) or as *reduced-form models* (also termed *intensity-based models*). In their original forms, the two approaches, structural and reduced-form, are extreme cases, in the sense that the default time is modelled either as a predictable stopping time (the first moment when the firm's value hits some barrier, as in Black and Cox [9]), or by a totally inaccessible stopping time (defined via its intensity, as in Jarrow and Turnbull [25]). However, as argued by several authors (see, for instance, Duffie and Lando [17], Giesecke [21]–[22], Jarrow and Protter [24], Jeanblanc and Valchev [28], or Guo *et al.* [23]), probabilistic properties of default time are directly related to the publicly available information (it is important, for instance, whether the value of the firm and/or the default triggering barrier are observed by the market with absolute accuracy).

In fact, in several structural models the default time is no longer predictable, as it was the case in classic models with deterministic default triggering barrier and full observation of the firm value process (see, Merton [33] or Black and Cox [9]). For this reason, we decided to refer to credit risk models considered in this work as models with totally unexpected default (the strict mathematical term, *totally inaccessible stopping time*, seems to be rather cumbersome for a frequent use). For a more exhaustive presentation of mathematical theory of credit risk, we refer to Arvanitis and Gregory [1], Bielecki and Rutkowski [3], Bielecki *et al.* [4], Cossin and Pirotte [15], Duffie and Singleton [18], Lando [30], or Schönbucher [36].

1. Totally Unexpected Default

In this section, we describe briefly the fundamental features of the credit risk models with unexpected default. Also, we collect here few technical results that are used in subsequent sections.

1.1 General set-up

We assume that we are given a probability space $(\Omega, \mathcal{G}, \mathbb{P})$ and a non-negative random variable τ on this space. We always postulate that τ is strictly positive with probability 1. Note that the probability measure $\mathbb{P}$ represents the historical probability reflecting the real-life dynamics of prices of primary traded assets, rather than some martingale measure for our financial model. We first focus on different definitions of default intensity encountered in the literature.

1.1.1 Intensity of a stopping time

Suppose that $(\Omega, \mathcal{G}, \mathbb{P})$ is endowed with some filtration $\widetilde{\mathbb{G}}$ such that τ is a $\widetilde{\mathbb{G}}$-stopping time. Let H be the *default process*, defined as $H_t = \mathbb{1}_{\{t \geq \tau\}}$ (note that H is a bounded $\widetilde{\mathbb{G}}$-submartingale). We say that τ admits a $\widetilde{\mathbb{G}}$-*intensity* if there exists a $\widetilde{\mathbb{G}}$-adapted, nonnegative process $\widetilde{\lambda}$ such that the process

$$M_t = H_t - \int_0^t \widetilde{\lambda}_u \, du = H_t - \int_0^{t \wedge \tau} \widetilde{\lambda}_u \, du \tag{1}$$

is a $\widetilde{\mathbb{G}}$-martingale (the second equality in (1) follows from the fact that the process H is stopped at τ). Then M is called the compensated $\widetilde{\mathbb{G}}$-martingale of the default process H. In order for a $\widetilde{\mathbb{G}}$-stopping time τ to admit a $\widetilde{\mathbb{G}}$-intensity $\widetilde{\lambda}$, it has to be *totally inaccessible* with respect to $\widetilde{\mathbb{G}}$, so that $\mathbb{P}(\tau = \theta) = 0$ for any $\widetilde{\mathbb{G}}$-predictable stopping time θ. The simplest example is the moment of the first jump a Poisson process. Note that the intensity $\widetilde{\lambda}$ necessarily vanishes after default.

Remark 1.1. Some authors define the intensity as the process $\widetilde{\lambda}$ such that $H_t - \int_0^{t \wedge \tau} \widetilde{\lambda}_u \, du$ is a $\widetilde{\mathbb{G}}$-martingale. In that case, the process $\widetilde{\lambda}$ is not uniquely defined after time τ.

1.1.2 $\mathbb{F}$-intensity of a random time

We change the perspective, and we no longer assume that the filtration $\widetilde{\mathbb{G}}$ is given a priori. We assume instead that τ is a positive random variable on some probability space $(\Omega, \mathcal{G}, \mathbb{P})$. Let $\mathbb{H} = (\mathcal{H}_t, \, t \geq 0)$ be the natural filtration generated by the default process $(H_t, \, t \geq 0)$, and let $\mathbb{F} = (\mathcal{F}_t, \, t \geq 0)$ be some *reference filtration* in $(\Omega, \mathcal{G}, \mathbb{P})$.

We assume throughout that the information available to an investor is modeled by the filtration $\mathbb{G} = \mathbb{F} \vee \mathbb{H}$. Consequently, we can reduce our study to the case where the default intensity (if it exists) is $\mathbb{G}$-adapted, meaning that the process M given by (1) is a $\mathbb{G}$-martingale for some $\mathbb{G}$-adapted process $\widetilde{\lambda}$. In this setting, there exists a process $\lambda = (\lambda_t, t \geq 0)$, called the $\mathbb{F}$-*intensity* of τ, which is $\mathbb{F}$-adapted and equal to $\widetilde{\lambda}$ before default, so that $\widetilde{\lambda}_t \mathbb{1}_{\{t \leq \tau\}} = \lambda_t \mathbb{1}_{\{t \leq \tau\}}$ for every $t \in \mathbb{R}_+$. The existence of $\widetilde{\lambda}$ (and its uniqueness under some technical conditions) follows from the following result (see Dellacherie *et al.* [16], Page 186).

Lemma 1.1. *Let $\mathbb{G} = \mathbb{F} \vee \mathbb{H}$. Then for any $\mathbb{G}$-predictable process ζ there exists an $\mathbb{F}$-predictable process $\widetilde{\zeta}$ such that*

$$\widetilde{\zeta}_t \mathbb{1}_{\{t \leq \tau\}} = \zeta_t \mathbb{1}_{\{t \leq \tau\}}, \quad \forall \, t \in \mathbb{R}_+. \tag{2}$$

If, in addition, the inequality $F_t := \mathbb{P}(\tau \leq t \,|\, \mathcal{F}_t) < 1$ holds for every $t \in \mathbb{R}_+$ then the process $\widetilde{\zeta}$ satisfying (2) *is unique.*

Of course, we have that

$$M_t = H_t - \int_0^{t \wedge \tau} \widetilde{\lambda}_u \, du = H_t - \int_0^{t \wedge \tau} \lambda_u \, du.$$

Suppose that the reference filtration is chosen in such a way that the default events $\{\tau \leq t\}$ are not in $\mathbb{F}$. Then the $\mathbb{F}$-intensity λ is uniquely defined after τ and, typically, does not vanish after τ.

1.2 Hypothesis (H)

In this section, we focus on the invariance property of the so-called hypothesis (H) under an equivalent change of a probability measure.

Definition 1.1. We say that filtrations $\mathbb{F}$ and $\mathbb{G}$, with $\mathbb{F} \subseteq \mathbb{G}$, satisfy the *hypothesis* (H) under $\mathbb{P}$ whenever any $\mathbb{F}$-local martingale L follows also a $\mathbb{G}$-local martingale.

Remark 1.2. We emphasize that, in general, an $\mathbb{F}$-martingale may fail to follow a $\mathbb{G}$-martingale. As a trivial example, consider a fixed date $T > 0$ and take $\mathcal{G}_t = \mathcal{F}_T$ for every $t \in [0, T]$. Then any $\mathbb{F}$-martingale L satisfies $\mathbb{E}_{\mathbb{P}}(L_t \,|\, \mathcal{G}_s) = L_t$ for $s \leq t$, and thus L is not a $\mathbb{G}$-martingale, in general. It is even possible, but more difficult, to produce an example of an $\mathbb{F}$-martingale, which is not a semi-martingale with respect to $\mathbb{G}$. For other counter-examples, in particular those involving progressive enlargement of filtrations, we refer interested reader to Protter [35], or Mansuy and Yor [32].

The original formulations of the hypothesis (H) refer to martingales (or even square-integrable martingales), rather than local martingales. We shall show that in our set-up the definition given above is equivalent to the original definition. In fact, the hypothesis (H) postulates a certain form of conditional independence of σ-fields associated with $\mathbb{F}$ and $\mathbb{G}$, rather than a specific property of $\mathbb{F}$-(local) martingales. In particular the following well known result is valid.

Lemma 1.2. *Assume that $\mathbb{G} = \mathbb{F} \vee \mathbb{H}$, where $\mathbb{F}$ is an arbitrary filtration and $\mathbb{H}$ is generated by the process $H_t = \mathbb{1}_{\{\tau \le t\}}$. Then the following conditions are equivalent to the hypothesis* (H).
(i) For any $t, h \in \mathbb{R}_+$, we have

$$\mathbb{P}(\tau \le t \,|\, \mathcal{F}_t) = \mathbb{P}(\tau \le t \,|\, \mathcal{F}_{t+h}). \tag{3}$$

(i′) For any $t \in \mathbb{R}_+$, we have

$$\mathbb{P}(\tau \le t \,|\, \mathcal{F}_t) = \mathbb{P}(\tau \le t \,|\, \mathcal{F}_\infty). \tag{4}$$

(ii) For any $t \in \mathbb{R}_+$, the σ-fields $\mathcal{F}_\infty$ and $\mathcal{G}_t$ are conditionally independent given $\mathcal{F}_t$ under $\mathbb{P}$, that is,

$$\mathbb{E}_{\mathbb{P}}(\xi\,\eta \,|\, \mathcal{F}_t) = \mathbb{E}_{\mathbb{P}}(\xi \,|\, \mathcal{F}_t)\, \mathbb{E}_{\mathbb{P}}(\eta \,|\, \mathcal{F}_t)$$

for any bounded, $\mathcal{F}_\infty$-measurable random variable ξ and bounded, $\mathcal{G}_t$-measurable random variable η.
(iii) For any $t \in \mathbb{R}_+$, and any $u \ge t$ the σ-fields $\mathcal{F}_u$ and $\mathcal{G}_t$ are conditionally independent given $\mathcal{F}_t$.
(iv) For any $t \in \mathbb{R}_+$ and any bounded, $\mathcal{F}_\infty$-measurable random variable ξ: $\mathbb{E}_{\mathbb{P}}(\xi \,|\, \mathcal{G}_t) = \mathbb{E}_{\mathbb{P}}(\xi \,|\, \mathcal{F}_t)$.
(v) For any $t \in \mathbb{R}_+$, and any bounded, $\mathcal{G}_t$-measurable random variable η: $\mathbb{E}_{\mathbb{P}}(\eta \,|\, \mathcal{F}_t) = \mathbb{E}_{\mathbb{P}}(\eta \,|\, \mathcal{F}_\infty)$.

Proof. The proof of equivalence of conditions (i′)–(v) can be found, for instance, in Section 6.1.1 of Bielecki and Rutkowski [3] (for related results, see Elliott *et al.* [20]). Using monotone class theorem it can be shown that conditions (i) and (i′) are equivalent. Hence, we shall only show that condition (iv) and the hypothesis (H) are equivalent.

Assume first that the hypothesis (H) holds. Consider any bounded, $\mathcal{F}_\infty$-measurable random variable ξ. Let $L_t = \mathbb{E}_{\mathbb{P}}(\xi \,|\, \mathcal{F}_t)$ be the martingale associated with ξ. Then, (H) implies that L is also a local martingale with respect to $\mathbb{G}$, and thus a $\mathbb{G}$-martingale, since L is bounded (recall that any bounded local martingale is a martingale). We conclude that $L_t = \mathbb{E}_{\mathbb{P}}(\xi \,|\, \mathcal{G}_t)$ and thus (iv) holds.

Suppose now that (iv) holds. First, we note that the standard truncation argument shows that the boundedness of ξ in (iv) can be replaced by the assumption that ξ is $\mathbb{P}$-integrable. Hence, any $\mathbb{F}$-martingale L is an $\mathbb{G}$-martingale, since L is clearly $\mathbb{G}$-adapted and we have, for every $t \leq s$,

$$L_t = \mathbb{E}_{\mathbb{P}}(L_s \,|\, \mathcal{F}_t) = \mathbb{E}_{\mathbb{P}}(L_s \,|\, \mathcal{G}_t).$$

Now, suppose that L is an $\mathbb{F}$-local martingale so that there exists an increasing sequence of $\mathbb{F}$-stopping times τ_n such that $\lim_{n\to\infty} \tau_n = \infty$, for any n the stopped process L^{τ_n} follows a uniformly integrable $\mathbb{F}$-martingale. Hence, L^{τ_n} is also a uniformly integrable $\mathbb{G}$-martingale, and this means that L follows a $\mathbb{G}$-local martingale. □

1.2.1 Hazard process

Let τ be a random time on a space $(\Omega, \mathcal{G}, \mathbb{P})$ such that the filtrations $\mathbb{F}$ and $\mathbb{G} = \mathbb{F} \vee \mathbb{H}$ satisfy the hypothesis (H). Then, from (4), the process $F_t = \mathbb{P}(\tau \leq t \,|\, \mathcal{F}_t)$ is increasing.

We make the standing assumption that $F_t < 1$ for every $t \in \mathbb{R}_+$, and we define the $\mathbb{F}$-*hazard process* Γ by setting $\Gamma_t = -\ln(1 - F_t)$. Let, in addition, the process F be absolutely continuous with respect to the Lebesgue measure, so that

$$F_t = \int_0^t f_u \, du, \quad \forall\, t \in \mathbb{R}_+,$$

for some $\mathbb{F}$-progressively measurable (or $\mathbb{F}$-predictable) process f. Then the $\mathbb{F}$-hazard process satisfies

$$\Gamma_t = \int_0^t \gamma_u \, du, \quad \forall\, t \in \mathbb{R}_+,$$

where the $\mathbb{F}$-*intensity* γ is given by

$$\gamma_t = \frac{f_t}{1 - F_t}, \quad \forall\, t \in \mathbb{R}_+. \tag{5}$$

From now on, we take (5) as the definition of the $\mathbb{F}$-intensity γ, and we make the standing assumption that the hypothesis (H) holds under $\mathbb{P}$. The following auxiliary result is standard (see, for instance, Elliott *et al.* [20] or Blanchet-Scalliet and Jeanblanc [10]).

Lemma 1.3. *For any $\mathbb{P}$-integrable, $\mathcal{F}_T$-measurable random variable X we have, for $t \in [0, T]$,*

$$\mathbb{E}_{\mathbb{P}}(X \mathbb{1}_{\{T<\tau\}} \,|\, \mathcal{G}_t) = \mathbb{1}_{\{t<\tau\}} e^{\Gamma_t} \mathbb{E}_{\mathbb{P}}(X e^{-\Gamma_T} \,|\, \mathcal{F}_t).$$

1.2.2 Canonical construction

We now describe the canonical construction of a random time with a given $\mathbb{F}$-hazard process. Let Ψ be an $\mathbb{F}$-adapted, increasing, nonnegative process with $\Psi_0 = 0$ and $\lim_{t\to\infty} \Psi_t = \infty$. We define a nonnegative random variable τ by setting

$$\tau = \inf\{\, t \in \mathbb{R}_+ : \Psi_t \geq \Theta\},$$

where Θ is a random variable independent of $\mathbb{F}$, with the exponential distribution of parameter 1. Of course, the existence of Θ on the original probability space $(\Omega, \mathcal{G}, \mathbb{P})$ is not guaranteed, so that we allow for an extension of the underlying probability space.

We shall now find the process $F_t = \mathbb{P}\{\tau \leq t \,|\, \mathcal{F}_t\}$. Since clearly $\{\tau > t\} = \{\Theta > \Psi_t\}$, we get

$$\mathbb{P}\{\tau > t \,|\, \mathcal{F}_\infty\} = \mathbb{P}\{\Theta > \Psi_t \,|\, \mathcal{F}_\infty\} = e^{-\Psi_t}.$$

Consequently,

$$1 - F_t = \mathbb{P}\{\tau > t \,|\, \mathcal{F}_t\} = \mathbb{E}_{\mathbb{P}}\Big(\mathbb{P}\{\tau > t \,|\, \mathcal{F}_\infty\} \,|\, \mathcal{F}_t\Big) = e^{-\Psi_t},$$

and so F is an $\mathbb{F}$-adapted, continuous, increasing process. We conclude that for every $t \in \mathbb{R}_+$

$$F_t = 1 - e^{-\Psi_t} = \mathbb{P}\{\tau \leq t \,|\, \mathcal{F}_\infty\} = \mathbb{P}\{\tau \leq t \,|\, \mathcal{F}_t\}, \tag{6}$$

and thus Ψ coincides with the $\mathbb{F}$-hazard process Γ of τ and the hypothesis (H) is valid. It is also not difficult to show that the process $M_t = H_t - \Gamma_{t\wedge\tau} = H_t - \Psi_{t\wedge\tau}$ follows a $\mathbb{G}$-martingale.

The following result shows that under the hypothesis (H), for any random time τ with continuous hazard process Γ, the auxiliary random variable Θ can be constructed on the original probability space, using τ and Γ (see El Karoui [19] or Blanchet-Scalliet and Jeanblanc [10]).

Lemma 1.4. *Let τ be a random time on a probability space $(\Omega, \mathcal{G}, \mathbb{P})$ such that the $\mathbb{F}$-hazard process Γ of τ under $\mathbb{P}$ is continuous and the hypothesis (H) holds. Then there exists a random variable Θ on $(\Omega, \mathcal{G}, \mathbb{P})$, independent of $\mathbb{F}$ and with the exponential distribution of parameter 1, such that*

$$\tau = \inf\{\, t \in \mathbb{R}_+ : \Gamma_t \geq \Theta\}. \tag{7}$$

Proof. It suffices to check that the random variable $\Theta = \Gamma_\tau$ has the desired properties. Indeed, we have, for every $t \in \mathbb{R}_+$,

$$\mathbb{P}(\Theta \geq t \,|\, \mathcal{F}_\infty) = \mathbb{P}(\Gamma_\tau \geq t \,|\, \mathcal{F}_\infty) = \mathbb{P}(\tau \geq A_t \,|\, \mathcal{F}_\infty) = \exp(-\Gamma_{A_t}) = e^{-t},$$

where A is the left inverse of Γ, so that $\Gamma_{A_t} = t$ for every $t \in \mathbb{R}_+$. □

1.3 Change of a probability measure

Kusuoka [29] shows, by means of a counter-example, that the hypothesis (H) is not invariant with respect to an equivalent change of the underlying probability measure, in general. It is worth noting that his counter-example is based on two filtrations, $\mathbb{H}^1$ and $\mathbb{H}^2$, generated by the two random times τ^1 and τ^2, and he chooses $\mathbb{H}^1$ to play the role of the reference filtration $\mathbb{F}$. We shall argue that in the case where $\mathbb{F}$ is generated by a Brownian motion (or, more generally, by some martingale orthogonal to M under $\mathbb{P}$), the above-mentioned invariance property is valid under mild technical assumptions.

1.3.1 Preliminary lemma

Let us first examine a general set-up in which $\mathbb{G} = \mathbb{F} \vee \mathbb{H}$, where $\mathbb{F}$ is an arbitrary filtration and $\mathbb{H}$ is generated by the default process H. We say that $\mathbb{Q}$ is locally equivalent to $\mathbb{P}$ if $\mathbb{Q}$ is equivalent to $\mathbb{P}$ on $(\Omega, \mathcal{G}_t)$ for every $t \in \mathbb{R}_+$. Then there exists the Radon-Nikodým density process η such that

$$d\mathbb{Q}\,|_{\mathcal{G}_t} = \eta_t \, d\mathbb{P}\,|_{\mathcal{G}_t}, \quad \forall\, t \in \mathbb{R}_+. \tag{8}$$

Part (i) in the next lemma is well known (see Jamshidian [27]). We assume that the hypothesis (H) holds under $\mathbb{P}$.

Lemma 1.5. *(i) Let $\mathbb{Q}$ be a probability measure equivalent to $\mathbb{P}$ on $(\Omega, \mathcal{G}_t)$ for every $t \in \mathbb{R}_+$, with the associated Radon-Nikodým density process η. If the density process η is $\mathbb{F}$-adapted then we have $\mathbb{Q}(\tau \le t \,|\, \mathcal{F}_t) = \mathbb{P}(\tau \le t \,|\, \mathcal{F}_t)$ for every $t \in \mathbb{R}_+$. Hence, the hypothesis* (H) *is also valid under $\mathbb{Q}$ and the $\mathbb{F}$-intensities of τ under $\mathbb{Q}$ and under $\mathbb{P}$ coincide.*
(ii) Assume that $\mathbb{Q}$ is equivalent to $\mathbb{P}$ on $(\Omega, \mathcal{G})$ and $d\mathbb{Q} = \eta_\infty \, d\mathbb{P}$, so that $\eta_t = \mathbb{E}_{\mathbb{P}}(\eta_\infty \,|\, \mathcal{G}_t)$. Then the hypothesis (H) *is valid under $\mathbb{Q}$ whenever we have, for every $t \in \mathbb{R}_+$,*

$$\frac{\mathbb{E}_{\mathbb{P}}(\eta_\infty H_t \,|\, \mathcal{F}_\infty)}{\mathbb{E}_{\mathbb{P}}(\eta_\infty \,|\, \mathcal{F}_\infty)} = \frac{\mathbb{E}_{\mathbb{P}}(\eta_t H_t \,|\, \mathcal{F}_\infty)}{\mathbb{E}_{\mathbb{P}}(\eta_t \,|\, \mathcal{F}_\infty)}. \tag{9}$$

Proof. To prove (i), assume that the density process η is $\mathbb{F}$-adapted. We have for each $t \le s \in \mathbb{R}_+$

$$\mathbb{Q}(\tau \le t \,|\, \mathcal{F}_t) = \frac{\mathbb{E}_{\mathbb{P}}(\eta_t \mathbb{1}_{\{\tau \le t\}} \,|\, \mathcal{F}_t)}{\mathbb{E}_{\mathbb{P}}(\eta_t \,|\, \mathcal{F}_t)} = \mathbb{P}(\tau \le t \,|\, \mathcal{F}_t) = \mathbb{P}(\tau \le t \,|\, \mathcal{F}_s) = \mathbb{Q}(\tau \le t \,|\, \mathcal{F}_s),$$

where the last equality follows by another application of the Bayes formula. The assertion now follows from part (i) in Lemma 1.2.

To prove part (ii), it suffices to establish the equality

$$\widehat{F}_t := \mathbb{Q}(\tau \le t \,|\, \mathcal{F}_t) = \mathbb{Q}(\tau \le t \,|\, \mathcal{F}_\infty), \quad \forall\, t \in \mathbb{R}_+. \tag{10}$$

Note that since the random variables $\eta_t \mathbb{1}_{\{\tau \le t\}}$ and η_t are $\mathbb{P}$-integrable and $\mathcal{G}_t$-measurable, using the Bayes formula, part (v) in Lemma 1.2, and assumed equality (9), we obtain the following chain of equalities

$$\begin{aligned}\mathbb{Q}(\tau \le t \,|\, \mathcal{F}_t) &= \frac{\mathbb{E}_{\mathbb{P}}(\eta_t \mathbb{1}_{\{\tau \le t\}} \,|\, \mathcal{F}_t)}{\mathbb{E}_{\mathbb{P}}(\eta_t \,|\, \mathcal{F}_t)} = \frac{\mathbb{E}_{\mathbb{P}}(\eta_t \mathbb{1}_{\{\tau \le t\}} \,|\, \mathcal{F}_\infty)}{\mathbb{E}_{\mathbb{P}}(\eta_t \,|\, \mathcal{F}_\infty)} \\ &= \frac{\mathbb{E}_{\mathbb{P}}(\eta_\infty \mathbb{1}_{\{\tau \le t\}} \,|\, \mathcal{F}_\infty)}{\mathbb{E}_{\mathbb{P}}(\eta_\infty \,|\, \mathcal{F}_\infty)} = \mathbb{Q}(\tau \le t \,|\, \mathcal{F}_\infty).\end{aligned}$$

We conclude that the hypothesis (H) holds under $\mathbb{Q}$ if and only if (9) is valid. □

Unfortunately, straightforward verification of condition (9) is rather cumbersome. For this reason, we shall provide alternative sufficient conditions for the preservation of the hypothesis (H) under a locally equivalent probability measure.

1.3.2 Case of the Brownian filtration

Let W be a Brownian motion under $\mathbb{P}$ with respect to its natural filtration $\mathbb{F}$. Since we work under the hypothesis (H), the process W is also a $\mathbb{G}$-martingale, where $\mathbb{G} = \mathbb{F} \vee \mathbb{H}$. Hence, W is a Brownian motion with respect to $\mathbb{G}$ under $\mathbb{P}$. Our goal is to show that the hypothesis (H) is still valid under $\mathbb{Q} \in \mathcal{Q}$ for a large class $\mathcal{Q}$ of (locally) equivalent probability measures on $(\Omega, \mathcal{G})$.

Let $\mathbb{Q}$ be an arbitrary probability measure locally equivalent to $\mathbb{P}$ on $(\Omega, \mathcal{G})$. Kusuoka [29] (see also Section 5.2.2 in Bielecki and Rutkowski [3]) proved that, under the hypothesis (H), any $\mathbb{G}$-martingale under $\mathbb{P}$ can be represented as the sum of stochastic integrals with respect to the Brownian motion W and the jump martingale M. In our set-up, Kusuoka's representation theorem implies that there exist $\mathbb{G}$-predictable processes θ and $\zeta > -1$, such that the Radon-Nikodým density η of $\mathbb{Q}$ with respect to $\mathbb{P}$ satisfies the following SDE

$$d\eta_t = \eta_{t-}\big(\theta_t \, dW_t + \zeta_t \, dM_t\big) \tag{11}$$

with the initial value $\eta_0 = 1$. More explicitly, the process η equals

$$\eta_t = \mathcal{E}_t\left(\int_0^{\cdot} \theta_u \, dW_u\right)\mathcal{E}_t\left(\int_0^{\cdot} \zeta_u \, dM_u\right) = \eta_t^{(1)}\eta_t^{(2)}, \tag{12}$$

where we write

$$\eta_t^{(1)} = \mathcal{E}_t\left(\int_0^{\cdot} \theta_u \, dW_u\right) = \exp\left(\int_0^t \theta_u \, dW_u - \frac{1}{2}\int_0^t \theta_u^2 \, du\right), \tag{13}$$

and

$$\eta_t^{(2)} = \mathcal{E}_t\left(\int_0^{\cdot} \zeta_u \, dM_u\right) = \exp\left(\int_0^t \ln(1+\zeta_u)\, dH_u - \int_0^{t\wedge\tau} \zeta_u \gamma_u \, du\right). \tag{14}$$

Moreover, by virtue of a suitable version of Girsanov's theorem, the following processes $\widehat{W}$ and $\widehat{M}$ are $\mathbb{G}$-martingales under $\mathbb{Q}$

$$\widehat{W}_t = W_t - \int_0^t \theta_u \, du, \quad \widehat{M}_t = M_t - \int_0^t \mathbb{1}_{\{u<\tau\}} \gamma_u \zeta_u \, du. \tag{15}$$

Proposition 1.1. *Assume that the hypothesis* (H) *holds under* $\mathbb{P}$. *Let* $\mathbb{Q}$ *be a probability measure locally equivalent to* $\mathbb{P}$ *with the associated Radon-Nikodým density process* η *given by formula* (12). *If the process* θ *is* $\mathbb{F}$*-adapted then the hypothesis* (H) *is valid under* $\mathbb{Q}$ *and the* $\mathbb{F}$*-intensity of* τ *under* $\mathbb{Q}$ *equals* $\widehat{\gamma}_t = (1+\widetilde{\zeta}_t)\gamma_t$, *where* $\widetilde{\zeta}$ *is the unique* $\mathbb{F}$*-predictable process such that the equality* $\widetilde{\zeta}_t \mathbb{1}_{\{t\le\tau\}} = \zeta_t \mathbb{1}_{\{t\le\tau\}}$ *holds for every* $t \in \mathbb{R}_+$.

Proof. Let $\widetilde{\mathbb{P}}$ be the probability measure locally equivalent to $\mathbb{P}$ on $(\Omega, \mathcal{G})$, given by

$$d\widetilde{\mathbb{P}}\,|_{\mathcal{G}_t} = \mathcal{E}_t\left(\int_0^{\cdot} \zeta_u \, dM_u\right) d\mathbb{P}\,|_{\mathcal{G}_t} = \eta_t^{(2)}\, d\mathbb{P}\,|_{\mathcal{G}_t}. \tag{16}$$

We claim that the hypothesis (H) holds under $\widetilde{\mathbb{P}}$. From Girsanov's theorem, the process W follows a Brownian motion under $\widetilde{\mathbb{P}}$ with respect to both $\mathbb{F}$ and $\mathbb{G}$. Moreover, from the predictable representation property of W under $\widetilde{\mathbb{P}}$, we deduce that any $\mathbb{F}$-local martingale L under $\widetilde{\mathbb{P}}$ can be written as a stochastic integral with respect to W. Specifically, there exists an $\mathbb{F}$-predictable process ξ such that

$$L_t = L_0 + \int_0^t \xi_u \, dW_u.$$

This shows that L is also a $\mathbb{G}$-local martingale, and thus the hypothesis (H) holds under $\widetilde{\mathbb{P}}$. Since

$$d\mathbb{Q}\,|_{\mathcal{G}_t} = \mathcal{E}_t\left(\int_0^{\cdot} \theta_u \, dW_u\right) d\widetilde{\mathbb{P}}\,|_{\mathcal{G}_t},$$

by virtue of part (i) in Lemma 1.5, the hypothesis (H) is valid under $\mathbb{Q}$ as well. The last claim in the statement of the lemma can be deduced from the fact that the hypothesis (H) holds under $\mathbb{Q}$ and, by Girsanov's theorem, the process

$$\widehat{M}_t = M_t - \int_0^t 1\!\!1_{\{u<\tau\}}\gamma_u\zeta_u\,du = H_t - \int_0^t 1\!\!1_{\{u<\tau\}}(1+\widetilde{\zeta}_u)\gamma_u\,du$$

is a $\mathbb{Q}$-martingale. □

We claim that the equality $\widetilde{\mathbb{P}} = \mathbb{P}$ holds on the filtration $\mathbb{F}$. Indeed, we have $d\widetilde{\mathbb{P}}\,|_{\mathcal{F}_t} = \widetilde{\eta}_t\,d\mathbb{P}\,|_{\mathcal{F}_t}$, where we write $\widetilde{\eta}_t = \mathbb{E}_{\mathbb{P}}(\eta^{(2)}_t\,|\,\mathcal{F}_t)$, and

$$\mathbb{E}_{\mathbb{P}}(\eta^{(2)}_t\,|\,\mathcal{F}_t) = \mathbb{E}_{\mathbb{P}}\left(\mathcal{E}_t\left(\int_0^{\cdot}\zeta_u\,dM_u\right)\,\Big|\,\mathcal{F}_\infty\right) = 1, \quad \forall\, t \in \mathbb{R}_+, \tag{17}$$

where the first equality follows from part (v) in Lemma 1.2.

To establish the second equality in (17), we first note that since the process M is stopped at τ, we may assume, without loss of generality, that $\zeta = \widetilde{\zeta}$ where the process $\widetilde{\zeta}$ is $\mathbb{F}$-predictable (see Lemma 1.1). Moreover, in view of (7) the conditional cumulative distribution function of τ given $\mathcal{F}_\infty$ has the form $1 - \exp(-\Gamma_t(\omega))$. Hence, for arbitrarily selected sample paths of processes ζ and Γ, the claimed equality can be seen as a consequence of the martingale property of the Doléans exponential.

Formally, it can be proved by following elementary calculations, where the first equality is a consequence of (14)),

$$\begin{aligned}
&\mathbb{E}_{\mathbb{P}}\left(\mathcal{E}_t\left(\int_0^{\cdot}\widetilde{\zeta}_u\,dM_u\right)\,\Big|\,\mathcal{F}_\infty\right) = \mathbb{E}_{\mathbb{P}}\left(\left(1 + 1\!\!1_{\{t\ge\tau\}}\widetilde{\zeta}_\tau\right)\exp\left(-\int_0^{t\wedge\tau}\widetilde{\zeta}_u\gamma_u\,du\right)\Big|\,\mathcal{F}_\infty\right)\\
&= \mathbb{E}_{\mathbb{P}}\left(\int_0^\infty\left(1 + 1\!\!1_{\{t\ge u\}}\widetilde{\zeta}_u\right)\exp\left(-\int_0^{t\wedge u}\widetilde{\zeta}_v\gamma_v\,dv\right)\gamma_u e^{-\int_0^u\gamma_v\,dv}du\,\Big|\,\mathcal{F}_\infty\right)\\
&= \mathbb{E}_{\mathbb{P}}\left(\int_0^t\left(1+\widetilde{\zeta}_u\right)\gamma_u\exp\left(-\int_0^u(1+\widetilde{\zeta}_v)\gamma_v\,dv\right)du\,\Big|\,\mathcal{F}_\infty\right)\\
&+ \exp\left(-\int_0^t\widetilde{\zeta}_v\gamma_v\,dv\right)\mathbb{E}_{\mathbb{P}}\left(\int_t^\infty\gamma_u e^{-\int_0^u\gamma_v\,dv}du\,\Big|\,\mathcal{F}_\infty\right)\\
&= \int_0^t\left(1+\widetilde{\zeta}_u\right)\gamma_u\exp\left(-\int_0^u(1+\widetilde{\zeta}_v)\gamma_v\,dv\right)du\\
&+ \exp\left(-\int_0^t\widetilde{\zeta}_v\gamma_v\,dv\right)\int_t^\infty\gamma_u e^{-\int_0^u\gamma_v\,dv}du\\
&= 1 - \exp\left(-\int_0^t(1+\widetilde{\zeta}_v)\gamma_v\,dv\right) + \exp\left(-\int_0^t\widetilde{\zeta}_v\gamma_v\,dv\right)\exp\left(-\int_0^t\gamma_v\,dv\right) = 1,
\end{aligned}$$

where the second last equality follows by an application of the chain rule.

1.3.3 Extension to orthogonal martingales

Equality (17) suggests that Proposition 1.1 can be extended to the case of arbitrary orthogonal local martingales. Such a generalization is convenient, if we wish to cover the situation considered in Kusuoka's counterexample.

Let N be a local martingale under $\mathbb{P}$ with respect to the filtration $\mathbb{F}$. It is also a $\mathbb{G}$-local martingale, since we maintain the assumption that the hypothesis (H) holds under $\mathbb{P}$. Let $\mathbb{Q}$ be an arbitrary probability measure locally equivalent to $\mathbb{P}$ on $(\Omega, \mathcal{G})$. We assume that the Radon-Nikodým density process η of $\mathbb{Q}$ with respect to $\mathbb{P}$ equals

$$d\eta_t = \eta_{t-}\big(\theta_t \, dN_t + \zeta_t \, dM_t\big) \tag{18}$$

for some $\mathbb{G}$-predictable processes θ and $\zeta > -1$ (the properties of the process θ depend, of course, on the choice of the local martingale N). The next result covers the case where N and M are orthogonal $\mathbb{G}$-local martingales under $\mathbb{P}$, so that the product MN follows a $\mathbb{G}$-local martingale.

Proposition 1.2. *Assume that the following conditions hold:*
(a) N and M are orthogonal $\mathbb{G}$-local martingales under $\mathbb{P}$,
(b) N has the predictable representation property under $\mathbb{P}$ with respect to $\mathbb{F}$, in the sense that any $\mathbb{F}$-local martingale L under $\mathbb{P}$ can be written as

$$L_t = L_0 + \int_0^t \xi_u \, dN_u, \quad \forall\, t \in \mathbb{R}_+,$$

for some $\mathbb{F}$-predictable process ξ,
(c) $\widetilde{\mathbb{P}}$ is a probability measure on $(\Omega, \mathcal{G})$ such that (16) *holds.*
Then we have:
(i) the hypothesis (H) *is valid under $\widetilde{\mathbb{P}}$,*
(ii) if the process θ is $\mathbb{F}$-adapted then the hypothesis (H) *is valid under $\mathbb{Q}$.*

The proof of the proposition hinges on the following simple lemma.

Lemma 1.6. *Under the assumptions of Proposition 1.2, we have:*
(i) N is a $\mathbb{G}$-local martingale under $\widetilde{\mathbb{P}}$,
(ii) N has the predictable representation property for $\mathbb{F}$-local martingales under $\widetilde{\mathbb{P}}$.

Proof. In view of (c), we have $d\widetilde{\mathbb{P}}|_{\mathcal{G}_t} = \eta^{(2)}_t \, d\mathbb{P}|_{\mathcal{G}_t}$, where the density process $\eta^{(2)}$ is given by (14), so that $d\eta^{(2)}_t = \eta^{(2)}_{t-}\zeta_t \, dM_t$. From the assumed orthogonality of N and M, it follows that N and $\eta^{(2)}$ are orthogonal $\mathbb{G}$-local martingales under $\mathbb{P}$, and thus $N\eta^{(2)}$ is a $\mathbb{G}$-local martingale under $\mathbb{P}$ as well. This means that N is a $\mathbb{G}$-local martingale under $\widetilde{\mathbb{P}}$, so that (i) holds.

To establish part (ii) in the lemma, we first define the auxiliary process $\widetilde{\eta}$ by setting $\widetilde{\eta}_t = \mathbb{E}_{\mathbb{P}}(\eta^{(2)}_t \,|\, \mathcal{F}_t)$. Then manifestly $d\widetilde{\mathbb{P}}|_{\mathcal{F}_t} = \widetilde{\eta}_t \, d\mathbb{P}|_{\mathcal{F}_t}$, and thus in order to show that any $\mathbb{F}$-local martingale under $\widetilde{\mathbb{P}}$ follows an $\mathbb{F}$-local martingale under $\mathbb{P}$, it suffices to check that $\widetilde{\eta}_t = 1$ for every $t \in \mathbb{R}_+$, so that $\widetilde{\mathbb{P}} = \mathbb{P}$ on $\mathbb{F}$. To this end, we note that

$$\mathbb{E}_{\mathbb{P}}(\eta_t^{(2)} \,|\, \mathcal{F}_t) = \mathbb{E}_{\mathbb{P}}\left(\mathcal{E}_t\left(\int_0^{\cdot} \zeta_u \, dM_u\right) \Big| \mathcal{F}_\infty\right) = 1, \quad \forall \, t \in \mathbb{R}_+,$$

where the first equality follows from part (v) in Lemma 1.2, and the second one can established similarly as the second equality in (17).

We are in a position to prove (ii). Let L be an $\mathbb{F}$-local martingale under $\widetilde{\mathbb{P}}$. Then it follows also an $\mathbb{F}$-local martingale under $\mathbb{P}$ and thus, by virtue of (b), it admits an integral representation with respect to N under $\mathbb{P}$ and $\widetilde{\mathbb{P}}$. This shows that N has the predictable representation property with respect to $\mathbb{F}$ under $\widetilde{\mathbb{P}}$. □

We now proceed to the proof of Proposition 1.2.

Proof of Proposition 1.2. We shall argue along the similar lines as in the proof of Proposition 1.1. To prove (i), note that by part (ii) in Lemma 1.6 we know that any $\mathbb{F}$-local martingale under $\widetilde{\mathbb{P}}$ admits the integral representation with respect to N. But, by part (i) in Lemma 1.6, N is a $\mathbb{G}$-local martingale under $\widetilde{\mathbb{P}}$. We conclude that L is a $\mathbb{G}$-local martingale under $\widetilde{\mathbb{P}}$, and thus the hypothesis (H) is valid under $\widetilde{\mathbb{P}}$. Assertion (ii) now follows from part (i) in Lemma 1.5. □

Remark 1.3. It should be stressed that Proposition 1.2 is not directly employed in what follows. We decided to present it here, since it sheds some light on specific technical problems arising in the context of modeling dependent default times through an equivalent change of a probability measure (see Kusuoka [29]).

Example 1.1. Kusuoka [29] presents a counter-example based on the two independent random times τ_1 and τ_2 given on some probability space $(\Omega, \mathcal{G}, \mathbb{P})$. We write $M_t^i = H_t^i - \int_0^{t \wedge \tau_i} \gamma_i(u)\, du$, where $H_t^i = \mathbb{1}_{\{t \geq \tau_i\}}$ and γ_i is the deterministic intensity function of τ_i under $\mathbb{P}$. Let us set $d\mathbb{Q}\,|_{\mathcal{G}_t} = \eta_t \, d\mathbb{P}\,|_{\mathcal{G}_t}$, where $\eta_t = \eta_t^{(1)} \eta_t^{(2)}$ and, for $i = 1, 2$ and every $t \in \mathbb{R}_+$,

$$\eta_t^{(i)} = 1 + \int_0^t \eta_{u-}^{(i)} \zeta_u^{(i)} \, dM_u^i = \mathcal{E}_t\left(\int_0^{\cdot} \zeta_u^{(i)} \, dM_u^i\right)$$

for some $\mathbb{G}$-predictable processes $\zeta^{(i)}$, $i = 1, 2$, where $\mathbb{G} = \mathbb{H}^1 \vee \mathbb{H}^2$. We set $\mathbb{F} = \mathbb{H}^1$ and $\mathbb{H} = \mathbb{H}^2$. Manifestly, the hypothesis (H) holds under $\mathbb{P}$.

Moreover, in view of Proposition 1.2, it is still valid under the equivalent probability measure $\widetilde{\mathbb{P}}$ given by

$$d\widetilde{\mathbb{P}}\,|_{\mathcal{G}_t} = \mathcal{E}_t\left(\int_0^{\cdot} \zeta_u^{(2)}\, dM_u^2\right) d\mathbb{P}\,|_{\mathcal{G}_t}.$$

It is clear that $\widetilde{\mathbb{P}} = \mathbb{P}$ on $\mathbb{F}$, since

$$\mathbb{E}_{\mathbb{P}}(\eta_t^{(2)}\,|\,\mathcal{F}_t) = \mathbb{E}_{\mathbb{P}}\left(\mathcal{E}_t\left(\int_0^{\cdot} \zeta_u^{(2)}\, dM_u^2\right)\Big|\,\mathcal{H}_t^1\right) = 1, \quad \forall\, t \in \mathbb{R}_+.$$

However, the hypothesis (H) is not necessarily valid under $\mathbb{Q}$ if the process $\zeta^{(1)}$ fails to be $\mathbb{F}$-adapted. In Kusuoka's counter-example, the process $\zeta^{(1)}$ was chosen to be explicitly dependent on both random times, and it was shown that the hypothesis (H) does not hold under $\mathbb{Q}$. For an alternative approach to Kusuoka's example, through an absolutely continuous change of a probability measure, the interested reader may consult Collin-Dufresne *et al.* [13].

2. Semimartingale Model with a Common Default

In what follows, we fix a finite horizon date $T > 0$. For the purpose of this work, it is enough to formally define a generic defaultable claim through the following definition.

Definition 2.1. A *defaultable claim* with maturity date T is represented by a triplet (X, Z, τ), where:
(i) the *default time* τ specifies the random time of default, and thus also the default events $\{\tau \leq t\}$ for every $t \in [0, T]$,
(ii) the *promised payoff* $X \in \mathcal{F}_T$ represents the random payoff received by the owner of the claim at time T, provided that there was no default prior to or at time T; the actual payoff at time T associated with X thus equals $X\mathbb{1}_{\{T<\tau\}}$,
(iii) the $\mathbb{F}$-adapted *recovery process* Z specifies the recovery payoff Z_τ received by the owner of a claim at time of default (or at maturity), provided that the default occurred prior to or at maturity date T.

In practice, hedging of a credit derivative after default time is usually of minor interest. Also, in a model with a single default time, hedging after default reduces to replication of a non-defaultable claim. It is thus natural to define the replication of a defaultable claim in the following way.

Definition 2.2. We say that a self-financing strategy ϕ replicates a defaultable claim (X, Z, τ) if its wealth process $V(\phi)$ satisfies $V_T(\phi)\mathbb{1}_{\{T<\tau\}} = X\mathbb{1}_{\{T<\tau\}}$ and $V_\tau(\phi)\mathbb{1}_{\{T\geq\tau\}} = Z_\tau\mathbb{1}_{\{T\geq\tau\}}$.

When dealing with replicating strategies, in the sense of Definition 2.2, we will always assume, without loss of generality, that the components of the process ϕ are $\mathbb{F}$-predictable processes.

2.1 Dynamics of asset prices

We assume that we are given a probability space $(\Omega, \mathcal{G}, \mathbb{P})$ endowed with a (possibly multi-dimensional) standard Brownian motion W and a random time τ admitting an $\mathbb{F}$-intensity γ under $\mathbb{P}$, where $\mathbb{F}$ is the filtration generated by W. In addition, we assume that τ satisfies (4), so that the hypothesis (H) is valid under $\mathbb{P}$ for filtrations $\mathbb{F}$ and $\mathbb{G} = \mathbb{F} \vee \mathbb{H}$. Since the default time admits an $\mathbb{F}$-intensity, it is not an $\mathbb{F}$-stopping time. Indeed, any stopping time with respect to a Brownian filtration is known to be predictable.

We interpret τ as the common default time for all defaultable assets in our model. For simplicity, we assume that only three primary assets are traded in the market, and the dynamics under the historical probability $\mathbb{P}$ of their prices are, for $i = 1, 2, 3$ and $t \in [0, T]$,

$$dY_t^i = Y_{t-}^i \big(\mu_{i,t}\, dt + \sigma_{i,t}\, dW_t + \kappa_{i,t}\, dM_t \big), \tag{19}$$

or equivalently,

$$dY_t^i = Y_{t-}^i \Big((\mu_{i,t} - \kappa_{i,t}\gamma_t \mathbb{1}_{\{t \le \tau\}})\, dt + \sigma_{i,t}\, dW_t + \kappa_{i,t}\, dH_t \Big). \tag{20}$$

The processes $(\mu_i, \sigma_i, \kappa_i) = (\mu_{i,t}, \sigma_{i,t}, \kappa_{i,t},\ t \ge 0)$, $i = 1, 2, 3$, are assumed to be $\mathbb{G}$-adapted, where $\mathbb{G} = \mathbb{F} \vee \mathbb{H}$. In addition, we assume that $\kappa_i \ge -1$ for any $i = 1, 2, 3$, so that Y^i are nonnegative processes, and they are strictly positive prior to τ.

Note that, according to Definition 2.2, replication refers to the behavior of the wealth process $V(\phi)$ on the random interval $[\![0, \tau \wedge T]\!]$ only. Hence, for the purpose of replication of defaultable claims of the form (X, Z, τ), it is sufficient to consider prices of primary assets stopped at $\tau \wedge T$. This implies that instead of dealing with $\mathbb{G}$-adapted coefficients in (19), it suffices to focus on $\mathbb{F}$-adapted coefficients of stopped price processes. However, for the sake of completeness, we shall also deal with T-maturity claims of the form $Y = G(Y_T^1, Y_T^2, Y_T^3, H_T)$ (see Section 5 below).

2.1.1 Pre-default values

As will become clear in what follows, when dealing with defaultable claims of the form (X, Z, τ), we will be mainly concerned with the so-called pre-default prices. The *pre-default price* $\widetilde{Y}^i$ of the ith asset is an $\mathbb{F}$-adapted, continuous process, given by the equation, for $i = 1, 2, 3$ and $t \in [0, T]$,

$$d\widetilde{Y}_t^i = \widetilde{Y}_t^i \Big((\mu_{i,t} - \kappa_{i,t}\gamma_t)\, dt + \sigma_{i,t}\, dW_t \Big) \tag{21}$$

with $\widetilde{Y}^i_0 = Y^i_0$. Put another way, $\widetilde{Y}^i$ is the unique $\mathbb{F}$-predictable process such that (see Lemma 1.1) $\widetilde{Y}^i_t \mathbb{1}_{\{t\leq\tau\}} = Y^i_t \mathbb{1}_{\{t\leq\tau\}}$ for $t \in \mathbb{R}_+$. When dealing with the pre-default prices, we may and do assume, without loss of generality, that the processes μ_i, σ_i and κ_i are $\mathbb{F}$-predictable.

It is worth stressing that the historically observed drift coefficient equals $\mu_{i,t} - \kappa_{i,t}\gamma_t$, rather than $\mu_{i,t}$. The drift coefficient denoted by $\mu_{i,t}$ is already credit-risk adjusted in the sense of our model, and it is not directly observed. This convention was chosen here for the sake of simplicity of notation. It also lends itself to the following intuitive interpretation: if ϕ^i is the number of units of the ith asset held in our portfolio at time t then the gains/losses from trades in this asset, prior to default time, can be represented by the differential

$$\phi^i_t \, d\widetilde{Y}^i_t = \phi^i_t \widetilde{Y}^i_t \big(\mu_{i,t}\, dt + \sigma_{i,t}\, dW_t\big) - \phi^i_t \widetilde{Y}^i_t \kappa_{i,t}\gamma_t \, dt.$$

The last term may be here separated, and formally treated as an effect of continuously paid dividends at the dividend rate $\kappa_{i,t}\gamma_t$. However, this interpretation may be misleading, since this quantity is not directly observed. In fact, the mere estimation of the drift coefficient in dynamics (21) is not practical.

Still, if this formal interpretation is adopted, it is sometimes possible make use of the standard results concerning the valuation of derivatives of dividend-paying assets. It is, of course, a delicate issue how to separate in practice both components of the drift coefficient. We shall argue below that although the dividend-based approach is formally correct, a more pertinent and simpler way of dealing with hedging relies on the assumption that only the effective drift $\mu_{i,t} - \kappa_{i,t}\gamma_t$ is observable. In practical approach to hedging, the values of drift coefficients in dynamics of asset prices play no essential role, so that they are considered as market observables.

2.1.2 Market observables

To summarize, we assume throughout that the *market observables* are: the pre-default market prices of primary assets, their volatilities and correlations, as well as the jump coefficients $\kappa_{i,t}$ (the financial interpretation of jump coefficients is examined in the next subsection). To summarize we postulate that under the statistical probability $\mathbb{P}$ we have

$$dY^i_t = Y^i_{t-}\big(\widetilde{\mu}_{i,t}\, dt + \sigma_{i,t}\, dW_t + \kappa_{i,t}\, dH_t\big) \tag{22}$$

where the drift terms $\widetilde{\mu}_{i,t}$ are not observable, but we can observe the volatilities $\sigma_{i,t}$ (and thus the assets correlations), and we have an a priori assessment of jump coefficients $\kappa_{i,t}$. In this general set-up, the most natural assumption is that the dimension of a driving Brownian motion W equals the number of tradable assets. However, for the sake of simplicity of presentation, we shall frequently assume that W is one-dimensional. One of our goals will be to derive closed-form solutions for replicating strategies for derivative securities in terms of market observables only (whenever replication of a given claim is actually feasible). To achieve this goal, we shall combine a general theory of hedging defaultable claims within a continuous semimartingale set-up, with a judicious specification of particular models with deterministic volatilities and correlations.

2.1.3 Recovery schemes

It is clear that the sample paths of price processes Y^i are continuous, except for a possible discontinuity at time τ. Specifically, we have that

$$\Delta Y^i_\tau := Y^i_\tau - Y^i_{\tau-} = \kappa_{i,\tau} Y^i_{\tau-},$$

so that $Y^i_\tau = Y^i_{\tau-}(1+\kappa_{i,\tau}) = \widetilde{Y}^i_{\tau-}(1+\kappa_{i,\tau})$.

A primary asset Y^i is termed a *default-free asset* (*defaultable asset*, respectively) if $\kappa_i = 0$ ($\kappa_i \neq 0$, respectively). In the special case when $\kappa_i = -1$, we say that a defaultable asset Y^i is subject to a *total default*, since its price drops to zero at time τ and stays there forever. Such an asset ceases to exist after default, in the sense that it is no longer traded after default. This feature makes the case of a total default quite different from other cases, as we shall see in our study below.

In market practice, it is common for a credit derivative to deliver a positive recovery (for instance, a *protection payment*) in case of default. Formally, the value of this recovery at default is determined as the value of some underlying process, that is, it is equal to the value at time τ of some $\mathbb{F}$-adapted recovery process Z.

For example, the process Z can be equal to δ, where δ is a constant, or to $g(t, \delta Y_t)$ where g is a deterministic function and $(Y_t, t \geq 0)$ is the price process of some default-free asset. Typically, the recovery is paid at default time, but it may also happen that it is postponed to the maturity date.

Let us observe that the case where a defaultable asset Y^i pays a predetermined recovery at default is covered by our set-up defined in (19). For instance, the case of a constant recovery payoff $\delta_i \geq 0$ at default time τ corresponds to the process $\kappa_{i,t} = \delta_i (Y^i_{t-})^{-1} - 1$. Under this convention, the price Y^i is governed under $\mathbb{P}$ by the SDE

$$dY^i_t = Y^i_{t-}\Big(\mu_{i,t}\,dt + \sigma_{i,t}\,dW_t + (\delta_i (Y^i_{t-})^{-1} - 1)\,dM_t\Big). \tag{23}$$

If the recovery is proportional to the pre-default value $Y^i_{\tau-}$, and is paid at default time τ (this scheme is known as the *fractional recovery of market value*), we have $\kappa_{i,t} = \delta_i - 1$ and

$$dY^i_t = Y^i_{t-}\Big(\mu_{i,t}\,dt + \sigma_{i,t}\,dW_t + (\delta_i - 1)\,dM_t\Big). \tag{24}$$

2.2 Risk-neutral valuation

To provide a partial justification for the postulated dynamics of the price of a defaultable asset delivering a recovery, let us study a toy example with two assets: a savings account with constant interest rate r and a defaultable asset Y represented by a defaultable claim (X, Z, τ). In this toy model, the only source of noise is the default time, hence, the only relevant filtration is $\mathbb{H}$ (in other words, the reference filtration $\mathbb{F}$ is trivial). We assume that by choosing today's prices of a large family liquidly traded defaultable assets, the market implicitly specifies a martingale measure $\mathbb{Q}$, equivalent to the historical probability $\mathbb{P}$. More precisely, the probability distribution of τ under an equivalent martingale measure (e.m.m.) $\mathbb{Q}$ can be inferred from market data. We are thus interested in the dynamics of the price process of (X, Z, τ) under $\mathbb{Q}$.

It is worth noting that in this subsection we adopt a totally different perspective than in the rest of the present paper. In fact, no attempt to replicate a defaultable claim is done in this section. We assume instead that the risk-neutral default intensity can be uniquely determined from prices of traded assets, and we postulate that the price of (X, Z, τ) is defined by the standard risk-neutral valuation formula. The argument that formally justifies the use of this pricing rule is that we obtain in this way an arbitrage-free market model in which $\mathbb{Q}$ is a martingale measure, and a defaultable claim can be considered to be an additional traded asset. Since we do not assume here that a defaultable claim is attainable, its spot price (that is, the price expressed in units of cash) depends explicitly on the risk-neutral default intensity. As was mentioned above, the arbitrage price of a defaultable claim, when expressed in terms of tradeable assets used for its replication, will be shown to not depend directly on real-world (or risk-neutral) default intensity.

To conclude, the rationale for the calculations given below, is that we strive here to justify the dynamics of prices of primary assets in our model. The risk-neutral valuation considered in this subsection is not supported by replication-based arguments, and thus it is not surprising that it exhibits specific features that are not present in the replication-based valuation.

We make the standing assumption that τ admits a continuous cumulative distribution function $\widehat{F}$ under $\mathbb{Q}$. Hence, the hazard function $\widehat{\Gamma}$ is also continuous, and the process $\widehat{M}_t = H_t - \widehat{\Gamma}(t \wedge \tau)$ is an $\mathbb{H}$-martingale under $\mathbb{Q}$. The following result is standard (see, e.g., Proposition 4.3.2 in Bielecki and Rutkowski [3]).

Proposition 2.1. *Assume that the cumulative distribution function F of τ is continuous. Let M^h be an $\mathbb{H}$-martingale given by $M^h_t = \mathbb{E}_{\mathbb{Q}}(h(\tau) \,|\, \mathcal{H}_t)$ for some Borel measurable function $h : \mathbb{R}_+ \to \mathbb{R}$ such that the random variable $h(\tau)$ is*

Q-integrable. Then

$$M^h_t = M^h_0 + \int_0^t (h(u) - g(u))\, d\widehat{M}_u = M^h_0 + \int_0^t (h(u) - M^h_{u-})\, d\widehat{M}_u, \tag{25}$$

where we write

$$g(t) = e^{\widehat{\Gamma}(t)}\, \mathbb{E}_{\mathbb{Q}}\big(\mathbb{1}_{\{t<\tau\}} h(\tau)\big).$$

Remark 2.1. Using the above proposition, it can be easily shown that on $(\Omega, \mathcal{G}_T)$ we have

$$d\mathbb{P} = \mathcal{E}_T\left(-\int_0^{\cdot} \zeta_u\, d\widehat{M}_u\right) d\mathbb{Q},$$

for some $\mathbb{H}$-predictable process ζ.

2.2.1 Price dynamics of a survival claim $(X, 0, \tau)$.

In what follows, we shall refer to a defaultable claim of the form $(X, 0, \tau)$ as a *survival claim*. By virtue of the risk-neutral valuation formula, the price of the payoff $\mathbb{1}_{\{T<\tau\}}X$ that settles at time T equals, for every $t \in [0, T]$,

$$Y_t = e^{rt}\, \mathbb{E}_{\mathbb{Q}}(\mathbb{1}_{\{T<\tau\}} e^{-rT} X \,|\, \mathcal{H}_t).$$

Note that X is $\mathcal{F}_T$-measurable, and thus constant since the σ-field $\mathcal{F}_T$ is trivial. To find the dynamics of the price process, it suffices to apply Proposition 2.1 to the function $h(u) = \mathbb{1}_{\{u>T\}} e^{-rT} X$. For the Q-martingale $M^h_t = e^{-rt} Y_t$, we thus get, for every $t \in [0, T]$,

$$e^{-rt} Y_t = Y_0 - \int_0^t e^{-ru} Y_{u-}\, d\widehat{M}_u.$$

Suppose that $\widehat{\Gamma}(t) = \int_0^t \widehat{\gamma}(u)\, du$. Then an application of Itô's formula yields

$$dY_t = rY_t\, dt - Y_{t-}\, d\widehat{M}_t = \big(r + \mathbb{1}_{\{t<\tau\}} \widehat{\gamma}(t)\big) Y_t\, dt - Y_{t-}\, dH_t. \tag{26}$$

We deal here with an example of a defaultable asset that is subject to the *total default*, meaning that its price vanishes at and after default.

2.2.2 Price dynamics of a recovery claim $(0, Z, \tau)$.

Recall that our standard convention stipulates that the recovery Z is paid at the time of default. Hence, the price process Y of $(0, Z, \tau)$ is given by the expression

$$Y_t = e^{rt}\, \mathbb{E}_{\mathbb{Q}}(\mathbb{1}_{\{T\geq\tau\}} e^{-r\tau} Z(\tau) \,|\, \mathcal{H}_t).$$

We now have $h(u) = \mathbb{1}_{\{u\leq T\}} e^{-ru} Z(u)$. Consequently,

$$e^{-rt} Y_t = Y_0 + \int_0^t \big(e^{-ru} Z(u) - e^{-ru} Y_{u-}\big)\, d\widehat{M}_u.$$

By applying Itô's formula, we conclude that the dynamics under $\mathbb{Q}$ of an asset that delivers $Z(\tau)$ at time τ are

$$\begin{aligned} dY_t &= rY_{t-}\,dt + (Z(t) - Y_{t-})\,d\widehat{M}_t \\ &= \big(r + \mathbb{1}_{\{t<\tau\}}\widehat{\gamma}(t)\big)Y_t\,dt - \mathbb{1}_{\{t<\tau\}}Z(t)\widehat{\gamma}(t)\,dt + (Z(t) - Y_{t-})\,dH_t. \end{aligned}$$

2.2.3 Price dynamics of a defaultable claim (X, Z, τ).

By combining the formula above with (26), and using Remark 2.1 together with Girsanov's theorem, we arrive at the following result.

Proposition 2.2. *The price process Y of a defaultable claim (X, Z, τ) satisfies under $\mathbb{Q}$*

$$dY_t = rY_{t-}\,dt + (Z(t) - Y_{t-})\,d\widehat{M}_t$$

with the initial condition

$$Y_0 = \mathbb{E}_{\mathbb{Q}}\big(\mathbb{1}_{\{T<\tau\}}e^{-rT}X + \mathbb{1}_{\{T\geq\tau\}}e^{-r\tau}Z(\tau)\big) = e^{-(rT+\widehat{\Gamma}(T))}X + \int_0^T Z(u)\widehat{\gamma}(u)e^{-\widehat{\Gamma}(u)}\,du.$$

Under the statistical probability $\mathbb{P}$, the price process Y satisfies

$$dY_t = \big(rY_{t-} + \mathbb{1}_{\{t<\tau\}}(Z(t) - Y_{t-})\widehat{\gamma}(t)\zeta(t)\big)\,dt + (Z(t) - Y_{t-})\,dM_t,$$

where the $\mathbb{G}$-martingale M under $\mathbb{P}$ equals

$$M_t = \widehat{M}_t + \int_0^t \mathbb{1}_{\{u<\tau\}}\widehat{\gamma}(u)\zeta(u)\,du.$$

Remark 2.2. Proposition 2.2 can be extended to the case when the recovery is random, and is given in the feedback form as $Z(t) = g(t, Y_{t-})$ for some function $g(t, y)$, which is Lipschitz continuous with respect to y. Assume, for instance, that the claim is subject to the fractional recovery of market value, so that $Z(t) = \delta Y_{t-}$ for some constant δ. If, in addition, ζ and $\widehat{\gamma}$ are constant, then we obtain (cf. (24))

$$dY_t = Y_{t-}\big((r + \mathbb{1}_{\{t<\tau\}}(\delta - 1)\widehat{\gamma}\zeta)\,dt + (\delta - 1)\,dM_t\big).$$

Note that here the drift coefficient $\mu_t = r + \mathbb{1}_{\{t<\tau\}}(\delta - 1)\widehat{\gamma}\zeta$ in dynamics of Y follows a $\mathbb{G}$-predictable process, but it is not $\mathbb{F}$-predictable. However, the drift of the pre-default value $\widetilde{Y}$ is simply r.

3. Trading Strategies in a Semimartingale Set-up

We consider trading within the time interval $[0, T]$ for some finite horizon date $T > 0$. For the sake of expositional clarity, we restrict our attention to the case where only three primary assets are traded. The general case of k traded assets was examined by Bielecki *et al.* [5]. We first recall some general properties, which do not depend on the choice of specific dynamics of asset prices.

In this section, we consider a fairly general set-up. In particular, processes Y^i, $i = 1, 2, 3$, are assumed to be nonnegative semi-martingales on a probability space $(\Omega, \mathcal{G}, \mathbb{P})$ endowed with some filtration $\mathbb{G}$. We assume that they represent spot prices of traded assets in our model of the financial market. Neither the existence of a savings account, nor the market completeness are assumed, in general.

Our goal is to characterize contingent claims which are *hedgeable*, in the sense that they can be replicated by continuously rebalanced portfolios consisting of primary assets. Here, by a contingent claim we mean an arbitrary $\mathcal{G}_T$-measurable random variable. We work under the standard assumptions of a frictionless market.

3.1 Unconstrained strategies

Let $\phi = (\phi^1, \phi^2, \phi^3)$ be a trading strategy; in particular, each process ϕ^i is predictable with respect to the filtration $\mathbb{G}$. The wealth of ϕ equals

$$V_t(\phi) = \sum_{i=1}^{3} \phi^i_t Y^i_t, \quad \forall\, t \in [0, T],$$

and a trading strategy ϕ is said to be *self-financing* if

$$V_t(\phi) = V_0(\phi) + \sum_{i=1}^{3} \int_0^t \phi^i_u \, dY^i_u, \quad \forall\, t \in [0, T].$$

Let Φ stand for the class of all self-financing trading strategies. We shall first prove that a self-financing strategy is determined by its initial wealth and the two components ϕ^2, ϕ^3. To this end, we postulate that the price of Y^1 follows a strictly positive process, and we choose Y^1 as a numéraire asset. We shall now analyze the relative values:

$$V^1_t(\phi) := V_t(\phi)(Y^1_t)^{-1}, \quad Y^{i,1}_t := Y^i_t (Y^1_t)^{-1}.$$

Lemma 3.1. (i) *For any $\phi \in \Phi$, we have*

$$V^1_t(\phi) = V^1_0(\phi) + \sum_{i=2}^{3} \int_0^t \phi^i_u \, dY^{i,1}_u, \quad \forall\, t \in [0, T].$$

(ii) *Conversely, let X be a $\mathcal{G}_T$-measurable random variable, and let us assume that there exists $x \in \mathbb{R}$ and $\mathbb{G}$-predictable processes ϕ^i, $i = 2,3$ such that*

$$(27) \qquad X = Y_T^1 \left(x + \sum_{i=2}^{3} \int_0^T \phi_u^i \, dY_u^{i,1} \right).$$

Then there exists a $\mathbb{G}$-predictable process ϕ^1 such that the strategy $\phi = (\phi^1, \phi^2, \phi^3)$ is self-financing and replicates X. Moreover, the wealth process of ϕ (i.e. the time-t price of X) satisfies $V_t(\phi) = V_t^1 Y_t^1$, where

$$(28) \qquad V_t^1 = x + \sum_{i=2}^{3} \int_0^t \phi_u^i \, dY_u^{i,1}, \quad \forall\, t \in [0,T].$$

Proof. The proof of part (i) is given, for instance, in Protter [34]. In the case of continuous semimartingales, this is a well-known result; for discontinuous processes, the proof is not much different. We reproduce it here for the reader's convenience.

Let us first introduce some notation. As usual, $[X,Y]$ stands for the *quadratic covariation* of the two semi-martingales X and Y, as defined by the integration by parts formula:

$$X_t Y_t = X_0 Y_0 + \int_0^t X_{u-} \, dY_u + \int_0^t Y_{u-} \, dX_u + [X,Y]_t.$$

For any càdlàg (i.e., RCLL) process Y, we denote by $\Delta Y_t = Y_t - Y_{t-}$ the size of the jump at time t. Let $V = V(\phi)$ be the value of a self-financing strategy, and let $V^1 = V^1(\phi) = V(\phi)(Y^1)^{-1}$ be its value relative to the numéraire Y^1. The integration by parts formula yields

$$dV_t^1 = V_{t-} d(Y_t^1)^{-1} + (Y_{t-}^1)^{-1} dV_t + d[(Y^1)^{-1}, V]_t.$$

From the self-financing condition, we have $dV_t = \sum_{i=1}^{3} \phi_t^i \, dY_t^i$. Hence, using elementary rules to compute the quadratic covariation $[X,Y]$ of the two semi-martingales X, Y, we obtain

$$\begin{aligned}
dV_t^1 &= \phi_t^1 Y_{t-}^1 \, d(Y_t^1)^{-1} + \phi_t^2 Y_{t-}^2 \, d(Y_t^1)^{-1} + \phi_t^3 Y_{t-}^3 \, d(Y_t^1)^{-1} \\
&\quad + (Y_{t-}^1)^{-1} \phi_t^1 \, dY_t^1 + (Y_{t-}^1)^{-1} \phi_t^2 \, dY_t^1 + (Y_{t-}^1)^{-1} \phi_t^3 \, dY_t^1 \\
&\quad + \phi_t^1 \, d[(Y^1)^{-1}, Y^1]_t + \phi_t^2 \, d[(Y^1)^{-1}, Y^2]_t + \phi_t^3 \, d[(Y^1)^{-1}, Y^1]_t \\
&= \phi_t^1 \Big(Y_{t-}^1 \, d(Y_t^1)^{-1} + (Y_{t-}^1)^{-1} \, dY_t^1 + d[(Y^1)^{-1}, Y^1]_t \Big) \\
&\quad + \phi_t^2 \Big(Y_{t-}^2 \, d(Y_t^1)^{-1} + (Y_{t-}^1)^{-1} \, dY_{t-}^1 + d[(Y^1)^{-1}, Y^2]_t \Big) \\
&\quad + \phi_t^3 \Big(Y_{t-}^3 \, d(Y_t^1)^{-1} + (Y_{t-}^1)^{-1} \, dY_{t-}^1 + d[(Y^1)^{-1}, Y^3]_t \Big).
\end{aligned}$$

We now observe that

$$Y^1_{t-}\,d(Y^1_t)^{-1}+(Y^1_{t-})^{-1}\,dY^1_t+d[(Y^1)^{-1},Y^1]_t=d(Y^1_t(Y^1_t)^{-1})=0$$

and

$$Y^i_{t-}\,d(Y^1_t)^{-1}+(Y^1_{t-})^{-1}\,dY^i_t+d[(Y^1)^{-1},Y^i]_t=d((Y^1_t)^{-1}Y^i_t).$$

Consequently,

$$dV^1_t=\phi^2_t\,dY^{2,1}_t+\phi^3_t\,dY^{3,1}_t,$$

as was claimed in part (i). We now proceed to the proof of part (ii). We assume that (27) holds for some constant x and processes ϕ^2,ϕ^3, and we define the process V^1 by setting (cf. (28))

$$V^1_t=x+\sum_{i=2}^{3}\int_0^t\phi^i_u\,dY^{i,1}_u,\quad\forall\,t\in[0,T].$$

Next, we define the process ϕ^1 as follows:

$$\phi^1_t=V^1_t-\sum_{i=2}^{3}\phi^i_tY^{i,1}_t=(Y^1_t)^{-1}\Big(V_t-\sum_{i=2}^{3}\phi^i_tY^i_t\Big),$$

where $V_t=V^1_tY^1_t$. Since $dV^1_t=\sum_{i=2}^{3}\phi^i_t\,dY^{i,1}_t$, we obtain

$$\begin{aligned}dV_t=d(V^1_tY^1_t)&=V^1_{t-}dY^1_t+Y^1_{t-}dV^1_t+d[Y^1,V^1]_t\\&=V^1_{t-}dY^1_t+\sum_{i=2}^{3}\phi^i_t\Big(Y^1_{t-}\,dY^{i,1}_t+d[Y^1,Y^{i,1}]_t\Big).\end{aligned}$$

From the equality

$$dY^i_t=d(Y^{i,1}_tY^1_t)=Y^{i,1}_{t-}dY^1_t+Y^1_{t-}dY^{i,1}_t+d[Y^1,Y^{i,1}]_t,$$

it follows that

$$dV_t=V^1_{t-}dY^1_t+\sum_{i=2}^{3}\phi^i_t\Big(dY^i_t-Y^{i,1}_{t-}dY^1_t\Big)=\Big(V^1_{t-}-\sum_{i=2}^{3}\phi^i_tY^{i,1}_{t-}\Big)dY^1_t+\sum_{i=2}^{3}\phi^i_t\,dY^i_t,$$

and our aim is to prove that $dV_t=\sum_{i=1}^{3}\phi^i_t\,dY^i_t$. The last equality holds if

$$\phi^1_t=V^1_t-\sum_{i=2}^{3}\phi^i_tY^{i,1}_t=V^1_{t-}-\sum_{i=2}^{3}\phi^i_tY^{i,1}_{t-},\tag{29}$$

i.e., if $\Delta V^1_t = \sum_{i=2}^3 \phi^i_t \Delta Y^{i,1}_t$, which is the case from the definition (28) of V^1. Note also that from the second equality in (29) it follows that the process ϕ^1 is indeed $\mathbb{G}$-predictable. Finally, the wealth process of ϕ satisfies $V_t(\phi) = V^1_t Y^1_t$ for every $t \in [0,T]$, and thus $V_T(\phi) = X$. □

We say that a self-financing strategy ϕ *replicates* a claim $X \in \mathcal{G}_T$ if

$$X = \sum_{i=1}^{3} \phi^i_T Y^i_T = V_T(\phi),$$

or equivalently,

$$X = V_0(\phi) + \sum_{i=1}^{3} \int_0^T \phi^i_t \, dY^i_t.$$

Suppose that there exists an e.m.m. for some choice of a numéraire asset, and let us restrict our attention to the class of all *admissible* trading strategies, so that our model is arbitrage-free.

Assume that a claim X can be replicated by some admissible trading strategy, so that it is *attainable* (or *hedgeable*). Then, by definition, the *arbitrage price* at time t of X, denoted as $\pi_t(X)$, equals $V_t(\phi)$ for any admissible trading strategy ϕ that replicates X.

In the context of Lemma 3.1, it is natural to choose as an e.m.m. a probability measure $\mathbb{Q}^1$ equivalent to $\mathbb{P}$ on $(\Omega, \mathcal{G}_T)$ and such that the prices $Y^{i,1}$, $i = 2,3$, are $\mathbb{G}$-martingales under $\mathbb{Q}^1$. If a contingent claim X is hedgeable, then its arbitrage price satisfies

$$\pi_t(X) = Y^1_t \mathbb{E}_{\mathbb{Q}^1}(X(Y^1_T)^{-1} \,|\, \mathcal{G}_t).$$

We emphasize that even if an e.m.m. $\mathbb{Q}^1$ is not unique, the price of any hedgeable claim X is given by this conditional expectation. That is to say, in case of a hedgeable claim these conditional expectations under various equivalent martingale measures coincide.

In the special case where $Y^1_t = B(t,T)$ is the price of a default-free zero-coupon bond with maturity T (abbreviated as ZC-bond in what follows), $\mathbb{Q}^1$ is called *T-forward martingale measure*, and it is denoted by $\mathbb{Q}_T$. Since $B(T,T) = 1$, the price of any hedgeable claim X now equals $\pi_t(X) = B(t,T)\,\mathbb{E}_{\mathbb{Q}_T}(X \,|\, \mathcal{G}_t)$.

3.2 Constrained strategies

In this section, we make an additional assumption that the price process Y^3 is strictly positive. Let $\phi = (\phi^1, \phi^2, \phi^3)$ be a self-financing trading strategy satisfying the following constraint:

$$\sum_{i=1}^{2} \phi^i_t Y^i_{t-} = Z_t, \quad \forall\, t \in [0,T], \tag{30}$$

for a predetermined, G-predictable process Z. In the financial interpretation, equality (30) means that a portfolio ϕ is rebalanced in such a way that the total wealth invested in assets Y^1, Y^2 matches a predetermined stochastic process Z. For this reason, the constraint given by (30) is referred to as the *balance condition*.

Our first goal is to extend part (i) in Lemma 3.1 to the case of constrained strategies. Let $\Phi(Z)$ stand for the class of all (admissible) self-financing trading strategies satisfying the balance condition (30). They will be sometimes referred to as *constrained strategies*. Since any strategy $\phi \in \Phi(Z)$ is self-financing, from $dV_t(\phi) = \sum_{i=1}^{3} \phi_t^i \, dY_t^i$, we obtain

$$\Delta V_t(\phi) = \sum_{i=1}^{3} \phi_t^i \Delta Y_t^i = V_t(\phi) - \sum_{i=1}^{3} \phi_t^i Y_{t-}^i.$$

By combining this equality with (30), we deduce that

$$V_{t-}(\phi) = \sum_{i=1}^{3} \phi_t^i Y_{t-}^i = Z_t + \phi_t^3 Y_{t-}^i.$$

Let us write $Y_t^{i,3} = Y_t^i (Y_t^3)^{-1}$, $Z_t^3 = Z_t (Y_t^3)^{-1}$. The following result extends Lemma 1.7 in Bielecki *et al.* [4] from the case of continuous semimartingales to the general case (see also [5]). It is apparent from Proposition 3.1 that the wealth process $V(\phi)$ of a strategy $\phi \in \Phi(Z)$ depends only on a single component of ϕ, namely, ϕ^2.

Proposition 3.1. *The relative wealth $V_t^3(\phi) = V_t(\phi)(Y_t^3)^{-1}$ of any trading strategy $\phi \in \Phi(Z)$ satisfies*

$$V_t^3(\phi) = V_0^3(\phi) + \int_0^t \phi_u^2 \left(dY_u^{2,3} - \frac{Y_{u-}^{2,3}}{Y_{u-}^{1,3}} \, dY_u^{1,3} \right) + \int_0^t \frac{Z_u^3}{Y_{u-}^{1,3}} \, dY_u^{1,3}. \tag{31}$$

Proof. Let us consider discounted values of price processes Y^1, Y^2, Y^3, with Y^3 taken as a numéraire asset. By virtue of part (i) in Lemma 3.1, we thus have

$$V_t^3(\phi) = V_0^3(\phi) + \sum_{i=1}^{2} \int_0^t \phi_u^i \, dY_u^{i,3}. \tag{32}$$

The balance condition (30) implies that

$$\sum_{i=1}^{2} \phi_t^i Y_{t-}^{i,3} = Z_t^3,$$

and thus

$$\phi^1_t = (Y^{1,3}_{t-})^{-1}\big(Z^3_t - \phi^2_t Y^{2,3}_{t-}\big). \tag{33}$$

By inserting (33) into (32), we arrive at the desired formula (31). □

The next result will prove particularly useful for deriving replicating strategies for defaultable claims.

Proposition 3.2. *Let a $\mathcal{G}_T$-measurable random variable X represent a contingent claim that settles at time T. We set*

$$dY^*_t = dY^{2,3}_t - \frac{Y^{2,3}_{t-}}{Y^{1,3}_{t-}}\, dY^{1,3}_t = dY^{2,3}_t - Y^{2,1}_{t-}\, dY^{1,3}_t, \tag{34}$$

where, by convention, $Y^*_0 = 0$. Assume that there exists a $\mathbb{G}$-predictable process ϕ^2, such that

$$X = Y^3_T\left(x + \int_0^T \phi^2_t\, dY^*_t + \int_0^T \frac{Z^3_t}{Y^{1,3}_{t-}}\, dY^{1,3}_t\right). \tag{35}$$

Then there exist $\mathbb{G}$-predictable processes ϕ^1 and ϕ^3 such that the strategy $\phi = (\phi^1, \phi^2, \phi^3)$ belongs to $\Phi(Z)$ and replicates X. The wealth process of ϕ equals, for every $t \in [0, T]$,

$$V_t(\phi) = Y^3_t\left(x + \int_0^t \phi^2_u\, dY^*_u + \int_0^t \frac{Z^3_u}{Y^{1,3}_{u-}}\, dY^{1,3}_u\right). \tag{36}$$

Proof. As expected, we first set (note that the process ϕ^1 is a $\mathbb{G}$-predictable process)

$$\phi^1_t = \frac{1}{Y^1_{t-}}\big(Z_t - \phi^2_t Y^2_{t-}\big) \tag{37}$$

and

$$V^3_t = x + \int_0^t \phi^2_u\, dY^*_u + \int_0^t \frac{Z^3_u}{Y^{1,3}_{u-}}\, dY^{1,3}_u.$$

Arguing along the same lines as in the proof of Proposition 3.1, we obtain

$$V^3_t = V^3_0 + \sum_{i=1}^{2} \int_0^t \phi^i_u\, dY^{i,3}_u.$$

Now, we define

$$\phi_t^3 = V_t^3 - \sum_{i=1}^{2} \phi_t^i Y_t^{i,3} = (Y_t^3)^{-1}\Big(V_t - \sum_{i=1}^{2} \phi_t^i Y_t^i\Big),$$

where $V_t = V_t^3 Y_t^3$. As in the proof of Lemma 3.1, we check that

$$\phi_t^3 = V_{t-}^3 - \sum_{i=1}^{2} \phi_t^i Y_{t-}^{i,3},$$

and thus the process ϕ^3 is $\mathbb{G}$-predictable. It is clear that the strategy $\phi = (\phi^1, \phi^2, \phi^3)$ is self-financing and its wealth process satisfies $V_t(\phi) = V_t$ for every $t \in [0, T]$. In particular, $V_T(\phi) = X$, so that ϕ replicates X. Finally, equality (37) implies (30), and thus ϕ belongs to the class $\Phi(Z)$. □

Note that equality (35) is a necessary (by Lemma 3.1) and sufficient (by Proposition 3.2) condition for the existence of a constrained strategy that replicates a given contingent claim X.

3.2.1 Synthetic asset

Let us take $Z = 0$, so that $\phi \in \Phi(0)$. Then the balance condition becomes $\sum_{i=1}^{2} \phi_t^i Y_{t-}^i = 0$, and formula (31) reduces to

$$dV_t^3(\phi) = \phi_t^2 \left(dY_t^{2,3} - \frac{Y_{t-}^{2,3}}{Y_{t-}^{1,3}}\, dY_t^{1,3} \right). \tag{38}$$

The process $\bar{Y}^2 = Y^3 Y^*$, where Y^* is defined in (34) is called a *synthetic asset*. It corresponds to a particular self-financing portfolio, with the long position in Y^2 and the short position of $Y_{t-}^{2,1}$ number of shares of Y^1, and suitably re-balanced positions in the third asset so that the portfolio is self-financing, as in Lemma 3.1.

It can be shown (see Bielecki *et al.* [5]) that trading in primary assets Y^1, Y^2, Y^3 is formally equivalent to trading in assets $Y^1, \bar{Y}^2, Y^3$. This observation supports the name synthetic asset attributed to the process $\bar{Y}^2$. Note, however, that the synthetic asset process may take negative values.

3.2.2 Case of continuous asset prices

In the case of continuous asset prices, the relative price $Y^* = \bar{Y}^2 (Y^3)^{-1}$ of the synthetic asset can be given an alternative representation, as the following result shows. Recall that the *predictable bracket* of the two continuous semi-martingales X and Y, denoted as $\langle X, Y \rangle$, coincides with their quadratic covariation $[X, Y]$.

Proposition 3.3. *Assume that the price processes Y^1 and Y^2 are continuous.*

Then the relative price of the synthetic asset satisfies

$$Y_t^* = \int_0^t (Y_u^{3,1})^{-1} e^{\alpha_u} \, d\widehat{Y}_u,$$

where $\widehat{Y}_t := Y_t^{2,1} e^{-\alpha_t}$ *and*

$$\alpha_t := \langle \ln Y^{2,1}, \ln Y^{3,1} \rangle_t = \int_0^t (Y_u^{2,1})^{-1} (Y_u^{3,1})^{-1} \, d\langle Y^{2,1}, Y^{3,1} \rangle_u. \tag{39}$$

In terms of the auxiliary process $\widehat{Y}$, *formula* (31) *becomes*

$$V_t^3(\phi) = V_0^3(\phi) + \int_0^t \widehat{\phi}_u \, d\widehat{Y}_u + \int_0^t \frac{Z_u^3}{Y_{u-}^{1,3}} \, dY_u^{1,3}, \tag{40}$$

where $\widehat{\phi}_t = \phi_t^2 (Y_t^{3,1})^{-1} e^{\alpha_t}$.

Proof. It suffices to give the proof for $Z = 0$. The proof relies on the integration by parts formula stating that for any two continuous semimartingales, say X and Y, we have

$$Y_t^{-1}\big(dX_t - Y_t^{-1} d\langle X, Y \rangle_t\big) = d(X_t Y_t^{-1}) - X_t \, dY_t^{-1},$$

provided that Y is strictly positive. An application of this formula to processes $X = Y^{2,1}$ and $Y = Y^{3,1}$ leads to

$$(Y_t^{3,1})^{-1}\big(dY_t^{2,1} - (Y_t^{3,1})^{-1} d\langle Y^{2,1}, Y^{3,1} \rangle_t\big) = d(Y_t^{2,1}(Y_t^{3,1})^{-1}) - Y_t^{2,1} \, d(Y^{3,1})_t^{-1}.$$

The relative wealth $V_t^3(\phi) = V_t(\phi)(Y_t^3)^{-1}$ of a strategy $\phi \in \Phi(0)$ satisfies

$$\begin{aligned} V_t^3(\phi) &= V_0^3(\phi) + \int_0^t \phi_u^2 \, dY_u^* \\ &= V_0^3(\phi) + \int_0^t \phi_u^2 (Y_u^{3,1})^{-1} e^{\alpha_u} \, d\widehat{Y}_u, \\ &= V_0^3(\phi) + \int_0^t \widehat{\phi}_u \, d\widehat{Y}_u \end{aligned}$$

where we denote $\widehat{\phi}_t = \phi_t^2 (Y_t^{3,1})^{-1} e^{\alpha_t}$.

Remark 3.1. The financial interpretation of the auxiliary process $\widehat{Y}$ will be studied in Sections 4.1.6 and 4.1.8 below. Let us only observe here that if Y^* is a local martingale under some probability $\mathbb{Q}^*$ then $\widehat{Y}$ is a

$\mathbb{Q}^*$-local martingale (and vice versa, if $\widehat{Y}$ is a $\widehat{\mathbb{Q}}$-local martingale under some probability $\widehat{\mathbb{Q}}$ then Y^* is a $\widehat{\mathbb{Q}}$-local martingale). Nevertheless, for the reader's convenience, we shall use two symbols $\mathbb{Q}^*$ and $\widehat{\mathbb{Q}}$, since this equivalence holds for continuous processes only.

It is thus worth stressing that we will apply Proposition 3.3 to pre-default values of assets, rather than directly to asset prices, within the set-up of a semimartingale model with a common default, as described in Section 2.1. In this model, the asset prices may have discontinuities, but their pre-default values follow continuous processes.

4. Martingale Approach to Valuation and Hedging

Our goal is to derive quasi-explicit conditions for replicating strategies for a defaultable claim in a fairly general set-up introduced in Section 2.1. In this section, we only deal with trading strategies based on the reference filtration $\mathbb{F}$, and the underlying price processes (that is, prices of default-free assets and pre-default values of defaultable assets) are assumed to be continuous. Hence, our arguments will hinge on Proposition 3.3, rather than on a more general Proposition 3.1. We shall also adapt Proposition 3.2 to our current purposes.

To simplify the presentation, we make a standing assumption that all coefficient processes are such that the SDEs appearing below admit unique strong solutions, and all stochastic exponentials (used as Radon-Nikodým derivatives) are true martingales under respective probabilities.

4.1 Defaultable asset with total default

In this section, we shall examine in some detail a particular model where the two assets, Y^1 and Y^2, are default-free and satisfy

$$dY_t^i = Y_t^i\big(\mu_{i,t}\,dt + \sigma_{i,t}\,dW_t\big), \quad i = 1, 2,$$

where W is a one-dimensional Brownian motion. The third asset is a defaultable asset with total default, so that

$$dY_t^3 = Y_{t-}^3\big(\mu_{3,t}\,dt + \sigma_{3,t}\,dW_t - dM_t\big).$$

Since we will be interested in replicating strategies in the sense of Definition 2.2, we may and do assume, without loss of generality, that the coefficients $\mu_{i,t}, \sigma_{i,t}$, $i = 1, 2$, are $\mathbb{F}$-predictable, rather than $\mathbb{G}$-predictable. Recall that, in general, there exist $\mathbb{F}$-predictable processes $\widetilde{\mu}_3$ and $\widetilde{\sigma}_3$ such that

$$\widetilde{\mu}_{3,t}\mathbb{1}_{\{t\le\tau\}} = \mu_{3,t}\mathbb{1}_{\{t\le\tau\}}, \quad \widetilde{\sigma}_{3,t}\mathbb{1}_{\{t\le\tau\}} = \sigma_{3,t}\mathbb{1}_{\{t\le\tau\}}. \tag{41}$$

We assume throughout that $Y_0^i > 0$ for every i, so that the price processes Y^1, Y^2 are strictly positive, and the process Y^3 is nonnegative, and has strictly positive pre-default value.

4.1.1 Default-free market

It is natural to postulate that the default-free market with the two traded assets, Y^1 and Y^2, is arbitrage-free. More precisely, we choose Y^1 as a numéraire, and we require that there exists a probability measure $\mathbb{P}^1$, equivalent to $\mathbb{P}$ on $(\Omega, \mathcal{F}_T)$, and such that the process $Y^{2,1}$ is a $\mathbb{P}^1$-martingale. The dynamics of processes $(Y^1)^{-1}$ and $Y^{2,1}$ are

$$d(Y_t^1)^{-1} = (Y_t^1)^{-1}\Big((\sigma_{1,t}^2 - \mu_{1,t})\,dt - \sigma_{1,t}\,dW_t\Big), \tag{42}$$

and

$$dY_t^{2,1} = Y_t^{2,1}\Big((\mu_{2,t} - \mu_{1,t} + \sigma_{1,t}(\sigma_{1,t} - \sigma_{2,t}))\,dt + (\sigma_{2,t} - \sigma_{1,t})\,dW_t\Big),$$

respectively. Hence, the necessary condition for the existence of an e.m.m. $\mathbb{P}^1$ is the inclusion $A \subseteq B$, where $A = \{(t,\omega) \in [0,T] \times \Omega : \sigma_{1,t}(\omega) = \sigma_{2,t}(\omega)\}$ and $B = \{(t,\omega) \in [0,T] \times \Omega : \mu_{1,t}(\omega) = \mu_{2,t}(\omega)\}$. The necessary and sufficient condition for the existence and uniqueness of an e.m.m. $\mathbb{P}^1$ reads

$$\mathbb{E}_{\mathbb{P}}\left\{\mathcal{E}_T\left(\int_0^{\cdot} \theta_u\,dW_u\right)\right\} = 1 \tag{43}$$

where the process θ is given by the formula (by convention, $0/0 = 0$)

$$\theta_t = \sigma_{1,t} - \frac{\mu_{1,t} - \mu_{2,t}}{\sigma_{1,t} - \sigma_{2,t}}, \quad \forall\, t \in [0,T]. \tag{44}$$

Note that in the case of constant coefficients, if $\sigma_1 = \sigma_2$ then the model is arbitrage-free only in the trivial case when $\mu_2 = \mu_1$.

Remark 4.1. Since the martingale measure $\mathbb{P}^1$ is unique, the default-free model (Y^1, Y^2) is complete. However, this is not a necessary assumption and thus it can be relaxed. As we shall see in what follows, it is typically more natural to assume that the driving Brownian motion W is multi-dimensional.

4.1.2 Arbitrage-free property

Let us now consider also a defaultable asset Y^3. Our goal is now to find a martingale measure $\mathbb{Q}^1$ (if it exists) for relative prices $Y^{2,1}$ and $Y^{3,1}$. Recall that we postulate that the hypothesis (H) holds under $\mathbb{P}$ for filtrations $\mathbb{F}$ and $\mathbb{G} = \mathbb{F} \vee \mathbb{H}$. The dynamics of $Y^{3,1}$ under $\mathbb{P}$ are

$$dY_t^{3,1} = Y_{t-}^{3,1}\Big\{\Big(\mu_{3,t} - \mu_{1,t} + \sigma_{1,t}(\sigma_{1,t} - \sigma_{3,t})\Big)dt + (\sigma_{3,t} - \sigma_{1,t})\,dW_t - dM_t\Big\}.$$

Let $\mathbb{Q}^1$ be any probability measure equivalent to $\mathbb{P}$ on $(\Omega, \mathcal{G}_T)$, and let η be the associated Radon-Nikodým density process, so that

$$d\mathbb{Q}^1\,|_{\mathcal{G}_t} = \eta_t\,d\mathbb{P}\,|_{\mathcal{G}_t}, \tag{45}$$

where the process η satisfies

$$d\eta_t = \eta_{t-}(\theta_t\, dW_t + \zeta_t\, dM_t) \tag{46}$$

for some $\mathbb{G}$-predictable processes θ and ζ, and η is a $\mathbb{G}$-martingale under $\mathbb{P}$.

From Girsanov's theorem, the processes $\widehat{W}$ and $\widehat{M}$, given by

$$\widehat{W}_t = W_t - \int_0^t \theta_u\, du, \quad \widehat{M}_t = M_t - \int_0^t \mathbb{1}_{\{u<\tau\}}\gamma_u\zeta_u\, du, \tag{47}$$

are $\mathbb{G}$-martingales under $\mathbb{Q}^1$. To ensure that $Y^{2,1}$ is a $\mathbb{Q}^1$-martingale, we postulate that (43) and (44) are valid. Consequently, for the process $Y^{3,1}$ to be a $\mathbb{Q}^1$-martingale, it is necessary and sufficient that ζ satisfies

$$\gamma_t\zeta_t = \mu_{3,t} - \mu_{1,t} - \frac{\mu_{1,t} - \mu_{2,t}}{\sigma_{1,t} - \sigma_{2,t}}(\sigma_{3,t} - \sigma_{1,t}).$$

To ensure that $\mathbb{Q}^1$ is a probability measure equivalent to $\mathbb{P}$, we require that $\zeta_t > -1$. The unique martingale measure $\mathbb{Q}^1$ is then given by the formula (45) where η solves (46), so that

$$\eta_t = \mathcal{E}_t\left(\int_0^\cdot \theta_u\, dW_u\right)\mathcal{E}_t\left(\int_0^\cdot \zeta_u\, dM_u\right).$$

We are in a position to formulate the following result.

Proposition 4.1. *Assume that the process θ given by (44) satisfies (43), and*

$$\zeta_t = \frac{1}{\gamma_t}\left(\mu_{3,t} - \mu_{1,t} - \frac{\mu_{1,t} - \mu_{2,t}}{\sigma_{1,t} - \sigma_{2,t}}(\sigma_{3,t} - \sigma_{1,t})\right) > -1. \tag{48}$$

Then the model $\mathcal{M} = (Y^1, Y^2, Y^3; \Phi)$ is arbitrage-free and complete. The dynamics of relative prices under the unique martingale measure $\mathbb{Q}^1$ are

$$\begin{aligned} dY_t^{2,1} &= Y_t^{2,1}(\sigma_{2,t} - \sigma_{1,t})\, d\widehat{W}_t, \\ dY_t^{3,1} &= Y_{t-}^{3,1}\big((\sigma_{3,t} - \sigma_{1,t})\, d\widehat{W}_t - d\widehat{M}_t\big). \end{aligned}$$

Since the coefficients $\mu_{i,t}, \sigma_{i,t}$, $i = 1,2$, are $\mathbb{F}$-adapted, the process $\widehat{W}$ is an $\mathbb{F}$-martingale (hence, a Brownian motion) under $\mathbb{Q}^1$. Hence, by virtue of Proposition 1.1, the hypothesis (H) holds under $\mathbb{Q}^1$, and the $\mathbb{F}$-intensity of default under $\mathbb{Q}^1$ equals

$$\widehat{\gamma}_t = \gamma_t(1 + \zeta_t) = \gamma_t + \left(\mu_{3,t} - \mu_{1,t} - \frac{\mu_{1,t} - \mu_{2,t}}{\sigma_{1,t} - \sigma_{2,t}}(\sigma_{3,t} - \sigma_{1,t})\right).$$

Example 4.1. We present an example where the condition (48) does not hold, and thus arbitrage opportunities arise. Assume the coefficients are constant and satisfy: $\mu_1 = \mu_2 = \sigma_1 = 0$, $\mu_3 < -\gamma$ for a constant default intensity $\gamma > 0$. Then

$$Y_t^3 = \mathbb{1}_{\{t<\tau\}} Y_0^3 \exp\Big(\sigma_3 W_t - \frac{1}{2}\sigma_3^2 t + (\mu_3 + \gamma)t\Big) \leq Y_0^3 \exp\Big(\sigma_3 W_t - \frac{1}{2}\sigma_3^2 t\Big) = V_t(\phi),$$

where $V(\phi)$ represents the wealth of a self-financing strategy $(\phi^1, \phi^2, 0)$ with $\phi^2 = \frac{\sigma_3}{\sigma_2}$. Hence, the arbitrage strategy would be to sell the asset Y^3, and to follow the strategy ϕ.

Remark 4.2. Let us stress once again, that the existence of an e.m.m. is a necessary condition for viability of a financial model, but the uniqueness of an e.m.m. is not always a convenient condition to impose on a model. In fact, when constructing a model, we should be mainly concerned with its flexibility and ability to reflect the pertinent risk factors, rather than with its mathematical completeness. In the present context, it is natural to postulate that the dimension of the underlying Brownian motion equals the number of tradeable risky assets. In addition, each particular model should be tailored to provide intuitive and handy solutions for a predetermined family of contingent claims, which will be priced and hedged within its framework.

4.1.3 Hedging a survival claim

We first focus on replication of a *survival claim* $(X, 0, \tau)$, that is, a defaultable claim represented by the terminal payoff $X\mathbb{1}_{\{T<\tau\}}$, where X is an $\mathcal{F}_T$-measurable random variable. For the moment, we maintain the simplifying assumption that W is one-dimensional. As we shall see in what follows, it may lead to certain pathological features of a model. If, on the contrary, the driving noise is multi-dimensional, most of the analysis remains valid, except that the model completeness is no longer ensured, in general.

Recall that $\widetilde{Y}^3$ stands for the pre-default price of Y^3, defined as (see (21))

$$d\widetilde{Y}_t^3 = \widetilde{Y}_t^3\big((\widetilde{\mu}_{3,t} + \gamma_t)\,dt + \widetilde{\sigma}_{3,t}\,dW_t\big) \tag{49}$$

with $\widetilde{Y}_0^3 = Y_0^3$. This strictly positive, continuous, $\mathbb{F}$-adapted process enjoys the property that $Y_t^3 = \mathbb{1}_{\{t<\tau\}}\widetilde{Y}_t^3$. Let us denote the pre-default values in the numéraire $\widetilde{Y}^3$ by $\widetilde{Y}_t^{i,3} = Y_t^i(\widetilde{Y}_t^3)^{-1}$, $i = 1, 2$, and let us introduce the pre-default relative price $\widetilde{Y}^*$ of the synthetic asset $\bar{Y}^2$ by setting

$$d\widetilde{Y}_t^* := d\widetilde{Y}_t^{2,3} - \frac{\widetilde{Y}_t^{2,3}}{\widetilde{Y}_t^{1,3}}\,d\widetilde{Y}_t^{1,3} = \widetilde{Y}_t^{2,3}\Big(\big(\mu_{2,t} - \mu_{1,t} + \sigma_{3,t}(\sigma_{1,t} - \sigma_{2,t})\big)\,dt + (\sigma_{2,t} - \sigma_{1,t})\,dW_t\Big),$$

and let us assume that $\sigma_{1,t} - \sigma_{2,t} \neq 0$. It is also useful to note that the process $\widehat{Y}$, defined in Proposition 3.3, satisfies

$$d\widehat{Y}_t = \widehat{Y}_t\Big(\big(\mu_{2,t} - \mu_{1,t} + \sigma_{3,t}(\sigma_{1,t} - \sigma_{2,t})\big)\,dt + (\sigma_{2,t} - \sigma_{1,t})\,dW_t\Big).$$

In Sections 4.1.6 and 4.1.8, we shall show that in the case, where α given by (39) is deterministic, the process $\widehat{Y}$ has a nice financial interpretation as a credit-risk adjusted forward price of Y^2 relative to Y^1. Therefore, it is more convenient to work with the process $\widetilde{Y}^*$ when dealing with the general case, but to use the process $\widehat{Y}$ when analyzing a model with deterministic volatilities.

Consider an $\mathbb{F}$-predictable self-financing strategy ϕ satisfying the balance condition $\phi^1_t Y^1_t + \phi^2_t Y^2_t = 0$, and the corresponding wealth process

$$V_t(\phi) := \sum_{i=1}^{3} \phi^i_t Y^i_t = \phi^3_t Y^3_t.$$

Let $\widetilde{V}_t(\phi) := \phi^3_t \widetilde{Y}^3_t$. Since the process $\widetilde{V}(\phi)$ is $\mathbb{F}$-adapted, we see that this is the *pre-default price* process of the portfolio ϕ, that is, we have $\mathbb{1}_{\{\tau>t\}} V_t(\phi) = \mathbb{1}_{\{\tau>t\}} \widetilde{V}_t(\phi)$; we shall call this process the *pre-default wealth* of ϕ. Consequently, the process $\widetilde{V}^3_t(\phi) := \widetilde{V}_t(\phi)(\widetilde{Y}^3_t)^{-1} = \phi^3_t$ is termed the relative pre-default wealth.

Using Proposition 3.1, with suitably modified notation, we find that the $\mathbb{F}$-adapted process $\widetilde{V}^3(\phi)$ satisfies, for every $t \in [0,T]$,

$$\widetilde{V}^3_t(\phi) = \widetilde{V}^3_0(\phi) + \int_0^t \phi^2_u \, d\widetilde{Y}^*_u.$$

Define a new probability on $(\Omega, \mathcal{F}_T)$ by setting

$$d\mathbb{Q}^* = \eta^*_T \, d\mathbb{P},$$

where $d\eta^*_t = \eta^*_t \theta^*_t \, dW_t$, and

$$\theta^*_t = \frac{\mu_{2,t} - \mu_{1,t} + \widetilde{\sigma}_{3,t}(\sigma_{1,t} - \sigma_{2,t})}{\sigma_{1,t} - \sigma_{2,t}}. \tag{50}$$

The process $\widetilde{Y}^*_t$, $t \in [0,T]$, is a (local) martingale under $\mathbb{Q}^*$. We shall require that this process is in fact a true martingale; a sufficient condition for this is that

$$\int_0^T \mathbb{E}_{\mathbb{Q}^*}\left(\widetilde{Y}^{2,3}_t(\sigma_{2,t} - \sigma_{1,t})\right)^2 dt < \infty.$$

From the predictable representation theorem, it follows that for any $X \in \mathcal{F}_T$, such that $X(\widetilde{Y}^3_T)^{-1}$ is square-integrable under $\mathbb{Q}^*$, there exists a constant x and an $\mathbb{F}$-predictable process ϕ^2 such that

$$X = \widetilde{Y}^3_T\left(x + \int_0^T \phi^2_u \, d\widetilde{Y}^*_u\right). \tag{51}$$

We now deduce from Proposition 3.2 that there exists a self-financing strategy ϕ with the pre-default wealth $\widetilde{V}_t(\phi) = \widetilde{Y}^3_t \widetilde{V}^3_t$ for every $t \in [0,T]$, where we set

$$\widetilde{V}^3_t = x + \int_0^t \phi^2_u \, d\widetilde{Y}^*_u. \tag{52}$$

Moreover, it satisfies the balance condition $\phi^1_t Y^1_t + \phi^2_t Y^2_t = 0$ for every $t \in [0,T]$. Since clearly $\widetilde{V}_T(\phi) = X$, we have that

$$V_T(\phi) = \phi^3_T Y^3_T = \mathbb{1}_{\{T<\tau\}} \phi^3_T \widetilde{Y}^3_T = \mathbb{1}_{\{T<\tau\}} \widetilde{V}_T(\phi) = \mathbb{1}_{\{T<\tau\}} X,$$

and thus this strategy replicates the survival claim $(X, 0, \tau)$. In fact, we have that $V_t(\phi) = 0$ on the random interval $[\![\tau, T]\!]$.

Definition 4.1. We say that a survival claim $(X, 0, \tau)$ is *attainable* if the process $\widetilde{V}^3$ given by (52) is a martingale under $\mathbb{Q}^*$.

The following result is an immediate consequence of (51) and (52).

Corollary 4.1. *Let $X \in \mathcal{F}_T$ be such that $X(\widetilde{Y}^3_T)^{-1}$ is square-integrable under $\mathbb{Q}^*$. Then the survival claim $(X, 0, \tau)$ is attainable. Moreover, the pre-default price $\widetilde{\pi}_t(X, 0, \tau)$ of the claim $(X, 0, \tau)$ is given by the conditional expectation*

$$\widetilde{\pi}_t(X, 0, \tau) = \widetilde{Y}^3_t \, \mathbb{E}_{\mathbb{Q}^*}(X(\widetilde{Y}^3_T)^{-1} \,|\, \mathcal{F}_t), \quad \forall\, t \in [0,T]. \tag{53}$$

The process $\widetilde{\pi}(X, 0, \tau)(\widetilde{Y}^3)^{-1}$ is an $\mathbb{F}$-martingale under $\mathbb{Q}^*$.

Proof. Since $X(\widetilde{Y}^3_T)^{-1}$ is square-integrable under $\mathbb{Q}^*$, we know from the predictable representation theorem that ϕ^2 in (51) is such that $\mathbb{E}_{\mathbb{Q}^*}\left(\int_0^T (\phi^2_t)^2 \, d\langle \widetilde{Y}^* \rangle_t\right) < \infty$, so that the process $\widetilde{V}^3$ given by (52) is a true martingale under $\mathbb{Q}^*$. We conclude that $(X, 0, \tau)$ is attainable.

Now, let us denote by $\pi_t(X, 0, \tau)$ the time-t price of the claim $(X, 0, \tau)$. Since ϕ is a hedging portfolio for $(X, 0, \tau)$ we thus have $V_t(\phi) = \pi_t(X, 0, \tau)$ for each $t \in [0,T]$. Consequently,

$$\begin{aligned}
\mathbb{1}_{\{\tau>t\}} \widetilde{\pi}_t(X, 0, \tau) &= \mathbb{1}_{\{\tau>t\}} \widetilde{V}_t(\phi) = \mathbb{1}_{\{\tau>t\}} \widetilde{Y}^3_t \, \mathbb{E}_{\mathbb{Q}^*}(\widetilde{V}^3_T \,|\, \mathcal{F}_t) \\
&= \mathbb{1}_{\{\tau>t\}} \widetilde{Y}^3_t \, \mathbb{E}_{\mathbb{Q}^*}(X(\widetilde{Y}^3_T)^{-1} \,|\, \mathcal{F}_t)
\end{aligned}$$

for each $t \in [0, T]$. This proves equality (53). □

In view of the last result, it is justified to refer to $\mathbb{Q}^*$ as the *pricing measure relative to* Y^3 for attainable survival claims.

Remark 4.3. It can be proved that there exists a unique absolutely continuous probability measure $\bar{\mathbb{Q}}$ on $(\Omega, \mathcal{G}_T)$ such that we have

$$Y_t^3 \, \mathbb{E}_{\bar{\mathbb{Q}}} \left(\frac{\mathbb{1}_{\{\tau > T\}} X}{Y_T^3} \,\Big|\, \mathcal{G}_t \right) = \mathbb{1}_{\{\tau > t\}} \widetilde{Y}_t^3 \, \mathbb{E}_{\mathbb{Q}^*} \left(\frac{X}{\widetilde{Y}_T^3} \,\Big|\, \mathcal{F}_t \right).$$

However, this probability measure is not equivalent to $\mathbb{Q}^*$, since its Radon-Nikodým density vanishes after τ (for a related result, see Collin-Dufresne *et al.* [13]).

Example 4.2. We provide here an explicit calculation of the pre-default price of a survival claim. For simplicity, we assume that $X = 1$, so that the claim represents a defaultable zero-coupon bond. Also, we set $\gamma_t = \gamma = const$, $\mu_{i,t} = 0$, and $\sigma_{i,t} = \sigma_i$, $i = 1, 2, 3$. Straightforward calculations yield the following pricing formula

$$\widetilde{\pi}_0(1, 0, \tau) = Y_0^3 e^{-(\gamma + \frac{1}{2}\sigma_3^2)T}.$$

We see that here the pre-default price $\widetilde{\pi}_0(1, 0, \tau)$ depends explicitly on the intensity γ, or rather, on the drift term in dynamics of pre-default value of defaultable asset. Indeed, from the practical viewpoint, the interpretation of the drift coefficient in dynamics of Y^2 as the real-world default intensity is questionable, since within our set-up the default intensity never appears as an independent variable, but is merely a component of the drift term in dynamics of pre-default value of Y^3.

Note also that we deal here with a model with three tradeable assets driven by a one-dimensional Brownian motion. No wonder that the model enjoys completeness, but as a downside, it has an undesirable property that the pre-default values of all three assets are perfectly correlated. Consequently, the drift terms in dynamics of traded assets are closely linked to each other, in the sense, that their behavior under an equivalent change of a probability measure is quite specific.

As we shall see later, if traded primary assets are judiciously chosen then, typically, the pre-default price (and hence the price) of a survival claim will not explicitly depend on the intensity process.

Remark 4.4. Generally speaking, we believe that one can classify a financial model as 'realistic' if its implementation does not require estimation of drift parameters in (pre-default) prices, at least for the purpose of hedging and valuation of a sufficiently large class of (defaultable) contingent claims of interest. It is worth recalling that the drift coefficients are not assumed to be market observables. Since the default intensity can formally interpreted as a component of the drift term in dynamics of pre-default prices, in a realistic model there is no need to estimate this quantity. From this perspective, the model considered in Example 4.2 may serve as an example of an 'unrealistic' model, since its implementation requires the knowledge of the drift parameter in the dynamics of Y^3. We do not pretend here that it is always possible to hedge derivative assets without using the drift coefficients in dynamics of tradeable assets, but it seems to us that a good idea is to develop models in which this knowledge is not essential.

Of course, a generic semimartingale model considered until now provides only a framework for a construction of realistic models for hedging of default risk. A choice of tradeable assets and specification of their dynamics should be examined on a case-by-case basis, rather than in a general semimartingale set-up. We shall address this important issue in the foregoing sections, in which we shall deal with particular examples of practically interesting defaultable claims.

4.1.4 Hedging a recovery process

Let us now briefly study the situation where the promised payoff equals zero, and the recovery payoff is paid at time τ and equals Z_τ for some $\mathbb{F}$-adapted process Z. Put another way, we consider a defaultable claim of the form $(0, Z, \tau)$. Once again, we make use of Propositions 3.1 and 3.2. In view of (35), we need to find a constant x and an $\mathbb{F}$-predictable process ϕ^2 such that

$$\psi_T := -\int_0^T \frac{Z_t}{Y_t^1}\, d\widetilde{Y}_t^{1,3} = x + \int_0^T \phi_t^2\, d\widetilde{Y}_t^{*}. \tag{54}$$

Similarly as in Section 4.1.3 we conclude that, under suitable integrability conditions on ψ_T, there exists ϕ^2 such that $d\psi_t = \phi_t^2\, dY_t^*$, where $\psi_t = \mathbb{E}_{\mathbb{Q}^*}(\psi_T \,|\, \mathcal{F}_t)$. We now set

$$\widetilde{V}_t^3 = x + \int_0^t \phi_u^2\, dY_u^* + \int_0^T \frac{\widetilde{Z}_u^3}{\widetilde{Y}_u^{1,3}}\, d\widetilde{Y}_u^{1,3},$$

so that, in particular, $\widetilde{V}_T^3 = 0$. Then it is possible to find processes ϕ^1 and ϕ^3 such that the strategy ϕ is self-financing and it satisfies: $\widetilde{V}_t(\phi) = \widetilde{V}_t^3 \widetilde{Y}_t^3$ and $V_t(\phi) = Z_t + \phi_t^3 Y_t^3$ for every $t \in [0, T]$. It is thus clear that $V_\tau(\phi) = Z_\tau$ on the set $\{\tau \le T\}$ and $V_T(\phi) = 0$ on the set $\{\tau > T\}$.

4.1.5 Bond market

For the sake of concreteness, we assume that $Y_t^1 = B(t,T)$ is the price of a default-free ZC-bond with maturity T, and $Y_t^3 = D(t,T)$ is the price of a defaultable ZC-bond with zero recovery, that is, an asset with the terminal payoff $Y_T^3 = \mathbb{1}_{\{T<\tau\}}$. We postulate that the dynamics under $\mathbb{P}$ of the default-free ZC-bond are

$$dB(t,T) = B(t,T)\Big(\mu(t,T)\,dt + b(t,T)\,dW_t\Big) \tag{55}$$

for some $\mathbb{F}$-predictable processes $\mu(t,T)$ and $b(t,T)$. We choose the process $Y_t^1 = B(t,T)$ as a numéraire. Since the prices of the other two assets are not given a priori, we may choose any probability measure $\mathbb{Q}$ equivalent to $\mathbb{P}$ on $(\Omega, \mathcal{G}_T)$ to play the role of $\mathbb{Q}^1$.

In such a case, an e.m.m. $\mathbb{Q}^1$ is referred to as the *forward martingale measure* for the date T, and is denoted by $\mathbb{Q}_T$. Hence, the Radon-Nikodým density of $\mathbb{Q}_T$ with respect to $\mathbb{P}$ is given by (46) for some $\mathbb{F}$-predictable processes θ and ζ, and the process

$$W_t^T = W_t - \int_0^t \theta_u\,du, \quad \forall\, t \in [0,T],$$

is a Brownian motion under $\mathbb{Q}_T$. Under $\mathbb{Q}_T$ the default-free ZC-bond is governed by

$$dB(t,T) = B(t,T)\Big(\widehat{\mu}(t,T)\,dt + b(t,T)\,dW_t^T\Big)$$

where $\widehat{\mu}(t,T) = \mu(t,T) + \theta_t b(t,T)$. Let $\widehat{\Gamma}$ stand for the $\mathbb{F}$-hazard process of τ under $\mathbb{Q}_T$, so that $\widehat{\Gamma}_t = -\ln(1-\widehat{F}_t)$, where $\widehat{F}_t = \mathbb{Q}_T(\tau \le t\,|\,\mathcal{F}_t)$. Assume that the hypothesis (H) holds under $\mathbb{Q}_T$ so that, in particular, the process $\widehat{\Gamma}$ is increasing. We define the price process of a defaultable ZC-bond with zero recovery by the formula

$$D(t,T) := B(t,T)\mathbb{E}_{\mathbb{Q}_T}(\mathbb{1}_{\{T<\tau\}}\,|\,\mathcal{G}_t) = \mathbb{1}_{\{t<\tau\}}B(t,T)\,\mathbb{E}_{\mathbb{Q}_T}\Big(e^{\widehat{\Gamma}_t-\widehat{\Gamma}_T}\,\Big|\,\mathcal{F}_t\Big),$$

where the second equality follows from Lemma 1.3. It is then clear that $Y_t^{3,1} = D(t,T)(B(t,T))^{-1}$ is a $\mathbb{Q}_T$-martingale, and the pre-default price $\widetilde{D}(t,T)$ equals

$$\widetilde{D}(t,T) = B(t,T)\,\mathbb{E}_{\mathbb{Q}_T}\Big(e^{\widehat{\Gamma}_t-\widehat{\Gamma}_T}\,\Big|\,\mathcal{F}_t\Big).$$

The next result examines the basic properties of the auxiliary process $\widehat{\Gamma}(t,T)$ given as, for every $t \in [0,T]$,

$$\widehat{\Gamma}(t,T) = \widetilde{Y}_t^{3,1} = \widetilde{D}(t,T)(B(t,T))^{-1} = \mathbb{E}_{\mathbb{Q}_T}\Big(e^{\widehat{\Gamma}_t-\widehat{\Gamma}_T}\,\Big|\,\mathcal{F}_t\Big).$$

The quantity $\widehat{\Gamma}(t,T)$ can be interpreted as the conditional probability (under $\mathbb{Q}_T$) that default will not occur prior to the maturity date T, given that we observe $\mathcal{F}_t$ and we know that the default has not yet happened. We will be more interested, however, in its volatility process $\beta(t,T)$ as defined in the following result.

Lemma 4.1. *Assume that the $\mathbb{F}$-hazard process $\widehat{\Gamma}$ of τ under $\mathbb{Q}_T$ is continuous. Then the process $\widehat{\Gamma}(t,T)$, $t \in [0,T]$, is a continuous $\mathbb{F}$-submartingale and*

$$d\widehat{\Gamma}(t,T) = \widehat{\Gamma}(t,T)\big(d\widehat{\Gamma}_t + \beta(t,T)\,dW_t^T\big) \tag{56}$$

for some $\mathbb{F}$-predictable process $\beta(t,T)$. The process $\widehat{\Gamma}(t,T)$ is of finite variation if and only if the hazard process $\widehat{\Gamma}$ is deterministic. In this case, we have $\widehat{\Gamma}(t,T) = e^{\widehat{\Gamma}_t - \widehat{\Gamma}_T}$.

Proof. We have

$$\widehat{\Gamma}(t,T) = \mathbb{E}_{\mathbb{Q}_T}\big(e^{\widehat{\Gamma}_t - \widehat{\Gamma}_T} \,|\, \mathcal{F}_t\big) = e^{\widehat{\Gamma}_t} L_t,$$

where we set $L_t = \mathbb{E}_{\mathbb{Q}_T}\big(e^{-\widehat{\Gamma}_T} \,|\, \mathcal{F}_t\big)$. Hence, $\widehat{\Gamma}(t,T)$ is equal to the product of a strictly positive, increasing, right-continuous, $\mathbb{F}$-adapted process $e^{\widehat{\Gamma}_t}$, and a strictly positive, continuous $\mathbb{F}$-martingale L. Furthermore, there exists an $\mathbb{F}$-predictable process $\widehat{\beta}(t,T)$ such that L satisfies

$$dL_t = L_t \widehat{\beta}(t,T)\,dW_t^T$$

with the initial condition $L_0 = \mathbb{E}_{\mathbb{Q}_T}\big(e^{-\widehat{\Gamma}_T}\big)$. Formula (56) now follows by an application of Itô's formula, by setting $\beta(t,T) = e^{-\widehat{\Gamma}_t}\widehat{\beta}(t,T)$. To complete the proof, it suffices to recall that a continuous martingale is never of finite variation, unless it is a constant process. □

Remark 4.5. It can be checked that $\beta(t,T)$ is also the volatility of the process

$$\Gamma(t,T) = \mathbb{E}_{\mathbb{P}}\big(e^{\Gamma_t - \Gamma_T} \,\big|\, \mathcal{F}_t\big).$$

Assume that $\widehat{\Gamma}_t = \int_0^t \widehat{\gamma}_u\,du$ for some $\mathbb{F}$-predictable, nonnegative process $\widehat{\gamma}$. Then we have the following auxiliary result, which gives, in particular, the volatility of the defaultable ZC-bond.

Corollary 4.2. *The dynamics under $\mathbb{Q}_T$ of the pre-default price $\widetilde{D}(t,T)$ equals*

$$d\widetilde{D}(t,T) = \widetilde{D}(t,T)\Big(\big(\widehat{\mu}(t,T) + b(t,T)\beta(t,T) + \widehat{\gamma}_t\big)\,dt + \big(b(t,T) + \beta(t,T)\big)\widetilde{d}(t,T)\,dW_t^T\Big).$$

Equivalently, the price $D(t,T)$ of the defaultable ZC-bond satisfies under $\mathbb{Q}_T$

$$dD(t,T) = D(t,T)\Big(\big(\widetilde{\mu}(t,T) + b(t,T)\beta(t,T)\big)\,dt + \widetilde{d}(t,T)\,dW_t^T - dM_t\Big).$$

where we set $\widetilde{d}(t,T) = b(t,T) + \beta(t,T)$.

Note that the process $\beta(t,T)$ can be expressed in terms of market observables, since it is simply the difference of volatilities $\widetilde{d}(t,T)$ and $b(t,T)$ of pre-default prices of tradeable assets.

4.1.6 Credit-risk-adjusted forward price

Assume that the price Y^2 satisfies under the statistical probability $\mathbb{P}$

$$dY_t^2 = Y_t^2\big(\mu_{2,t}\,dt + \sigma_t\,dW_t\big) \tag{57}$$

with $\mathbb{F}$-predictable coefficients μ and σ. Let $F_{Y^2}(t,T) = Y_t^2(B(t,T))^{-1}$ be the forward price of Y_T^2. For an appropriate choice of θ (see 50), we shall have that

$$dF_{Y^2}(t,T) = F_{Y^2}(t,T)\big(\sigma_t - b(t,T)\big)\,dW_t^T.$$

Therefore, the dynamics of the pre-default synthetic asset $\widetilde{Y}_t^*$ under $\mathbb{Q}^T$ are

$$d\widetilde{Y}_t^* = \widetilde{Y}_t^{2,3}\big(\sigma_t - b(t,T)\big)\big(dW_t^T - \beta(t,T)\,dt\big),$$

and the process $\widehat{Y}_t = Y_t^{2,1}e^{-\alpha_t}$ satisfies

$$d\widehat{Y}_t = \widehat{Y}_t\big(\sigma_t - b(t,T)\big)\big(dW_t^T - \beta(t,T)\,dt\big).$$

Let $\widehat{\mathbb{Q}}$ be an equivalent probability measure on $(\Omega,\mathcal{G}_T)$ such that $\widehat{Y}$ (or, equivalently, $\widetilde{Y}^*$) is a $\widehat{\mathbb{Q}}$-martingale. By virtue of Girsanov's theorem, the process $\widehat{W}$ given by the formula

$$\widehat{W}_t = W_t^T - \int_0^t \beta(u,T)\,du, \quad \forall\, t \in [0,T],$$

is a Brownian motion under $\widehat{\mathbb{Q}}$. Thus, the forward price $F_{Y^2}(t,T)$ satisfies under $\widehat{\mathbb{Q}}$

$$dF_{Y^2}(t,T) = F_{Y^2}(t,T)\big(\sigma_t - b(t,T)\big)\big(d\widehat{W}_t + \beta(t,T)\,dt\big). \tag{58}$$

It appears that the valuation results are easier to interpret when they are expressed in terms of forward prices associated with vulnerable forward contracts, rather than in terms of spot prices of primary assets. For this reason, we shall now examine credit-risk-adjusted forward prices of default-free and defaultable assets.

Definition 4.2. Let Y be a $\mathcal{G}_T$-measurable claim. An $\mathcal{F}_t$-measurable random variable K is called the *credit-risk-adjusted forward price* of Y if the pre-default value at time t of the vulnerable forward contract represented by the claim $\mathbb{1}_{\{T<\tau\}}(Y-K)$ equals 0.

Lemma 4.2. *The credit-risk-adjusted forward price $\widehat{F}_Y(t,T)$ of an attainable survival claim $(X,0,\tau)$, represented by a $\mathcal{G}_T$-measurable claim $Y = X\mathbb{1}_{\{T<\tau\}}$, equals $\widetilde{\pi}_t(X,0,\tau)(\widetilde{D}(t,T))^{-1}$, where $\widetilde{\pi}_t(X,0,\tau)$ is the pre-default price of $(X,0,\tau)$. The process $\widehat{F}_Y(t,T)$, $t\in[0,T]$, is an $\mathbb{F}$-martingale under $\widehat{\mathbb{Q}}$.*

Proof. The forward price is defined as an $\mathcal{F}_t$-measurable random variable K such that the claim

$$\mathbb{1}_{\{T<\tau\}}(X\mathbb{1}_{\{T<\tau\}}-K) = X\mathbb{1}_{\{T<\tau\}} - KD(T,T)$$

is worthless at time t on the set $\{t<\tau\}$. It is clear that the pre-default value at time t of this claim equals $\widetilde{\pi}_t(X,0,\tau)-K\widetilde{D}(t,T)$. Consequently, we obtain $\widetilde{F}_Y(t,T) = \widetilde{\pi}_t(X,0,\tau)(\widetilde{D}(t,T))^{-1}$. □

Let us now focus on default-free assets. Manifestly, the credit-risk-adjusted forward price of the bond $B(t,T)$ equals 1. To find the credit-risk-adjusted forward price of Y^2, let us write

$$\widehat{F}_{Y^2}(t,T) := F_{Y^2}(t,T)\,e^{\alpha_T-\alpha_t} = Y_t^{2,1}e^{\alpha_T-\alpha_t},$$

where α is given by (see (39))

$$(59)\quad \alpha_t = \int_0^t \big(\sigma_u - b(u,T)\big)\beta(u,T)\,du = \int_0^t \big(\sigma_u - b(u,T)\big)\big(\widetilde{d}(u,T)-b(u,T)\big)\,du.$$

Lemma 4.3. *Assume that α given by* (59) *is a deterministic function. Then the credit-risk-adjusted forward price of Y^2 equals $\widehat{F}_{Y^2}(t,T)$ for every $t\in[0,T]$.*

Proof. According to Definition 4.2, the price $\widehat{F}_{Y^2}(t,T)$ is an $\mathcal{F}_t$-measurable random variable K, which makes the forward contract represented by the claim $D(T,T)(Y_T^2-K)$ worthless on the set $\{t<\tau\}$. Assume that the claim Y_T^2-K is attainable.[1] Since $\widetilde{D}(T,T)=1$, from equation (53) it follows that the pre-default value of this claim is given by the conditional expectation

$$\widetilde{D}(t,T)\,\mathbb{E}_{\widehat{\mathbb{Q}}}\big(Y_T^2-K\,\big|\,\mathcal{F}_t\big).$$

[1] Attainability of this claim can be shown in a similar way as the attainability of a vulnerable call option considered in Section 4.1.7.

Consequently,

$$\widehat{F}_{Y^2}(t,T) = \mathbb{E}_{\widehat{\mathbb{Q}}}\big(Y_T^2 \,\big|\, \mathcal{F}_t\big) = \mathbb{E}_{\widehat{\mathbb{Q}}}\big(F_{Y^2}(T,T) \,\big|\, \mathcal{F}_t\big) = F_{Y^2}(t,T)\, e^{\alpha_T - \alpha_t},$$

as was claimed. □

It is worth noting that the process $\widehat{F}_{Y^2}(t,T)$ is a (local) martingale under the pricing measure $\widehat{\mathbb{Q}}$, since it satisfies

$$(60) \qquad d\widehat{F}_{Y^2}(t,T) = \widehat{F}_{Y^2}(t,T)(\sigma_t - b(t,T))\, d\widehat{W}_t.$$

Under the present assumptions, the auxiliary process $\widehat{Y}$ introduced in Proposition 3.3 and the credit-risk-adjusted forward price $\widehat{F}_{Y^2}(t,T)$ are closely related to each other. Indeed, we have $\widehat{F}_{Y^2}(t,T) = \widehat{Y}_t e^{\alpha_T}$, so that the two processes are proportional.

4.1.7 Vulnerable option on a default-free asset

We shall now analyze a vulnerable call option with the payoff

$$C_T^d = \mathbb{1}_{\{T<\tau\}}(Y_T^2 - K)^+.$$

Our goal is to find a replicating strategy for this claim, interpreted as a survival claim $(X, 0, \tau)$ with the promised payoff $X = C_T = (Y_T^2 - K)^+$, where C_T is the payoff of an equivalent non-vulnerable option. The method presented below is quite general, however, so that it can be applied to any survival claim with the promised payoff $X = G(Y_T^2)$ for some function $G : \mathbb{R} \to \mathbb{R}$ satisfying the usual integrability assumptions.

We assume that $Y_t^1 = B(t,T)$, $Y_t^3 = D(t,T)$ and the price of a default-free asset Y^2 is governed by (57). Then

$$C_T^d = \mathbb{1}_{\{T<\tau\}}(Y_T^2 - K)^+ = \mathbb{1}_{\{T<\tau\}}(Y_T^2 - KY_T^1)^+.$$

We are going to apply Proposition 3.3. In the present set-up, we have $Y_t^{2,1} = F_{Y^2}(t,T)$ and $\widehat{Y}_t = F_{Y^2}(t,T)e^{-\alpha_t}$. Since a vulnerable option is an example of a survival claim, in view of Lemma 4.2, its credit-risk-adjusted forward price satisfies $\widehat{F}_{C^d}(t,T) = \widetilde{C}_t^d(\widetilde{D}(t,T))^{-1}$.

Proposition 4.2. *Suppose that the volatilities σ, b and β are deterministic functions. Then the credit-risk-adjusted forward price of a vulnerable call option written on a default-free asset Y^2 equals*

$$(61) \qquad \widehat{F}_{C^d}(t,T) = \widehat{F}_{Y^2}(t,T)N(d_+(\widehat{F}_{Y^2}(t,T),t,T)) - KN(d_-(\widehat{F}_{Y^2}(t,T),t,T))$$

where

$$d_\pm(\widehat{f},t,T) = \frac{\ln \widehat{f} - \ln K \pm \frac{1}{2}v^2(t,T)}{v(t,T)}$$

and

$$v^2(t,T) = \int_t^T (\sigma_u - b(u,T))^2\,du.$$

The replicating strategy ϕ in the spot market satisfies for every $t \in [0,T]$, on the set $\{t < \tau\}$,

$$\phi_t^1 B(t,T) = -\phi_t^2 Y_t^2,\ \phi_t^2 = \widetilde{D}(t,T)(B(t,T))^{-1} N(d_+(t,T))e^{\alpha_T - \alpha_t},\ \phi_t^3 \widetilde{D}(t,T) = \widetilde{C}_t^d,$$

where $d_+(t,T) = d_+(\widehat{F}_{Y^2}(t,T),t,T)$.

Proof. In the first step, we establish the valuation formula. Assume for the moment that the option is attainable. Then the pre-default value of the option equals, for every $t \in [0,T]$,

$$\widetilde{C}_t^d = \widetilde{D}(t,T)\,\mathbb{E}_{\widehat{\mathbb{Q}}}\big((F_{Y^2}(T,T)-K)^+ \,\big|\, \mathcal{F}_t\big) = \widetilde{D}(t,T)\,\mathbb{E}_{\widehat{\mathbb{Q}}}\big((\widehat{F}_{Y^2}(T,T)-K)^+ \,\big|\, \mathcal{F}_t\big). \tag{62}$$

In view of (60), the conditional expectation above can be computed explicitly, yielding the valuation formula (61).

To find the replicating strategy, and establish attainability of the option, we consider the Itô differential $d\widehat{F}_{C^d}(t,T)$ and we identify terms in (52). It appears that

$$\begin{aligned} d\widehat{F}_{C^d}(t,T) &= N(d_+(t,T))\,d\widehat{F}_{Y^2}(t,T) = N(d_+(t,T))e^{\alpha_T}\,d\widehat{Y}_t \\ &= N(d_+(t,T))\widetilde{Y}_t^{3,1} e^{\alpha_T-\alpha_t}\,d\widetilde{Y}_t^*, \end{aligned} \tag{63}$$

so that the process ϕ^2 in (51) equals

$$\phi_t^2 = \widetilde{Y}_t^{3,1} N(d_+(t,T))e^{\alpha_T-\alpha_t}.$$

Moreover, ϕ^1 is such that $\phi_t^1 B(t,T) + \phi_t^2 Y_t^2 = 0$ and $\phi_t^3 = \widetilde{C}_t^d(\widetilde{D}(t,T))^{-1}$. It is easily seen that this proves also the attainability of the option. □

Let us examine the financial interpretation of the last result.

First, equality (63) shows that it is easy to replicate the option using vulnerable forward contracts. Indeed, we have

$$\widehat{F}_{C^d}(T,T) = X = \frac{\widetilde{C}_0^d}{\widetilde{D}(0,T)} + \int_0^T N(d_+(t,T))\,d\widehat{F}_{Y^2}(t,T)$$

and thus it is enough to invest the premium $\widetilde{C}_0^d = C_0^d$ in defaultable ZC-bonds of maturity T, and take at any instant t prior to default $N(d_+(t,T))$ positions in vulnerable forward contracts. It is understood that if default occurs prior to T, all outstanding vulnerable forward contracts become void.

Second, it is worth stressing that neither the arbitrage price, nor the replicating strategy for a vulnerable option, depend explicitly on the default intensity. This remarkable feature is due to the fact that the default risk of the writer of the option can be completely eliminated by trading in defaultable zero-coupon bond with the same exposure to credit risk as a vulnerable option.

In fact, since the volatility β is invariant with respect to an equivalent change of a probability measure, and so are the volatilities σ and $b(t,T)$, the formulae of Proposition 4.2 are valid for any choice of a forward measure $\mathbb{Q}_T$ equivalent to $\mathbb{P}$ (and, of course, they are valid under $\mathbb{P}$ as well). The only way in which the choice of a forward measure $\mathbb{Q}_T$ makes an impact on these results is through the pre-default value of a defaultable ZC-bond.

We conclude that we deal here with the volatility based relative pricing a defaultable claim. This should be contrasted with the more popular intensity-based risk-neutral pricing, which is commonly used to produce an arbitrage-free model of tradeable defaultable assets. Recall, however, that if tradeable assets are not chosen carefully for a given class of survival claims, then both hedging strategy and pre-default price may depend explicitly on values of drift parameters, which can be linked in our set-up to the default intensity (see Example 4.2).

Remark 4.6. Assume that $X = G(Y^2_T)$ for some function $G : \mathbb{R} \to \mathbb{R}$. Then the credit-risk-adjusted forward price of a survival claim satisfies $\widehat{F}_X(t,T) = v(t, \widehat{F}_{Y^2}(t,T))$, where the pricing function v solves the PDE

$$\partial_t v(t,\widehat{f}) + \frac{1}{2}(\sigma_t - b(t,T))^2 \widehat{f}^2 \partial_{\widehat{f}\widehat{f}} v(t,\widehat{f}) = 0$$

with the terminal condition $v(T,\widehat{f}) = G(\widehat{f})$. The PDE approach is studied in Section 5 below.

Remark 4.7. Proposition 4.2 is still valid if the driving Brownian motion is two-dimensional, rather than one-dimensional. In an extended model, the volatilities $\sigma_t, b(t,T)$ and $\beta(t,T)$ take values in $\mathbb{R}^2$ and the respective products are interpreted as inner products in $\mathbb{R}^2$. Equivalently, one may prefer to deal with real-valued volatilities, but with correlated one-dimensional Brownian motions.

4.1.8 Vulnerable swaption

In this section, we relax the assumption that Y^1 is the price of a default-free bond. We now let Y^1 and Y^2 to be arbitrary default-free assets, with dynamics

$$dY_t^i = Y_t^i\big(\mu_{i,t}\,dt + \sigma_{i,t}\,dW_t\big), \quad i = 1,2.$$

We still take $D(t,T)$ to be the third asset, and we maintain the assumption that the model is arbitrage-free, but we no longer postulate its completeness. In other words, we postulate the existence an e.m.m. $\mathbb{Q}^1$, as defined in Section 4.1.2, but not the uniqueness of $\mathbb{Q}^1$.

We take the first asset as a numéraire, so that all prices are expressed in units of Y^1. In particular, $Y_t^{1,1} = 1$ for every $t \in \mathbb{R}_+$, and the relative prices $Y^{2,1}$ and $Y^{3,1}$ satisfy under $\mathbb{Q}^1$ (cf. Proposition 4.1)

$$\begin{aligned} dY_t^{2,1} &= Y_t^{2,1}(\sigma_{2,t} - \sigma_{1,t})\,d\widehat{W}_t, \\ dY_t^{3,1} &= Y_{t-}^{3,1}\big((\sigma_{3,t} - \sigma_{1,t})\,d\widehat{W}_t - d\widehat{M}_t\big). \end{aligned}$$

It is natural to postulate that the driving Brownian noise is two-dimensional. Under this assumption, we may represent the joint dynamics of $Y^{2,1}$ and $Y^{3,1}$ under $\mathbb{Q}^1$ as follows

$$\begin{aligned} dY_t^{2,1} &= Y_t^{2,1}(\sigma_{2,t} - \sigma_{1,t})\,dW_t^1, \\ dY_t^{3,1} &= Y_{t-}^{3,1}\big((\sigma_{3,t} - \sigma_{1,t})\,dW_t^2 - d\widehat{M}_t\big), \end{aligned}$$

where W^1, W^2 are one-dimensional Brownian motions under $\mathbb{Q}^1$, such that $d\langle W^1, W^2\rangle_t = \rho_t\,dt$ for a deterministic instantaneous correlation coefficient ρ taking values in $[-1,1]$.

We assume from now on that the volatilities σ_i, $i = 1,2,3$ are deterministic. Let us set

$$\alpha_t = \langle \ln \widetilde{Y}^{2,1}, \ln \widetilde{Y}^{3,1}\rangle_t = \int_0^t \rho_u(\sigma_{2,u} - \sigma_{1,u})(\sigma_{3,u} - \sigma_{1,u})\,du, \tag{64}$$

and let $\widehat{\mathbb{Q}}$ be an equivalent probability measure on $(\Omega, \mathcal{G}_T)$ such that the process $\widehat{Y}_t = Y_t^{2,1}e^{-\alpha_t}$ is a $\widehat{\mathbb{Q}}$-martingale. To clarify the financial interpretation of the auxiliary process $\widehat{Y}$ in the present context, we introduce the concept of credit-risk-adjusted forward price relative to the numéraire Y^1.

Definition 4.3. Let Y be a $\mathcal{G}_T$-measurable claim. An $\mathcal{F}_t$-measurable random variable K is called the time-t *credit-risk-adjusted* Y^1*-forward price* of Y if the pre-default value at time t of a vulnerable forward contract, represented by the claim

$$\mathbb{1}_{\{T<\tau\}}(Y_T^1)^{-1}(Y - KY_T^1) = \mathbb{1}_{\{T<\tau\}}(Y(Y_T^1)^{-1} - K),$$

equals 0.

The credit-risk-adjusted Y^1-forward price of Y is denoted by $\widehat{F}_{Y|Y^1}(t,T)$, and it is also interpreted as an abstract defaultable swap rate. The following auxiliary results are easy to establish, along the same lines as Lemmas 4.2 and 4.3.

Lemma 4.4. *The credit-risk-adjusted Y^1-forward price of a survival claim $Y = (X, 0, \tau)$ equals*

$$\widehat{F}_{Y|Y^1}(t,T) = \widetilde{\pi}_t(X^1, 0, \tau)(\widetilde{D}(t,T))^{-1}$$

where $X^1 = X(Y^1_T)^{-1}$ is the price of X in the numéraire Y^1, and $\widetilde{\pi}_t(X^1, 0, \tau)$ is the pre-default value of a survival claim with the promised payoff X^1.

Proof. It suffices to note that for $Y = \mathbb{1}_{\{T<\tau\}}X$, we have

$$\mathbb{1}_{\{T<\tau\}}(Y(Y^1_T)^{-1} - K) = \mathbb{1}_{\{T<\tau\}}X^1 - KD(T,T),$$

where $X^1 = X(Y^1_T)^{-1}$, and to consider the pre-default values. □

Lemma 4.5. *The credit-risk-adjusted Y^1-forward price of the asset Y^2 equals*

$$\widehat{F}_{Y^2|Y^1}(t,T) = Y^{2,1}_t\, e^{\alpha_T - \alpha_t} = \widehat{Y}_t e^{\alpha_T}, \tag{65}$$

where α is given by (64).

Proof. It suffices to find an $\mathcal{F}_t$-measurable random variable K for which

$$\widetilde{D}(t,T)\, \mathbb{E}_{\widehat{\mathbb{Q}}}\Big(Y^2_T(Y^1_T)^{-1} - K \,\Big|\, \mathcal{F}_t\Big) = 0.$$

Consequently, $K = \widehat{F}_{Y^2|Y^1}(t,T)$, where

$$\widehat{F}_{Y^2|Y^1}(t,T) = \mathbb{E}_{\widehat{\mathbb{Q}}}\Big(Y^{2,1}_T \,\Big|\, \mathcal{F}_t\Big) = Y^{2,1}_t\, e^{\alpha_T - \alpha_t} = \widehat{Y}_t\, e^{\alpha_T},$$

where we have used the facts that $\widehat{Y}_t = Y^{2,1}_t e^{-\alpha_t}$ is a $\widehat{\mathbb{Q}}$-martingale, and α is deterministic. □

We are in a position to examine a vulnerable option to exchange default-free assets with the payoff

$$C^d_T = \mathbb{1}_{\{T<\tau\}}(Y^1_T)^{-1}(Y^2_T - KY^1_T)^+ = \mathbb{1}_{\{T<\tau\}}(Y^{2,1}_T - K)^+. \tag{66}$$

The last expression shows that the option can be interpreted as a vulnerable swaption associated with the assets Y^1 and Y^2. It is useful to observe that

$$\frac{C^d_T}{Y^1_T} = \frac{\mathbb{1}_{\{T<\tau\}}}{Y^1_T}\left(\frac{Y^2_T}{Y^1_T} - K\right)^+,$$

so that, when expressed in the numéraire Y^1, the payoff becomes

$$C_T^{1,d} = D^1(T,T)(Y_T^{2,1} - K)^+,$$

where $C_t^{1,d} = C_t^d(Y_t^1)^{-1}$ and $D^1(t,T) = D(t,T)(Y_t^1)^{-1}$ stand for the prices relative to Y^1.

It is clear that we deal here with a model analogous to the model examined in Sections 4.1.5 and 4.1.7 in which, however, all prices are now relative to the numéraire Y^1. This observation allows us to directly derive the valuation formula below from Proposition 4.2.

Proposition 4.3. *The credit-risk-adjusted Y^1-forward price of a vulnerable call option written with the payoff given by* (66) *equals*

$$\widehat{F}_{C^d|Y^1}(t,T) = \widehat{F}_{Y^2|Y^1}(t,T)N\Big(d_+(\widehat{F}_{Y^2|Y^1}(t,T),t,T)\Big) - KN\Big(d_-(\widehat{F}_{Y^2|Y^1}(t,T),t,T)\Big)$$

where

$$d_\pm(\widehat{f},t,T) = \frac{\ln \widehat{f} - \ln K \pm \frac{1}{2}v^2(t,T)}{v(t,T)}$$

and

$$v^2(t,T) = \int_t^T (\sigma_{2,u} - \sigma_{1,u})^2\, du.$$

The replicating strategy ϕ in the spot market satisfies for every $t \in [0,T]$, on the set $\{t < \tau\}$,

$$\phi_t^1 Y_t^1 = -\phi_t^2 Y_t^2, \quad \phi_t^2 = \widetilde{D}(t,T)(Y_t^1)^{-1}N(d_+(t,T))e^{\alpha_T - \alpha_t}, \quad \phi_t^3 \widetilde{D}(t,T) = \widetilde{C}_t^d,$$

where $d_+(t,T) = d_+\big(\widehat{F}_{Y^2}(t,T),t,T\big)$.

Proof. The proof is analogous to that of Proposition 4.2, and thus it is omitted. □

It is worth noting that the payoff (66) was judiciously chosen. Suppose instead that the option payoff is not specified by (66), but it is given by an apparently simpler expression

$$C_T^d = \mathbb{1}_{\{T<\tau\}}(Y_T^2 - KY_T^1)^+. \tag{67}$$

Since the payoff C_T^d can be represented as follows

$$C_T^d = \widehat{G}(Y_T^1, Y_T^2, Y_T^3) = Y_T^3(Y_T^2 - KY_T^1)^+,$$

where $\widehat{G}(y_1, y_2, y_3) = y_3(y_2 - Ky_1)^+$, the option can be seen an option to exchange the second asset for K units of the first asset, but with the payoff expressed in units of the defaultable asset. When expressed in relative prices, the payoff becomes

$$C_T^{1,d} = \mathbb{1}_{\{T<\tau\}}(Y_T^{2,1} - K)^+,$$

where $\mathbb{1}_{\{T<\tau\}} = D^1(T,T)Y_T^1$. It is thus rather clear that it is not longer possible to apply the same method as in the proof of Proposition 4.2.

4.2 Two defaultable assets with total default

We shall now assume that we have only two tradeable assets, and both are defaultable assets with total default. This case is also examined by Carr [12], who studies the imperfect hedging of digital options. Note that here we present results for the perfect hedging.

We shall briefly outline the analysis of hedging of a survival claim. Under the present assumptions, we have, for $i = 1,2$,

$$dY_t^i = Y_{t-}^i\Big(\mu_{i,t}\,dt + \sigma_{i,t}\,dW_t - dM_t\Big), \tag{68}$$

where W is a one-dimensional Brownian motion, so that

$$Y_t^1 = \mathbb{1}_{\{t<\tau\}}\widetilde{Y}_t^1, \quad Y_t^2 = \mathbb{1}_{\{t<\tau\}}\widetilde{Y}_t^2,$$

with the pre-default prices governed by the SDEs

$$d\widetilde{Y}_t^i = \widetilde{Y}_t^i\Big((\mu_{i,t} + \gamma_t)\,dt + \sigma_{i,t}\,dW_t\Big). \tag{69}$$

The wealth process V associated with the self-financing trading strategy (ϕ^1,ϕ^2) satisfies, for every $t \in [0,T]$,

$$V_t = Y_t^1\left(V_0^1 + \int_0^t \phi_u^2\,d\widetilde{Y}_u^{2,1}\right),$$

where $\widetilde{Y}_t^{2,1} = \widetilde{Y}_t^2/\widetilde{Y}_t^1$. Since both primary traded assets are subject to total default, it is clear that the present model is incomplete, in the sense, that not all defaultable claims can be replicated. We shall check in Section 4.2.1 that, under the assumption that the driving Brownian motion W is one-dimensional, all survival claims satisfying natural technical conditions are hedgeable, however. In the more realistic case of a two-dimensional noise, we will still be able to hedge a large class of survival claims, including options on a defaultable asset (see Section 4.2.2) and options to exchange defaultable assets (see Section 4.2.3).

4.2.1 Hedging a survival claim

For the sake of expositional simplicity, we assume in this section that the driving Brownian motion W is one-dimensional. This is definitely not the right choice, since we deal here with two risky assets, and thus their prices will be perfectly correlated. However, this assumption is convenient for the expositional purposes, since it ensures the model completeness with respect to survival claims, and it will be later relaxed anyway.

We shall argue that in a model with two defaultable assets governed by (68), replication of a survival claim $(X, 0, \tau)$ is in fact equivalent to replication of the promised payoff X using the pre-default processes.

Lemma 4.6. *If a strategy ϕ^i, $i = 1, 2$, based on pre-default values $\widetilde{Y}^i$, $i = 1, 2$, is a replicating strategy for an $\mathcal{F}_T$-measurable claim X, that is, if ϕ is such that the process $\widetilde{V}_t(\phi) = \phi_t^1 \widetilde{Y}_t^1 + \phi_t^2 \widetilde{Y}_t^2$ satisfies, for every $t \in [0, T]$,*

$$\begin{aligned} d\widetilde{V}_t(\phi) &= \phi_t^1 \, d\widetilde{Y}_t^1 + \phi_t^2 \, d\widetilde{Y}_t^2, \\ \widetilde{V}_T(\phi) &= X, \end{aligned}$$

then for the process $V_t(\phi) = \phi_t^1 Y_t^1 + \phi_t^2 Y_t^2$ we have, for every $t \in [0, T]$,

$$\begin{aligned} dV_t(\phi) &= \phi_t^1 \, dY_t^1 + \phi_t^2 \, dY_t^2, \\ V_T(\phi) &= X \mathbb{1}_{\{T < \tau\}}. \end{aligned}$$

This means that a strategy ϕ replicates a survival claim $(X, 0, \tau)$.

Proof. It is clear that $V_t(\phi) = \mathbb{1}_{\{t<\tau\}} V_t(\phi) = \mathbb{1}_{\{t<\tau\}} \widetilde{V}_t(\phi)$. From

$$\phi_t^1 \, dY_t^1 + \phi_t^2 \, dY_t^2 = -(\phi_t^1 \widetilde{Y}_t^1 + \phi_t^2 \widetilde{Y}_t^2) \, dH_t + (1 - H_{t-})(\phi_t^1 \, d\widetilde{Y}_t^1 + \phi_t^2 \, d\widetilde{Y}_t^2),$$

it follows that

$$\phi_t^1 \, dY_t^1 + \phi_t^2 \, dY_t^2 = -\widetilde{V}_t(\phi) \, dH_t + (1 - H_{t-}) d\widetilde{V}_t(\phi),$$

that is,

$$\phi_t^1 \, dY_t^1 + \phi_t^2 \, dY_t^2 = d(\mathbb{1}_{\{t<\tau\}} \widetilde{V}_t(\phi)) = dV_t(\phi).$$

It is also obvious that $V_T(\phi) = X \mathbb{1}_{\{T<\tau\}}$. □

Combining the last result with Lemma 3.1, we see that a strategy (ϕ^1, ϕ^2) replicates a survival claim $(X, 0, \tau)$ whenever we have

$$\widetilde{Y}_T^1 \Big(x + \int_0^T \phi_t^2 \, d\widetilde{Y}_t^{2,1} \Big) = X$$

for some constant x and some $\mathbb{F}$-predictable process ϕ^2, where, in view of (69),

$$d\widetilde{Y}_t^{2,1} = \widetilde{Y}_t^{2,1} \Big(\big(\mu_{2,t} - \mu_{1,t} + \sigma_{1,t}(\sigma_{1,t} - \sigma_{2,t}) \big) dt + (\sigma_{2,t} - \sigma_{1,t}) \, dW_t \Big).$$

We introduce a probability measure $\widetilde{\mathbb{Q}}$, equivalent to $\mathbb{P}$ on $(\Omega, \mathcal{G}_T)$, and such that $\widetilde{Y}^{2,1}$ is an $\mathbb{F}$-martingale under $\widetilde{\mathbb{Q}}$. It is easily seen that the Radon-Nikodým density η satisfies, for $t \in [0,T]$,

$$d\widetilde{\mathbb{Q}}\,|_{\mathcal{G}_t} = \eta_t \, d\mathbb{P}\,|_{\mathcal{G}_t} = \mathcal{E}_t\left(\int_0^{\cdot} \theta_s \, dW_s\right) d\mathbb{P}\,|_{\mathcal{G}_t} \tag{70}$$

with

$$\theta_t = \frac{\mu_{2,t} - \mu_{1,t} + \sigma_{1,t}(\sigma_{1,t} - \sigma_{2,t})}{\sigma_{1,t} - \sigma_{2,t}},$$

provided, of course, that the process θ is well defined and satisfies suitable integrability conditions. We shall show that a survival claim is attainable if the random variable $X(\widetilde{Y}^1_T)^{-1}$ is $\widetilde{\mathbb{Q}}$-integrable. Indeed, the pre-default value $\widetilde{V}_t$ at time t of a survival claim equals

$$\widetilde{V}_t = \widetilde{Y}^1_t \, \mathbb{E}_{\widetilde{\mathbb{Q}}}\big(X(\widetilde{Y}^1_T)^{-1} \,|\, \mathcal{F}_t\big),$$

and from the predictable representation theorem, we deduce that there exists a process ϕ^2 such that

$$\mathbb{E}_{\widetilde{\mathbb{Q}}}\big(X(\widetilde{Y}^1_T)^{-1} \,|\, \mathcal{F}_t\big) = \mathbb{E}_{\widetilde{\mathbb{Q}}}\big(X(\widetilde{Y}^1_T)^{-1}\big) + \int_0^t \phi^2_u \, d\widetilde{Y}^{2,1}_u.$$

The component ϕ^1 of the self-financing trading strategy $\phi = (\phi^1, \phi^2)$ is then chosen in such a way that

$$\phi^1_t \widetilde{Y}^1_t + \phi^2_t \widetilde{Y}^2_t = \widetilde{V}_t, \quad \forall\, t \in [0,T].$$

To conclude, by focusing on pre-default values, we have shown that the replication of survival claims can be reduced here to classic results on replication of (non-defaultable) contingent claims in a default-free market model.

4.2.2 Option on a defaultable asset

In order to get a complete model with respect to survival claims, we postulated in the previous section that the driving Brownian motion in dynamics (68) is one-dimensional. This assumption is questionable, since it implies the perfect correlation of risky assets. However, we may relax this restriction, and work instead with the two correlated one-dimensional Brownian motions. The model will no longer be complete, but options on a defaultable assets will be still attainable. The payoff of a (non-vulnerable) call option written on the defaultable asset Y^2 equals

$$C_T = (Y^2_T - K)^+ = \mathbb{1}_{\{T<\tau\}}(\widetilde{Y}^2_T - K)^+,$$

so that it is natural to interpret this contract as a survival claim with the promised payoff $X = (\widetilde{Y}^2_T - K)^+$.

To deal with this option in an efficient way, we consider a model in which

$$dY^i_t = Y^i_{t-}\big(\mu_{i,t}\,dt + \sigma_{i,t}\,dW^i_t - dM_t\big), \tag{71}$$

where W^1 and W^2 are two one-dimensional correlated Brownian motions with the instantaneous correlation coefficient ρ_t. More specifically, we assume that $Y^1_t = D(t,T) = \mathbb{1}_{\{t<\tau\}}\widetilde{D}(t,T)$ represents a defaultable ZC-bond with zero recovery, and $Y^2_t = \mathbb{1}_{\{t<\tau\}}\widetilde{Y}^2_t$ is a generic defaultable asset with total default. Within the present set-up, the payoff can also be represented as follows

$$C_T = G(Y^1_T, Y^2_T) = (Y^2_T - KY^1_T)^+,$$

where $g(y_1, y_2) = (y_2 - Ky_1)^+$, and thus it can also be seen as an option to exchange the second asset for K units of the first asset.

The requirement that the process $\widetilde{Y}^{2,1}_t = \widetilde{Y}^2_t(\widetilde{Y}^1_t)^{-1}$ follows an $\mathbb{F}$-martingale under $\widetilde{\mathbb{Q}}$ implies that

$$d\widetilde{Y}^{2,1}_t = \widetilde{Y}^{2,1}_t\Big((\sigma_{2,t}\rho_t - \sigma_{1,t})\,d\widetilde{W}^1_t + \sigma_{2,t}\sqrt{1-\rho^2_t}\,d\widetilde{W}^2_t\Big), \tag{72}$$

where $\widetilde{W} = (\widetilde{W}^1, \widetilde{W}^2)$ follows a two-dimensional Brownian motion under $\widetilde{\mathbb{Q}}$. Since $\widetilde{Y}^1_T = 1$, replication of the option reduces to finding a constant x and an $\mathbb{F}$-predictable process ϕ^2 satisfying

$$x + \int_0^T \phi^2_t\,d\widetilde{Y}^{2,1}_t = (\widetilde{Y}^2_T - K)^+.$$

To obtain closed-form expressions for the option price and replicating strategy, we postulate that the volatilities $\sigma_{1,t}, \sigma_{2,t}$ and the correlation coefficient ρ_t are deterministic. Let $\widehat{F}_{Y^2}(t,T) = \widetilde{Y}^2_t(\widetilde{D}(t,T))^{-1}$ ($\widehat{F}_C(t,T) = \widetilde{C}_t(\widetilde{D}(t,T))^{-1}$, respectively) stand for the credit-risk-adjusted forward price of the second asset (the option, respectively). The proof of the following valuation result is fairly standard, and thus it is omitted.

Proposition 4.4. *The credit-risk-adjusted forward price of the option written on Y^2 equals*

$$\widehat{F}_C(t,T) = \widehat{F}_{Y^2}(t,T)N\big(d_+(\widehat{F}_{Y^2}(t,T),t,T)\big) - KN\big(d_-(\widehat{F}_{Y^2}(t,T),t,T)\big).$$

Equivalently, the pre-default price of the option equals

$$\widetilde{C}_t = \widetilde{Y}^2_t N\big(d_+(\widehat{F}_{Y^2}(t,T),t,T)\big) - K\widetilde{D}(t,T)N\big(d_-(\widehat{F}_{Y^2}(t,T),t,T)\big),$$

where

$$d_{\pm}(\widetilde{f},t,T) = \frac{\ln \widetilde{f} - \ln K \pm \frac{1}{2}v^2(t,T)}{v(t,T)}$$

and

$$v^2(t,T) = \int_t^T (\sigma_{1,u}^2 + \sigma_{2,u}^2 - 2\rho_u\sigma_{1,u}\sigma_{2,u})\, du.$$

Moreover the replicating strategy ϕ in the spot market satisfies for every $t \in [0,T]$, on the set $\{t < \tau\}$,

$$\phi_t^1 = -KN\big(d_-(\widehat{F}_{Y^2}(t,T),t,T)\big), \quad \phi_t^2 = N\big(d_+(\widehat{F}_{Y^2}(t,T),t,T)\big).$$

4.2.3 Option to exchange defaultable assets

We work here with the two correlated one-dimensional Brownian motions, so that

(73) $$dY_t^i = Y_{t-}^i\big(\mu_{i,t}\,dt + \sigma_{i,t}\,dW_t^i - dM_t\big), \quad i = 1,2,$$

where $d\langle W^1, W^2\rangle_t = \rho_t\,dt$ for some function ρ with values in $[-1,1]$. The model is no longer complete, but it is still not difficult to establish a direct counterpart of Proposition 4.4 for the exchange option with the payoff $(Y_T^2 - KY_T^1)^+$. In fact, the next result shows that the pricing formula expressed in terms of pre-default prices has the same shape as the standard formula for the option to exchange non-defaultable assets with dynamics (68). It is notable that we do not need to make any assumption about the behavior of the default intensity.

We only assume that the coefficients in (73) are such that there exist an e.m.m. for the process $\widetilde{Y}^{2,1}$, where

(74) $$d\widetilde{Y}_t^i = \widetilde{Y}_t^i\big((\mu_{i,t} + \gamma_t)\,dt + \sigma_{i,t}\,dW_t^i\big), \quad i = 1,2,$$

so that we implicitly impose mild technical conditions on drift coefficients.

Proposition 4.5. *Assume that the volatilities σ_1, σ_2 and the instantaneous correlation coefficient ρ are deterministic. Then the pre-default price of the exchange option equals*

$$\widetilde{C}_t = \widetilde{Y}_t^2 N\big(d_+(\widetilde{Y}_t^{2,1},t,T)\big) - K\widetilde{Y}_t^1 N\big(d_-(\widetilde{Y}_t^{2,1},t,T)\big),$$

where

$$d_{\pm}(\widetilde{y},t,T) = \frac{\ln \widetilde{y} - \ln K \pm \frac{1}{2}v^2(t,T)}{v(t,T)}$$

and

$$v^2(t,T) = \int_t^T \left(\sigma_{1,u}^2 + \sigma_{2,u}^2 - 2\rho_u \sigma_{1,u}\sigma_{2,u}\right) du.$$

Moreover the replicating strategy ϕ in the spot market satisfies for every $t \in [0,T]$, on the set $\{t < \tau\}$,

$$\phi_t^1 = -KN\Big(d_-(\widetilde{Y}_t^{2,1}, t, T)\Big), \quad \phi_t^2 = N\Big(d_+(\widetilde{Y}_t^{2,1}, t, T)\Big).$$

The pricing formula for the option on a defaultable asset (see Proposition 4.4) can be seen as a special case of the formula established in Proposition 4.5.

Similarly as in Sections 4.1.7 and 4.1.8, we conclude that the pricing and hedging of any attainable survival claim with the promised payoff $X = g(\widetilde{Y}_T^1, \widetilde{Y}_T^2)$ depends on the choice of a default intensity only through the pre-default prices $\widetilde{Y}_t^1$ and $\widetilde{Y}_t^2$. This property shows that we have correctly specified the hedging instruments for a claim at hand. Of course, the model considered in this section is not complete, even if the concept of completeness is reduced to survival claims. Basically, a survival claim can be hedged if its promised payoff can be represents as $X = \widetilde{Y}_T^1 h(Y_T^{2,1})$.

5. PDE Approach to Valuation and Hedging

In the remaining part of the paper, we take a different perspective, and we assume that trading occurs on the time interval $[0,T]$ and our goal is to replicate a contingent claim of the form

$$Y = \mathbb{1}_{\{T \geq \tau\}} g_1(Y_T^1, Y_T^2, Y_T^3) + \mathbb{1}_{\{T < \tau\}} g_0(Y_T^1, Y_T^2, Y_T^3) = G(Y_T^1, Y_T^2, Y_T^3, H_T),$$

which settles at time T. We do not need to assume here that the coefficients in dynamics of primary assets are $\mathbb{F}$-predictable. Since our goal is to develop the PDE approach, it will be essential, however, to postulate a Markovian character of a model. For the sake of simplicity, we assume that the coefficients are constant, so that

$$dY_t^i = Y_{t-}^i\Big(\mu_i\, dt + \sigma_i\, dW_t + \kappa_i\, dM_t\Big), \quad i = 1,2,3.$$

The assumption of constancy of coefficients is rarely, if ever, satisfied in practically relevant models of credit risk. It is thus important to note that it was postulated here mainly for the sake of notational convenience, and the general results established in this section can be easily extended to a non-homogeneous Markov case in which $\mu_{i,t} = \mu_i(t, Y_{t-}^1, Y_{t-}^2, Y_{t-}^3, H_{t-}), \sigma_{i,t} = \sigma_i(t, Y_{t-}^1, Y_{t-}^2, Y_{t-}^3, H_{t-})$, etc.

5.1 Defaultable asset with total default

We first assume that Y^1 and Y^2 are default-free, so that $\kappa_1 = \kappa_2 = 0$, and the third asset is subject to total default, i.e. $\kappa_3 = -1$,

$$dY_t^3 = Y_{t-}^3\big(\mu_3\,dt + \sigma_3\,dW_t - dM_t\big).$$

We work throughout under the assumptions of Proposition 4.1. This means that any $\mathbb{Q}^1$-integrable contingent claim $Y = G(Y_T^1, Y_T^2, Y_T^3; H_T)$ is attainable, and its arbitrage price equals

$$\pi_t(Y) = Y_t^1\,\mathbb{E}_{\mathbb{Q}^1}(Y(Y_T^1)^{-1}\,|\,\mathcal{G}_t), \quad \forall\, t \in [0,T]. \tag{75}$$

The following auxiliary result is thus rather obvious.

Lemma 5.1. *The process (Y^1, Y^2, Y^3, H) has the Markov property with respect to the filtration $\mathbb{G}$ under the martingale measure $\mathbb{Q}^1$. For any attainable claim $Y = G(Y_T^1, Y_T^2, Y_T^3; H_T)$ there exists a function $v : [0,T] \times \mathbb{R}^3 \times \{0,1\} \to \mathbb{R}$ such that $\pi_t(Y) = v(t, Y_t^1, Y_t^2, Y_t^3; H_t)$.*

We find it convenient to introduce the *pre-default* pricing function $v(\cdot\,;0) = v(t, y_1, y_2, y_3; 0)$ and the *post-default* pricing function $v(\cdot\,;1) = v(t, y_1, y_2, y_3; 1)$. In fact, since $Y_t^3 = 0$ if $H_t = 1$, it suffices to study the post-default function $v(t, y_1, y_2; 1) = v(t, y_1, y_2, 0; 1)$. Also, we write

$$\alpha_i = \mu_i - \sigma_i \frac{\mu_1 - \mu_2}{\sigma_1 - \sigma_2}, \qquad b = (\mu_3 - \mu_1)(\sigma_1 - \sigma_2) - (\mu_1 - \mu_3)(\sigma_1 - \sigma_3).$$

Let $\gamma > 0$ be the constant default intensity under $\mathbb{P}$, and let $\zeta > -1$ be given by formula (48).

Proposition 5.1. *Assume that the functions $v(\cdot\,;0)$ and $v(\cdot\,;1)$ belong to the class $C^{1,2}([0,T] \times \mathbb{R}_+^3, \mathbb{R})$. Then $v(t, y_1, y_2, y_3; 0)$ satisfies the PDE*

$$\begin{aligned} &\partial_t v(\cdot\,;0) + \sum_{i=1}^{2} \alpha_i y_i \partial_i v(\cdot\,;0) + (\alpha_3 + \zeta) y_3 \partial_3 v(\cdot\,;0) + \frac{1}{2} \sum_{i,j=1}^{3} \sigma_i \sigma_j y_i y_j \partial_{ij} v(\cdot\,;0) \\ &\quad - \alpha_1 v(\cdot\,;0) + \left(\gamma - \frac{b}{\sigma_1 - \sigma_2}\right)\Big[v(t, y_1, y_2; 1) - v(t, y_1, y_2, y_3; 0)\Big] = 0 \end{aligned}$$

subject to the terminal condition $v(T, y_1, y_2, y_3; 0) = G(y_1, y_2, y_3; 0)$, and $v(t, y_1, y_2; 1)$ satisfies the PDE

$$\partial_t v(\cdot\,;1) + \sum_{i=1}^{2} \alpha_i y_i \partial_i v(\cdot\,;1) + \frac{1}{2} \sum_{i,j=1}^{2} \sigma_i \sigma_j y_i y_j \partial_{ij} v(\cdot\,;1) - \alpha_1 v(\cdot\,;1) = 0$$

subject to the terminal condition $v(T, y_1, y_2; 1) = G(y_1, y_2, 0; 1)$.

Proof. For simplicity, we write $C_t = \pi_t(Y)$. Let us define

$$\Delta v(t, y_1, y_2, y_3) = v(t, y_1, y_2; 1) - v(t, y_1, y_2, y_3; 0).$$

Then the jump $\Delta C_t = C_t - C_{t-}$ can be represented as follows:

$$\Delta C_t = \mathbb{1}_{\{\tau=t\}}\Big(v(t, Y_t^1, Y_t^2; 1) - v(t, Y_t^1, Y_t^2, Y_{t-}^3; 0)\Big) = \mathbb{1}_{\{\tau=t\}}\,\Delta v(t, Y_t^1, Y_t^2, Y_{t-}^3).$$

We write ∂_i to denote the partial derivative with respect to the variable y_i, and we typically omit the variables $(t, Y_{t-}^1, Y_{t-}^2, Y_{t-}^3, H_{t-})$ in expressions $\partial_t v$, $\partial_i v$, Δv, etc. We shall also make use of the fact that for any Borel measurable function g we have

$$\int_0^t g(u, Y_u^2, Y_{u-}^3)\, du = \int_0^t g(u, Y_u^2, Y_u^3)\, du$$

since Y_u^3 and Y_{u-}^3 differ only for at most one value of u (for each ω). Let $\xi_t = \mathbb{1}_{\{t<\tau\}}\gamma$. An application of Itô's formula yields

$$\begin{aligned}
dC_t &= \partial_t v\, dt + \sum_{i=1}^{3} \partial_i v\, dY_t^i + \frac{1}{2}\sum_{i,j=1}^{3} \sigma_i \sigma_j Y_{t-}^i Y_{t-}^j \partial_{ij} v\, dt \\
&\quad + \Big(\Delta v + Y_{t-}^3 \partial_3 v\Big)\, dH_t \\
&= \partial_t v\, dt + \sum_{i=1}^{3} \partial_i v\, dY_t^i + \frac{1}{2}\sum_{i,j=1}^{3} \sigma_i \sigma_j Y_{t-}^i Y_{t-}^j \partial_{ij} v\, dt \\
&\quad + \Big(\Delta v + Y_{t-}^3 \partial_3 v\Big)\Big(dM_t + \xi_t\, dt\Big),
\end{aligned}$$

and this in turn implies that

$$\begin{aligned}
dC_t &= \partial_t v\, dt + \sum_{i=1}^{3} Y_{t-}^i \partial_i v\Big(\mu_i\, dt + \sigma_i\, dW_t\Big) + \frac{1}{2}\sum_{i,j=1}^{3} \sigma_i \sigma_j Y_{t-}^i Y_{t-}^j \partial_{ij} v\, dt \\
&\quad + \Delta v\, dM_t + \Big(\Delta v + Y_{t-}^3 \partial_3 v\Big)\xi_t\, dt \\
&= \left\{ \partial_t v + \sum_{i=1}^{3} \mu_i Y_{t-}^i \partial_i v + \frac{1}{2}\sum_{i,j=1}^{3} \sigma_i \sigma_j Y_{t-}^i Y_{t-}^j \partial_{ij} v + \Big(\Delta v + Y_{t-}^3 \partial_3 v\Big)\xi_t \right\} dt \\
&\quad + \Big(\sum_{i=1}^{3} \sigma_i Y_{t-}^i \partial_i v\Big)\, dW_t + \Delta v\, dM_t.
\end{aligned}$$

We now use the integration by parts formula together with (42) to derive dynamics of the relative price $\widehat{C}_t = C_t(Y_t^1)^{-1}$. We find that

$$d\widehat{C}_t = \widehat{C}_{t-}\big((-\mu_1 + \sigma_1^2)\,dt - \sigma_1\,dW_t\big)$$
$$+ (Y_{t-}^1)^{-1}\left\{\partial_t v + \sum_{i=1}^{3}\mu_i Y_{t-}^i\partial_i v + \frac{1}{2}\sum_{i,j=1}^{3}\sigma_i\sigma_j Y_{t-}^i Y_{t-}^j\partial_{ij}v + \big(\Delta v + Y_{t-}^3\partial_3 v\big)\xi_t\right\}dt$$
$$+ (Y_{t-}^1)^{-1}\sum_{i=1}^{3}\sigma_i Y_{t-}^i\partial_i v\,dW_t + (Y_{t-}^1)^{-1}\Delta v\,dM_t - (Y_{t-}^1)^{-1}\sigma_1\sum_{i=1}^{3}\sigma_i Y_{t-}^i\partial_i v\,dt.$$

Hence, using (47), we obtain

$$d\widehat{C}_t = \widehat{C}_{t-}\big(-\mu_1 + \sigma_1^2\big)\,dt + \widehat{C}_{t-}\big(-\sigma_1\,d\widehat{W}_t - \sigma_1\theta\,dt\big)$$
$$+ (Y_{t-}^1)^{-1}\left\{\partial_t v + \sum_{i=1}^{3}\mu_i Y_{t-}^i\partial_i v + \frac{1}{2}\sum_{i,j=1}^{3}\sigma_i\sigma_j Y_{t-}^i Y_{t-}^j\partial_{ij}v + \big(\Delta v + Y_{t-}^3\partial_3 v\big)\xi_t\right\}dt$$
$$+ (Y_{t-}^1)^{-1}\sum_{i=1}^{3}\sigma_i Y_{t-}^i\partial_i v\,d\widehat{W}_t + (Y_{t-}^1)^{-1}\sum_{i=1}^{3}\sigma_i Y_{t-}^i\theta\partial_i v\,dt$$
$$+ (Y_{t-}^1)^{-1}\Delta v\,d\widehat{M}_t + (Y_{t-}^1)^{-1}\zeta\xi_t\Delta v\,dt - (Y_{t-}^1)^{-1}\sigma_1\sum_{i=1}^{3}\sigma^i Y_{t-}^i\partial_i v\,dt.$$

This means that the process $\widehat{C}$ admits the following decomposition under $\mathbb{Q}^1$

$$d\widehat{C}_t = \widehat{C}_{t-}\big(-\mu_1 + \sigma_1^2 - \sigma_1\theta\big)\,dt$$
$$+ (Y_{t-}^1)^{-1}\left\{\partial_t v + \sum_{i=1}^{3}\mu_i Y_{t-}^i\partial_i v + \frac{1}{2}\sum_{i,j=1}^{3}\sigma_i\sigma_j Y_{t-}^i Y_{t-}^j\partial_{ij}v + \big(\Delta v + Y_{t-}^3\partial_3 v\big)\xi_t\right\}dt$$
$$+ (Y_{t-}^1)^{-1}\sum_{i=1}^{3}\sigma_i Y_{t-}^i\theta\partial_i v\,dt + (Y_{t-}^1)^{-1}\zeta\xi_t\Delta v\,dt$$
$$- (Y_{t-}^1)^{-1}\sigma_1\sum_{i=1}^{3}\sigma_i Y_{t-}^i\partial_i v\,dt + \text{a } \mathbb{Q}^1\text{-martingale}.$$

From (75), it follows that the process $\widehat{C}$ is a martingale under $\mathbb{Q}^1$. Therefore, the continuous finite variation part in the above decomposition necessarily vanishes, and thus we get

$$
\begin{aligned}
0 &= C_{t-}(Y^1_{t-})^{-1}\Big(-\mu_1+\sigma_1^2-\sigma_1\theta\Big)\\
&+(Y^1_{t-})^{-1}\left\{\partial_t v+\sum_{i=1}^{3}\mu_i Y^i_{t-}\partial_i v+\frac{1}{2}\sum_{i,j=1}^{3}\sigma_i\sigma_j Y^i_{t-}Y^j_{t-}\partial_{ij}v+\Big(\Delta v+Y^3_{t-}\partial_3 v\Big)\xi_t\right\}\\
&+(Y^1_{t-})^{-1}\sum_{i=1}^{3}\sigma_i Y^i_{t-}\theta\partial_i v+(Y^1_{t-})^{-1}\zeta\xi_t\Delta v-(Y^1_{t-})^{-1}\sigma_1\sum_{i=1}^{3}\sigma_i Y^i_{t-}\partial_i v.
\end{aligned}
$$

Consequently, we have that

$$
\begin{aligned}
0 &= C_{t-}\Big(-\mu_1+\sigma_1^2-\sigma_1\theta\Big)\\
&+\partial_t v+\sum_{i=1}^{3}\mu_i Y^i_{t-}\partial_i v+\frac{1}{2}\sum_{i,j=1}^{3}\sigma_i\sigma_j Y^i_{t-}Y^j_{t-}\partial_{ij}v+\Big(\Delta v+Y^3_{t-}\partial_3 v\Big)\xi_t\\
&+\sum_{i=1}^{3}\sigma_i Y^i_{t-}\theta\partial_i v+\zeta\xi_t\Delta v-\sigma_1\sum_{i=1}^{3}\sigma_i Y^i_{t-}\partial_i v.
\end{aligned}
$$

Finally, we conclude that

$$
\begin{aligned}
&\partial_t v+\sum_{i=1}^{2}\alpha_i Y^i_{t-}\partial_i v+(\alpha_3+\xi_t)\,Y^3_{t-}\partial_3 v+\frac{1}{2}\sum_{i,j=1}^{3}\sigma_i\sigma_j Y^i_{t-}Y^j_{t-}\partial_{ij}v\\
&-\alpha_1 C_{t-}+(1+\zeta)\xi_t\Delta v=0.
\end{aligned}
$$

Recall that $\xi_t = \mathbb{1}_{\{t<\tau\}}\gamma$. It is thus clear that the pricing functions $v(\cdot,0)$ and $v(\cdot;1)$ satisfy the PDEs given in the statement of the proposition. □

The next result deals with a replicating strategy for Y.

Proposition 5.2. *The replicating strategy ϕ for the claim Y is given by formulae*

$$
\begin{aligned}
\phi^3_t Y^3_{t-} &= -\Delta v(t,Y^1_t,Y^2_t,Y^3_{t-}) = v(t,Y^1_t,Y^2_t,Y^3_{t-};0)-v(t,Y^1_t,Y^2_t;1),\\
\phi^2_t Y^2_t(\sigma_2-\sigma_1) &= -(\sigma_1-\sigma_3)\Delta v-\sigma_1 v+\sum_{i=1}^{3}Y^i_{t-}\sigma_i\partial_i v,\\
\phi^1_t Y^1_t &= v-\phi^2_t Y^2_t-\phi^3_t Y^3_t.
\end{aligned}
$$

Proof. As a by-product of our computations, we obtain

$$
d\widehat{C}_t = -(Y^1_t)^{-1}\sigma_1 v\,d\widehat{W}_t+(Y^1_t)^{-1}\sum_{i=1}^{3}\sigma_i Y^i_{t-}\partial_i v\,d\widehat{W}_t+(Y^1_t)^{-1}\Delta v\,d\widehat{M}_t.
$$

The self-financing strategy that replicates Y is determined by two components ϕ^2, ϕ^3 and the following relationship:

$$d\widehat{C}_t = \phi^2_t\, dY^{2,1}_t + \phi^3_t\, dY^{3,1}_t = \phi^2_t Y^{2,1}_t (\sigma_2 - \sigma_1)\, d\widehat{W}_t + \phi^3_t Y^{3,1}_{t-} \left((\sigma_3 - \sigma_1)\, d\widehat{W}_t - d\widehat{M}_t \right).$$

By identification, we obtain $\phi^3_t Y^{3,1}_{t-} = (Y^1_t)^{-1} \Delta v$ and

$$\phi^2_t Y^2_t (\sigma_2 - \sigma_1) - (\sigma_3 - \sigma_1)\Delta v = -\sigma_1 C_t + \sum_{i=1}^{3} Y^i_{t-} \sigma_i \partial_i v.$$

This yields the claimed formulae. □

Corollary 5.1. *In the case of a total default claim, the hedging strategy satisfies the balance condition.*

Proof. A total default corresponds to the assumption that $G(y_1, y_2, y_3, 1) = 0$. We now have $v(t, y_1, y_2; 1) = 0$, and thus $\phi^3_t Y^3_{t-} = v(t, Y^1_t, Y^2_t, Y^3_{t-}; 0)$ for every $t \in [0, T]$. Hence, the equality $\phi^1_t Y^1_t + \phi^2_t Y^2_t = 0$ holds for every $t \in [0, T]$. The last equality is the balance condition for $Z = 0$. Recall that it ensures that the wealth of a replicating portfolio jumps to zero at default time. □

5.1.1 Hedging with the savings account

Let us now study the particular case where Y^1 is the savings account, i.e.,

$$dY^1_t = rY^1_t\, dt, \quad Y^1_0 = 1,$$

which corresponds to $\mu_1 = r$ and $\sigma_1 = 0$. Let us write $\widehat{r} = r + \widehat{\gamma}$, where

$$\widehat{\gamma} = \gamma(1 + \zeta) = \gamma + \mu_3 - r + \frac{\sigma_3}{\sigma_2}(r - \mu_2)$$

stands for the intensity of default under $\mathbb{Q}^1$. The quantity $\widehat{r}$ has a natural interpretation as the risk-neutral *credit-risk adjusted* short-term interest rate. Straightforward calculations yield the following corollary to Proposition 5.1.

Corollary 5.2. *Assume that* $\sigma_2 \neq 0$ *and*

$$\begin{aligned} dY^1_t &= rY^1_t\, dt, \\ dY^2_t &= Y^2_t\left(\mu_2\, dt + \sigma_2\, dW_t\right), \\ dY^3_t &= Y^3_{t-}\left(\mu_3\, dt + \sigma_3\, dW_t - dM_t\right). \end{aligned}$$

Then the function $v(\cdot\,;0)$ satisfies

$$\partial_t v(t,y_2,y_3;0) + ry_2\partial_2 v(t,y_2,y_3;0) + \widehat{r}y_3\partial_3 v(t,y_2,y_3;0) - \widehat{r}v(t,y_2,y_3;0) + \frac{1}{2}\sum_{i,j=2}^{3}\sigma_i\sigma_j y_i y_j \partial_{ij} v(t,y_2,y_3;0) + \widehat{\gamma}v(t,y_2;1) = 0$$

with $v(T,y_2,y_3;0) = G(y_2,y_3;0)$, and the function $v(\cdot\,;1)$ satisfies

$$\partial_t v(t,y_2;1) + ry_2\partial_2 v(t,y_2;1) + \frac{1}{2}\sigma_2^2 y_2^2 \partial_{22} v(t,y_2;1) - rv(t,y_2;1) = 0$$

with $v(T,y_2;1) = G(y_2,0;1)$.

In the special case of a survival claim, the function $v(\cdot\,;1)$ vanishes identically, and thus the following result can be easily established.

Corollary 5.3. *The pre-default pricing function $v(\cdot\,;0)$ of a survival claim $Y = \mathbb{1}_{\{T<\tau\}}G(Y_T^2, Y_T^3)$ is a solution of the following PDE:*

$$\partial_t v(t,y_2,y_3;0) + ry_2\partial_2 v(t,y_2,y_3;0) + \widehat{r}y_3\partial_3 v(t,y_2,y_3;0) + \frac{1}{2}\sum_{i,j=2}^{3}\sigma_i\sigma_j y_i y_j \partial_{ij} v(t,y_2,y_3;0) - \widehat{r}v(t,y_2,y_3;0) = 0$$

with the terminal condition $v(T,y_2,y_3;0) = G(y_2,y_3)$. The components ϕ^2 and ϕ^3 of the replicating strategy satisfy

$$\phi_t^2\sigma_2 Y_t^2 = \sum_{i=2}^{3}\sigma_i Y_{t-}^i \partial_i v(t,Y_t^2,Y_{t-}^3;0) + \sigma_3 v(t,Y_t^2,Y_{t-}^3;0),$$
$$\phi_t^3 Y_{t-}^3 = v(t,Y_t^2,Y_{t-}^3;0).$$

Example 5.1. Consider a survival claim $Y = \mathbb{1}_{\{T<\tau\}}g(Y_T^2)$, that is, a vulnerable claim with default-free underlying asset. Its pre-default pricing function $v(\cdot\,;0)$ does not depend on y_3, and satisfies the PDE (y stands here for y_2 and σ for σ_2)

$$\partial_t v(t,y;0) + ry\partial_2 v(t,y;0) + \frac{1}{2}\sigma^2 y^2 \partial_{22} v(t,y;0) - \widehat{r}v(t,y;0) = 0 \tag{76}$$

with the terminal condition $v(T,y;0) = \mathbb{1}_{\{t<\tau\}}g(y)$. The solution to (76) is

$$v(t,y) = e^{(\widehat{r}-r)(t-T)}\,v^{r,g,2}(t,y) = e^{\widehat{\gamma}(t-T)}\,v^{r,g,2}(t,y),$$

where the function $v^{r,g,2}$ is the Black-Scholes price of $g(Y_T)$ in a Black-Scholes model for Y_t with interest rate r and volatility σ_2.

5.2 Defaultable asset with non-zero recovery

We now assume that

$$dY_t^3 = Y_{t-}^3(\mu_3\,dt + \sigma_3\,dW_t + \kappa_3\,dM_t)$$

with $\kappa_3 > -1$ and $\kappa_3 \neq 0$. We assume that $Y_0^3 > 0$, so that $Y_t^3 > 0$ for every $t \in \mathbb{R}_+$. We shall briefly describe the same steps as in the case of a defaultable asset with total default.

5.2.1 Arbitrage-free property

As usual, we need first to impose specific constraints on model coefficients, so that the model is arbitrage-free. Indeed, an e.m.m. $\mathbb{Q}^1$ exists if there exists a pair (θ, ζ) such that

$$\theta_t(\sigma_i - \sigma_1) + \zeta_t\xi_t\frac{\kappa_i - \kappa_1}{1+\kappa_1} = \mu_1 - \mu_i + \sigma_1(\sigma_i - \sigma_1) + \xi_t(\kappa_i - \kappa_1)\frac{\kappa_1}{1+\kappa_1}, \quad i = 2,3.$$

To ensure the existence of a solution (θ, ζ) on the set $\tau < t$, we impose the condition

$$\sigma_1 - \frac{\mu_1 - \mu_2}{\sigma_1 - \sigma_2} = \sigma_1 - \frac{\mu_1 - \mu_3}{\sigma_1 - \sigma_3},$$

that is,

$$\mu_1(\sigma_3 - \sigma_2) + \mu_2(\sigma_1 - \sigma_3) + \mu_3(\sigma_2 - \sigma_1) = 0.$$

Now, on the set $\tau \geq t$, we have to solve the two equations

$$\theta_t(\sigma_2 - \sigma_1) = \mu_1 - \mu_2 + \sigma_1(\sigma_2 - \sigma_1),$$
$$\theta_t(\sigma_3 - \sigma_1) + \zeta_t\gamma\kappa_3 = \mu_1 - \mu_3 + \sigma_1(\sigma_3 - \sigma_1).$$

If, in addition, $(\sigma_2 - \sigma_1)\kappa_3 \neq 0$, we obtain the unique solution

$$\theta = \sigma_1 - \frac{\mu_1 - \mu_2}{\sigma_1 - \sigma_2} = \sigma_1 - \frac{\mu_1 - \mu_3}{\sigma_1 - \sigma_3},$$
$$\zeta = 0 > -1,$$

so that the martingale measure $\mathbb{Q}^1$ exists and is unique.

5.2.2 Pricing PDE and replicating strategy

We are in a position to derive the pricing PDEs. For the sake of simplicity, we assume that Y^1 is the savings account, so that Proposition 5.3 is a counterpart of Corollary 5.2. For the proof of Proposition 5.3, the interested reader is referred to Bielecki *et al.* [7].

Proposition 5.3. *Let $\sigma_2 \neq 0$ and let Y^1, Y^2, Y^3 satisfy*

$$dY_t^1 = rY_t^1\,dt,$$
$$dY_t^2 = Y_t^2\big(\mu_2\,dt + \sigma_2\,dW_t\big),$$
$$dY_t^3 = Y_{t-}^3\big(\mu_3\,dt + \sigma_3\,dW_t + \kappa_3\,dM_t\big).$$

Assume, in addition, that $\sigma_2(r-\mu_3)=\sigma_3(r-\mu_2)$ *and* $\kappa_3\neq 0$, $\kappa_3>-1$. *Then the price of a contingent claim* $Y=G(Y_T^2,Y_T^3,H_T)$ *can be represented as* $\pi_t(Y)=v(t,Y_t^2,Y_t^3,H_t)$, *where the pricing functions* $v(\cdot\,;0)$ *and* $v(\cdot\,;1)$ *satisfy the following PDEs*

$$\partial_t v(t,y_2,y_3;0)+ry_2\partial_2 v(t,y_2,y_3;0)+y_3\,(r-\kappa_3\gamma)\,\partial_3 v(t,y_2,y_3;0)-rv(t,y_2,y_3;0)$$
$$+\frac{1}{2}\sum_{i,j=2}^{3}\sigma_i\sigma_j y_i y_j \partial_{ij}v(t,y_2,y_3;0)+\gamma\Big(v(t,y_2,y_3(1+\kappa_3);1)-v(t,y_2,y_3;0)\Big)=0$$

and

$$\partial_t v(t,y_2,y_3;1)+ry_2\partial_2 v(t,y_2,y_3;1)+ry_3\partial_3 v(t,y_2,y_3;1)-rv(t,y_2,y_3;1)$$
$$+\frac{1}{2}\sum_{i,j=2}^{3}\sigma_i\sigma_j y_i y_j \partial_{ij}v(t,y_2,y_3;1)=0$$

subject to the terminal conditions

$$v(T,y_2,y_3;0)=G(y_2,y_3;0),\quad v(T,y_2,y_3;1)=G(y_2,y_3;1).$$

The replicating strategy ϕ *equals*

$$\phi_t^2=\frac{1}{\sigma_2 Y_t^2}\sum_{i=2}^{3}\sigma_i y_i\partial_i v(t,Y_t^2,Y_{t-}^3,H_{t-})$$
$$-\frac{\sigma_3}{\sigma_2\kappa_3 Y_t^2}\Big(v(t,Y_t^2,Y_{t-}^3(1+\kappa_3);1)-v(t,Y_t^2,Y_{t-}^3;0)\Big),$$
$$\phi_t^3=\frac{1}{\kappa_3 Y_{t-}^3}\Big(v(t,Y_t^2,Y_{t-}^3(1+\kappa_3);1)-v(t,Y_t^2,Y_{t-}^3;0)\Big),$$

and ϕ_t^1 *is given by* $\phi_t^1Y_t^1+\phi_t^2Y_t^2+\phi_t^3Y_t^3=C_t$.

5.2.3 Hedging of a survival claim

We shall illustrate Proposition 5.3 by means of examples. First, consider a survival claim of the form

$$Y=G(Y_T^2,Y_T^3,H_T)=\mathbb{1}_{\{T<\tau\}}g(Y_T^3).$$

Then the post-default pricing function $v^g(\cdot\,;1)$ vanishes identically, and the pre-default pricing function $v^g(\cdot\,;0)$ solves the PDE

$$\partial_t v^g(\cdot\,;0)+ry_2\partial_2 v^g(\cdot\,;0)+y_3\,(r-\kappa_3\gamma)\,\partial_3 v^g(\cdot\,;0)$$
$$+\frac{1}{2}\sum_{i,j=2}^{3}\sigma_i\sigma_j y_i y_j\partial_{ij}v^g(\cdot\,;0)-(r+\gamma)v^g(\cdot\,;0)=0$$

with the terminal condition $v^g(T, y_2, y_3; 0) = g(y_3)$. Denote $\alpha = r - \kappa_3\gamma$ and $\beta = \gamma(1 + \kappa_3)$.

It is not difficult to check that $v^g(t, y_2, y_3; 0) = e^{\beta(T-t)} v^{\alpha,g,3}(t, y_3)$ is a solution of the above equation, where the function $w(t, y) = v^{\alpha,g,3}(t, y)$ is the solution of the standard Black-Scholes PDE equation

$$\partial_t w + y\alpha \partial_y w + \frac{1}{2}\sigma_3^2 y^2 \partial_{yy} w - \alpha w = 0$$

with the terminal condition $w(T, y) = g(y)$, that is, the price of the contingent claim $g(Y_T)$ in the Black-Scholes framework with the interest rate α and the volatility parameter equal to σ_3.

Let C_t be the current value of the contingent claim Y, so that

$$C_t = \mathbb{1}_{\{t<\tau\}} e^{\beta(T-t)} v^{\alpha,g,3}(t, Y_t^3).$$

The hedging strategy of the survival claim is, on the event $\{t < \tau\}$,

$$\begin{aligned} \phi_t^3 Y_t^3 &= -\frac{1}{\kappa_3} e^{-\beta(T-t)} v^{\alpha,g,3}(t, Y_t^3) = -\frac{1}{\kappa_3} C_t, \\ \phi_t^2 Y_t^2 &= \frac{\sigma_3}{\sigma_2} \left(Y_t^3 e^{-\beta(T-t)} \partial_y v^{\alpha,g,3}(t, Y_t^3) - \phi_t^3 Y_t^3 \right). \end{aligned}$$

5.2.4 Hedging of a recovery payoff

As another illustration of Proposition 5.3, we shall now consider the contingent claim $G(Y_T^2, Y_T^3, H_T) = \mathbb{1}_{\{T\geq\tau\}} g(Y_T^2)$, that is, we assume that recovery is paid at maturity and equals $g(Y_T^2)$. Let v^g be the pricing function of this claim. The post-default pricing function $v^g(\cdot\,; 1)$ does not depend on y_3. Indeed, the equation (we write here $y_2 = y$)

$$\partial_t v^g(\cdot\,; 1) + ry\partial_y v^g(\cdot\,; 1) + \frac{1}{2}\sigma_2^2 y^2 \partial_{yy} v^g(\cdot\,; 1) - rv^g(\cdot\,; 1) = 0,$$

with $v^g(T, y; 1) = g(y)$, admits a unique solution $v^{r,g,2}$, which is the price of $g(Y_T)$ in the Black-Scholes model with interest rate r and volatility σ_2.

Prior to default, the price of the claim can be found by solving the following PDE

$$\begin{aligned} &\partial_t v^g(\cdot; 0) + ry_2 \partial_2 v^g(\cdot; 0) + y_3 \left(r - \kappa_3\gamma\right) \partial_3 v^g(\cdot; 0) \\ &+ \frac{1}{2} \sum_{i,j=2}^{3} \sigma_i \sigma_j y_i y_j \partial_{ij} v^g(\cdot; 0) - (r + \gamma) v^g(\cdot; 0) = -\gamma v^g(t, y_2; 1) \end{aligned}$$

with $v^g(T, y_2, y_3; 0) = 0$. It is not difficult to check that

$$v^g(t, y_2, y_3; 0) = (1 - e^{\gamma(t-T)}) v^{r,g,2}(t, y_2).$$

The reader can compare this result with the one of Example 5.1.

5.3 Two defaultable assets with total default

We shall now assume that we have only two assets, and both are defaultable assets with total default. We shall briefly outline the analysis of this case, leaving the details and the study of other relevant cases to the reader. We postulate that

$$dY_t^i = Y_{t-}^i\big(\mu_i\,dt + \sigma_i\,dW_t - dM_t\big),\ i = 1,2, \tag{77}$$

so that

$$Y_t^1 = \mathbb{1}_{\{t<\tau\}}\widetilde{Y}_t^1, \quad Y_t^2 = \mathbb{1}_{\{t<\tau\}}\widetilde{Y}_t^2,$$

with the pre-default prices governed by the SDEs

$$d\widetilde{Y}_t^i = \widetilde{Y}_t^i\big((\mu_i + \gamma)\,dt + \sigma_i\,dW_t\big),\ i = 1,2.$$

In the case where the promised payoff X is path-independent, so that

$$X\mathbb{1}_{\{T<\tau\}} = G(Y_T^1, Y_T^2)\mathbb{1}_{\{T<\tau\}} = G(\widetilde{Y}_T^1, \widetilde{Y}_T^2)\mathbb{1}_{\{T<\tau\}}$$

for some function G, it is possible to use the PDE approach in order to value and replicate survival claims prior to default (needless to say that the valuation and hedging after default are trivial here).

We know already from the martingale approach that hedging of a survival claim $X\mathbb{1}_{\{T<\tau\}}$ is formally equivalent to replicating the promised payoff X using the pre-default values of tradeable assets

$$d\widetilde{Y}_t^i = \widetilde{Y}_t^i\big((\mu_i + \gamma)\,dt + \sigma_i\,dW_t\big), \quad i = 1,2.$$

We need not to worry here about the balance condition, since in case of default the wealth of the portfolio will drop to zero, as it should in view of the equality $Z = 0$.

We shall find the pre-default pricing function $v(t, y_1, y_2)$, which is required to satisfy the terminal condition $v(T, y_1, y_2) = G(y_1, y_2)$, as well as the hedging strategy (ϕ^1, ϕ^2). The replicating strategy ϕ is such that for the pre-default value $\widetilde{C}$ of our claim we have $\widetilde{C}_t := v(t, \widetilde{Y}_t^1, \widetilde{Y}_t^2) = \phi_t^1\widetilde{Y}_t^1 + \phi_t^2\widetilde{Y}_t^2$, and

$$d\widetilde{C}_t = \phi_t^1\,d\widetilde{Y}_t^1 + \phi_t^2\,d\widetilde{Y}_t^2. \tag{78}$$

Proposition 5.4. *Assume that $\sigma_1 \neq \sigma_2$. Then the pre-default pricing function v satisfies the PDE*

$$\begin{aligned}\partial_t v &+ y_1\left(\mu_1 + \gamma - \sigma_1\frac{\mu_2 - \mu_1}{\sigma_2 - \sigma_1}\right)\partial_1 v + y_2\left(\mu_2 + \gamma - \sigma_2\frac{\mu_2 - \mu_1}{\sigma_2 - \sigma_1}\right)\partial_2 v \\ &+ \frac{1}{2}\big(y_1^2\sigma_1^2\partial_{11}v + y_2^2\sigma_2^2\partial_{22}v + 2y_1y_2\sigma_1\sigma_2\partial_{12}v\big) = \left(\mu_1 + \gamma - \sigma_1\frac{\mu_2 - \mu_1}{\sigma_2 - \sigma_1}\right)v\end{aligned}$$

with the terminal condition $v(T, y_1, y_2) = G(y_1, y_2)$.

Proof. We shall merely sketch the proof. By applying Itô's formula to $v(t, \widetilde{Y}^1_t, \widetilde{Y}^2_t)$, and comparing the diffusion terms in (78) and in the Itô differential $dv(t, \widetilde{Y}^1_t, \widetilde{Y}^2_t)$, we find that

$$y_1\sigma_1\partial_1 v + y_2\sigma_2\partial_2 v = \phi^1 y_1\sigma_1 + \phi^2 y_2\sigma_2, \tag{79}$$

where $\phi^i = \phi^i(t, y_1, y_2)$. Since $\phi^1 y_1 = v(t, y_1, y_2) - \phi^2 y_2$, we deduce from (79) that

$$y_1\sigma_1\partial_1 v + y_2\sigma_2\partial_2 v = v\sigma_1 + \phi^2 y_2(\sigma_2 - \sigma_1),$$

and thus

$$\phi^2 y_2 = \frac{y_1\sigma_1\partial_1 v + y_2\sigma_2\partial_2 v - v\sigma_1}{\sigma_2 - \sigma_1}.$$

On the other hand, by identification of drift terms in (79), we obtain

$$\begin{aligned}
&\partial_t v + y_1(\mu_1 + \gamma)\partial_1 v + y_2(\mu_2 + \gamma)\partial_2 v \\
&\quad + \frac{1}{2}\left(y_1^2\sigma_1^2\partial_{11} v + y_2^2\sigma_2^2\partial_{22} v + 2y_1 y_2\sigma_1\sigma_2\partial_{12} v\right) \\
&= \phi^1 y_1(\mu_1 + \gamma) + \phi^2 y_2(\mu_2 + \gamma).
\end{aligned}$$

Upon elimination of ϕ^1 and ϕ^2, we arrive at the stated PDE. □

Recall that the historically observed drift terms are $\widehat{\mu}_i = \mu_i + \gamma$, rather than μ_i. The pricing PDE can thus be simplified as follows:

$$\begin{aligned}
&\partial_t v + y_1\left(\widehat{\mu}_1 - \sigma_1\frac{\widehat{\mu}_2 - \widehat{\mu}_1}{\sigma_2 - \sigma_1}\right)\partial_1 v + y_2\left(\widehat{\mu}_2 - \sigma_2\frac{\widehat{\mu}_2 - \widehat{\mu}_1}{\sigma_2 - \sigma_1}\right)\partial_2 v \\
&+ \frac{1}{2}\left(y_1^2\sigma_1^2\partial_{11} v + y_2^2\sigma_2^2\partial_{22} v + 2y_1 y_2\sigma_1\sigma_2\partial_{12} v\right) = v\left(\widehat{\mu}_1 - \sigma_1\frac{\widehat{\mu}_2 - \widehat{\mu}_1}{\sigma_2 - \sigma_1}\right).
\end{aligned}$$

The pre-default pricing function v depends on the market observables (drift coefficients, volatilities, and pre-default prices), but not on the (deterministic) default intensity.

To make one more simplifying step, we make an additional assumption about the payoff function. Suppose, in addition, that the payoff function is such that $G(y_1, y_2) = y_1 g(y_2/y_1)$ for some function $g : \mathbb{R}_+ \to \mathbb{R}$ (or equivalently, $G(y_1, y_2) = y_2 h(y_1/y_2)$ for some function $h : \mathbb{R}_+ \to \mathbb{R}$). Then we may focus on relative pre-default prices $\widehat{C}_t = \widetilde{C}_t(\widetilde{Y}^1_t)^{-1}$ and $\widetilde{Y}^{2,1} = \widetilde{Y}^2_t(\widetilde{Y}^1_t)^{-1}$. The corresponding pre-default pricing function $\widehat{v}(t, z)$, such that $\widehat{C}_t = \widehat{v}(t, Y^{2,1}_t)$ will satisfy the PDE

$$\partial_t\widehat{v} + \frac{1}{2}(\sigma_2 - \sigma_1)^2 z^2\partial_{zz}\widehat{v} = 0$$

with terminal condition $\widehat{v}(T,z) = g(z)$. If the price processes Y^1 and Y^2 in (68) are driven by the correlated Brownian motions W and $\widehat{W}$ with the constant instantaneous correlation coefficient ρ, then the PDE becomes

$$\partial_t \widehat{v} + \frac{1}{2}(\sigma_2^2 + \sigma_1^2 - 2\rho\sigma_1\sigma_2)z^2 \partial_{zz}\widehat{v} = 0.$$

Consequently, the pre-default price $\widetilde{C}_t = \widetilde{Y}_t^1 \widehat{v}(t, \widetilde{Y}_t^{2,1})$ will not depend directly on the drift coefficients $\widehat{\mu}_1$ and $\widehat{\mu}_2$, and thus, in principle, we should be able to derive an expression the price of the claim in terms of market observables: the prices of the underlying assets, their volatilities and the correlation coefficient. Put another way, neither the default intensity nor the drift coefficients of the underlying assets appear as independent parameters in the pre-default pricing function.

Before we conclude this work, let us stress once again that the martingale approach can be used in a fairly general set-up. By contrast, the PDE methodology is only suitable when dealing with a Markovian framework. In a forthcoming paper [8], we analyze a more general situation where a traded defaultable asset is a credit default swap, so that its dynamics involve also a continuous dividend stream.

Acknowledgments. Some results of this work were presented by Monique Jeanblanc at the "International Workshop on Stochastic Processes and Applications to Mathematical Finance" held at Ritsumeikan University on March 3–6, 2005. She deeply thanks the participants for questions and comments. The first version of this paper was written during her stay at Nagoya City University on the invitation by Professor Yoshio Miyahara, whose the warm hospitality is gratefully acknowledged. The work was completed during our visit to the Isaac Newton Institute for Mathematical Sciences in Cambridge. We thank the organizers of the programme *Developments in Quantitative Finance* for the kind invitation.

References

1. Arvanitis A. and J. Gregory, *Credit: The Complete Guide to Pricing, Hedging and Risk Management*, Risk Publications, 2001.
2. Ayache, E., P. Henrotte, S. Nassar and X. Wang, Can anyone solve the smile problem? *Wilmott* (2004), 78–96.
3. Bielecki, T. R. and M. Rutkowski, *Credit Risk: Modelling, Valuation and Hedging*. Springer-Verlag, Berlin Heidelberg New York, 2002.
4. Bielecki, T. R., M. Jeanblanc and M. Rutkowski, Hedging of defaultable claims, *Paris-Princeton Lectures on Mathematical Finance 2003*, R. A. Carmona, E. Çinlar, I. Ekeland, E. Jouini, J. E. Scheinkman, N. Touzi, eds., Springer-Verlag, Berlin Heidelberg New York, pp. 1–132, 2004.
5. Bielecki, T. R., M. Jeanblanc and M. Rutkowski, Completeness of a general semimartingale market under constrained trading, to appear, Proceedings of International Lisbonn Conference, Springer,2005.

6. Bielecki, T. R., M. Jeanblanc and M. Rutkowski, Completeness of a reduced-form credit risk model with discontinuous asset prices, to appear, 2005.
7. Bielecki, T. R., M. Jeanblanc and M. Rutkowski, PDE approach to valuation and hedging of credit derivatives, *Quantitative Finance* **5**, (2005), 257–270.
8. Bielecki, T. R., M. Jeanblanc and M. Rutkowski, Pricing and trading credit default swaps, working paper, 2005.
9. Black, F. and J. C. Cox, Valuing corporate securities: some effects of bond indenture provisions, *Journal of Finance* **31** (1976), 351–367.
10. Blanchet-Scalliet, C. and M. Jeanblanc, Hazard rate for credit risk and hedging defaultable contingent claims, *Finance and Stochastics* **8** (2004), 145–159.
11. Brémaud, P., *Point Processes and Queues. Martingale Dynamics*, Springer-Verlag, Berlin Heidelberg New York, 1981.
12. Carr, P., Dynamic replication of a digital default claim, working paper, 2005.
13. Collin-Dufresne, P., R. S. Goldstein and J.-N. Hugonnier, A general formula for valuing defaultable securities, *Econometrica* **72** (2004), 1377–1407.
14. Collin-Dufresne, P. and J.-N. Hugonnier, On the pricing and hedging of contingent claims in the presence of extraneous risks, working paper, 1999.
15. Cossin, D. and H. Pirotte, *Advanced Credit Risk Analysis*, J. Wiley, Chichester, 2000.
16. Dellacherie, C., B. Maisonneuve and P.-A. Meyer, *Probabilités et potentiel, chapitres XVII-XXIV*, Hermann, Paris, 1992.
17. Duffie, D. and D. Lando, The term structure of credit spreads with incomplete accounting information, *Econometrica* **69** (2001), 633–664.
18. Duffie, D. and K. Singleton, *Credit Risk: Pricing, Measurement and Management*, Princeton University Press, Princeton, 2003.
19. El Karoui, N., Modélisation de l'information, CEA-EDF-INRIA, Ecole d'été, unpublished manuscript, 1999.
20. Elliott, R. J., M. Jeanblanc and M. Yor, On models of default risk, *Mathematical Finance* **10** (2000), 179–195.
21. Giesecke, K., Default and information, working paper, Cornell University, 2001.
22. Giesecke, K., Correlated default with incomplete information, *Journal of Banking and Finance* **28** (2004), 1521–1545.
23. Guo, X., R. A. Jarrow and Y. Zheng, Information reduction in credit risk models, working paper, Cornell University, 2005.
24. Jarrow, R. A. and P. Protter, Structural versus reduced form models: A new information based perspective, *Journal of Investment Management* **2/2** (2004), 1–10.
25. Jarrow, R. A. and S. M. Turnbull, Pricing derivatives on financial securities subject to credit risk, *Journal of Finance* **50** (1995), 53–85.
26. Jacod, J. and A. N. Shiryaev, *Limit Theorems for Stochastic Processes*, Springer-Verlag, Berlin Heidelberg New York, 1987.
27. Jamshidian, F., Valuation of credit default swap and swaptions, *Finance and Stochastics* **8** (2004), 343–371.
28. Jeanblanc, M. and S. Valchev, Partial information, default hazard process, and default-risky bonds, *IJTAF* **8** (2005), 807-838.
29. Kusuoka, S., A remark on default risk models, *Advances in Mathematical Eco-*

nomics **1** (1999), 69–82.

30. Lando, D., *Credit Risk Modeling,* Princeton University Press, Princeton, 2004.
31. Laurent, J.-P., Applying hedging techniques to credit derivatives, Credit Risk Conference, London, 2001.
32. Mansuy, R. and Yor, M., *Random Times and Enlargement of Filtrations in a Brownian Setting*, forthcoming, Springer-Verlag, Berlin Heidelberg New York, 2005.
33. Merton, R. C., On the pricing of corporate debt: The risk structure of interest rates, *Journal of Finance* **29** (1974), 449–470.
34. Protter, P., A partial introduction to financial asset pricing theory, *Stochastic Processes and Their Applications* **91** (2001), 169–203.
35. Protter, P., *Stochastic Integration and Differential Equations*, 2nd edition, version 2.1, Springer-Verlag, Berlin Heidelberg New York, 2005.
36. Schönbucher, P. J., *Credit Derivatives Pricing Models,* J. Wiley, Chichester, 2003.
37. Vaillant, N., A beginner's guide to credit derivatives, working paper, Nomura International, 2001.

A Large Trader-Insider Model

Arturo Kohatsu-Higa and Agnès Sulem

INRIA-Rocquencourt
Domaine de Voluceau, Rocquencourt, B.P. 105,
F–78153 Le Chesnay Cedex, France

We give some remarks on the anticipating approach to insider modelling introduced by the authors recently. In particular, we define forward integrals by using limits of Riemmann sums. This definition is well adapted to financial applications.
As an application, we consider a portfolio maximization problem of a large trader with insider information. We show that the forward integral is a natural tool to handle such problems and we compute the optimal portfolios for an insider and a small trader.
Key words: Anticipating Calculus, Information asymmetry, large traders.

1. Introduction

In this article, we would like to explain the anticipating approach to insider information. The section on the forward integrals properties relies on Chapter 3 of Nualart (1995). Nevertheless, as we have not found a standard reference for this material in the form of the forward integral we will do it here in detail. For this we need to introduce the basic tools of differentiation on the Wiener space.

Consider the interval $[0,T]$ and a complete probability space $(\Omega, \mathcal{F}, P)$ on which a standard one dimensional Brownian motion W is defined; $\{\mathcal{F}_t\}_{t\in[0,T]}$ denotes the filtration generated by W, augmented with the P–null sets and made right continuous. Since all the results in the paper rely heavily on Malliavin calculus, we introduce some of its terminology briefly.

We denote by $\mathbb{C}_b^\infty(\mathbb{R}^n)$ the set of $\mathbb{C}^\infty$ bounded functions f from $\mathbb{R}^n$ to $\mathbb{R}$, with bounded derivatives of all orders. If S is the class of real random variables F that can be represented as $f(W_{t_1},\ldots,W_{t_n})$ for some $n \in \mathbf{N}$, $t_1,\ldots,t_n \in [0,T]$ and $f \in \mathbb{C}_b^\infty(\mathbb{R}^n)$, we can complete this space under the

Sobolev norm $\|\cdot\|_{1,p}$ given by

$$\|F\|_{1,p}^p = E(|F|^p) + E\left(\left(\int_0^T |D_sF|^2 ds\right)^{\frac{p}{2}}\right),$$

where D is defined as $D_sF = \sum_{i=1}^{n} \frac{\partial f}{\partial x_i}(W_{t_1}, \ldots, W_{t_n}) 1_{[0,t_i]}(s)$, obtaining a Banach space, usually indicated with $\mathrm{D}^{1,p}$. Analogously, we can construct the space $\mathrm{D}^{k,p}$ by completing S under the Sobolev norm

$$\|F\|_{k,p}^p = E(|F|^p) + \sum_{i=1}^{k} E\Big[\Big(\int_0^T \cdots \int_0^T |D^i_{s_i..s_1}F|^2 ds_1 \ldots ds_i\Big)^{\frac{p}{2}}\Big],$$

where $D^i_{s_i\ldots s_1}F = D_{s_i}\ldots D_{s_1}F$. Finally, we denote $\mathrm{D}^\infty = \bigcap_{p\geq 1}\bigcap_{k\geq 1} \mathrm{D}^{k,p}$.

We denote the adjoint of the closable unbounded operator

$$D : \mathrm{D}^{1,2} \subseteq L^2(\Omega) \longrightarrow L^2([0,T]\times\Omega)$$

by δ_0^T. This operator is called the Skorohod integral. The domain of δ_0^T is the set of all processes u in $L^2([0,T]\times\Omega)$ such that

$$\left|E\left(\int_0^T D_tFu_t dt\right)\right| \leq C\,\|F\|_2 \qquad \forall\, F \in S,$$

for some constant C possibly depending on u and $\|F\|_2 = E(|F|^2)^{1/2}$.

If $u \in Dom(\delta_0^T)$, then $\delta_0^T(u)$ is the square integrable random variable determined by the duality relation

$$E(\delta_0^T(u)F) = E(\int_0^T D_tFu_t dt) \qquad \forall\, F \in \mathrm{D}^{1,2}.$$

Note that the above construction can be carried through for any fixed time interval $[s,S]$, in the space $L^2([s,S]\times\Omega)$. We will also use the notation

$$\delta_0^T(u) = \int_0^T u(s)\delta W(s).$$

For a stochastic process ϕ, we say that $\phi \in \mathbf{L}^{1,2}$ if the following norm is finite:

$$\|\phi\|_{1,2}^2 = E\left[\int_0^T |\phi(s)|^2 ds\right] + E\left[\int_0^T\int_0^T |D_u\phi(s)|^2 dsdu\right].$$

2. The Forward Integral

Consider an insider, that is an agent that has sensible information about the future values of a stock, who may also have an influence on the evolution of the stock price. This is called a large trader-insider.

In general one would like to study models of the type

$$S(t) = S(0) + \int_0^t \mu(s, \pi(s))S(s)ds + \int_0^t \sigma(s, \pi(s))S(s)d^- W(s).$$

Here π represents the insider's strategy which is adapted to a filtration $\mathcal{G}$, which may be bigger (or just different) than the filtration generated by the Wiener process W with natural filtration $\mathcal{F}$. Therefore S is also adapted to $\mathcal{G}$ (if $\mathcal{F} \subseteq \mathcal{G}$) and the above stochastic integral will be an anticipating integral commonly known as the forward integral of Russo-Vallois.

Next, we define the forward integral. For this, given any partition $0 = t_0 < ... < t_n = T$ such that $\max\{t_{i+1} - t_i; i = 0, ..., n-1\} \to 0$ as $n \to \infty$, let

$$\eta(s) := \max\{t_i; t_i \leq s\}.$$

Then we can define the forward integral as follows:

Definition 2.1. Let $\phi : [0, T] \times \Omega \to$ be a measurable continuous process. The *forward integral* of ϕ with respect to $W(.)$ is defined by

$$\int_0^T \phi(t)d^- W(t) = \lim_{n \to +\infty} \sum_{i=0}^{n-1} \phi(t_i)(W(t_{i+1}) - W(t_i)), \tag{1}$$

if the limit exists in probability and is independent of the partition sequence taken.

This definition does not coincide exactly with the original definition of Russo-Vallois, unless we put some additional assumptions.

Note that the above definition is local. That is, let ϕ be forward integrable such that for a measurable set $A \subset \Omega$ we have that $\phi 1_A = 0$, Then $\int_0^T \phi(t)1_A d^- W(t) = 0$. In that sense, as in Nualart, (1995), page 45 we will use the local defintion of all the spaces to appear below.

First let us start proving that the expectation of this integral is not zero and therefore the usual rules of calculus do not apply. In particular, usual martingale properties are not true. For parallel martingale properties of anticipating integrals, see the interesting articles of Tudor (2004) and Pecatti-Theieullen-Tudor (2005).

Definition 2.2. Let $\phi : [0, T] \times \Omega \to \mathbb{R}$ be a measurable process such that

$\phi(t) \in \mathbb{L}^{1,2}$. We say that $\phi \in \mathbb{L}^{1,2}_{+}$ if the following stability property is satisfied: for any sequence of partitions $0 = t_0 < \dots < t_n = T$ such that its norm tends to zero as $n \to \infty$, there exists the trace process $D_{s+}\phi \in L^2([0,T] \times \Omega)$ such that

$$\left\|\phi(\eta(\cdot)) - \phi(\cdot)\right\|_{1,2} + E\left[\int_0^T \left|D_s\phi(\eta(s)) - D_{s+}\phi\right|^2 ds\right] \to 0$$

In such a case we say that $\phi \in \mathbb{L}^{1,2}_{+}$ and we define

$$\left\|\phi\right\|_{1,2,+}^2 := \left\|\phi\right\|_{1,2}^2 + E\left[\int_0^T \left|D_{s+}\phi\right|^2 ds\right].$$

This norm will serve to control the variance of the forward integral as it is shown in the next Theorem.

Theorem 2.1. *Suppose that $\phi \in \mathbb{L}^{1,2}_{+}$. Then the forward integral of ϕ exists, the limit in the definition 2.1 being satisfied in $L^1(\Omega)$ and furthermore*

$$\int_0^T \phi(t)d^-W(t) = \int_0^T \phi(t)\delta W(t) + \int_0^T D_{t+}\phi dt,$$

where δ denotes the Skorohod integral. Furthermore,

$$E\left[\left(\int_0^T \phi(t)d^-W(t)\right)^2\right] \leq 2\left\|\phi\right\|_{1,2,+}^2.$$

Proof. In order to prove that the integral exists we use the following formula (see formula (1.12) in page 130 in Nualart (1995a))

$$\phi(t_i)(W(t_{i+1}) - W(t_i)) = \int_{t_i}^{t_{i+1}} \phi(t_i)\delta W(s) + \int_{t_i}^{t_{i+1}} D_s\phi(t_i)ds.$$

Then the existence of the forward integral follows from Definition 2.2. Furthermore we have that each element in this expression belongs to $L^2(\Omega)$ and therefore we have that

$$\sum_{i=0}^{n-1} E\left[\phi(t_i)(W(t_{i+1}) - W(t_i))\right] = E\left[\int_0^T D_s\phi(\eta(s))ds\right]$$
$$\to E\left[\int_0^T D_{s+}\phi ds\right].$$

The last estimate is obtained similarly. We have

$$\sum_{i=0}^{n-1} \phi(t_i)(W(t_{i+1}) - W(t_i)) = \int_0^T \phi(\eta(s))\delta W(s) + \int_0^T D_s\phi(\eta(s))ds.$$

Therefore,

$$E\left[\left(\sum_{i=0}^{n-1} \phi(t_i)(W(t_{i+1}) - W(t_i))\right)^2\right]$$
$$\leq 2\left\{E\left[\left(\int_0^T \phi(\eta(s))\delta W(s)\right)^2\right] + E\left[\left(\int_0^T D_s\phi(\eta(s))ds\right)^2\right]\right\}.$$

Then the Riemmann sum sequence is bounded in $L^2(\Omega)$ and therefore converges in $L^2(\Omega)$ as it converges in $L^1(\Omega)$. Then taking limits in the above inequality we obtained the desired result. □

Next we prove that the integral process is a continuous process.

Theorem 2.2. *Suppose that* $\phi \in \mathbb{L}_+^{1,2}$, *such that* $E\left[\int_0^T \left(\int_0^T |D_s\phi(u)|^2 ds\right)^{p/2} du\right] < \infty$ *for some* $p > 2$ *then the process* $\{\int_0^t \phi(s)d^-W(s); t \in [0,T]\}$ *has a continuous version.*

Proof. Use Proposition 5.1.1 in Nualart (1995a). □

Now we give the formula for the quadratic variation.

Theorem 2.3. *Given any sequence of partitions of the interval* $[0,t]$, $\pi_n : 0 = t_0 < \dots < t_n = t$ *such that* $\max\{t_{i+1} - t_i; i = 0, \dots, n-1\} \to 0$ *as* $n \to \infty$, *we have that*

$$\sum_{i=0}^{n-1}\left(\int_{t_i}^{t_{i+1}} \phi(s)d^-W(s)\right)^2 \to \int_0^t |\phi(s)|^2 ds \text{ a.s.}$$

for $\phi \in \mathbb{L}_+^{1,2}$.

Proof. First suppose the simple case that there exists a fixed partition $0 = s_0 < \dots < s_m = t$ such that

$$\phi(s) = \sum_{i=0}^{m-1} F_i 1(s_i < s \leq s_{i+1}),$$

where $F_i \in \mathbb{D}^{1,2}$. In such a case we obviously have that ϕ is forward integrable and furthermore

$$\int_0^t \phi(s)d^-W(s) = \sum_{i=0}^{m-1} F_i(W(s_{i+1}) - W(s_i)).$$

We then also have that for the sequence of partitions $\pi_n' = \{t_i; i = 0, \ldots, n\} \cup \{s_j; j = 0, \ldots, m\}$ then

$$\sum_{t_i \in \pi_n} \left(\int_{t_i}^{t_{i+1}} \phi(s)d^-W(s) \right)^2 - \sum_{s_j \in \pi_n'} \left(\int_{s_j}^{s_{j+1}} \phi(s)d^-W(s) \right)^2 \to 0,$$

as $n \to \infty$ because the partition $\{s_j; j = 0, \ldots, m\}$ is fixed and the forward integrals are L^2- continuous in the time variable. Therefore without loss of generality we will suppose that $\{s_j; j = 0, \ldots, m\} \subset \pi_n$. Then we have that

$$\sum_{i=0}^{n-1} \left(\int_{t_i}^{t_{i+1}} \phi(s)d^-W(s) \right)^2 = \sum_{j=0}^{m-1} F_j^2 \sum_{s_j < t_i < s_{j+1}} (W(t_{i+1}) - W(t_i))^2.$$

As the partition $\{s_j; j = 0, \ldots, m\}$ is fixed we have that

$$\sum_{s_j < t_i < s_{j+1}} (W(t_{i+1}) - W(t_i))^2 \to s_{j+1} - s_j$$

as $n \to \infty$. Therefore

$$\sum_{i=0}^{n-1} \left(\int_{t_i}^{t_{i+1}} \phi(s)d^-W(s) \right)^2 \to \sum_{j=0}^{m-1} F_j^2 \left(s_{j+1} - s_j \right).$$

Finally the result follows from the following density argument:

$$E\left[\left| \sum_{i=0}^{n-1} \left(\int_{t_i}^{t_{i+1}} \phi(s)d^-W(s) \right)^2 - \sum_{i=0}^{n-1} \left(\int_{t_i}^{t_{i+1}} \psi(s)d^-W(s) \right)^2 \right| \right]$$
$$\leq \left\{ E\left[\sum_{i=0}^{n-1} \left(\int_{t_i}^{t_{i+1}} (\phi - \psi)(s)d^-W(s) \right)^2 \right] \right\}^{1/2} \times \left\{ E\left[\sum_{i=0}^{n-1} \left(\int_{t_i}^{t_{i+1}} (\phi + \psi)(s)d^-W(s) \right)^2 \right] \right\}^{1/2}$$
$$\leq 2 \left\| \phi - \psi \right\|_{1,2,+} \left\| \phi + \psi \right\|_{1,2,+}.$$

□

Now we give the Itô formula that is necessary for our calculations. Before we need a preliminary Lemma.

Lemma 2.1. *Suppose that* $\phi \in \mathbb{L}_+^{1,2} \cap \mathbb{L}^{2,4}$ *with* $D_{s+}\phi \in \mathbb{L}^{1,2}$ *and* b *is a stochastic process with* $b \in \mathbb{L}^{1,2}$*. Define the process*

$$X(t) = x + \int_0^t b(s)ds + \int_0^t \phi(s)d^-W(s).$$

Then $f(\cdot, X)\phi \in \mathbb{L}_+^{1,2}$ *for any* $f \in C_b^{1,2}([0,T] \times \mathbb{R})$.

Proof. First, note that $\int_0^t b(s)ds \in \mathbb{L}_+^{1,2}$. In fact, $D_u \int_0^t b(s)ds = \int_0^t D_u b(s)ds$. Furthermore, one clearly has that

$$E\left[\int_0^T \left|D_u \int_0^{\eta(u)} b(s)ds - \int_0^u D_u b(s)ds\right|^2 du\right] \to 0.$$

The other properties being clear, the assertion $\int_0^t b(s)ds \in \mathbb{L}_+^{1,2}$ follows. Next, consider

$$\begin{aligned} D_u\left(\int_0^t \phi(s)d^-W(s)\right) &= D_u\left(\int_0^t \phi(s)\delta W(s) + \int_0^t D_{s+}\phi ds\right) \\ &= \phi(u)1(u \le \eta(t)) + \int_0^t D_u\phi(s)\delta W(s) + \int_0^t D_u D_{s+}\phi ds. \end{aligned}$$

Therefore we have that

$$D_{u+}\left(\int_0^\cdot \phi(s)d^-W(s)\right) = \int_0^u D_u\phi(s)\delta W(s) + \int_0^u D_u D_{s+}\phi ds.$$

Finally by the chain rule and product rule, we have that

$$D_s\left(f(t, X(t))\phi(t)\right) = \frac{\partial f}{\partial x}(t, X(t))D_s X(t)\phi(t) + f(t, X(t))D_s\phi(t)$$

and

$$D_{s+}\left(f(\cdot, X)\phi\right) = \frac{\partial f}{\partial x}(\cdot, X)D_{s+}X\phi + f(s, X(s))D_{s+}\phi.$$

□

Theorem 2.4. *Suppose that* $\phi \in \mathbb{L}_+^{1,2} \cap \mathbb{L}^{2,4}$ *with* $D_{s+}\phi \in \mathbb{L}^{1,2}$ *and* b *is a stochastic process with* $b \in \mathbb{L}^{1,2}$ *then for any* $f \in C_b^{1,2}([0,T] \times \mathbb{R})$ *we have that*

$$\begin{aligned} f(t, X(t)) = f(0, x) &+ \int_0^t \frac{\partial f}{\partial t}(s, X(s)) + \frac{\partial f}{\partial x}(s, X(s))b(s) + \frac{1}{2}\frac{\partial^2 f}{\partial x^2}(s, X(s))\phi(s)^2 ds \\ &+ \int_0^t \frac{\partial f}{\partial x}(s, X(s))\phi(s)d^-W(s), \end{aligned}$$

for

$$X(t) = x + \int_0^t b(s)ds + \int_0^t \phi(s)d^- W(s).$$

Proof. In order to prove that the integral exists we find first a smooth approximation of the process ϕ of the type

$$\phi^n(s) = \sum_{i=0}^{n-1} F_i^n 1(s_i < s \leq s_{i+1})$$

where $F_i^n = \phi(s_i) \in \mathbb{D}^{1,2}$ and $0 = s_0 < ... < s_n = T$ is a fixed partition and

$$\|\phi^n - \phi\|_{1,2,+} \to 0$$

as $n \to \infty$. Note that in this case one has that

$$\int_0^t \phi^n(s)d^- W(s) = \sum_{i=0}^{n-1} F_i^n (W(s_{i+1}) - W(s_i)).$$

Now define $\eta_1(s) = \inf\{s_i; s_i > s\}$ and $\eta_2(s) = \sup\{s_i; s_i \leq s\}$. We define similarly the approximation process

$$X^n(t) = x + \int_0^t \frac{1}{\eta_1(s) - \eta_2(s)} \int_{\eta_2(s)}^{\eta_1(s)} b(u)duds + \int_0^t \phi^n(s)d^- W(s).$$

Consider any partition $0 = t_0 < ... < t_m = t$ such that it contains all the points s_j, $j = 0, ..., n$. Using the Taylor expansion we have

$$f(t, X^n(t)) = f(0, x) + \sum_{i=0}^{m-1} (\partial_x f(t_i, X^n(t_i))(X^n(t_{i+1}) - X^n(t_i)) + \partial_t f(\bar{t}_i, X^n(t_i))(t_{i+1} - t_i))$$

$$+ \frac{1}{2} \sum_{i=0}^{m-1} \partial_{xx} f(t_i, \overline{X}^n(t_i))(X^n(t_{i+1}) - X^n(t_i))^2.$$

Here $\overline{X}^n(t_i)$ denotes a value between $X^n(t_i)$ and $X^n(t_{i+1})$ and $\bar{t}_i$ a value between t_i and t_{i+1}. Obviously, $f(t, X^n(t))$ converges a.s. to $f(t, X(t))$ as $n \to \infty$. The last term above, as in the previous Theorem 2.3 converges to

$$\frac{1}{2} \int_0^t \partial_{xx} f(s, X(s))\phi(s)^2 ds.$$

In fact, one can easily reduce the problem to the calculation of the limit of

$$\sum_{i=0}^{m-1} \partial_{xx} f(t_i, \overline{X}^n(t_i)) \left(\int_{t_i}^{t_{i+1}} \phi^n(s) d^- W(s) \right)^2 .$$

$$= \sum_{i=0}^{m-1} \left(\partial_{xx} f(t_i, \overline{X}^n(t_i)) - \partial_{xx} f(\eta_2(t_i), X^n(\eta_2(t_i))) \right) \left(\int_{t_i}^{t_{i+1}} \phi^n(s) d^- W(s) \right)^2$$

$$+ \sum_{i=0}^{n-1} \partial_{xx} f(s_j, X^n(s_j)) \phi^n(s_j) \sum_{s_j \le t_i < s_{j+1}} (W(t_{i+1}) - W(t_i))^2 .$$

The first term converges to zero as $n \to \infty$ and the second converges first as $m \to \infty$ to

$$\int_0^t \partial_{xx} f(\eta_2(s), X^n(\eta_2(s))) \phi^n(\eta_2(s)) ds,$$

and to $\int_0^t \partial_{xx} f(s, X(s)) \phi(s)^2 ds$ as $n \to \infty$. The other terms converge clearly to

$$\int_0^t \left(\frac{\partial f}{\partial t}(s, X(s)) + \frac{\partial f}{\partial x}(s, X(s)) b(s) \right) ds.$$

So we only have to consider the last term which is $\sum_{i=0}^{m-1} \partial_x f(t_i, X^n(t_i)) \int_{t_i}^{t_{i-1}} \phi^n(s) d^- W(s)$. First, as $m \to \infty$ this term converges a.s. as all the other terms converge. Therefore this limit is the forward integral $\int_0^t \partial_x f(s, X^n(s)) \phi^n(s) d^- W(s)$. The forward integral $\int_0^t \partial_x f(s, X(s)) \phi(s) d^- W(s)$ exists due to Lemma 2.1. The rest of the argument follows using a subsequence that converges at a fast speed. That is, take a uniform partition of the interval $[0, T]$, say $s_i = Ti/n$, then consider the sequence ϕ^n such that $\sup_{t \le T} |X(t) - X^n(t)| \le n^{-\epsilon}$ for $\epsilon > 1/2$. We will then have that for the same sequence $t_i = Ti/m$

$$\int_0^t \left(\partial_x f(s, X(s)) \phi(s) - \partial_x f(s, X^n(s)) \phi^n(s) \right) d^- W(s)$$

$$= \lim_{m \to \infty} \sum_{i=0}^{m-1} \left(\partial_x f(t_i, X(t_i)) \phi(t_i) - \partial_x f(t_i, X^n(t_i)) \phi^n(t_i) \right) (W(t_{i+1}) - W(t_i)).$$

Now we consider the subsequence for which $n = m$ to obtain that the above limit converges to zero. Then the result follows. □

Remark 2.1. 1.The previous proof also gives a sense to the integral $\int_0^t \partial_x f(s, X(s)) d^- X(s)$.

2. In fact the original definition of the forward integral by Russo-Vallois is somewhat different to the one given here. In general, their definition is more general. Nevertheless, once one wants that this integral becomes the limit of Riemman sums then one is forced to the above framework. Still, we remark that the above conditions can be somewhat relaxed but the general idea remains.
3. For example, the above proof is also satisfied in local form. That is, the result is also satisfied if $\phi \in \mathbb{L}^{1,2}_{+,loc} \cap \mathbb{L}^{2,4}_{loc}$ with $D_{s+}\phi \in \mathbb{L}^{1,2}_{loc}$ and b is a stochastic process with $b \in \mathbb{L}^{1,2}_{loc}$ and $f \in C^{1,2}([0,T] \times \mathbb{R})$. For the definition of these spaces see Nualart [25].
4. The fact that the above Itô formula demands an extra condition ($D_{s+}\phi \in \mathbb{L}^{1,2}$) in comparison with its counterpart in Skorohod integral form is well documented in the literature. In particular, in the case of the Stratonovich-Skorohod integral. Nevertheless as our restriction comes from the financial interpretation of the models to be used we accept them as natural.

3. A First Toy Example

Rather than following the general theory exposed in Kohatsu-Sulem (2006), we will give some examples in order to illustrate the theory. In this section, we consider a first toy model where the dynamics of the prices are given by

$$dS(t) = S(t)(\mu + bW(T))dt + \sigma S(t)d^{-}W(t) \tag{2}$$

where μ and b are real numbers, $\sigma > 0$. We suppose moreover that $\rho(t) = \rho =$ constant. The interpretation of this model when $b \geq 0$ is that the insider introduces a higher appreciation rate in the stock price if $W(T) > 0$. Given the linearity of the equation of S this indicates that the higher the final stock price the bigger the value of the drift in the equation driving S. Some cases of negative values for b can also be studied but the practical interpretation of such a study is dubious.

Furthermore we remark that usually in this model we assume that the trades of the insider are not revealed to the public. This is also an interesting modelling issue which is also assumed by Kyle and Back. They assume that the cumulative trades of the insider plus a Wiener process in the insider's filtration are public information. The Wiener process is interpreted as the effect of the so-called noise traders.

This interpretation can also be applied in any of the cases studied with the enlargement of filtration approach and as we will see it can also be applied here.

The difference here is that we will introduce large trader-insider models with finite utility where there can also be small traders that act rationally.

In order to compare with the theory given in our previous article, we decide to first give an approach which is easier to introduce at this stage but that later will not be possible to apply. This is the set-up of enlargement of filtration. For this, consider the filtration $\mathcal{G}_t = \mathcal{F}_t \vee \sigma(W(T))$. In this filtration it is well known that W is a semimartingale and its semimartingale decomposition is given by

$$W(t) = \hat{W}(t) + \int_0^t \frac{W(T) - W(s)}{T - s} ds,$$

where $\hat{W}$ is a Wiener process in $\mathcal{G}$. Therefore in this case, as the forward integral becomes a semimartingale integral we have that the model for S is

$$dS(t) = S(t)\left(\mu + bW(T) + \sigma \frac{W(T) - W(t)}{T - t}\right) dt + \sigma S(t) d\widehat{W}(t).$$

$$S(t) = S(0) \exp\left(\left(\mu - \frac{\sigma^2}{2}\right) t + bW(T)t + \sigma W(t)\right).$$

Therefore the optimization of the logarithmic utility for this model is done through classical methods. Briefly, one has that the wealth process associated with this price process is given by

$$V(t) = V(0) + \int_0^t \frac{\pi(s)V(s)}{S(s)} dS(s) + \int_0^t (1 - \pi(s))V(s) r e^{rs} ds.$$

Then the discounted wealth, $\hat{V}(t) = e^{-rt} V(t)$ can be written as

$$\hat{V}(t) = V(0) + \int_0^t \left(\mu - r + bW(T) + \sigma \frac{W(T) - W(s)}{T - s}\right)$$
$$\cdot \pi(s) \hat{V}(s) ds + \int_0^t \sigma \pi(s) \hat{V}(s) d\widehat{W}(s).$$

The solution to the above equation is

$$\hat{V}(t) = V(0) \exp\left(\int_0^t \left(\mu - r + bW(T) + \sigma \frac{W(T) - W(s)}{T - s}\right)\right.$$
$$\left.\cdot \pi(s) - \frac{\sigma^2}{2} \pi(s)^2 ds + \int_0^t \sigma \pi(s) d\widehat{W}(s)\right).$$

Therefore if we consider the optimization of the logarithmic utility we have the following problem

$$\max_{\pi \in \mathcal{A}_{\mathcal{H}}} J(\pi)$$

where

$$J(\pi) \equiv J_{\mathcal{H}}(t,\pi) = E\left[\int_0^t \left(\mu - r + bW(T) + \sigma\frac{W(T)-W(s)}{T-s}\right)\cdot\pi(s) - \frac{\sigma^2}{2}\pi(s)^2 ds + \int_0^t \sigma\pi(s)d\widehat{W}(s)\right],$$

and for any filtration $\mathcal{H} \subseteq \mathcal{G}$ satisfying the usual conditions we define

$$\mathcal{A}_{\mathcal{H}}(t) = \left\{\pi \text{ is } \mathcal{H} \text{ adapted}; \int_0^t |\pi(s)|^2\, ds < \infty\right\}.$$

We then have the following theorem

Theorem 3.1. *Assume that $\mathcal{H}$ is any filtration included in $\mathcal{G}$. Then the optimal portfolio for the above problem is given by*

$$\hat{\pi}(s) = \frac{\mu - r}{\sigma^2} + E\left[\frac{b}{\sigma^2}W(T) + \sigma^{-1}\frac{W(T)-W(s)}{T-s}\Big/\mathcal{H}_s\right]$$

and the optimal value is given by

$$\frac{(\mu-r)^2\,t}{2\sigma^2} + \frac{1}{2\sigma^2}\int_0^t E\left[E\left[bW(T) + \sigma\frac{W(T)-W(s)}{T-s}\Big/\mathcal{H}_s\right]^2\right]ds.$$

In particular,

$$\lim_{t\to T} J_{\mathcal{G}}(t,\hat{\pi}) = \infty, \tag{3}$$

while

$$\lim_{t\to T} J_{\mathcal{H}}(t,\hat{\pi}) < \infty$$

for $\mathcal{H}_t = \sigma(S(s); s \le t)$. Furthermore the functions $J_{\mathcal{G}}(t,\hat{\pi})$ and $J_{\mathcal{H}}(t,\hat{\pi})$ are increasing in b.

A far more general theorem was given in Kohatsu-Sulem (2006).

Proof. In order to obtain the result first note that given that $\pi \in \mathcal{A}_{\mathcal{H}}(t)$, we have that

$$E\left[\int_0^t \sigma\pi(s)d\widehat{W}(s)\right] = 0.$$

Next the function

$$f_s(\pi) = \left(\mu - r + E\left[bW(T) + \sigma\frac{W(T)-W(s)}{T-s}\Big/\mathcal{H}_s\right]\right)\pi - \frac{\sigma^2}{2}\pi^2$$

is a strictly convex function adapted to the filtration $\mathcal{H}$. Therefore the maximal value is obtained for the value $\hat{\pi}$ given in the statement of the theorem. The limit wealth for the full insider is infinite because

$$E\left[E\left[bW(T) + \sigma\frac{W(T) - W(s)}{T - s}\Big/\mathcal{G}_s\right]^2\right] = b^2T + +2b\sigma + \frac{\sigma^2}{T-s}.$$

The last result follows by noting that

$$\mathcal{H}_t = \sigma\left(bW(T)s + \sigma W(s); s \leq t\right).$$

By using a formula for conditional expectations of Gaussian random variables (see Kohatsu-Sulem (2006)) one obtains that

$$\begin{aligned} E[W(T) - W(t)|\mathcal{H}_t] &= \frac{b(T-t)}{(b^2T + 2b\sigma)t + \sigma^2}(bW(T)t + \sigma W(t)), \\ E[W(T)|\mathcal{H}_t] &= \frac{(bT + \sigma)}{(b^2T + 2b\sigma)t + \sigma^2}(bW(T)t + \sigma W(t)). \end{aligned}$$

Therefore the result follows because

$$E\left[\left(\frac{E[W(T) - W(t)|\mathcal{H}_t]}{T - t}\right)^2\right] = \frac{b^2t}{(b^2T + 2b\sigma)t + \sigma^2}.$$

To finish one only needs to note that

$$\begin{aligned} J_{\mathcal{H}}(t, \hat{\pi}) &= \frac{(\mu - r)^2 t}{2\sigma^2} + \frac{1}{2\sigma^2}\int_0^t E\left[E\left[bW(T) + \sigma\frac{W(T) - W(s)}{T - s}\Big/\mathcal{H}_s\right]^2\right] ds \\ &= \frac{(\mu - r)^2 t}{2\sigma^2} + \frac{b^2}{2\sigma^2}\int_0^t s\frac{((bT + \sigma) + \sigma(T - s))^2}{(b^2T + 2b\sigma)s + \sigma^2} ds. \end{aligned}$$

Finally differentiating with respect to b it follows that $J_{\mathcal{H}}(t, \hat{\pi})$ is increasing. □

There are various other interesting remarks that are made in Kohatsu-Sulem (2006) with respect to the interpretation of this result. This result says that in various situations the insider which acts as a large trader may have effects in the market and the small trader only uses a projection of this market in order to optimize its utility.

This projection does not transfer the information from the insider to the small investor. This example also reflects the fact that there is not only one insider but various insiders that may act depending on the filtration that one takes between $\mathcal{H}_t = \sigma(S(s); s \le t)$ and $\mathcal{G}_t = \mathcal{F}_t \vee \sigma(W(T))$. Finding examples where the calculations can be done explicitely will be an interesting subject of future research.

This toy example, which can be solved using the simple technique showed here was solved in Kohatsu-Sulem (2006) using a powerful technique consisting on optimization in an anticipating framework. We will show in the next section an example which can be considered as a non-trivial application which cannot be solved using the previous technique.

Before that we will discuss another issue related with (3). In fact with a small modification we can obtain that the optimal logarithmic utility of the insider is finite.

Theorem 3.2. *Consider the filtration* $\mathcal{G}'_t = \mathcal{F}_t \vee \sigma\left(W(T) + W'((T-s)^\theta); s \le t\right)$ *where* W' *is another Wiener process independent of* W *and* $\theta \in (0,1)$. *Then we have that*

$$\lim_{t \to T} J_{\mathcal{G}'}(t, \hat{\pi}) < \infty.$$

Proof. First note that the first part of Theorem 3.1 can be applied to the filtration $\bar{\mathcal{G}}_t = \mathcal{F}_t \vee \sigma(W(T)) \vee \sigma(W'(s); s \le T^\theta)$. Therefore we only need to compute

$$E\left(\frac{W(T) - W(s)}{T-s} \Big/ \mathcal{G}'_s\right) = \frac{W(T) - W(s) + W'((T-s)^\theta)}{T - s + (T-s)^\theta}.$$

From here it follows that the logarithmic utility is finite if $\theta < 1$ □

To finish we prove a theorem that can be interpreted as the non-existence of arbitrage or the issue of non-conspicuous insider trader.

Theorem 3.3. *For any filtration* $\mathcal{H}$ *included in* $\mathcal{G}$ *such that* S *is* $\mathcal{H}$*-adapted, suppose that there exists an* $\mathcal{H}$*-optimal portfolio* $\hat{\pi} \in \mathbb{L}^{1,2}_+$ *which leads to a finite logarithmic utility. Then there exists an* $\mathcal{H}$*-Wiener process* $W_{\mathcal{H}}$ *such that*

$$\log(S(t)/S(0)) = \int_0^t \left(r + \sigma^2 \hat{\pi}(s)\right) ds + \sigma W_{\mathcal{H}}(t).$$

Proof. Just to avoid explicit notation let $\mu_s(\omega) = \mu + bW(T)$. If there exists an optimal portfolio $\hat{\pi}$ then it minimizes the logarithmic utility of this trader which is

$$E\left(\int_0^t (\mu_s - r)\pi(s) - \frac{\sigma^2}{2}\pi(s)^2 ds + \int_0^t \sigma\pi(s) dW(s)\right).$$

Applying variational calculus to the above expression we obtain that

$$E\left(\int_u^t (\mu_s - r) - \sigma^2 \hat{\pi}(s)ds + \sigma(W(t) - W(u)) \Big/ \mathcal{H}_u\right) = 0.$$

Furthermore note that

$$\log(S(t)/S(0)) = \int_0^t \mu_s - \frac{\sigma^2}{2} ds + \sigma W(t).$$

Then

$$E\left(\log(S(t)/S(u))/\mathcal{H}_u\right) = r(t-u) + \int_u^t \sigma^2 \hat{\pi}(s)ds.$$

Therefore by Lévy's characterization of the Wiener process we have the result. □

Note that in the classical Merton model $\hat{\pi}(s) = \frac{\mu - r}{\sigma^2}$. Therefore the previous theorem states that the small trader will not find any anomaly in his trading of the stock even if this is influenced by an insider.

This result also says that if we interpret $W_{\mathcal{H}}$ as the effect of $\mathcal{H}$-noise traders then the market maker will only see the information in the stock price itself.

4. Continuous Stream of Information

In this section, we consider for $\delta > T$ fixed

$$S(t) = S(0) + \int_0^t (\mu + bW(s+\delta))\, S(s)ds + \int_0^t \sigma S(s) d^- W(s). \tag{4}$$

In this model, the insider has an effect on the drift of the diffusion through information that is δ units of time in the future. This continuous deformation of information may be used to model streams of information rather than one single piece of information. In this case, it is difficult to see what is the information held by the insider but his/her effect on the market is known. One first important remark is the following proposition.

Proposition 4.1. *W is not a semimartingale on the filtration* $(\mathcal{F}_{t+\delta})_{t \in [0,T]}$.

Proof. Consider the definition of semimartingale as given in Protter's book page 52. If W is a $(\mathcal{F}_{t+\delta})$-semimartingale, then for any partition whose norm tends to zero and always smaller than δ, consider the process

$$H(t) = \sum_{i=0}^{n-1} (W(t_{i+1}) - W(t_i)) 1_{(t_i, t_{i+1}]}(t).$$

This process is then $(\mathcal{F}_{t+\delta})$-adapted and converges uniformly to zero but its stochastic integral converges to the quadratic variation of W leading to a contradiction. □

This shows that the insider filtration does not even correspond to $(\mathcal{F}_{t+\delta})_{t\in[0,T]}$. The definition for the insider's filtration in the particular case that $\delta \geq T$ is

$$\mathcal{G}_t = \mathcal{F}_t \vee \sigma(W(T)) \vee \sigma(W(s+\delta) - W(T); s \leq t).$$

Then the calculations can be carried out as in the previous section. Nevertheless, we need to be more precise here in the general case. We do this here.

In such a situation, we have to clearly use the anticipative set-up given in the first section. Therefore we have to find the solution for the equation of the prices.

Proposition 4.2.

$$S(t) = S(0)\exp\left(\left(\mu - \frac{1}{2}\sigma^2\right)t + b\int_\delta^{t+\delta} W(s)ds + \sigma W(t)\right)$$

is the unique solution of equation (4) *in the space* $\mathbb{L}^{1,2}_{+,loc}$.

The proof of this result follows directly from the Itô formula given in theorem 2.4. We are interested in computing the optimal policy of the small investor with filtration $\mathcal{H}_t = \sigma(S_s; s \leq t)$. From the previous proposition, we have that

$$\mathcal{H}_t = \sigma\left(Y(s); s \leq t\right),$$

where $Y(s) = b\int_\delta^{s+\delta} W(r)dr + \sigma W(s)$. Now we study the wealth process associated with this price process. The wealth process is defined as the solution of

$$V(t) = V(0) + \int_0^t \frac{\pi(s)V(s)}{S(s)} d^- S(s) + \int_0^t (1 - \pi(s))V(s)re^{rs}ds$$

where the interpretation of $d^-S(t)$ is as in Definition 2.1. Note that in order that this equation among others has a sensible financial interpretation we introduced in Section 2 the forward integral as a limit of Riemmann sums.

Then the discounted wealth, $\hat{V}(t) = e^{-rt}V(t)$ can be written as

$$\hat{V}(t) = V(0) + \int_0^t (\mu - r + bW(t+\delta))\pi(s)\hat{V}(s)ds + \int_0^t \sigma\pi(s)\hat{V}(s)d^- W(s).$$

As before the solution to the above equation is

$$\hat{V}(t) = V(0)\exp\left(\int_0^t (\mu - r + bW(t+\delta))\pi(s) - \frac{\sigma^2}{s}\pi(s)^2 ds + \int_0^t \sigma\pi(s)d^-W(s)\right).$$

We will later show that the optimal portfolios proposed satisfy the conditions stated in Section 3. With these assumptions, we have that the limit of the logarithmic wealth process can be written as

$$\begin{aligned} J(\pi) = J_{\mathcal{H}}(t,\pi) &:= E\log(\hat{V}(t)) - \log(V_0) \\ &= E\left[\int_0^t \left(\pi(s)(\mu - r + bW(s+\delta)) - \frac{1}{2}\sigma^2\pi(s)^2\right)ds + \sigma\int_0^t \pi(s)d^-W(s)\right]. \end{aligned}$$

The class of admissible portfolios is given by

$$\mathcal{A} = \{\pi \text{ is } \mathcal{H} \text{ adapted}; \pi \in \mathbb{L}_+^{1,2}\}.$$

Theorem 4.1. *Define the following portfolio*

$$\hat{\pi}(s) = \frac{\mu - r}{\sigma^2} + \sigma^{-2}E\left(bW(s+\delta)/\mathcal{H}_s\right) + \sigma^{-1}a(s)$$

where

$$a(s) = L^1(\Omega) - \lim_{h\to 0} E\left[\frac{W(s+h) - W(s)}{h}\Big/\mathcal{H}_s\right].$$

If $\hat{\pi} \in \mathbb{L}_+^{1,2}$ then $\hat{\pi}$ is the optimal portfolio for the above problem for any filtration $\mathcal{H}$ and the optimal value is given by

$$J(t,\pi^*) = \frac{\sigma^2}{2}E\left[\int_0^t \hat{\pi}(s)^2 ds\right]$$

A more general theorem was proved in Kohatsu-Sulem (2006).

Proof. In order to obtain the result we have to prove first that the functional J is strictly convex. For this, let π_0 and $\pi_1 \in \mathcal{A}$. Then we have that for any $\alpha \in (0,1)$

$$J(\alpha\pi_0 + (1-\alpha)\pi_1) < \alpha J(\pi_0) + (1-\alpha)J(\pi_1).$$

This property clearly comes from the factor $-\frac{\sigma^2}{2}\pi(s)^2$ in the expression for J. Next, we find the first directional derivative of J.

Consider for $\pi, v \in \mathcal{A}$, then

$$\begin{aligned} D_\pi J(\pi) &:= \lim_{\epsilon\to 0}\frac{J(\pi + \epsilon v) - J(\pi)}{\epsilon} \\ &= E\left[\int_0^t (\mu - r + bW(s+\delta))v(s) - \sigma^2\pi(s)v(s)ds + \int_0^t \sigma v(s)d^-W(s)\right]. \end{aligned}$$

If we set the above equation equal to zero for all $v \in \mathcal{A}$ and in particular for $v = X1_{[s_0,t_0]}$ for $X \in \mathbb{D}^{1,2}$ we have by a density argument that

$$E\left[\int_{s_0}^{t_0} (\mu - r + bW(s+\delta))ds - \sigma^2 \pi(s) ds + \sigma (W(t_0) - W(s_0)) \Big/ \mathcal{H}_{s_0}\right] = 0.$$

Now note that $\hat{\pi}$ satisfies the above equation. In fact, replacing $\hat{\pi}$ in the above equation, we have

$$\begin{aligned} &E\left[\int_{s_0}^{t_0} -\sigma \lim_{h\to 0} E\left[\frac{W(s+h)-W(s)}{h} \Big/ \mathcal{H}_s\right] ds + \sigma (W(t_0) - W(s_0)) \Big/ \mathcal{H}_{s_0}\right] \\ &= -\sigma \lim_{h\to 0} E\left[\int_{s_0}^{t_0} \frac{W(s+h)-W(s)}{h} ds + (W(t_0) - W(s_0)) \Big/ \mathcal{H}_{s_0}\right] = 0, \end{aligned}$$

by continuity of the paths of the Wiener process. Therefore $\hat{\pi}$ has to be optimal. In fact, for all $\beta \in \mathcal{A}$ and $\varepsilon \in (0,1)$, we have

$$\begin{aligned} J(\hat{\pi} + \varepsilon\beta) - J(\hat{\pi}) &= J((1-\varepsilon)\frac{\hat{\pi}}{1-\varepsilon} + \varepsilon\beta) - J(\hat{\pi}) \\ &\geq (1-\varepsilon)J(\frac{\hat{\pi}}{1-\varepsilon}) + \varepsilon J(\beta) - J(\hat{\pi}) \\ &= J(\frac{\hat{\pi}}{1-\varepsilon}) - J(\hat{\pi}) + \varepsilon(J(\beta) - J(\frac{\hat{\pi}}{1-\varepsilon})). \end{aligned}$$

Now, with $\frac{1}{1-\varepsilon} = 1 + \eta$ we have

$$\lim_{\varepsilon\to 0} \frac{1}{\varepsilon}(J(\frac{\hat{\pi}}{1-\varepsilon}) - J(\hat{\pi})) = \lim_{\eta\to 0} \frac{1+\eta}{\eta}(J(\hat{\pi} + \eta\hat{\pi}) - J(\hat{\pi})) = D_{\hat{\pi}} J(\hat{\pi}).$$

Then we get

$$D_\beta J(\hat{\pi}) = \lim_{\varepsilon\to 0} \frac{1}{\varepsilon}(J(\hat{\pi} + \varepsilon\beta) - J(\hat{\pi})) \geq D_{\hat{\pi}} J(\hat{\pi}) + J(\beta) - J(\hat{\pi}).$$

We conclude that

$$J(\beta) - J(\hat{\pi}) \leq D_\beta J(\hat{\pi}) - D_{\hat{\pi}} J(\hat{\pi}) \;\; ; \hat{\pi}, \beta \in \mathcal{A}.$$

In particular, using that $D_\beta J(\pi^*) = 0$, we get

$$J(\beta) - J(\pi^*) \leq 0,$$

which proves that π^* is optimal.

To find the optimal expression for the utility it is enough to note that

$$E\left[\int_0^t (\mu - r + bW(s+\delta))\hat{\pi}(s) - \sigma^2 \pi(s)\hat{\pi}(s)ds + \int_0^t \sigma\hat{\pi}(s)d^- W(s)\right] = 0,$$

therefore the optimal utility is

$$E\left[\int_0^t \left(\hat{\pi}(s)(\mu - r + bW(s+\delta)) - \frac{1}{2}\sigma^2\hat{\pi}(s)^2\right)ds - \int_0^t (\mu - r + bW(s+\delta))\hat{\pi}(s) - \sigma^2\pi(s)\hat{\pi}(s)ds)\right].$$

From here the result follows. □

A very useful property is that the optimal portfolios in a smaller filtration is just a projection.

Proposition 4.3. *Let $\mathcal{H}^1 \subset \mathcal{H}^2 \subset \mathcal{G}$ be two filtrations satifying the usual conditions such that there is an optimal portfolio $\hat{\pi}_2$ in $\mathcal{H}^2$ within a class of protfolios $\mathcal{A}_{\mathcal{H}^2}$. If $\mathcal{A}_{\mathcal{H}^1} \subset \mathcal{A}_{\mathcal{H}^2}$, then there is an optimal portfolio $\hat{\pi}_1$ in $\mathcal{H}^1$ which satisfies*

$$\hat{\pi}_1(s) = E\left[\hat{\pi}_2(s)/\mathcal{H}^1_s\right],$$
$$J_{\mathcal{H}^1}(t, \widehat{\pi}_1) \leq J_{\mathcal{H}^2}(t, \widehat{\pi}_2).$$

Therefore in order to prove the existence of the optimal portfolio it is essential to compute a or at least obtain its existence and some regularity properties. We do this, first in the case that $\delta \geq T$. This is done in the next proposition.

Proposition 4.4. *Suppose that $\delta \geq T$. The optimal logarithmic utility portfolio in the filtration $\mathcal{H} \subset \mathcal{G}$ is given by*

$$\hat{\pi}(s) = \frac{\mu - r}{\sigma^2} + \sigma^{-2}E\left[bW(s+\delta) + \sigma\frac{W(T) - W(s)}{T - s}\Big/\mathcal{H}_s\right].$$

The optimal value is given by

$$\frac{(\mu - r)^2 t}{2\sigma^2} + \frac{1}{2\sigma^2}\int_0^t E\left[E\left[bW(s+\delta) + \sigma\frac{W(T) - W(s)}{T - s}\Big/\mathcal{H}_s\right]^2\right]ds.$$

In particular,

$$\lim_{t \to T} J_{\mathcal{G}}(t, \hat{\pi}) = \infty,$$

while

$$\lim_{t \to T} J_{\mathcal{H}}(t, \hat{\pi}) < \infty$$

for $\mathcal{H}_t = \sigma(S(s); s \leq t)$.*Furthermore the functions* $J_{\mathcal{G}}(t, \hat{\pi})$ *and* $J_{\mathcal{H}}(t, \hat{\pi})$ *are increasing in b.*

Proof. Define $Y(t) = b \int_\delta^{t+\delta} W(r)dr + \sigma W(t)$. Then for $\delta \geq T$

$$\lim_{s \downarrow t} E\left[\frac{W(s) - W(t)}{s - t} \Big/ \mathcal{H}_t \right] = bM \int_0^t g(t, u) dY(u).$$

$$E[W(t + \delta) / \mathcal{H}_t] = (b(t + \delta) + \sigma) M \int_0^t g(t, u) dY(u)$$

where $M \equiv M_t = \sigma^{-1} \left((b\delta + 2\sigma) \left(e^{\frac{2bt}{\sigma}} - 1 \right) + \sigma \left(e^{\frac{2bt}{\sigma}} + 1 \right) \right)^{-1}$ and $g(t, u) = e^{\frac{b}{\sigma}(2t-u)} + e^{\frac{b}{\sigma}u}$.

In fact, note that Y is a Gaussian process. Therefore $E[W(s)/\mathcal{H}_t] = \int_0^t h(s, t, u) dY(u)$ for a deterministic function h. To compute h we compute the covariances between $W(s)$ and the stochastic integral and $Y(v)$ for some $v \leq t \leq s \leq T$. First

$$E[W(s)Y(v)] = bsv + \sigma(s \wedge v).$$

Also

$$\begin{aligned} E\left[\int_0^t h(s, t, u) dY(u) Y(v) \right] = \quad & b^2 \int_0^t \int_0^v h(s, t, \theta_1)(\theta_1 \wedge \theta_2 + \delta) d\theta_2 d\theta_1 \\ & + 2b\sigma v \int_0^t h(s, t, \theta) d\theta + \sigma^2 \int_0^v h(s, t, \theta) d\theta. \end{aligned} \tag{5}$$

Therefore the above two expressions have to be equal. After differentiation of the equality with respect to $v \leq t$ three times, we obtain

$$-b^2 h(s, t, u) + \sigma^2 \frac{\partial^2 h}{\partial u^2}(s, t, u) = 0.$$

Solving this differential equation gives

$$h(s, t, u) = C_1(s, t) e^{-\frac{b}{\sigma}u} + C_2(s, t) e^{\frac{b}{\sigma}u}.$$

Next one verifies that for the following constants, the covariances coincide.

$$C_2(s, t) = \sigma^{-1}(bs + \sigma) \left((b\delta + 2\sigma) \left(e^{\frac{2bt}{\sigma}} - 1 \right) + \sigma \left(e^{\frac{2bt}{\sigma}} + 1 \right) \right)^{-1}$$

$$C_1(s, t) = e^{\frac{2bt}{\sigma}} C_2(s, t).$$

Therefore, we have that

$$E\left(\frac{W(s)-W(t)}{s-t}\Big/\mathcal{H}_t\right)=\int_0^t \frac{h(s,t,u)-h(t,t,u)}{s-t}dY(u).$$

Then the result follows.

Next, using Theorem 4.1, we have that the possible optimal portfolio π^* defined by

$$\pi^*(t)=\frac{\mu-r}{\sigma^2}+bM_t\left(\frac{b(t+\delta)+2\sigma}{\sigma^2}\right)\int_0^t g(t,u)dY(u)$$

is in $\mathbb{L}_+^{1,2}$. In fact, all the properties are obtained through the process Y. We do not give the details of this verification.

Then the optimal utility is finite as it is given by

$$J(t,\pi^*)=\log(V_0)+\frac{\sigma^2}{2}E\left[\int_0^t \pi^*(s)^2 ds\right].$$

□

Remark 4.1. When $s \leq T$, we have that

$$E\left[W(s)/\mathcal{H}_T\right]=\int_0^s h(s,T,u)dY(u)+\int_s^t \bar{h}(s,T,u)dY(u),$$

where

$$\bar{h}(s,t,u)=\bar{C}_1(s,t)e^{-\frac{b}{\sigma}u}+\bar{C}_2(s,t)e^{\frac{b}{\sigma}u}$$

$$\bar{C}_2(s,t)=\sigma^{-1}\left(1+\sigma h(s,T,s)\right)\left(e^{\frac{b(2T-s)}{\sigma}}+e^{\frac{2bs}{\sigma}}\right)^{-1}$$

$$\bar{C}_1(s,t)=e^{\frac{2bt}{\sigma}}\bar{C}_2(s,t).$$

This shows that even the information on all the prices of the interval $[0,T]$ does not reveal the information held by the insider to the small trader.

As before we can also show that the insider's utility is finite if we use the filtration $\mathcal{G}'_t=\mathcal{F}_t\vee\sigma\left(W(s+\delta)+W'((T-t)^\theta);s\leq t\right)$ for $\theta<1$. Similarly we can also obtain a representation theorem such as Theorem 3.3. Instead we will take a look at the case $\delta<T$. We use a different shortcut through the anticipating Girsanov's theorem. For details and notation we refer to Chapter 4 in [25].

Theorem 4.2. *Consider the case $\delta<T$. Then there is no arbitrage for the filtration $\mathcal{H}_t=\sigma(S(s);s\leq t)$ and the logarithmic utility for the optimal portfolio value for this investor is finite.*

Proof. We apply Theorem 4.1.2 in [25] in the interval $[0, T + \delta]$ with the transformation

$$T(\omega) = \omega + b1(\cdot \leq T) \int_0^{\cdot} \omega(s + \delta) ds,$$

defined in $C[0, T + \delta]$. Then we have that if $T(\omega) = 0$ then $\omega(t) = 0$ for all $t \in [T, T + \delta]$. Therefore

$$T(\omega) = \omega + b1(\cdot \leq T - \delta) \int_0^{\cdot} \omega(s + \delta) ds.$$

That is, by finite induction we have that T is an injection. To prove that it is surjective one follows a similar pattern.

Next we have that

$$\det_2 (I + Du) > 0$$

for $u_s(\omega) = b1(s \leq T)\omega(s + \delta)$ and that under the change of measure

$$\frac{dQ}{dP} = \det_2 (I + Du) \exp\left(- \int_0^T bW(s + \delta) dW(s) - \frac{b^2}{2} \int_0^T W(s + \delta)^2 ds \right)$$

then $\hat{W} = T(W)$ has the law of a Wiener process under Q. Therefore there exists an equivalent martingale measure for this problem.

In order to compute the optimal portfolio one uses the dual method. That is, denote $m = \sigma^{-2}(\mu - r)$ and define

$$\frac{dQ'}{dP} = \det_2 (I + Du) \exp$$
$$\cdot \left(- \int_0^T (bW(s + \delta) - m)\, dW(s) - \frac{1}{2} \int_0^T (bW(s + \delta) + m)^2\, ds \right).$$

Then the optimal portfolio value is

$$\hat{V}_T = V_0 \frac{dQ'}{dP}.$$

The optimal portfolio value is finite because $E\left[\log\left(\frac{dQ'}{dP}\right)\right] < \infty$. □

Off course an interesting problem is to compute explicitely the optimal portfolio for the case $\delta < T$.

Although one may consider that the large trader effect is somewhat hidden in this paper through the process appearing in the drift. We remark that this may be considered as a first learning step towards more complex models. Some of these models were presented in Kohatsu-Sulem (2006) or Kohatsu (2005).

References

1. Amendinger, J., Imkeller, P., and Schweizer, M., 1998. Additional logaritmic utility of an insider. *Stochastic Proc. Appl.* **75**, 263–286.
2. Amendinger, J., 2000. Martingale representations theorems for initially enlarged filtrations. *Stochastic Proc. Appl.* **89**, 101–116.
3. Amendinger, J., Becherer, D., and Schweizer, M., 2003. A monetary value for initial information in portfolio optimization. *Finance and Stochastics* **7**, 29–46.
4. Baudoin, F., 2003. Modelling anticipations in financial markets. In *Paris-Princeton Lectures on Mathematical Finance 2002*. Lect. Notes in Maths. 1814, Springer-Verlag. Berlin.
5. Baudoin, F., 2002. *Conditioned Stochastic Differential Equations and Application to Finance, Stochastic Processes and their Applications*, Vol. 100, 109–145.
6. Back, K., 1992. Insider Trading in Continuous Time. *Review of Financial Studies* **5**, 387–409.
7. Biagini, F. and Øksendal, B.: A general stochastic calculus approach to insider trading. Preprint Series, Dept. of Mathematics, Univ. of Oslo, 17/2002.
8. Corcuera, J. M., Imkeller, P., Kohatsu-Higa, A., and Nualart, D., 2004. Additional utility of insiders with imperfect dynamical information. *Finance and Stochastics* **8**, 437–450.
9. Chaumont, L. and Yor, M., 2004. *Exercises in Probability*, Cambridge University Press, 2004.
10. Elliot, R. J., Geman, H., and Korkie, B. M., 1997. Portfolio optimization and contingent claim pricing with differential information. Stochastics and Stochastics Reports 60, 185–203.
11. Grorud, A., 2000. Asymmetric information in a financial market with jumps. *International Journal of Theoretical and Applied Finance* **3**, 641–659.
12. Grorud, A. and Pointier, M., 1998. Insider Trading in a continuous Time Market Model. *International Journal of Theoretical and Applied Finance* **1**, 331–347.
13. Imkeller, P., 1996. Enlargement of the Wiener filtration by an absolutely continuous random variable via Malliavin's calculus. *Probab. Th. Rel. Fields* **106**, 105–135.
14. Imkeller, P., 1997. Enlargement of the Wiener filtration by a manifold valued random element via Malliavin's calculus. In *Statistics and Control of Stochastic Processes. The Lipster Festschrift*, Y. M. Kabanov, B. L. Rosovskii, and A. N. Shiryaev (eds.) World Scientific, Singapore.
15. Imkeller, P., 2002. Random times at which insiders can have free lunches. *Stochastics and Stochastics Reports* **74**, 465–487.
16. Imkeller, P., Pontier, M., Weisz, F., 2001. Free lunch and arbitrage possibilities in a financial market with an insider. *Stochastic Proc. Appl.* **92**, 103–130.
17. Jacod, J., 1985. Grossissement initial, hypothèse (H'), et théorème de Girsanov. In *Grossissements de Filtrations: Exemples et Applications*, T. Jeulin, and M. Yor (eds.) Lect. Notes in Maths. 1118. Springer-Verlag. Berlin.
18. Jeulin, T., 1980. *Semi-Martingales et Groissessement de Filtration*. Lect. Notes in Maths. 833. Springer-Verlag, Berlin.
19. Karatzas, I. and Pikovsky, I., 1996. Anticipative portfolio optimization. *Adv. Appl. Prob.* **28**, 1095–1122.

20. Kohatsu-Higa, A., 2005. Insider models with finite utility. Lecture Notes.
21. Kohatsu-Higa, A. and Sulem, A., 2006. Utility maximization in an insider influenced market, *Mathematical Finance* **16**, 153-179.
22. Kyle, A., 1985. Continuous Auctions and Insider Trading. *Econometrica* **53**, 1315–1335.
23. Liptser, R. S. and Shiryaev, A. N., 1997. *Statistics of Random Processes I. General Theory*. Springer-Verlag. New York.
24. Mansuy, R. and Yor, M., 2004. Harnesses, Lévy processes and Monsieur Jourdain. to appear in *Stochastic Process. Appl.*
25. Nualart, D., 1995. *The Malliavin Calculus and Related Topics*. Springer-Verlag. Berlin.
26. Nualart, D., 1995a. Analysis on Wiener space and anticipating calculus. In *Lectures on Probability Theory and Statistics. Ecole d'eté de Probabilités de Saint-Flour XXV*. Lect. Notes in Maths. 1690. Springer-Verlag.
27. Øksendal, B. and Sulem, A.: Partial observation in an anticipative environment. Preprint University Oslo 31/2003.
28. Peccati, G., Thieullen, M., and Tudor, C., 2005. Martingale structure for Skorohod integral processes. to appear in *The Annals of Probability*.
29. Protter, P., 2004. *Stochastic Integration and Differential Equations. A New Approach*. Springer-Verlag. New York.
30. Russo, F. and Vallois, P., 1993. Forward, backward and symmetric stochastic integration. *Probab. Th. Rel. Fields* **97**, 403–421.
31. Russo, F. and Vallois, P., 2000. Stochastic calculus with respect to continuous finite quadratic variation processes. *Stochastics and Stochastics Reports* **70**, 1–40.
32. Russo, F. and Vallois, P., 1995. The generalized covariation process and Itô formula. *Stochastic Process. Appl.* **59**, 81–104.
33. Séminaire de Calcul Stochastique 1982/83, Université Paris VI, 1985. *Grossisements de Fitrations: Exemples et Applications*. T. Jeulin and M. Yor (eds.) Lect. Notes in Maths. 1118. Springer-Velag. Berlin.
34. Tudor, C., 2004. Martingale type stochastic calculus for anticipating integrals, *Bernoulli* **10**(2), 313–325.
35. Yor, M., 1985. Grossissement de filtrations et absolue continuite de noyaux. In *Grossissements de Filtrations: Exemples et Applications*. T. Jeulin and M. Yor (eds.) Lect. Notes in Maths. 1118. Springer-Verlag. Berlin.

[GLP & MEMM] Pricing Models and Related Problems

Yoshio Miyahara

Graduate School of Economics, Nagoya City University
Mizuhochou Mizuhoku, Nagoya, 467-8501, Japan

The [GLP & MEMM] pricing model (= [Geometric Lévy Process & Minimal Entropy Martingale Measure] pricing model) has been introduced as a pricing model for the incomplete financial market. This model has many good properties and is applicable to very wide classes of underlying asset price processes including the geometric stable processes. We explain those good properties and see several examples of this model. After that we investigate the calibration problems of [GLP & MEMM] model.

Key words: Geometric Lévy Process, Relative entropy, Minimal entropy martingale measure, Stable process, Calibration

1. Introduction

The [Geometric Lévy Process & MEMM] pricing model was first introduced in [36]. This model is one of the incomplete markets, and is based on the geometric Lévy process and the minimal entropy martingale measure (= MEMM).

We assume that the value process of bond is given by $B_t = \exp\{rt\}$, where r is a positive constant. The price process of the underlying asset is denoted by S_t.

1.1 Black-Scholes model

The explicit form of Black-Scholes model (Geometric Brownian motion model) is given by

$$S_t = S_0 e^{(\mu - \frac{1}{2}\sigma^2)t + \sigma W_t}, \tag{1}$$

and the stochastic differential equation (SDE) form is given by

$$dS_t = S_t\left(\mu dt + \sigma dW_t\right), \tag{2}$$

where W_t is a standard Wiener process.

The risk neutral measure Q is uniquely determined by the Girsanov's lemma. Under the Q the process $\tilde{W}_t = W_t + (\mu - r)\sigma^{-1}t$ is a Wiener process and the price process S_t is expressed in the form of

$$(3) \qquad S_t = S_0 e^{(r-\frac{1}{2}\sigma^2)t+\sigma\tilde{W}_t} \quad \text{or} \quad dS_t = S_t\left(rdt + \sigma d\tilde{W}_t\right).$$

The price of an option X is given by $e^{-rT}E_Q[X]$. The theoretical B-S price of the European call option, $C(S_0, K, T)$, with the strike price K and the fixed maturity T is given by the following formula

$$(4) \qquad C_K = C(S_0, K, T) = e^{-rT}E_Q[(S_T - K)^+] = S_0\Phi(d_1) - e^{-rT}K\Phi(d_2),$$

where $\Phi(d)$ is the normal distribution function and

$$(5) \qquad d_1 = \frac{\log\frac{S_0}{K} + (r + \frac{\sigma^2}{2})T}{\sigma\sqrt{T}}, \quad d_2 = \frac{\log\frac{S_0}{K} + (r - \frac{\sigma^2}{2})T}{\sigma\sqrt{T}} = d_1 - \sigma\sqrt{T}.$$

1.2 Properties of B-S models

1.2.1 Distribution of log returns

The log return is the increment of the logarithm of S_t,

$$(6) \qquad \triangle\log S_t = \log S_{t+\triangle t} - \log S_t = (\mu - \frac{1}{2}\sigma^2)\triangle t + \sigma\triangle W_t,$$

and the log return process is $(\mu - \frac{1}{2}\sigma^2)t + \sigma W_t$.

The distribution of the log return (or the log return process) of the B-S model is normal. This is convenient for the calculation of the option prices. For example we have obtained the explicit formula of the price of European call options. But it is said that the distributions of the log returns in the real market usually have the fat tail and the asymmetry. These facts suggest us the necessity to consider another models.

1.2.2 Historical volatility and implied volatility

Under the setting of the B-S model, the historical volatility of the process is defined as the estimated value of σ based on the sequential data of the price process S_t. We denote it by $\widehat{\sigma}$. On the other hand the implied volatility is defined as what follows. Suppose that the market price of the European call option with the strike K, say $C_K^{(m)}$, were given. Then the value of σ which satisfies the following equation

$$(7) \qquad S_0\Phi(d_1) - e^{-rT}K\Phi(d_2) = C_K^{(m)},$$

is the implied volatility, and this value is denoted by $\sigma_K^{(im)}$. We remark here that the implied volatility $\sigma_K^{(im)}$ depends on the strike value K, and that on the contrary the historical volatility $\widehat{\sigma}$ does not depend on K.

We first consider the case where the market value of options obey to the Black-Scholes model, and so the market price $C_K^{(m)}$ is equal to the theoretical B-S price C_K . In this case the solution of the equation for the implied volatility is equal to the original σ and it holds true that $\sigma_K^{(im)} = \sigma = constant$. This means that if the market obeys exactly to the Black-Scholes model, then the implied volatility $\sigma_K^{(im)}$ should be equal to the historical volatility $\widehat{\sigma}$.

But in the real world this is not true. It is well-known that the implied volatility is not equal to the historical volatility, and the implied volatility $\sigma_K^{(im)}$ is sometimes a convex function of K, and sometimes the combination of convex part and concave part. These properties are so-called volatility smile or smirk properties.

1.3 Generalization of B-S model

1.3.1 Geometric Lévy Process models

We start from the explicit form of Geometric Brownian motion: $S_t = S_0 e^{(\mu - \frac{1}{2}\sigma^2)t + \sigma W_t}$. It may be a natural idea to replace the Wiener process with the more general Lévy processes Z_t and set

$$S_t = S_0 e^{Z_t}. \tag{8}$$

This type processes called the Geometric Lévy Processes (GLP). The [GLP & MEMM] pricing model is one of this type of generalisation of B-S model.

The class of Lévy processes are very wide and the distributions of S_t may have the fat tail property and may be asymmetric.

1.3.2 Stochastic volatility models

We start from the SDE form $dS_t = S_t (\mu dt + \sigma dW_t)$. When we replace the Brownian motion with a Lévy process, we obtain the equation described in the previous subsection (see §2).

When we randomize the volatility σ as follows

$$dS_t = S_t (\mu dt + \tilde{\sigma}_t dW_t), \tag{9}$$

where $\tilde{\sigma}_t$ is a stochastic process, then we obtain the so-called stochastic volatility models.

1.4 Our Goal

The purposes of this lecture are, 1) we introduce the [GLP & MEMM] pricing model and see that this model has many good properties, and next 2) we review some relating problems of this model, in theoretical sense and (or) in practical fence (for example, the fitness analysis and calibration analysis).

2. Geometric Lévy Process Pricing Models

We assume that the value process of bond is given by

$$B_t = \exp\{rt\}, \tag{10}$$

where r is a positive constant.

A pricing model consists of the following two parts:
(A) The price process S_t of the underlying asset.
(B) The rule to compute the prices of options.

For the part (A) we adopt the geometric Lévy processes, so the part (A) is reduced to the selecting problem of a suitable class of the geometric Lévy processes.

For the part (B) we adopt the martingale measure method, so the part (B) is reduced to the selecting problem of a suitable martingale measure Q, and then the price of an option X is given by $e^{-rT}E_Q[X]$.

2.1 Geometric Lévy processes

The price process S_t of a stock is assumed to be defined as what follows. We suppose that a probability space $(\Omega, \mathcal{F}, P)$ and a filtration $\{\mathcal{F}_t, 0 \leq t \leq T\}$ are given. We also suppose that a Lévy process Z_t is defined on this probability space and that the price process S_t of a stock is given in the form

$$S_t = S_0 e^{Z_t}, \quad 0 \leq t \leq T. \tag{11}$$

Throughout this paper we assume that $\mathcal{F}_t = \sigma(S_s, 0 \leq s \leq t) = \sigma(Z_s, 0 \leq s \leq t)$ and $\mathcal{F} = \mathcal{F}_T$.

We give here the definition of Lévy process and the characterization of it (see [45]).

Definition 2.1. A stochastic process $\{Z_t\}$ on R^d is a Lévy process if the following conditions are satisfied.
1) For any choice of $n \geq 1$ and $0 \leq t_0 < t_1 \cdots < t_n$, random variables $Z_{t_0}, Z_{t_1} - Z_{t_0}, Z_{t_2} - Z_{t_1}, \ldots, Z_{t_n} - Z_{t_0 n-1}$, are independent (independent increments property).
2) $Z_0 = 0$ *a.s.*
3) The distribution of $Z_{s+t} - Z_s$ does not depend on s (temporal homogeneity or stationary increments property).
4) It is stochastically continuous.
5) There is $\Omega_0 \in \mathcal{F}$ with $P(\Omega_0) = 1$ such that for every $\omega \in \Omega_0$, $Z_t(\omega)$ is right-continuous in $t \geq 0$ and has left limits in $t > 0$.

In this lecture we discuss the case of $d = 1$. The Lévy process Z_t is characterized by the generating triplet $(\sigma^2, \nu(dx), b)$, where σ^2 is a nonnegative constant, $\nu(dx)$ is a measure such that

$$\nu(\{0\}) = 0, \quad \int_{-\infty}^{\infty} (|x|^2 \wedge 1)\nu(dx) < \infty \tag{12}$$

and b is a constant. By the use of this generating triplet, the characteristic function of Z_t is

$$\phi_t(u) = E[e^{iuZ_t}] = \exp\{t\psi(u)\} \tag{13}$$

$$\psi(u) = -\frac{1}{2}\sigma^2 u^2 + ibu + \int_{-\infty}^{\infty} (e^{iux} - 1 - iux1_{\{|x|\leq 1\}})\nu(dx) \tag{14}$$

Using Ito formula, we know that S_t satisfies the following stochastic differential equation

$$dS_t = S_{t-}d\tilde{Z}_t, \tag{15}$$

where $\tilde{Z}_t$ is another Lévy process given by

$$\tilde{Z}_t = Z_t + \frac{1}{2}\langle Z^c \rangle_t + \sum_{0<s\leq t} (e^{\triangle Z_s} - 1 - \triangle Z_s). \tag{16}$$

And the price process S_t has the following expression

$$S_t = S_0\mathcal{E}(\tilde{Z})_t \tag{17}$$

where $\mathcal{E}(\tilde{Z})_t$ is the Doléans-Dade exponential (or stochastic exponential) of $\tilde{Z}_t$.

$$\mathcal{E}(\tilde{Z})_t = e^{\tilde{Z}_t - \tilde{Z}_0 - \frac{1}{2}<\tilde{Z}^c, \tilde{Z}^c>_t} \prod_{s\leq t} (1 + \triangle\tilde{Z}_s)e^{-\triangle\tilde{Z}_s} \tag{18}$$

The generating triplet of $\tilde{Z}_t$, say $(\tilde{\sigma}^2, \tilde{\nu}(dx), \tilde{b})$, is

$$\tilde{\sigma}^2 = \sigma^2 \tag{19}$$

$$\tilde{\nu}(A) = \int 1_A(e^x - 1)\nu(dx), \tag{20}$$

$$\tilde{b} = b + \frac{1}{2}\sigma^2 + \int ((e^x - 1)1_{\{|e^x-1|\leq 1\}}(x) - x1_{\{|x|\leq 1\}}(x))\, \nu(dx). \tag{21}$$

Remark 2.1. (i) It holds that *supp* $\tilde{\nu} \subset (-1, \infty)$.
(ii) If $\nu(dx)$ has the density $n(x)$, then $\tilde{\nu}(dx)$ has the density $\tilde{n}(x)$ and $\tilde{n}(x)$ is given by

$$\tilde{n}(x) = \frac{1}{1+x} n(\log(1+x)). \tag{22}$$

(iii) The relations between Z_t and $\tilde{Z}_t$ are more precisely discussed in [30] , where the stochastic logarithm of X_t, $\mathcal{L}(X)_t$, is defined and the following relations are obtained.

$$Z_t = \log \mathcal{E}(\tilde{Z})_t, \quad \tilde{Z}_t = \mathcal{L}(e^{Z.})_t \tag{23}$$

Many candidates for the suitable Lévy process have been proposed. We give some examples below.
1) Stable process (Mandelbrot and Fama (1963))
2) Jump diffusion process (Merton (1973))
3) Variance Gamma process (Madan (1990))
4) Generalized Hyperbolic process (Eberlein (1995))
5) CGMY process (Carr-Geman-Madam-Yor (2002))
6) Normal inverse Gaussian process (Barndorff-Nielsen (1995, 1977))

2.2 Equivalent martingale measures

A probability measure Q on $(\Omega, \mathcal{F})$ is called an equivalent martingale measure of S_t if $Q \sim P$ and $e^{-rt}S_t$ is $(\mathcal{F}_t, Q)$-martingale. Since the geometric Lévy process model is incomplete in general, there are many equivalent martingale measures. For the part of (B) of the pricing model we have to select a special martingale measure.

Many candidates for the equivalent martingale measure have been proposed as follows.
1) Minimal Martingale Measure (MMM) (Föllmer-Schweizer (1991))
2) Variance Optimal Martingale Measure (VOMM) (Schweizer (1995))
3) Esscher Martingale Measure (ESMM) (Gerber-Shiu (1994), B-D-E-S (1996))
4) Minimal Entropy Martingale Measure (MEMM) (Miyahara (1996), Frittelli (2000))
5) Utility Martingale Measure (Utility-MM)
6) Mean Correcting Martingale Measure (MCMM)

3. Esscher Transformed Martingale Measures

3.1 Esscher transforms

The Esscher transform is very popular and thought to be very important method in the actuary theory (see [24]). Esscher has introduced the risk function and the transformed risk function for the calculation of collective risk. His idea has been developed in his work [18] and by many authors, and played very important roles in the option pricing theory.

3.1.1 Esscher transforms and risk processes

We give several definitions.

Definition 3.1. Let R be a random variable and h be a constant. Then the probability measure $P^{(ESS)}_{R,h}$ defined by

$$\frac{dP^{(ESS)}_{R,h}}{dP}|_{\mathcal{F}} = \frac{e^{hR}}{E[e^{hR}]} \tag{24}$$

is called the Esscher transformed measure of P by the risk variable R and the index h, and this measure transformation is called the Esscher transform by the risk variable R and the index h.

Definition 3.2. Let $R_t, 0 \leq t \leq T$, be a stochastic process. Then the Esscher transformed measure of P by the risk process R_t and the index process h_s is the probability measure $P^{(ESS)}_{R_{[0,T]},h}$ defined by

$$\frac{dP^{(ESS)}_{R_{[0,T]},h_{[0,T]}}}{dP}|_{\mathcal{F}} = \frac{e^{\int_0^T h_s dR_s}}{E[e^{\int_0^T h_s dR_s}]} \tag{25}$$

This measure transformation is called the Esscher transform by the risk process R_t and the index process h_s.

Definition 3.3. In the above definitions, if the index index process is chosen so that the $P^{(ESS)}_{R_{[0,T]},h_{[0,T]}}$ is a martingale measure of S_t, then $P^{(ESS)}_{R_{[0,T]},h_{[0,T]}}$ is called the Esscher transformed martingale measure of S_t by the risk process R_t, and it is denoted by $P^{(ESS)}_{R_{[0,T]}}$ or simply $P^{(ESS)}_{R.}$.

3.1.2 Simple return process and compound return process

When we give a certain risk process R_t, we obtain a corresponding Esscher transformed martingale measure if it exists. As we have seen in the previous section, the GLP has two kinds of representation such that $S_t = S_0 e^{Z_t} = S_0 \mathcal{E}(\tilde{Z})_t$. So the processes Z_t and $\tilde{Z}_t$ are both candidates for the risk process.

We shall see the economical meaning of them. For this purpose, we will review the discrete time approximation of geometric Lévy processes. Set

$$S^{(n)}_k = S_{k/2^n}, \quad k = 1, 2, \ldots. \tag{26}$$

According to the above two kinds of expression of S_t, we obtain two kinds of approximation formula. First one is

$$S^{(n)}_k = S_0 e^{Z^{(n)}_k}, \quad Z^{(n)}_k = Z_{k/2^n}, \quad k = 1, 2, \ldots. \tag{27}$$

Second approximation is

$$S_k^{(n)} = S_0\mathcal{E}(Y^{(n)})_k = \prod_{j=1}^{k}\left(1 + (Y_j^{(n)} - Y_{j-1}^{(n)})\right), \quad k = 1, 2, \ldots, \tag{28}$$

where $\mathcal{E}(Y^{(n)})_k$ is the discrete time Doléans-Dade exponential of $Y_k^{(n)}$, and $Y_k^{(n)}$ is defined from the following relations

$$e^{Z_k^{(n)}} = \prod_{j=1}^{k}\left(1 + (Y_j^{(n)} - Y_{j-1}^{(n)})\right), \quad k = 1, 2, \ldots. \tag{29}$$

So we obtain

$$e^{\triangle Z_k^{(n)}} = e^{Z_k^{(n)} - Z_{k-1}^{(n)}} = \left(1 + (Y_k^{(n)} - Y_{k-1}^{(n)})\right) = 1 + \triangle Y_k^{(n)}. \tag{30}$$

From this we obtain

$$\triangle Y_k^{(n)} = e^{\triangle Z_k^{(n)}} - 1 = \frac{S_k^{(n)}}{S_{k-1}^{(n)}} - 1 = \frac{\triangle S_k^{(n)}}{S_{k-1}^{(n)}} \tag{31}$$

and we know that $\triangle Y_k^{(n)}$ is the simple return process of $S_k^{(n)}$.

On the other hand, we obtain from the definition of $\triangle Z_k^{(n)}$

$$\triangle Z_k^{(n)} = \log S_k^{(n)} - \log S_{k-1}^{(n)} = \log\left(1 + \frac{\triangle S_k^{(n)}}{S_{k-1}^{(n)}}\right), \tag{32}$$

and we know that $\triangle Z_k^{(n)}$ is the increment of log-return and it is called the compound return process of $S_k^{(n)}$.

Remark 3.1. The terms "simple return" and "compound return" were introduced in [1] , and well known in economics (See also [47]).

For $t \in (\frac{k-1}{2^n}, \frac{k}{2^n})$ we define $Z_t^{(n)} = Z_k^{(n)}, \quad Y_t^{(n)} = Y_k^{(n)}$. Then it is easy to see that the process $Z_t^{(n)}$ converges to the process Z_t when n goes to ∞. On the other hand the process $Y_t^{(n)}$ converges to the process $\tilde{Z}_t$ as follows. The process S_t satisfies the stochastic differential equation $dS_t = S_{t-}d\tilde{Z}_t$. From this it follows that $d\tilde{Z}_t = \frac{dS_t}{S_{t-}}$. (For the justification of this formula, see [30] .) Comparing this formula with (31), we know that the process $Y_t^{(n)}$ is the approximation process in the procedure of solving the stochastic differential equation for $\tilde{Z}_t$. This fact means that the process $Y_t^{(n)}$ converges to the process $\tilde{Z}_t$.

Based on the above observation, it is natural for us to give the following definition.

Definition 3.4. The process $\tilde{Z}_t$ is called the "simple return process" of S_t, and the process Z_t is called the "compound return process" of S_t.

3.2 Esscher transformed martingale measures for geometric Lévy processes

We next study the existence and uniqueness of Esscher transformed martingale measures.

3.2.1 Esscher Martingale measure (ESMM)

Suppose that Z_t is adopted as the risk process. Then, if the corresponding Esscher transformed martingale measure $P^{(ESS)}_{Z_{[0,T]}}$ is well defined, then it should be called the *'compound return Esscher transformed martingale measure'*. This is the Gerber-Shiu's Esscher martingale measure introduced in [24], and the term *'Esscher martingale measure'* is usually suggesting this compound return Esscher transformed martingale measure $P^{(ESS)}_{Z_{[0,T]}}$.

We suppose that the expectations which appear in what follows exist. Then the martingale condition for an Esscher transformed probability measure $Q = P^{(ESS)}_{Z_{[0,T]},h}$ is

$$E_Q[e^{-r}S_1] = e^{-r}S_0E_Q[e^{Z_1}] = e^{-r}S_0\frac{E_P[e^{(h+1)Z_1}]}{E_P[e^{hZ_1}]} = S_0. \tag{33}$$

This condition is equal to the following condition

$$E_P[e^{(h+1)Z_1}] = e^r E_P[e^{hZ_1}], \tag{34}$$

and this is also equivalent to the following expression,

$$\phi(-i(h+1)) = e^r\phi(-ih), \tag{35}$$

where $\phi(u)$ is the characteristic function of Z_1 ($\phi(u) = E_P[e^{iuZ_1}]$).

To formulate the existence theorem, we set

$$f^{(ESMM)}(h) = b + (\frac{1}{2} + h)\sigma^2 + \int_{-\infty}^{\infty}\left((e^x - 1)e^{hx} - x1_{\{|x|\le 1\}}\right)\nu(dx). \tag{36}$$

Then we obtain

Theorem 3.1. *If the equation*

$$f^{(ESMM)}(h) = r, \tag{37}$$

has a solution h^*, then the ESMM of S_t, $P^{(ESMM)}$, exists and

$$P^{(ESMM)} = P^{(ESS)}_{Z_{[0,T]},h^*} = P^{(ESS)}_{Z_T,h^*} \tag{38}$$

The process Z_t is also a Lévy process under $P^{(ESMM)}$ and the generating triplet of Z_t under $P^{(ESMM)}$, say $(\sigma^{(ESMM)2}, \nu^{(ESMM)}(dx), b^{(ESMM)})$, is

$$\sigma^{(ESMM)2} = \sigma^2, \tag{39}$$

$$\nu^{(ESMM)}(dx) = e^{h^* x}\nu(dx), \tag{40}$$

$$b^{(ESMM)} = b + h^*\sigma^2 + \int_{\{|x|\leq 1\}} x(e^{h^* x} - 1)\nu(dx). \tag{41}$$

(Proof) The equation (37) is equivalent to the condition (3.16). Therefore $P^{(ESS)}_{Z_{[0,T]},h^*}$ is a martingale measure of S_t.

The characteristic function of Z_t under $P^{(ESMM)} = P^{(ESS)}_{Z_{[0,T]},h^*}$, $\phi_t^{(ESMM)}(u)$, is by definition

$$\phi_t^{(ESMM)}(u) = E_{P^{(ESMM)}}[e^{iuZ_t}] = \frac{E_P[e^{iuZ_t}e^{h^* Z_T}]}{E_P[e^{h^* Z_T}]}. \tag{42}$$

And this is equal to

$$\frac{E_P[e^{(iu+h^*)Z_t}]}{E_P[e^{h^* Z_t}]} = \frac{\phi_t(u - ih^*)}{\phi_t(-ih^*)}. \tag{43}$$

By simple calculation we obtain

$$\begin{aligned} &\phi_t^{(ESMM)}(u) \\ &= \exp\left\{t\left(-\tfrac{1}{2}\sigma^2u^2 + i(b + h^*\sigma^2 + \textstyle\int_{\{|x|\leq 1\}} x(e^{h^* x} - 1)\nu(dx))u \right.\right. \\ &\qquad \left.\left. + \textstyle\int_{-\infty}^{\infty}(e^{iux} - 1 - iux1_{\{|x|\leq 1\}})e^{h^* x}\nu(dx)\right)\right\}. \end{aligned} \tag{44}$$

Using this formula, we can see that the martingale condition (3.16) is reduced to the equation (3.18). (Q.E.D.)

3.2.2 Minimal Entropy Martingale Measure (MEMM)

Next we consider the case where $\tilde{Z}_t$ is adopted as the risk process. If the corresponding Esscher transformed martingale measure $P^{(ESS)}_{\tilde{Z}_{[0,T]}}$ exists, then it should be called the *"simple return Esscher transformed martingale measure"*.

We will see here the relation between the Esscher transform and the minimal entropy martingale measure (MEMM). We first give the definition of the MEMM.

Definition 3.5. If an equivalent martingale measure P^* satisfies

(45) $\quad H(P^*|P) \leq H(Q|P) \quad \forall Q: \ \textit{equivalent martingale measure},$

then P^* is called the "minimal entropy martingale measure (MEMM)" of S_t. Where $H(Q|P)$ is the relative entropy of Q with respect to P

(46)
$$H(Q|P) = \begin{Bmatrix} \int_\Omega \log[\frac{dQ}{dP}]dQ, & if \quad Q \ll P, \\ \infty, & otherwise, \end{Bmatrix}.$$

From the proof of [23] it follows that

Proposition 3.1. *The simple return Esscher transformed martingale measure* $P^{(ESS)}_{\tilde{Z}_{[0,T]}}$ *of* S_t *is the minimal entropy martingale measure* (MEMM) *of* S_t.

Based on the above results, we give the following definition.

Definition 3.6. (i) The compound return Esscher transformed martingale measure $P^{(ESS)}_{Z_{[0,T]}}$ is called the "Esscher martingale measure (ESMM)" and denoted by $P^{(ESMM)}$.
(ii) The simple return Esscher transformed martingale measure $P^{(ESS)}_{\tilde{Z}_{[0,T]}}$ is called the "minimal entropy martingale measure (MEMM)" and denoted by P^* (or $P^{(MEMM)}$).

As we have mentioned above, the MEMM, P^*, is the simple return Esscher transformed martingale measure. ($P^* = P^{(ESS)}_{\tilde{Z}_{[0,T]}}$). The existence theorem of the MEMM is obtained in [23].

Set

(47)
$$f^{(MEMM)}(\theta) = b + (\frac{1}{2} + \theta)\sigma^2 + \int_{-\infty}^{\infty} \left((e^x - 1)e^{\theta(e^x-1)} - x1_{\{|x|\leq 1\}}(x)\right) \nu(dx)$$

Then the following result is obtained ([23]).

Theorem 3.2. *If the equation*

(48)
$$f^{(MEMM)}(\theta) = r$$

has a solution θ^*, *then the MEMM of* S_t, P^*, *exists and*

(49)
$$P^* = P^{(MEMM)} = P^{(ESS)}_{\tilde{Z}_{[0,T]},\theta^*} = P^{(ESS)}_{\tilde{Z}_T,\theta^*}$$

The process Z_t is also a Lévy process under P^* and the generating triplet of Z_t under P^*, say $(\sigma^{*2}, \nu^*(dx), b^*)$, is

$$\sigma^{*2} = \sigma^2, \tag{50}$$

$$\nu^*(dx) = e^{\theta^*(e^x-1)}\nu(dx), \tag{51}$$

$$b^* = b + \theta^*\sigma^2 + \int_{\{|x|\leq 1\}} x(e^{\theta^*(e^x-1)} - 1)\nu(dx). \tag{52}$$

(Proof) The results of this theorem follows directly from the proof of [23] . (Q.E.D.)

Remark 3.2. (i) The above result is improved to the multi dimensional cases (see [22] or [17]).
(ii)The function $f^{(MEMM)}(\theta)$ is also expressed as follows

$$f^{(MEMM)}(\theta) = \tilde{b} + \sigma^2\theta + \int_{-\infty}^{\infty} \left(xe^{\theta x} - x1_{\{|x|\leq 1\}}(x)\right) \tilde{\nu}(dx) \tag{53}$$

And the process $\tilde{Z}_t$ is also a Lévy process under P^* and the generating triplet of $\tilde{Z}_t$ under P^* is

$$\tilde{\sigma}^{*2} = \tilde{\sigma}^2 = \sigma^2, \tag{54}$$

$$\tilde{\nu}^*(dx) = e^{\theta^* x}\tilde{\nu}(dx), \tag{55}$$

$$\tilde{b}^* = \tilde{b} + \theta^*\tilde{\sigma}^2 + \int_{\{|x|\leq 1\}} x(e^{\theta^* x} - 1)\tilde{\nu}(dx). \tag{56}$$

(iii)The function $f^{(MEMM)}(\theta)$ is a non-decreasing function of θ.
(iv) If S_t is integrable, then it holds that

$$E(S_t) = S_0 \exp(t f^{(MEMM)}(0)). \tag{57}$$

(vi) If the condition that $f^{(MEMM)}(0) > r$ is satisfied, then the solution θ^* of (48) is negative ($\theta^* < 0$) if it exists. Such cases occur very often.

3.2.3 Mean Correcting Martingale Measure (MCMM)

For the jump-diffusion type models, the Brownian motion can be adopted as the risk process. In that case the corresponding Esscher transformed martingale measure is the mean correcting martingale measure (MCMM) (see [46] or [8]).

We suppose that the expectations which appear in what follows exist. Then the martingale condition for an Esscher transformed probability measure $Q = P^{(ESS)}_{W_{[0,T]},h}$ is

$$E_Q[e^{-r}S_1] = e^{-r}S_0 E_Q[e^{Z_1}] = e^{-r}S_0 \frac{E_P[e^{Z_1+hW_1}]}{E_P[e^{hW_1}]} = S_0. \tag{58}$$

This condition is equal to the following condition

$$E_P[e^{Z_1+hW_1}] = e^r E_P[e^{hW_1}], \tag{59}$$

and this is also equivalent to the following expression,

$$\psi(-i) + \frac{1}{2}h^2 + \sigma h = r + \frac{1}{2}h^2. \tag{60}$$

So the martingale condition is

$$\frac{1}{2}\sigma^2 + b + \int_{-\infty}^{\infty} (e^x - 1 - x1_{\{|x|\leq 1\}}(x))\ \nu(dx) + \sigma h = r. \tag{61}$$

To formulate the existence theorem, we set

$$f^{(MCMM)}(h) = b + \frac{1}{2}\sigma^2 + h\sigma + \int_{-\infty}^{\infty} (e^x - 1 - x1_{\{|x|\leq 1\}}(x))\ \nu(dx) \tag{62}$$

Then we obtain

Theorem 3.3. *If the equation*

$$f^{(MCMM)}(h) = r, \tag{63}$$

has a solution h^*, then the MCMM of S_t, $P^{(MCMM)}$, exists and

$$P^{(MCMM)} = P^{(ESS)}_{W_{[0,T]},h^*} = P^{(ESS)}_{W_T,h^*} \tag{64}$$

The process Z_t is also a Lévy process under $P^{(MCMM)}$ and the generating triplet of Z_t under $P^{(MCMM)}$, say $(\sigma^{(MCMM)2}, \nu^{(MCMM)}(dx), b^{(MCMM)})$, is

$$\sigma^{(MCMM)2} = \sigma^2, \tag{65}$$

$$\nu^{(MCMM)}(dx) = \nu(dx), \tag{66}$$

$$\begin{aligned} b^{(MCMM)} &= b + h^*\sigma \\ &= r - \frac{1}{2}\sigma^2 - \int_{-\infty}^{\infty} (e^x - 1 - x1_{\{|x|\leq 1\}}(x))\ \nu(dx). \end{aligned} \tag{67}$$

3.3 Comparison of ESMM and MEMM

The ESMM, MEMM and MCMM are all obtained by Esscher transforms, but they have slightly different properties. The MCMM is supported in only the case of $\sigma > 0$. So we restrict our attention on the ESMM and MEMM, and we will survey the properties and the differences of them.

3.3.1 Corresponding risk process

The risk process corresponding to the ESMM is the compound return process, and the risk process corresponding to the MEMM is the simple return process. The simple return process seems to be more essential in the relation to the original process rather than the compound return process. In this sense we can say that the MEMM is more reasonable martingale measure than the ESMM.

3.3.2 Existence condition

As we have seen in the previous section, for the existence of ESMM, $P^{(ESMM)}$, the following condition

$$\int_{\{|x|>1\}} |(e^x - 1)e^{h^* x}|\, \nu(dx) < \infty \tag{68}$$

is necessary. On the other hand, for the existence of MEMM, P^*, the corresponding condition is

$$\int_{\{|x|>1\}} |(e^x - 1)e^{\theta^*(e^x-1)}|\, \nu(dx) < \infty. \tag{69}$$

This condition is satisfied for wide class of Lévy measures, if $\theta^* < 0$. Namely, the former condition is strictly stronger than the latter condition.

This means that the MEMM may be applied to the wider class of models than the ESMM. This difference does work in the stable process cases. In fact the MEMM method can be applied to the geometric stable model but the ESMM method can not be applied to this same model.

3.3.3 Corresponding utility function

The ESMM is corresponding to power utility function or logarithm utility function (see [24] or [25]). On the other hand the MEMM is corresponding to the exponential utility function (see [21] or [25]).

We remark here that, in the case of ESMM, the power parameter of the utility function depends on the parameter value h^* of the Esscher transform.

3.3.4 Properties special to MEMM

a) Minimal distance to the original probability:

The relative entropy is very popular in the field of information theory, and it is called Kullback-Leibler Information Number(see [27]) or Kullback-Leibler distance (see [12]). Therefore we can state that the MEMM is the nearest equivalent martingale measure to the original probability P in the sense of Kullback-Leibler distance. Recently the idea of minimal distance martingale measure is studied. In [25] it is mentioned that the relative entropy is the typical example of the distance in their theory.

b) Large deviation property:

The large deviation theory is closely related to the minimum relative entropy analysis, and the Sanov's theorem or Sanov property is well-known (see, e.g. [14]). Applying this theorem on the paths spaces of the price process, we can conclude that that the MEMM is the most possible empirical probability measure of paths of price process in the class of the all equivalent martingale measures. In this sense the MEMM should be considered to be the exceptional measure in the class of all equivalent martingale measures.

The Sanov's theorem is

Theorem (Sanov) [14] Let μ be a probability measure on the Polish space Σ and let $\tilde{\mu}_n \in M_1(M_1(\Sigma))$ be the distribution under μ^n of the $\frac{1}{n}\sum_{j=1}^{n}\delta_{\sigma_j}$. Then $H(\cdot|\mu)$ is a good, convex rate function on $M_1(\Sigma)$ and $\{\tilde{\mu}_n : n \geq 1\}$ satisfies the full large deviation principle with rate function $H(\cdot|\mu)$.

c) Indifference Price: The MEMM is related to the indifference prices with the exponential utility in the following sense.

Let $p_\gamma(X)$ be the indifference price of X, where γ is the parameter for the absolute risk aversion. Then it holds that

$$\lim_{\gamma \to 0} p_\gamma(X) = e^{-rT} E_{P^*}[X]$$

We also remark that, in the case of MEMM, the relation of the MEMM to the utility indifference price is known (see [44] , [23] and [48]).

As the result of the above discussions, we can say that the MEMM has many better properties than the ESMM in the theoretical sense.

4. [GLP & MEMM] Pricing Model

In this section we explain the [GLP & MEMM] pricing model and see examples of the model.

4.1 Model

Now we give the definition of the [GLP & MEMM] Pricing Model. The [GLP & MEMM] Pricing Model is such a model (see [36]):

(A) The price process S_t is a geometric Lévy process (GLP).
(B) The price of an option X is defined to be $e^{-rT}E_{P^*}[X]$, where P^* is the MEMM.

Of course this model can be considered for the cases where the MEMM exists. The existence condition is given in Theorem 2 of Section 3.

4.2 Examples of [GLP & MEMM] Pricing Model

In this section we see several examples of [GLP & MEMM] Pricing Models. To do this, we have to check the existence condition of the MEMM, i.e. we have to examine that the given geometric Lévy process $S_t = S_0 \exp Z_t$ satisfies the conditions of Theorem 2.

We denote the function $f^{(MEMM)}(\theta)$ by $f(\theta)$ for simplicity, namely

$$f(\theta) = b + (\frac{1}{2} + \theta)\sigma^2 + \int_{-\infty}^{\infty} \left((e^x - 1)e^{\theta(e^x-1)} - x1_{\{|x|\leq 1\}}(x)\right) \nu(dx). \tag{70}$$

Then the condition is that the following equation has a solution θ^*.

$$f(\theta) = r. \tag{71}$$

4.2.1 Geometric Variance Gamma Model

The Lévy measure of Variance Gamma process is of the following form (see [31]).

$$\nu(dx) = C\,(I_{\{x<0\}} \exp(-c_1|x|) + I_{\{x>0\}} \exp(-c_2|x|))\,|x|^{-1}dx, \tag{72}$$

where C, c_1, c_2 are positive constants.

The following results are obtained (see [23]).

Proposition 4.1. 1) *If $c_2 \leq 1$, then the equation $f(\theta) = r$ has a unique solution θ^*, and the solution is negative.*
2) *If $c_2 > 1$ and $f(0) \geq r$, then the equation $f(\theta) = r$ has a unique solution θ^*, and the solution is non-positive.*
3) *If $c_2 > 1$ and $f(0) < r$, then the equation $f(\theta) = r$ has no solution.*

4.2.2 Geometric CGMY Model

The Lévy measure of the CGMY process is

$$\nu(dx) = C\,(I_{\{x<0\}} \exp(-G|x|) + I_{\{x>0\}} \exp(-M|x|))\,|x|^{-(1+Y)}dx, \tag{73}$$

where $C > 0, G \geq 0, M \geq 0, Y < 2$ (see [3]). If $Y \leq 0$, then $G > 0$ and $M > 0$ are assumed. We mention here that the case $Y = 0$ is the VG process case, and the case $G = M = 0$ and $0 < Y < 2$ is the symmetric stable process case. In the sequel we assume that $G, M > 0$.

For this model the following results are obtained (see [41]).

Proposition 4.2. 1) *If $M \leq 1$, then the equation $f(\theta) = r$ has a unique solution θ^*, and the solution is negative.*
2) *If $M > 1$ and $f(0) \geq r$, then the equation $f(\theta) = r$ has a unique solution θ^*, and the solution is non-positive.*
3) *If $M > 1$ and $f(0) < r$, then the equation $f(\theta) = r$ has no solution.*

4.2.3 Geometric Stable Model

We consider the stable model. Suppose that Z_t is a stable process and let $(0, \nu(dx), b)$ be its generating triplet. The Lévy measure is

$$\nu(dx) = c_1 I_{\{x<0\}}|x|^{-(\alpha+1)}dx + c_2 I_{\{x>0\}}|x|^{-(\alpha+1)}dx, \tag{74}$$

where $0 < \alpha < 2$ and we assume that

$$c_1 \geq 0, \quad c_2 \geq 0, \quad c = c_1 + c_2 > 0. \tag{75}$$

The following results are obtained (see [23]).

Proposition 4.3. *Under the assumption $c_1, c_2 > 0$, the equation $f(\theta) = r$ has a unique solution θ^*, and the solution θ^* is negative.*

Remark 4.1. Consider the case where $c_1, c_2 > 0$. Under the original measure P, $S_t, t > 0$ is not integrable. But under the MEMM P^*, any moments $E_{P^*}[|S_t|^k], k = 1, 2, \ldots,$ of S_t are finite. This fact follows easily from the result that θ^* is negative, and this property is very useful for the study of option pricing of this model.

4.3 Option Pricing and Volatility Smile/Smirk Properties

In order to apply the [GLP & MEMM] Pricing Models to the financial problems, we have to establish the methods to compute the option prices. Namely we have to compute the expectations $E_{P^*}[F(\omega)]$, where $F(\omega)$ is a functional of Lévy process.

4.3.1 European Type Options

If a contingent claim C is depending only on the terminal value of the stock price $S_T = S_0 e^{Z_T}$, then we can compute the price of C as what follows.

Let $C = f(S_T) = f(S_0 e^{Z_T}) = F(Z_T)$, $(F(z) = f(S_0 e^z))$, and set $C(t, y) = E_{P^*}[e^{-r(T-t)} f(S_T)|S_t = y]$ and $\tilde{C}(t, z) = E_{P^*}[e^{-r(T-t)} F(Z_T)|Z_t = z] = E_{P^*}[e^{-r(T-t)} f(S_T)|S_t = S_0 e^z]$. (Remark that $\tilde{C}(t, z) = C(t, S_0 e^z)$.) Since the process Z_t is a Lévy process with the generating triplet $(\sigma^2, \nu^*(dx), b^*)$, $\tilde{C}(t, z)$ satisfies the following equation under the assumption of the smoothness of $\tilde{C}(t, z)$.

$$\begin{aligned} -\frac{\partial \tilde{C}(t,z)}{\partial t} &= \frac{1}{2}\sigma^2 \frac{\partial^2 \tilde{C}(t,z)}{\partial z^2} + b\frac{\partial \tilde{C}(t,z)}{\partial z} \\ &\quad + \int_{-\infty}^{\infty} \left(\tilde{C}(t, z + \tilde{z}) - \tilde{C}(t, z) - \tilde{z}\frac{\partial \tilde{C}(t,z)}{\partial z} 1_{\{|\tilde{z}|<1\}}(\tilde{z}) \right) \nu(d\tilde{z}) \\ &\quad - r\tilde{C}(t, z), \quad 0 \leq t < T, \end{aligned} \tag{76}$$

$$\tilde{C}(T, z) = F(z). \tag{77}$$

Solving this equation, we obtain the option price $C(0, S_0) = \tilde{C}(0, 0)$.

4.3.2 FFT Method for European Call Options

The fast Fourier transform method (FFT method) is very useful for the computation of option prices. We need to compute the such an expectation $E_{P^*}[F(\omega)]$, and in the case of European type options such type of expectations $E_{P^*}[G(S_T)]$. If we know the distribution function $p_T^*(z)$ of Z_T under P^*, then $E_{P^*}[G(S_T)] = \int_{-\infty}^{\infty} G(z)p_T^*(z)dz$. Lévy process is characterized by the generating triplet, and the generating triplet is given explicitly in the characteristic function. So we can assume that the characteristic function $\phi_T^*(u)$ of Z_T under P^* is given and the density function $p_T^*(z)$ is obtained as the inverse Fourier transform of $\phi_T^*(u)$.

For the computer simulation of the theoretical prices of European call options, the FFT method is very useful. Carr and Madan have introduced their idea in [4] , and their method has been improved by Cont and Tankov in [10] . We rearrange their ideas in such a form that we can easily apply the formula to our [GLP & MEMM] pricing models.

The characteristic function $\phi_t^*(u)$ of Z_t under the MEMM P^* is

$$\phi_t^*(u) = \phi_{Z_t}^*(u) = E_{P^*}[e^{iuZ_t}] = \exp(\psi_t^*(u)) = \exp(t\psi^*(u)), \quad i = \sqrt{-1}, \tag{78}$$

where $\psi^*(u) = \psi_1^*(u)$. Let $\mu_t^*(dz)$ be the distribution of Z_t under the MEMM P^*, and assume that $\mu_t^*(dz) = p_t^*(z)dz$. Then

$$\phi_t^*(u) = \phi_{Z_t}^*(u) = E_{P^*}[e^{iuZ_t}] = \int_{-\infty}^{\infty} e^{iuz}p_t^*(z)dz, \tag{79}$$

The price of European call option is

$$C(S_0, K, T) = e^{-rT}E_{P^*}[(S_T - K)^+] = e^{-rT}\int_{-\infty}^{\infty}(S_0e^z - K)^+p_T^*(z)dz. \tag{80}$$

Set $K/S_0 = e^k$, and define $c(k; S_0, T) = C(S_0, S_0e^k, T)$. Then

$$c(k; S_0, T) = S_0e^{-rT}\int_{-\infty}^{\infty}(e^z - e^k)^+p_T^*(z)dz \tag{81}$$

We introduce the so-called time value of option

$$\tilde{c}(k; S_0, T) = c(k; S_0, T) - (S_0 - e^{-rT}K)^+ = c(k; S_0, T) - S_0(1 - e^{k-rT})^+ \tag{82}$$

and let $\zeta(v; S_0, T)$ be the Fourier transform of $\tilde{c}(k; S_0, T)$

$$\zeta(v; S_0, T) = \int_{-\infty}^{\infty} e^{ivk}\tilde{c}(k; S_0, T)dk. \tag{83}$$

Using (4.12)

$$\zeta(v; S_0, T) = S_0 e^{-rT} \int_{-\infty}^{\infty} p_T^*(z)dz \int_z^{rT} e^{ivk}(e^k - e^z)dk. \tag{84}$$

and

$$\zeta(v; S_0, T) = S_0 \frac{e^{-rT}\phi_T^*(v-i) - e^{ivrT}}{iv(1+iv)}. \tag{85}$$

The characteristic function $\phi_T^*(u)$ is computed directly from the generating triplet $(\sigma^2, b^*, \nu^*(dx))$, so $\zeta(v; S_0, T)$ is obtained from the above formula. Next, by (4.14), $\tilde{c}(k; S_0, T)$ is obtained by the inverse Fourier transform

$$\tilde{c}(k; S_0, T) = \frac{1}{2\pi} \int_{-\infty}^{\infty} e^{-ikv} \zeta(v; S_0, T)dv \tag{86}$$

and

$$c(k; S_0, T) = \tilde{c}(k; S_0, T) + (S_0 - e^{-rT}K)^+, \quad K = S_0 e^k. \tag{87}$$

Finally we obtain the price of the European call option $C(S_0, K, T)$ as

$$C(S_0, K, T) = c(\log(K/S_0); S_0, T) = \tilde{c}(\log(K/S_0); S_0, T) + (S_0 - e^{-rT}K)^+. \tag{88}$$

4.3.3 Volatility Smile/Smirk Properties

The volatility smile/smirk properties are reported for many market prices of options. This fact tells us that the Black-Scholes model is not necessarily best model, and that we should study other models which may have the volatility simile/smirk properties. It is known that the [GLP & MEMM] models have those properties (see [39]).

5. Physical World and MEMM World

The behavior of the price process S_t is governed by the original probability P, and the movement of S_t is observable. This is the real world (=Physical world).

On the other hand the price of an option X is computed as the expectation $e^{-rT}E_{P^*}[X]$, namely the process S_t is supposed to obey the MEMM P^*. This world is differ from the real world, and this world should be called the imaginary world (=MEMM world).

5.1 From Physical World to MEMM World

Suppose that the price process $S_t = S_0 e^{Z_t}$ is given and the generating triplet of Z_t is (σ^2, ν, b). Let θ^* is the solution of $f(\theta) = r$, where the function $f(\theta)$ is defined by (4.6). Then, by Theorem 3 in §4.2, the generating triplet $(\sigma^{*2}, \nu^*, b^*)$ of Z_t under P^* is

$$(89) \qquad \sigma^{*2} = \sigma^2,$$

$$(90) \qquad \nu^*(dx) = e^{\theta^*(e^x-1)}\nu(dx),$$

$$(91) \qquad b^* = b + \theta^*\sigma^2 + \int_{\{|x|\le 1\}} xd(\nu^* - \nu)$$

$$= b + \theta^*\sigma^2 + \int_{\{|x|\le 1\}} x\left(e^{\theta^*(e^x-1)} - 1\right)\nu(dx).$$

This triplet determines the prices of options in the framework of [GLP & MEMM] pricing model.

Remark 5.1. The existence condition for θ^* (i.e. θ^* is the solution of $f(\theta) = r$) is equivalent to the following martingale condition (M*)

$$(\mathrm{M}^*) \quad b^* + \frac{1}{2}\sigma^{*2} + \int_{-\infty}^{\infty} (e^x - 1 - x1_{\{|x|\le 1\}}(x))\nu^*(dx) = r.$$

We should notice that the θ^* does not appear explicitly in this formula, and that this formula is just the same condition that P^* is a martingale measure of the price process S_t.

Concerning to the martingale condition for more general cases of semi-martingales, see [47] .

5.2 From MEMM World to Physical World

We study the inverse problem of the previous subsection. Suppose that the generating triplet $(\sigma^{*2}, \nu^*, b^*)$ of Z_t under P^* is given. Since we assume that P^* is martingale measure, the condition (M*) is satisfied.

We try to construct a probability $\bar{P}$ such that under $\bar{P}$ the price process $S_t = S_0e^{Z_t}$ is geometric Lévy process and the MEMM of $S_t = S_0e^{Z_t}$ with $\bar{P}$ is P^*.

Let θ^* be any real number (it is usually supposed that $\theta^* < 0$) and set

$$(93) \qquad \bar{\sigma}^2_{\theta^*} = \sigma^{*2}$$

$$(94) \qquad \bar{\nu}_{\theta^*}(x) = e^{-\theta^*(e^x-1)}\nu^*(dx)$$

$$(95) \qquad \bar{b}_{\theta^*} = b^* - \theta^*\sigma^2 + \int_{\{|x|\le 1\}} x\left(e^{-\theta^*(e^x-1)} - 1\right)\nu^*(dx),$$

where we assume that all integrals are converge. Then suppose that we could construct the probability measure $\bar{P}_{\theta^*}$ such that under $\bar{P}_{\theta^*}$ the process Z_t is a Lévy process with the generating triplet $(\bar{\sigma}^2_{\theta^*}, \bar{\nu}_{\theta^*}, \bar{b}_{\theta^*})$.

It is easy to see that P^* is the MEMM of $S_t = S_0e^{Z_t}$ with $\bar{P}_{\theta^*}$. We remark here that there are many geometric Lévy processes whose MEMM is just the same P^*.

5.3 Example: Geometric Stable Process Case

• Parameters in the physical world: (α, c_1, c_2, b), $0 < \alpha < 2$, $c_1, c_2 \geq 0$, $c_1 + c_2 > 0$, $-\infty < b < \infty$.

The triplet is $(0, \nu, b)$, where

$$\nu(dx) = \frac{c_1 I_{\{x<0\}}(x) + c_2 I_{\{x>0\}}(x)}{|x|^{(\alpha+1)}} dx. \tag{96}$$

• Parameters in the MEMM world: $(\theta^*, \alpha^*, c_1^*, c_2^*, b^*)$, $\theta^* < 0$, $0 < \alpha^* < 2$, $c_1^*, c_2^* \geq 0$, $c_1^* + c_2^* > 0$, $-\infty < b^* < \infty$, where

$$\nu^*(dx) = e^{\theta^*(e^x-1)} \frac{c_1^* I_{\{x<0\}}(x) + c_2^* I_{\{x>0\}}(x)}{|x|^{(\alpha^*+1)}} dx. \tag{97}$$

and the following martingale condition

$$(\mathrm{M}^*): \quad b^* + \int_{-\infty}^{\infty} (e^x - 1 - x 1_{\{|x|\leq 1\}}(x))\nu^*(dx) = r$$

must be satisfied. So, if we have given the values of $(\theta^*, \alpha^*, c_1^*, c_2^*,)$, then the value of b^* is determined by the above condition (M*).

5.4 Diagram of Physical World and MEMM World

Physical World		MEMM World
P		P^*
	$S_t = S_0 e^{Z_t} = S_0 \mathcal{E}(\tilde{Z})_t$	
(σ^2, ν, b) under P	Z_t	$(\sigma^{*2}, \nu^*, b^*)$ under P^*
$(\tilde{\sigma}^2, \tilde{\nu}, \tilde{b})$ under P	$\tilde{Z}_t$	$(\tilde{\sigma}^{*2}, \tilde{\nu}^*, \tilde{b}^*)$ under P^*

Z_t: log-return process,
$\tilde{Z}_t$: simple return process

6. Estimation of Lévy Processes in the Physical World

Usually this procedure is carried on under the restriction of the class of Lévy processes, for example the stable process class, VG process class, etc. Therefore the estimation problem of the process is reduced to the parameter estimation problems.

There are many papers on this subject. (see [37] for example). If the MLE is possible and easy, then this method may be good. But this method is not easy to apply our cases.

6.1 Characteristic Function Method of Moments

1) Characteristic Function and Moment Generating Function:

The characteristic function (in the sense of distribution) $\phi(u)$ of X is defined by

(98) $$\phi(u) = \phi(u; X) = E[e^{iuX}] = \exp\{\psi(u)\}, \quad i = \sqrt{-1}.$$

The sample characteristic function $\hat{\phi}_n(u)$ is given by

(99) $$\hat{\phi}_n(u) = \frac{1}{n}\sum_{j=1}^{n} e^{iu\xi_j}, \quad -\infty < u < \infty.$$

Note that $\hat{\phi}_n(u)$ is a consistent estimator of $\phi(u)$:

(100) $$\lim_{n\to\infty} \hat{\phi}_n(u) = \phi(u), \quad -\infty < u < \infty.$$

The moment generating function $M(u)$ of X is defined, if it exists, by

(101) $$M(u) = M(u; X) = E[e^{uX}], \quad -\infty < u < \infty,$$

and the sample moment generating function $\hat{M}(u)$ is given by

(102) $$\hat{M}_n(u) = \frac{1}{n}\sum_{j=1}^{n} e^{u\xi_j}, \quad -\infty < u < \infty.$$

The sample moment generating function $\hat{M}(u)$ is a consistent estimator of $M(u)$.

2) Moment Equations for Characteristic Function:

When we take the function e^{iuX} as the function $f_u(X)$ for the generalized method of moments, then the generalized moment equations are

(103) $$\phi(u) = \hat{\phi}_n(u), \quad -\infty < u < \infty.$$

If the moment $E[X^k]$ exists, then it is well-known that

(104) $$m_k = E[X^k] = \frac{1}{i^k}E[(iX)^k] = \frac{1}{i^k}\frac{d^k\phi}{du^k}(0),$$

and the classical moment equations are

(105) $$\frac{1}{i^k}\frac{d^k\phi}{du^k}(0) = \hat{m}_k, \quad k = 1, 2, \ldots.$$

3) Moment Equations for Moment Generating Function:

Suppose that the moment generating function $M(u)$ exists. In such cases we can take the function e^{uX} as the function $f_u(X)$ for the generalized method of moments, and then the generalized moment equations are

(106) $$M(u) = \hat{M}_n(u), \quad -\infty < u < \infty.$$

6.2 Estimation of Lévy Processes

Set $Z = Z_1$. The corresponding characteristic function $\phi(u)$ is

$$\begin{aligned}(107)\qquad \phi(u) &= E[e^{iuZ}] = \exp\{\psi(u)\}\\ &= \exp\left\{-\frac{\sigma^2}{2}u^2 + ibu + \int_{-\infty}^{\infty}\left(e^{iux} - 1 - iux1_{\{|x|\le 1\}}(x)\right)\nu(dx)\right\}\end{aligned}$$

What we have to do is to estimate the generating triplet $(\sigma^2, \nu(dx), b)$ of the distribution of Z. These parameters explicitly contained in the characteristic functions. So it is natural for us to apply the characteristic function method to those estimation problems.

Set

$$(108)\qquad I_j = Z_j - Z_{j-1}, j = 1, 2, \ldots,$$

then $\{I_j, j = 1, 2, \ldots\}$ is i.i.d. with the same distribution as Z_1, since Lévy process has temporally homogeneous independent increment. So, if we are given a sequential data of a Lévy process Z_t, then we can apply the method described above to estimate the distribution of $Z = Z_1$, namely we can apply the generalized moment equation or the classical moment equation when it exists (see [37] or [2]).

7. Fitness Analysis of the Models

Suppose that the sequential data of the price process S_t of underlying asset and the data of market prices of options. From these data, we have to select a model which is most fitting to the given data. This is the fitness analysis of the models.

7.1 Procedure of fitness analysis

- Collecting data: the sequential data of the price process S_t, and the data of market prices of options.
- Selection of the most fitting model to the obtained data.

To solve this problem, we first fix a type of model (for example, [G-stable-P & MEMM]), and we take the following steps.

1) Estimation the price process of the underlying asset in the physical world from the sequential data of it.
2) Calculation of the MEMM from the estimated parameter, and computation of the theoretical prices of options in the estimated MEMM world.
3) Analyzing the fitness of the theoretical prices to the market prices.

We carry on the above procedures for several types of models, and the final step is

4) Determination of the most reasonable model.

7.2 Fitting error of the model

Denote the estimated probability by $\widehat{P}$, (or equivalently, the estimated generating triplet by $(\widehat{\sigma^2}, \widehat{\nu}, \widehat{b})$), and let $\widehat{P^*}$ be the corresponding MEMM. Then the theoretical price of option C is $E_{\widehat{P^*}}[Ce^{-rT}]$. We denote this value by $\widehat{C^*}$.

Suppose that the data, $\eta_l, l = 1, 2, \ldots, L$, of market prices of options C_l are given. Then we define the fitting error of the model by

$$\epsilon^* = \frac{1}{L}\sum_{l=1}^{L} \frac{|\widehat{C_l}^* - \eta_l|}{\eta_l} \tag{109}$$

Procedure to obtain the fitting error:

Physical World		MEMM World
Data: $\{\xi_j\}$ (time series data)		
Estimated: $\widehat{P}$		Transformed: $\widehat{P^*}$
$(\widehat{\sigma^2}, \widehat{\nu}, \widehat{b})$	$\widehat{\theta^*}$	$(\widehat{\sigma^*}^2, \widehat{\nu^*}, \widehat{b^*})$
	(European Call Options)	Theoretical prices: $\widehat{C_l}^*$

Data: $\{\eta_l\}$ (European Call Options, in the Market)

$$\text{Fitting error}: \quad \epsilon^* = \frac{1}{L}\sum_{l=1}^{L} \frac{|\widehat{C_l}^* - \eta_l|}{\eta_l}$$

Remark 7.1. the second candidate for the fitting error is

$$\sqrt{\frac{1}{L}\sum_{l=1}^{L} |\widehat{C_l}^* - \eta_l|^2} \tag{110}$$

7.3 Example: Geometric Stable Process Case

Parameter in the physical world: (α, c_1, c_2, b)

$$\nu(dx) = \frac{c_1 1_{\{x<0\}}(x) + c_2 1_{\{x>0\}}(x)}{|x|^{(\alpha+1)}} dx. \tag{111}$$

Estimators: $(\widehat{\alpha}, \widehat{c_1}, \widehat{c_2}, \widehat{b})$

$$\widehat{\nu}(dx) = \frac{\widehat{c_1} 1_{\{x<0\}}(x) + \widehat{c_2} 1_{\{x>0\}}(x)}{|x|^{(\widehat{\alpha}+1)}} dx. \tag{112}$$

$\widehat{\theta^*}$ is determined by

$$\widehat{b}+\int_{-\infty}^{\infty}\left((e^x-1)e^{\widehat{\theta^*}(e^x-1)}-x1_{\{|x|\leq 1\}}(x)\right)\widehat{\nu}(dx)=r. \tag{113}$$

The process which determines the theoretical option prices is

$$\widehat{\nu^*}(dx)=e^{\widehat{\theta^*}(e^x-1)}\frac{\widehat{c_1}1_{\{x<0\}}(x)+\widehat{c_2}1_{\{x>0\}}(x)}{|x|^{(\widehat{\alpha}+1)}}dx \tag{114}$$

$$\widehat{b^*}=\widehat{b}+\int_{-\infty}^{\infty}x1_{\{|x|\leq 1\}}(x)d(\widehat{\nu^*}-\widehat{\nu}). \tag{115}$$

7.4 Fitness analysis of the estimated model

As the results of the above procedure, if the value ϵ^*

$$\epsilon^*=\frac{1}{L}\sum_{l=1}^{L}\frac{|\widehat{C_l}^*-\eta_l|}{\eta_l}$$

is small, then the fitness of the model to the data is good.

The value ϵ^* depends on the model, namely on the selected class of the process (for example, class of stable processes, class of CGMY processes, etc.) We can conclude that the class whose fitness error is the smallest is the best class (see [40] .)

8. Calibration

The calibration is similar to the fitness analysis, but usually the calibration is done based on only the data of option prices in the market (see [11]). Namely the calibration solves the following minimization problem

$$\min_{\gamma^*}\epsilon^{(\gamma^*)}=\frac{1}{L}\sum_{l=1}^{L}\frac{|\widehat{C_l}^{(\gamma^*)}-\eta_l|}{\eta_l} \tag{116}$$

where γ^* is the parameter of a model in the risk neutral world. For example, in the geometric stable process case, the calibration means the estimation of $\gamma^*=(\theta^*,\alpha^*,c_1^*,c_2^*,b^*)$ in the MEMM world.

Suppose that the above minimization were attained at $\gamma^{(*)}=\gamma^{(cal)}$, then the calibration error is

$$\text{ARPE}^{(cal)}=\frac{1}{L}\sum_{l=1}^{L}\frac{|\widehat{C_l}^{(cal)}-\eta_l|}{\eta_l} \tag{117}$$

The second candidate for the calibration error is the root mean-square error (RMSE) given by

$$\text{RMSE}^{(cal)} = \min_{\gamma^*} \sqrt{\frac{1}{L}\sum_{l=1}^{L} |\widehat{C_l}^{(\gamma^*)} - \eta_l|^2} \tag{118}$$

9. Calculation of the Option Prices: Expectation of Functionals of Lévy Processes

In order to apply the [GLP & MEMM] Pricing Models to the financial problems, we have to establish the methods to compute the option prices. Namely we have to compute the expectations $E_{P^*}[F(\omega)]$, where $F(\omega)$ is a functional of Lévy process.

9.1 Asian Option

Let $A(S, K)$ be the Asian option, namely

$$A(S,K) = \max\{\frac{1}{T}\int_0^T S_t dt - K, 0\}. \tag{119}$$

Then the price of $A(S_., K)$ is $e^{-rT}E_{P^*}[A(S_., K)]$.

For the computation of the above value, in [49] the following interest results have been obtained.

Set $q_t = \frac{1}{rT}\left(1 - e^{-r(T-t)}\right)$, and let X_t be a stochastic process determined by the following equation

$$dX_t = q_t dS_t + r(X_{t-} - q_t S_{t-})dt, 0 \le t \le T, \tag{120}$$

$$X_0 = q_0 S_0 - e^{-rT}K. \tag{121}$$

Then it holds that

$$X_T = A(S, K) \tag{122}$$

and so the calculation problem of the price $A^*(S_0, K) = e^{-rT}E_{P^*}[A(S(\omega), K)]$ is reduced to the calculation of $e^{-rT}E_{P^*}[X_T]$. Next we set

$$Y_t = \frac{X_t}{S_t} \tag{123}$$

and define a new probability measure Q^* by

$$\frac{dQ^*}{dP^*}|_{\mathcal{F}_t} = \frac{e^{-rt}S_t}{S_0} = e^{-rt+Z_t} \tag{124}$$

then the process Y_t is a Markov process under Q^* and solves the following equation
(125)

$$dY_t = (q_t - Y_{t-})\left(\sigma dW_t + (b^* - \frac{1}{2}\sigma^2)dt + \int_{-\infty}^{\infty} x\widetilde{N^*}_p(dudx) - \int_{-\infty}^{\infty} \frac{x^2}{1+x} N_p^*(dudx)\right),$$

Next we define $V(t,y)$ as the solution of the following equation

$$\begin{aligned}(126)\qquad & \frac{\partial V(t,y)}{\partial t} + \frac{1}{2}\sigma^2 \frac{\partial^2 V(t,y)}{\partial y^2}(q_t - y)^2 \\ & + \int_{-\infty}^{\infty}\left(V(t, y + (q_t - y)\tfrac{e^x-1}{e^x}) - V(t,y) - \frac{\partial V(t,y)}{\partial y}(q_t - y)\tfrac{e^x-1}{e^x}\right)\bar{\nu}^*(dx) \\ & = 0, \qquad 0 \le t < T, \\ (127)\qquad & V(T,y) = S_0 \max\{y, 0\},\end{aligned}$$

where $\bar{\nu}^*(dx)$ is the Lévy measure of $\bar{Z}_t$ under the probability P^*, and $\bar{Z}_t$ is the Lévy process which appears when S_t is expresses in the form of $S = S_0\mathcal{E}(\bar{Z})_t$ (Doléans-Dade exponential). Then $V(0, \frac{X_0}{S_0}) = V(0, q_0 - e^{-rT}\frac{K}{S_0})$ is the price of the Asian option $A(S,K)$.

The points of this result is that the problem to obtain the price of Asian option is reduced to the problem to obtain the price of European type option of Markov process.

10. Utility Indifference Prices and Risk Measure

We consider the exponential utility function

$$(128)\qquad U_\alpha(x) = 1 - e^{-\alpha x},$$

and set

$$\begin{aligned}(129)\qquad J_\alpha(c,B) &:= \sup_{\theta\in\Theta} E_P[U_\alpha(c + G(\theta)_T - B)] \\ &= \sup_{\theta\in\Theta} E_P[1 - \exp\{-\alpha(c + G(\theta) - B\}]\end{aligned}$$

where Θ is a suitable set of strategies, $G(\theta)$ is the gain of a strategy θ, B is a contingent claim.

We give the definition of the utility indifference price $p_\alpha(c,B)$ (see [13] §4.2, or [26]).

Definition 10.1. The value $p_\alpha(c,B)$ which satisfies the following equation

$$(130)\qquad J_\alpha(c + p_\alpha(c,B), B) = J_\alpha(c, 0)$$

is called the utility indifference price of B.

It is easy to see that the value $p_\alpha(c, B)$ does not depend on c, so we use the notation $p_\alpha(B)$.

The quantity $J_\alpha(c, B)$ is related to the relative entropy by the following duality relation

$$J_\alpha(c, B) = 1 - \exp\{-\inf_{Q \in \mathcal{M}} \Big(H(Q|P) + \alpha c - E_Q[\alpha B]\Big)\} \tag{131}$$
$$= 1 - e^{-\alpha c} \exp\{\alpha \sup_{Q \in \mathcal{M}} \Big(E_Q[B] - \frac{1}{\alpha} H(Q|P)\Big)\},$$

where $\mathcal{M}$ is a convex subset of local martingale measures corresponding to Θ (see [22]). And the utility indifference price $p_\alpha(B)$ has the following property

$$\lim_{\alpha \downarrow 0} p_\alpha(B) = E_{P^*}[B]. \tag{132}$$

(See [44] , [23] and [48] .)

This result suggests us that $-E_{P^*}[B]$ may be an example of the reasonable coherent risk measure.

11. Generalization of the [GLP & MEMM] Pricing Model

11.1 Multi dimensional cases

The multidimensional cases are very important in the practical sense, because of the fact that many options are based on the index, for example Nikkei 225, and the index is the combination of the multi dimensional price processes of underlying assets.

Suppose that the price processes are given by

$$S_t^j = S_0^j e^{Z_t^j}, \quad j = 1, \ldots, d \tag{133}$$

where $Z_t = (Z_t^1, \ldots, Z_t^d)^T$ is a d-dimensional Lévy process.

This process is equivalent to

$$dS_t^j = S_{t-}^j d\tilde{Z}_t^j, \quad j = 1, \ldots, d \tag{134}$$

where $\tilde{Z}_t = (\tilde{Z}_t^1, \ldots, \tilde{Z}_t^d)^T$ is the corresponding d-dimensional Lévy process, and the price processes S_t^j have the following expression

$$S_t^j = S_0^j \mathcal{E}(\tilde{Z}^j)_t, \quad j = 1, \ldots, d \tag{135}$$

where $\mathcal{E}(\tilde{Z}^j)_t$ is the Doléans-Dade exponential of $\tilde{Z}_t^j$.

The results described in the previous sections for 1-dimensional case are generalized to the multi dimensional cases (See [22] and [17]).

The points are the following two.
1) The MEMM P^* is obtained by the Esscher transform by $\tilde{Z}$.
2) The processes Z_t and $\tilde{Z}_t$ are also Lévy processes under P^*.

Remark 11.1. The Esscher transform by $\tilde{Z}_t$ is unique if it exists, but the Esscher transform by Z_t is not necessary unique in the multi dimensional cases (See [30]).

11.2 [GLP & MEMM] models with defaultable risk

We have started from the following type models $S_t = S_0 e^{Z_t}, \quad 0 \le t \le T$, or the following SDE $dS_t = S_{t-} d\tilde{Z}_t$, where $\tilde{Z}_t$ is a Lévy process such that *supp* $\tilde{\nu} \subset (-1, \infty)$. In this case the model is without defaultable risk.

If we permit the case *supp* $\tilde{\nu} \subset [-1, \infty)$, then the defaultable risk is in mind. To do this we introduce a new Lévy process $\tilde{Z}_t^{(\lambda)}$ $(\lambda > 0)$, defined on a new probability space $(\Omega^{(\lambda)} \mathcal{F}^{(\lambda)}, P^{(\lambda)})$, whose Lévy measure $\tilde{\nu}^{(\lambda)}(dx)$ is

$$\tilde{\nu}^{(\lambda)}(dx) = \tilde{\nu}(dx) + \lambda \delta_{\{-1\}}(dx). \tag{136}$$

And let $S_t^{(\lambda)}$ be the solution of $dS_t^{(\lambda)} = S_{t-}^{(\lambda)} d\tilde{Z}_t^{(\lambda)}$, namely $S_t^{(\lambda)} = S_0 \mathcal{E}(\tilde{Z}^{(\lambda)})_t$.

We can assume naturally that the original processes S_t, Z_t and $\tilde{Z}_t$ are defined on the same probability space $(\Omega^{(\lambda)}, \mathcal{F}^{(\lambda)}, P^{(\lambda)})$, where the Lévy measure of $\tilde{Z}_t$ is $\tilde{\nu}(dx)$.

Next we define the following stopping time

$$\tau^{(\lambda)} = \inf\{t > 0; S_t^{(\lambda)} = 0)\} = \inf\{t > 0; \tilde{Z}_t^{(\lambda)} = -1\}. \tag{137}$$

It is easy to see that $\tau^{(\lambda)}$ is independent of S_t, Z_t and $\tilde{Z}_t$, and that

$$P(\tau^{(\lambda)} \le t) = \lambda \int_0^t e^{-\lambda s} ds = 1 - e^{-\lambda t}. \tag{138}$$

It is obvious that $S_t^{(\lambda)} = S_t 1_{\{\tau^{(\lambda)} > t\}}$.

Suppose that the MEMM of $S_t^{(\lambda)}$, $P^{(\lambda)^*}$, exists. Then $\tilde{Z}_t^{(\lambda)}$ is Lévy process under $P^{(\lambda)^*}$ and the Lévy measure of $\tilde{Z}_t^{(\lambda)}$ under $P^{(\lambda)^*}$ is

$$\tilde{\nu}^{(\lambda)^*}(dx) = e^{\theta^{(\lambda)^*} x} \tilde{\nu}^{(\lambda)}(dx), \tag{139}$$

where $\theta^{(\lambda)^*}$ is the solution of the following equation for θ

$$\tilde{b} + \sigma^2 \theta + \int_{-\infty}^{\infty} \left(x e^{\theta x} - x 1_{\{|x| \le 1\}}(x) \right) \tilde{\nu}^{(\lambda)}(dx) = r. \tag{140}$$

We can see that $\tilde{Z}_t$ is also a Lévy process under $P^{(\lambda)^*}$ and the Lévy measure of it is $e^{\theta^{(\lambda)^*} x} \tilde{\nu}(dx)$. Therefore Z_t is also a Lévy process under $P^{(\lambda)^*}$.

Remark 11.2. The Lévy measure of $\tilde{Z}_t$ under $P^{(\lambda)^*}$ $(= e^{\theta^{(\lambda)^*}x}\tilde{\nu}(dx))$ is different from the Lévy measure of $\tilde{Z}_t$ under P^*, which is $e^{\theta^* x}\tilde{\nu}(dx)$.

From these results we know that $\tau^{(\lambda)}$ is independent of S_t and $\tilde{Z}_t$ under $P^{(\lambda)^*}$, and

$$P^{(\lambda)^*}(\tau^{(\lambda)} \leq t) = \lambda e^{-\theta^{(\lambda)^*}} \int_0^t e^{-\lambda e^{-\theta^{(\lambda)^*}} s} ds = 1 - e^{-\lambda e^{-\theta^{(\lambda)^*}} t}. \tag{141}$$

The theoretical prices of options can be computed as the expectation with respect to $P^{(\lambda)^*}$. In particular, The prices of European type options are easily computed, using the above properties of $\tau^{(\lambda)}$.

Remark 11.3. The arguments of this subsection are possible in the MEMM setting, but not possible in the ESMM setting because the process $Z_t^{(\lambda)}$ such that $S_t^{(\lambda)} = S_0\mathcal{E}(\tilde{Z}^{(\lambda)})_t = S_0 e^{Z_t^{(\lambda)}}$ is not well-defined.

11.3 semimartingale process model

The martingale theory is established in the framework of semimartingale process. So, in the theoretical or mathematical sense, it is natural to study the semimartingale process models. In fact many articles are studied under the semimartingale setting.

Among them the generalization of the [GLP & MEMM] Pricing Model is discussed in [9] , where the entropy-Hellinger martingale measure is introduced.

Acknowledgments. The author likes to thank professor Albert Shiryaev for valuable comments.

References

1. Bühlmann, H., Delbaen, F., Embrechts, P., and Shiryaev, A. N. (1996), No-arbitrage, Change of Measure and Conditional Esscher Transforms, *CWI Quarterly* **9**(4), 291–317.
2. Carrasco, M. and Florens, J.-P. (2002), Efficient GMM Estimation Using the Empirical Characteristic Function (preprint).
3. Carr, P., Geman, H., Madan, D. B., and Yor, M. (2002), The Fine Structure of Asset Returns: An Empirical Investigation, *J. of Business,* **2**, 305–332.
4. Carr, P. and Madan, D. (1999), Option valuation using the fast Fourier transform. *Journal of Computational Finance* **2**, 61–73.
5. Carr, P. and Wu, L. (2003), "The Finite Moment Log Stable Processes and Option Pricing," The Journal of Finance **58**(2), 753–777.
6. Carr, P. and Wu, L. (2004), "Time-changed Lévy Processes and Option Pricing," *Journal of Financial Economics* **71**, 113–141.
7. Chan, T. (1999), Pricing Contingent Claims on Stocks Derived by Lévy Processes. *The Annals of Applied Probability* **9**(2), 504–528.

8. Cheang, Gerald H. L. (2003), "A simple approach to pricing options with jumps" (preprint)
9. Choulli, T. and Stricker, C. (2004), Minimal entropy-Hellinger martingale measure in incomplete markets, to appear in *Mathematical Finance.*
10. Cont, R. and Tankov, P. (2004a), Non-Parametric Calibration of Jump-Diffusion Option-Pricing Models, J. of Computational Finance **7**(3), 1–49.
11. Cont, R. and Tankov, P. (2004b), *Financial Modelling with Jump Processes*, CHAPMAN & Hall/CRC.
12. Cover, T. and Thomas, J. (1991), *Elements of Information Theory*, Wiley.
13. Delbaen, F., Grandits, P., Rheinländer, T., Samperi, D., Schweizer, M., and Stricker, C. (2002), "Exponential Hedging and Entropic Penalties," *Mathematical Finance*, **12**(2), 99–124.
14. Deuschel, J.-D. and Stroock, D. W. (1989), *Large Deviations*, Academic Press.
15. Eberlein, E. and Keller, U. (1995), Hyperbolic distributions in finance. *Bernoulli* **1**, 281–299.
16. Edelman, D. (1995), A Note: Natural Generalization of Black-Scholes in the Presence of Skewness, Using Stable Processes, *ABACUS* **31**(1), 113–119.
17. Esche, F. and Schweizer, M. (2005), Minimal Entropy Preserves the Lévy Property: How and Why, *Stochastic Processes and their Applications* **115**, 299–327.
18. Esscher, F. (1932), On the Probability Function in the Collective Theory of Risk, *Skandinavisk Aktuarietidskrift* **15**, 175–195.
19. Fama, E. F. (1963), Mandelbrot and the Stable Paretian Hypothesis. *J. of Business*, 36, 420–429.
20. Föllmer, H. and Schweizer, M. (1991), Hedging of Contingent Claims under Incomplete Information. In M. H. A. Davis and R. J. Elliot (ed.): *Applied Stochastic Analysis*, Gordon and Breach, 389–414.
21. Frittelli, M. (2000), The Minimal Entropy Martingale Measures and the Valuation Problem in Incomplete Markets, *Mathematical Finance* **10**, 39–52.
22. Fujiwara, T. (2002), The Minimal Entropy Martingale Measures for multi-dimensional Geometric Lévy Processes and the optimal strategies for the exponential utility maximization (preprint)
23. Fujiwara, T. and Miyahara, Y. (2003), The Minimal Entropy Martingale Measures for Geometric Lévy Processes, *Finance and Stochastics* **7**, 509–531.
24. Gerber, H. U. and Shiu, E. S. W. (1994), Option Pricing by Esscher Transforms, *Transactions of the Society of Actuaries* **XLVI**, 99–191.
25. Goll, T. and Rüschendorf, L. (2001), Minimax and minimal distance martingale measures and their relationship to portfolio optimization, *Finance and Stochastics*, **5**(4), 557–581.
26. Hodges, S. D. and Neuberger, A. (1989), "Optimal Replication of Contingent Claims under Transaction Costs," *Review of Futures Markets* **8**, 222–239.
27. Ihara, S. (1993), *Information Theory for Continuous System*, World Scientific.
28. Jacod, J. and Shiryaev, A. (1987), *Limit Theorems for Stochastic Processes*, Springer-Verlag.
29. Kallsen, J. (2002), Utility-Based Derivative Pricing in Incomplete Markets, in *Mathematical Finance - Bachelier Congress 2000*, Springer, 313–338.
30. Kallsen, J. and Shiryaev, A. N. (2002), The Cumulant Process and Esscher's

Change of Measure, *Finance and Stochastics* **6**(4), 397–428.
31. Madan, D. and Seneta, E. (1990), The variance gamma (vg) model for share market returns. *Journal of Business* **63**(4), 511–524.
32. Madan, D., Carr, P., and Chang, E. (1998), The variance gamma process and option pricing. *European Finance Review* **2**, 79–105.
33. Mandelbrot, B. (1963), The variation of certain speculative prices. *J. of Business* **36**, 394–419.
34. Miyahara, Y. (1996), Canonical Martingale Measures of Incomplete Assets Markets, in *Probability Theory and Mathematical Statistics: Proceedings of the Seventh Japan-Russia Symposium, Tokyo 1995 (eds. S. Watanabe et al.)*, pp. 343–352.
35. Miyahara, Y. (1999), Minimal Entropy Martingale Measures of Jump Type Price Processes in Incomplete Assets Markets. *Asian-Pacific Financial Markets* **6**(2), 97–113.
36. Miyahara, Y. (2001), [Geometric Lévy Process & MEMM] Pricing Model and Related Estimation Problems, *Asia-Pacific Financial Markets* **8**(1), 45–60.
37. Miyahara, Y. (2002), Estimation of Lévy Processes, *Discussion Papers in Economics, Nagoya City University* No. 318, pp. 1–36.
38. Miyahara, Y. (2004), A Note on Esscher Transformed Martingale Measures for Geometric Lévy Processes, *Discussion Papers in Economics, Nagoya City University* No. 379, pp. 1–14.
39. Miyahara, Y. and Moriwaki, N. (2004), Volatility Smile/Smirk Properties of [GLP & MEMM] Pricing Models, *Discussion Papers in Economics, Nagoya City University* No. 405, pp. 1–16.
40. Miyahara, Y. and Moriwaki, N. (2005), Application of [GLP & MEMM] Model to Nikkei 225 Option (preprint).
41. Miyahara, Y. and Novikov, A. (2002), Geometric Lévy Process Pricing Model, *Proceedings of Steklov Mathematical Institute*, **237**, pp. 176–191.
42. Overhaus, M., Ferraris, A., Knudsen, T., Milward, R., Nguyen-Ngoc, L., and Schindlmayr, G. (2002), *Equity derivatives: Theory and Applications*, Wiley.
43. Rachev, S. and Mittnik, S. (2000), "*Stable Paretian Models in Finance*", Wiley.
44. Rouge, R. and El Karoui, N. (2000), Pricing via Utility Maximization and Entropy, *Mathematical Finance* **10**(2), 259–276.
45. Sato, K. (1999), "*Lévy Processes and Infinitely Divisible Distributions*", Cambridge University Press.
46. Schoutens, W, (2003), *Lévy Processes in Finance: Pricing Financial Derivatives*, Wiley.
47. Shiryaev, A. N. (1999), *Essentials of Stochastic Finance: Facts, Models, Theory*, World Scientific.
48. Stricker, C. (2004), Indifference Pricing with Exponential Utility, *Progress in Probability*, **58**, 323–328. Birkhäuser 2004.
49. Večeř, J. and Xu, M. (2001), "Pricing Asian Options in a Semimartingale Model" (preprint)
50. Xiao, K., Misawa, T., and Miyahara, Y. (2000), Computer Simulation of [Geometric Lévy Process & MEMM] Pricing Model, *Discussion Papers in Economics, Nagoya City University* No. 266, pp. 1–16.

Topics Related to Gamma Processes

Makoto Yamazato

Department of Mathematical Sciences
University of the Ryukyus,
Senbaru 1, Nishihara-cho, Okinawa, 903-0213 Japan

The aim of this paper is to explain important but not popular properties related to gamma processes and show the applicability of these properties. We define subclasses (**CME** and its subclasses) of the class of infinitely divisible distributions, which are generated by mixtures and convolutions from gamma distributions, and study their properties. Then we apply the obtained results to the unimodality of the distributions in the above classes, the boundedness in space-time parameters of transition densities of subordinators generated by **CME** distributions and the determination of the class of hitting time distributions of 1-dimensional generalized diffusion processes. Finally, we remark that some subclasses of the class **CME** and the class of selfdecomposable distributions are often used in mathematical finance.
Key words: gamma process, convolution, mixture

1. Introduction

Gamma process appears in various fields both theoretical and practical. Its 1-dimensional distribution - gamma distribution has various remarkable properties (refer [23]). The aim of this paper is to introduce and explain important but not popular (in Japan) properties related to gamma processes and show the applicability of these properties.

We restrict ourselves to 1-dimensional processes. We define subclasses **ME, CE, CG, CEME** and **CME** of the class of infinitely divisible distributions, which are generated by mixtures and convolutions from gamma distributions, and study their properties. For this purpose, we mainly use the fact that the distributions in these classes can be decomposed into exponential distributions and gamma distributions (Sections 3-6).

Then we apply the results obtained in Sections 3-6 to unimodality of the distributions in the above classes (Section 7), boundedness in space-time parameters of transition densities of **CME**$_+$ processes (Lévy processes generated by **CME** distributions on $[0,\infty)$) in Section 8 and determination of the class of hitting time distributions of 1-dimensional generalized diffusion processes (Section 9). In the final section, we remark that distributions of some subclasses of the class **CME** and, class **L** (class of selfdecomposable distributions) and related stochastic processes are often used in mathematical finance.

Almost all facts described in this paper are already known. However, some results are new (Theorems 3.3, 3.4, 6.3, 8.2, 8.3, Example 8.1) and the proof of Theorem 3.5 is simplified.

2. Infinitely divisible distributions and Lévy processes

Definition 2.1. A probability measure μ on $\mathbb{R}$ is said to be infinitely divisible if for every $n \geq 1$, there is a probability measure μ_n on $\mathbb{R}$ such that $\mu = (\mu_n)^{n*}$ where $(\mu_n)^{n*}$ is the n-fold convolution of μ_n.

Theorem 2.1. *A probability measure μ on $\mathbb{R}$ is said to be infinitely divisible if and only if its characteristic function is represented as follows:*

$$\mathcal{F}\mu(z) = \int e^{izx}\mu(dx) = \exp[i\gamma z - \frac{\sigma^2}{2}z^2 + \int_{\mathbb{R}\setminus\{0\}}(e^{izu} - 1 - izu1_{\{|u|\leq 1\}})\nu(du)]$$

where, $\gamma \in \mathbb{R}$, $\sigma \geq 0$ and ν is a measure, called the Lévy measure of μ, satisfying

$$\int 1 \wedge |x|^2\nu(dx) < \infty.$$

We denote by **I** the class of infinitely divisible distributions on $\mathbb{R}$.

Definition 2.2. A probablity measure μ on $\mathbb{R}$ is said to be selfdecomposable if for any $0 < c < 1$, there is a probablity measure μ_c on $\mathbb{R}$ such that

$$\mathcal{F}\mu(z) = \mathcal{F}\mu(cz)\mathcal{F}\mu_c(z).$$

We denote by **L** the class of selfdecomposable distributions. The following characterization of the class **L** is well known.

Theorem 2.2. *A probability measure μ on $\mathbb{R}$ is selfcecomposable if and only if it is infinitely divisible and its characteristic function is represented as*

$$\mathcal{F}\mu(z) = \exp[i\gamma z - \frac{\sigma^2}{2}z^2 + \int_{\mathbb{R}\setminus\{0\}}(e^{izu} - 1 - izu1_{\{|u|\leq 1\}})|u|^{-1}k(u)du]$$

where, $\gamma \in \mathbb{R}$, $\sigma \geq 0$ and $k(u)$ is a nonnegative measurable function nondecreasing on $(-\infty, 0)$ and nonincreasing on $(0,\infty)$ satisfying $\int(|x|^{-1} \wedge |x|)k(x)dx < \infty$.

Characteristic function of an exponential distribution μ with density $\lambda e^{-\lambda x}$ is represented as

$$\mathcal{F}\mu(z) = \frac{\lambda}{\lambda - iz} = \exp[\int_0^\infty (e^{izu} - 1)u^{-1}e^{-\lambda u}du].$$

Hence the exponential distribution is selfdecomposable with Lévy measure $u^{-1}e^{-\lambda u}$. The class of exponential distributions is regarded as a subclass of the class of gamma distributions. A probability measure μ is said to be a gamma distribution with parameter (α, β) if it is absolutely continuous with density

$$f(x) = \begin{cases} \frac{\beta^\alpha}{\Gamma(\alpha)} x^{\alpha-1}e^{-\beta x} & \textit{for } x > 0 \\ 0 & \textit{otherwise.} \end{cases}$$

The parameter α is called the shape parameter and β is called the scale parameter. The gamma distribution is selfdecomposable with Lévy measure $\alpha u^{-1}e^{-\beta u}$.

Definition 2.3. A stochastic process $\{X_t : t \geq 0\}$ on $\mathbb{R}$ is a Lévy process if the following conditions are satisfied:
(1) For any choice of $n \geq 1$ and $0 \leq t_0 < t_1 < \cdots < t_n$, random variables $X_{t_0}, X_{t_1} - X_{t_0}, \ldots, X_{t_n} - X_{t_{n-1}}$ are independent
(2) $X_0 = 0$ a.s.
(3) The distribution of $X_{s+t} - X_s$ does not depend on s.
(4) $\lim_{t\downarrow 0} P(|X_t| > \epsilon) = 0$ for $\epsilon > 0$.
(5) X_t is right-continuous in $t \geq 0$ and has left limits in $t > 0$.

Theorem 2.3. *Let $\{X_t\}$ be a Lévy process on $\mathbb{R}$. Then the distribution of X_t is infinitely divisible for every $t \geq 0$. If the characteristic function of X_1 is represented as*

$$\begin{aligned} (1) \qquad & E(e^{izX_1}) \\ & = \exp[i\gamma z - \frac{\sigma^2}{2}z^2 + \int_{\mathbb{R}\setminus\{0\}} (e^{izu} - 1 - izu1_{\{|u|\leq 1\}})\nu(du)] =: e^{\Psi(z)}, \end{aligned}$$

then the characteristic function of X_t is represented as

$$E(e^{izX_t}) = e^{t\Psi(z)}.$$

Let $\{X_t\}$ be a Lévy process. If its distribution at $t = t_0 > 0$ is an exponential distribution with density $\beta e^{-\beta x}$, $(\beta > 0)$, then the distribution at $t > 0$ is a gamma distribution with density $\frac{\beta^{t/t_0}}{\Gamma(t/t_0)} x^{(t/t_0 - 1)}e^{-\beta x}$.

3. Mixtures of exponential distributions

We say that a probability measure μ is an $\mathbf{ME}_+$ distribution if there is a probability measure G on $(0, \infty]$ such that

$$\mu([0, x]) = \int_{(0,\infty]} (1 - e^{-xu})G(du) \quad \text{for } x > 0. \tag{2}$$

The measure μ may have a point mass $G(\{\infty\})$ at the origin and the distribution function of μ is infinitely differentiable on $(0, \infty)$. Similarly, we say that a probability measure μ is an $\mathbf{ME}_-$ distribution if there is a probability measure G on $[-\infty, 0)$ such that

$$\mu([x, 0]) = \int_{[-\infty,0)} (1 - e^{-xu})G(du) \quad \text{for } x < 0.$$

Theorem 3.1. ([21]) *A probability measure μ on $[0, \infty)$ is an $\mathbf{ME}_+$ distribution if and only if there is a nonnegative and absolutely continuous measure Q on $(0, \infty)$ with density bounded by 1 a.e. satisfying $\int_0^1 u^{-1}Q(du) < \infty$ such that, for $s \in \mathbb{R}$,*

$$\begin{aligned} \mathcal{L}\mu(s) &= \exp[\int_{(0,\infty)} (e^{-sx} - 1)(\int_{(0,\infty)} e^{-xu}Q(du))dx] \\ &= \exp \int_{(0,\infty)} (\frac{1}{u+s} - \frac{1}{u})Q(du), \end{aligned} \tag{3}$$

where $\mathcal{L}\mu$ stands for the Laplace transform of μ.

We have

$$G(\{\infty\}) = \lim_{s\to\infty} \mathcal{L}\mu(s) = \exp(-\int_0^\infty \frac{1}{u}Q(du)).$$

We call measures G appearing in (2) and Q appearing in (3) G-measure and Q-measure of $\mu \in \mathbf{ME}_+$, respectively.

We denote $\mathbf{ME} = \mathbf{ME}_+ * \mathbf{ME}_-$.

Theorem 3.2. ([29]) *Let $\mu_+ \in \mathbf{ME}_+$ and $\mu_- \in \mathbf{ME}_-$ and let $\mu = \mu_+ * \mu_- \in \mathbf{ME}$. Let G_+ and G_- be mixing distributions of μ_+ and μ_-, respectively. Denote*

$$d_- = \sup\{v < 0 : G_-([v, 0)) > 0\}$$

and

$$d_+ = \inf\{v > 0 : G_+((0, v]) > 0\}.$$

If $d_- < d_+$, then the Laplace transform $\mathcal{L}\mu(s)$ of μ exists for $-d_+ < s < -d_-$ and is represented as

$$\mathcal{L}\mu(s) = \int_{(-\infty,0)} \mathcal{L}\mu_+(-v)\frac{v}{s+v}G_-(dv) + \int_{(0,\infty)} \mathcal{L}\mu_-(-u)\frac{u}{s+u}G_+(du) + G_+(\{\infty\})G_-(\{-\infty\}).$$

Theorem 3.3. *Let $\mu \in \mathbf{ME}_+$ and let $q(u)du$ be its Q-measure. Assume that $\int_0^\infty \frac{1}{u}q(u)du = \infty$, equivalently, μ is absolutely continuous w.r.t. Lebesgue measure. Let f be the density of $\mu(dx)$. Then f is bounded if and only if $\int_1^\infty \frac{1}{u}(1-q(u))du < \infty$.*

Proof. Let G be the G-measure of μ. Since $f(x)$ is bounded iff $\int_0^\infty uG(du) < \infty$ and since

$$s\mathcal{L}\mu(s) = \int_0^\infty \frac{su}{s+u}G(du) \to \int_0^\infty uG(du)$$

as $s \to \infty$, we have that $f(x)$ is bounded if and only if $\lim_{s\to\infty} s\mathcal{L}\mu(s) < \infty$. The quantity $s\mathcal{L}\mu(s)$ is written as

$$\begin{aligned} s\mathcal{L}\mu(s) &= \exp\Big[\int_0^\infty (\frac{1}{u+s} - \frac{1}{u})q(u)du + \int_1^s \frac{1}{u}du\Big] \\ &= -\int_0^1 \frac{s}{u(u+s)}q(u)du + \int_1^s \frac{1}{u}(1-q(u))du \\ &\quad + \int_1^s \frac{1}{u+s}q(u)du - \int_s^\infty \frac{s}{u(u+s)}q(u)du. \end{aligned}$$

The first and the second terms tend to

$$-\int_0^1 \frac{1}{u}q(u)du > -\infty \text{ and } \int_1^\infty \frac{1}{u}(1-q(u))du \le \infty,$$

respectively, by monotone convergence theorem. The third term minus fourth term is equal to

$$\int_0^1 \frac{1}{u+1}q(su)du - \int_1^\infty \frac{1}{u(u+1)}q(su)du,$$

which is bounded in s. If $\int_1^\infty \frac{1}{u}(1-q(u))du < \infty$, then since $\lim_{s\to\infty} s\mathcal{L}\mu(s) \le \infty$ exists,

$$\lim_{s\to\infty}\Big(\int_0^1 \frac{1}{u+1}q(su)du - \int_1^\infty \frac{1}{u(u+1)}q(su)du\Big)$$

exists and finite. Hence, $\lim_{s\to\infty} s\mathcal{L}\mu(s) < \infty$. Converse is obvious. □

Theorem 3.4. *Let $\mu \in$ **ME**$_+$ and let $q(u)du$ be its Q-measure. Assume that $\int_0^\infty \frac{1}{u}q(u)du < \infty$, equivalently, μ has a point mass $\exp(-\int_0^\infty \frac{1}{u}q(u)du)$ at the origin and is absolutely continuous w.r.t. Lebesgue measure on $(0,\infty)$. Let f be the density of the absolutely continuous part of $\mu(dx)$. Then $\lim_{x\downarrow 0} f(x) = \int_0^\infty q(u)du \exp(-\int_0^\infty \frac{1}{u}q(u)du)$ and hence f is bounded if and only if $\int_1^\infty q(u)du < \infty$.*

Proof. Let G be the G-measure of μ. Since $\mu(dx) = G(\{\infty\})\delta_0(dx) + \int_{(0,\infty)} ue^{-ux}G(du)dx$,

$$s(\mathcal{L}\mu(s) - G(\{\infty\})) = \int_{(0,\infty)} \frac{us}{u+s}G(du) \to \int_{(0,\infty)} uG(du)$$

as $s \to \infty$. Hence we have

$$\lim_{s\to\infty} s(\mathcal{L}\mu(s) - G(\{\infty\})) = f(0+).$$

The left hand side of the above equality is written as

$$\exp(-\int_0^\infty \frac{1}{u}q(u)du)\{\exp(\int_0^\infty \frac{1}{u+s}q(u)du) - 1\}s.$$

By l'Hôspital's rule,

$$s\{\exp(\int_0^\infty \frac{1}{u+s}q(u)du) - 1\} \sim \int_0^\infty \frac{s^2}{(u+s)^2}q(u)du \uparrow \int_0^\infty q(u)du$$

as $s \to \infty$. □

In order to solve some problems, it is necessary to extend the domain of the Laplace transform of an **ME**$_+$ distribution to the complex number plane $\mathbb{C}$. We denote by $\mathbb{C}_+$ the open upper half plane in $\mathbb{C}$. A function analytic on $\mathbb{C}_+$ with positive imaginary part is called a Pick function ([8]). By $P(a,b)$, we denote the subclass of Pick functions which admit an analytic continuation by reflection across the interval (a,b) into the lower half-plane. Pick functions are related to moment generating functions. In order to apply to Laplace transforms, we consider other classes of functions dual to Pick functions : Let

$$\overline{P}(a,b) = \{\psi(s) : \overline{\psi(-\bar{s})} \in P(-b,-a)\}.$$

Proposition 3.1. ([5], [8]) *In order that a function h belongs to $\overline{P}(0,\infty)$, it is necessary and sufficient that there are $\alpha \in \mathbb{R}$, $\beta \le 0$ and a measure σ on $[0,\infty)$ satisfying $\int_{[0,\infty)}(1+u^2)^{-1}\sigma(du) < \infty$ such that*

$$h(s) = \alpha + \beta s + \int_{[0,\infty)} \left(\frac{1}{u+s} - \frac{u}{u^2+1}\right)\sigma(du).$$

The measure σ is called the spectral measure of h.

Since the imaginary part of the principal value Log $h(s)$ ($s \in \mathbb{C}_+$) for $h \in \overline{P}(0,\infty)$ belongs to $[-\pi, 0]$, we have another representation (exponential representation).

Proposition 3.2. ([5], [8]) *In order that a nonnegative function h on $(0,\infty)$ not identically 0 belongs to $\overline{P}(0,\infty)$, it is necessary and sufficient that there are a $\gamma \in \mathbb{R}$ and a measurable function $q(u)$ satisfying $0 \le q(u) \le 1$ a.s. such that*

$$h(s) = \exp\{\gamma + \int_0^\infty \left(\frac{1}{u+s} - \frac{u}{u^2+1}\right)q(u)du\}$$

for $s \in \mathbb{C}_+$.

This proposition shows that the Laplace transform of an $\mathbf{ME}_+$-distribution belongs to $\overline{P}(0,\infty)$ with $\gamma = -\int_0^\infty \frac{1}{u(u^2+1)}q(u)du$ and $\sigma(du) = uG(du)$.

Remark 3.1. For $s \in \mathbb{C}_+ \cup (0,\infty)$

$$s + a = \begin{cases} \exp\{-\int_a^\infty (\frac{1}{u+s} - \frac{1}{u})du + \log a\} & if\ a > 0, \\ \exp\{-\int_0^\infty (\frac{1}{u+s} - \frac{u}{u^2+1})du\} & if\ a = 0. \end{cases}$$

The imaginary part of the quantity in the braces belongs to $[0,\pi]$.

Lemma 3.1. *Let $\mu \in \mathbf{ME}_+$ with exponential representation* (3). *Then, for $s \in \mathbb{C}_+ \cup (0,\infty)$*

$$(s+a)\mathcal{L}\mu(s) = \begin{cases} \exp\{\log a + \int_0^\infty (\frac{1}{u+s} - \frac{1}{u})(q(u) - 1_{[a,\infty)}(u))du\} & if\ a > 0, \\ \exp\{\gamma + \int_0^\infty (\frac{1}{u+s} - \frac{u}{u^2+1})(q(u) - 1)du\} & if\ a = 0. \end{cases}$$

where $\gamma = -\int_0^\infty \frac{1}{u(u^2+1)}q(u)du$. The imaginary part of the quantity in the braces belongs to $[-\pi,\pi]$. It belongs to $[0,\pi]$, in particular, if $a = 0$.

Remark 3.2. Although the reciprocal $\frac{1}{s\mathcal{L}\mu(s)}$ does not belong to the class $\mathbf{ME}_+$, it has a representation in Proposition 3.2 and hence $\frac{1}{s\mathcal{L}\mu(s)}$ belongs to $\overline{P}(0,\infty)$.

Lemma 3.2. *Let $\mu \in \mathbf{ME}_+$ and let G be its G-measure. Let $a > 0$. Then for $s = -a + ib$,*

$$ib\mathcal{L}\mu(s) \to aG(\{a\}) \text{ as } b \downarrow 0.$$

Proof. Write $\mathcal{L}\mu(s)$ as

$$\mathcal{L}\mu(s) = \frac{aG(\{a\})}{s+a} + \int_{(0,a)\cup(a,\infty]} \frac{1}{s+u} uG(du).$$

Thus,

$$ib\mathcal{L}\mu(s) = aG(\{a\}) + \int_{(0,a)\cup(a,\infty)} \frac{b^2 + ib(u-a)}{(u-a)^2 + b^2} uG(du)$$

The second term in the right hand side of the above equality tends to 0 as $b \downarrow 0$, since $b^2 u\{(u-a)^2 + b^2\}^{-1}$ is bounded in $b \in (0,1]$, $b^2\{(u-a)^2+b^2\}^{-1} \downarrow 0$ as $b \downarrow 0$ if $u \neq a$ and $|bu(u-a)\{(u-a)^2+b^2\}^{-1}| \le |bu(u-a)^{-1}| \wedge \frac{1}{2}u$. □

The following result is obtained by Aronszajn and Donoghue Jr. ([1]) under more general setting and hence the proof is complicated. The proof in our restricted setting is simplified in [28].

Theorem 3.5. ([1], [28]) *Let $\mu \in \mathbf{ME}_+$ and let G and $q(u)du$ be its G-measure and Q-measure, respectively. Let $a > 0$. Then $G(\{a\}) > 0$ if and only if there is an $\epsilon > 0$ such that*

$$\int_{a-\epsilon}^{a+\epsilon} \frac{1}{u-a}\Big(1_{[a,\infty)}(u) - q(u)\Big)du < \infty. \tag{4}$$

Moreover,

$$G(\{a\}) = \exp\left\{-\int_0^\infty \left(\frac{1}{u-a} - \frac{1}{u}\right)\Big(1_{[a,\infty)}(u) - q(u)\Big)du\right\}$$

Proof. By Lemma 3.1, we have

$$ib\mathcal{L}\mu(s) = \exp\left\{\log a + \int_0^\infty \frac{a(u-a) - b^2 - ibu}{u\{(u-a)^2+b^2\}}\Big(q(u) - 1_{[a,\infty)}(u)\Big)du\right\}$$

for $s = -a + ib$ $(a, b > 0)$. By Lemma 3.2, $ib\mathcal{L}\mu(s)$ approaches to $aG(\{a\})$ as $b \downarrow 0$. Assume that $G(\{a\}) > 0$. Then,

$$\mathrm{Arg}(ib\mathcal{L}\mu(s)) \to 0, \ \log|ib\mathcal{L}\mu(s)| \to \log(aG(\{a\}))$$

as $b \to 0$. Here, Arg denotes the principal value of the argument. Thus,

$$\int_0^\infty \frac{a(u-a) - b^2}{u\{(u-a)^2+b^2\}}\Big(q(u) - 1_{[a,\infty)}(u)\Big)du \to \log G(\{a\})$$

and

$$\int_0^\infty \frac{b}{(u-a)^2+b^2}\Big(q(u)-1_{[a,\infty)}(u)\Big)du \to 0 \tag{5}$$

as $b \downarrow 0$. By the monotone covergence theorem, we have

$$\begin{aligned}&\lim_{b\downarrow 0}\int_0^\infty \frac{a(u-a)}{u\{(u-a)^2+b^2\}}\Big(q(u)-1_{[a,\infty)}(u)\Big)du\\ &=\int_0^\infty \frac{a}{u(u-a)}\Big(q(u)-1_{[a,\infty)}(u)\Big)du \geq -\infty.\end{aligned}$$

Also, we have

$$\lim_{b\downarrow 0}\int_0^\infty \frac{b^2}{u\{(u-a)^2+b^2\}}\Big(q(u)-1_{[a,\infty)}(u)\Big)du = 0.$$

Thus

$$0 \geq \int_0^\infty \frac{a}{u(u-a)}\Big(q(u)-1_{[a,\infty)}(u)\Big)du = -\log G(\{a\}) > -\infty.$$

Hence for $0 < \epsilon < a$

$$\int_{a-\epsilon}^{a+\epsilon} \frac{1}{a-u}\Big(1_{[a,\infty)}(u)-q(u)\Big)du < \infty.$$

Conversely, assume (4) for some $\epsilon > 0$. Then, by the same argument as above, we have

$$\begin{aligned}&\lim_{b\downarrow 0}\int_0^\infty \frac{a(u-a)-b^2}{u\{(u-a)^2+b^2\}}\Big(q(u)-1_{[a,\infty)}(u)\Big)du\\ &=\int_0^\infty \frac{a}{u(u-a)}\Big(q(u)-1_{[a,\infty)}(u)\Big)du > -\infty.\end{aligned}$$

Since $b|u-a|\{|u-a|^2+b^2\}^{-1} \leq \frac{1}{2}$, we have (5) by the assumption (4). Therefore,

$$G(\{a\}) = \exp\Big\{-\int_0^\infty (\frac{1}{u-a}-\frac{1}{u})\Big(1_{[a,\infty)}(u)-q(u)\Big)du\Big\} > 0. \qquad \square$$

Remark 3.3. Lemma 3.2 and Theorem 3.5 are valid for $\overline{P}(0,\infty)$ function with $\sigma(du) = uG(du)$.

4. Convolutions of exponential distributions

Let

$$\mathbf{CE}_+^f = \{\mu \in \mathcal{P}([0,\infty)) : \mathcal{L}\mu(s) = \prod_{k=1}^{m} a_k(s+a_k)^{-1},$$
$$\text{for } m \geq 1 \text{ and } a_1, \dots, a_m > 0\}$$

and let $\mathbf{CE}_-^f$ be the mirror image of $\mathbf{CE}_+^f$. We denote by $\mathbf{CE}_+$, $\mathbf{CE}_-$ and $\mathbf{CE}$ the closures in the weak convergence sense of $\mathbf{CE}_+^f$, $\mathbf{CE}_-^f$ and $\mathbf{CE}_+^f * \mathbf{CE}_-^f$, respectively (CE = Convolutions of Exponential distributions) .

Theorem 4.1. *Let $\mu \in \mathcal{P}([0,\infty))$. Then, μ is a $\mathbf{CE}_+$ distribution if and only if $\mu \in \mathbf{I}$ and*

$$\mathcal{L}\mu(s) = \exp\Big\{-\gamma' s + \int_0^\infty (e^{-sx}-1)n(x)dx\Big\}$$
$$= \exp\Big\{-\gamma' s + \int_0^\infty (\frac{1}{u+s} - \frac{1}{u})q(u)(du)\Big\}$$

where, $\gamma' \geq 0$, $q(u) = \sum_k 1_{[a_k,\infty)}(u)$ for $u > 0$ and $n(x) = (x^{-1}\sum_k e^{-a_k x})$ for $x > 0$ with $(0,\infty)$-valued finite or infinite sequence $\{a_k\}$ satisfying $\sum_k a_k^{-1} < \infty$.

Theorem 4.2. *Let $\mu \in \mathcal{P}(\mathbb{R})$. Then, μ is a $\mathbf{CE}$ distribution if and only if $\mu \in \mathbf{I}$ and there is an $\mathbb{R}\backslash\{0\}$-valued finite or infinite non-decreasing sequence $\{a_k\}$ such that*

$$\cdots \leq a_k \leq a_{k+1} \leq \cdots \leq a_{-1} < 0 < a_1 \leq \cdots \leq a_\ell \leq a_{\ell+1} \leq \cdots,$$
$$\sum a_k^{-2} < \infty$$

and the Lévy measure ν of μ is represented as

$$\nu(dx) = \begin{cases} (x^{-1}\sum_{k>0} e^{-a_k x})dx & \text{for } x > 0, \\ (|x|^{-1}\sum_{k<0} e^{-a_k x})dx & \text{for } x < 0. \end{cases}$$

5. Pólya frequency functions

Definition 5.1. A function $f(x)$ defined on $(-\infty,\infty)$ is said to be a $\mathbf{PF_r}$-function (a Pólya frequency function of order r) or an r-times positive-function if for $n = 2,3,\dots,r$ and for $x_1 < x_2 \cdots < x_n$, $y_1 < y_2 < \cdots < y_n$,

$$\det\Big(f(x_i-y_j)\Big) := \begin{vmatrix} f(x_1-y_1) & f(x_1-y_2) & \cdots & f(x_1-y_n) \\ f(x_2-y_1) & f(x_2-y_2) & \cdots & f(x_2-y_n) \\ \vdots & \vdots & \ddots & \vdots \\ f(x_n-y_1) & f(x_n-y_2) & \cdots & f(x_n-y_n) \end{vmatrix} \geq 0.$$

Definition 5.2. A function $f(x)$ is said to be a **PF** (or totally positive) function if it is r-times positive for any $r \geq 2$.

PF (or $\mathbf{PF}_r$) function is said to be a **PF** (or $\mathbf{PF}_r$) density if it is a probability density. The concept **PF**-function is considered by Pólya et al. and $\mathbf{PF_r}$-function is introduced by Schoenberg ([19]). **PF** and $\mathbf{PF_r}$ densities have sign diminishing property. This fact is used in hypothesis test of number of modes using kernel density estimate ([14]). For an increasing sequence $x_1 < x_2 < \cdots < x_n$, $v(x_1, \cdots, x_n)$ denotes the number of sign changes of the sequence. For a function f on $\mathbb{R}$, we define

$$v(f) = \sup v(f(x_1), \ldots, f(x_n))$$

for all n and all choices of the sequence $\{x_k\}_{k=1}^n$.

Theorem 5.1. *Let f be a* **PF** *density and let g be a probability density on $\mathbb{R}$. Let $h(x) = \int f(x-y)g(y)dy$. Then*

$$v(h) \leq v(g).$$

Theorem 5.2. *Let f be a* $\mathbf{PF_r}$ *density and let g be a probability density on $\mathbb{R}$. Let $h(x) = \int f(x-y)g(y)dy$. If $v(g) \leq r$, then*

$$v(h) \leq v(g).$$

Theorem 5.3. *A necessary and sufficient condition that a probability density on $\mathbb{R}$ be* **PF** *is that its Laplace transform is represented as*

$$\phi(s) = \exp\{\gamma s + \sigma^2 s^2/2\} \prod_{j=1}^{\infty} a_j(a_j + s)^{-1} e^{s/a_j}$$

where $\sigma^2 \geq 0$, $\gamma \in \mathbb{R}$, a_j are real and $0 < \sigma^2/2 + \sum_{j=1}^{\infty} a_j^{-2} < \infty$.

This theorem and Theorem 4.1 show that the class of **PF** densities coincides with the class of nondegenerate **CE** distributions. It is known that a density function is a $\mathbf{PF_2}$ density if and only if its logarithm is concave.

Example 5.1. Normal densities and exponential densities are **PF** densities. Rather a nontrivial example is $\frac{1}{\pi}\frac{1}{e^{x/2}+e^{-x/2}}$ ([11]). A gamma density with parameter (α, β), $\alpha \geq 1$, is a $\mathbf{PF_2}$ density.

Total positivity is related to Riemann hypothesis. Let $\zeta(s)$ be Riemann's zeta function. Let $\xi(s) = \frac{1}{2}s(s-1)\pi^{-\frac{1}{2}s}\Gamma(\frac{1}{2}s)\zeta(s)$ and let $\Xi(z) = \xi(\frac{1}{2} + iz)$. The function Ξ is written as

$$\Xi(z) = 2\int_0^{\infty} \Phi(u) \cos 2uz du,$$

where

$$\Phi(u) = 4\sum_{n=1}^{\infty}(2n^4\pi^2e^{9u} - 3n^2\pi e^{5u})e^{-n^2\pi e^{4u}},$$

(refer [22]). Riemann hypothesis is equivalent to that $\frac{\Xi(0)}{\Xi(z)}$ is the Laplace transform of a **PF** density (Thorin, private communication with Bondesson ([5] p.124), and [16] et al.).

6. Classes CME, CEME and CG

Definition 6.1. We say that a probability measure μ on $\mathbb{R}$ is a **CME** distribution if $\mu \in \mathbf{I}$ and its Lévy measure ν is absolutely continuous with density ℓ represented as

$$\ell(y) = \begin{cases} \int_{(0,\infty)} e^{-yu}Q(du) & for\ y > 0, \\ \int_{(-\infty,0)} e^{-yu}Q(du) & for\ y < 0, \end{cases}$$

where Q is a measure on $\mathbb{R} \setminus \{0\}$ satisfying

$$\int_{\mathbb{R}\setminus\{0\}} |u|^{-1} \wedge |u|^{-3}Q(dy) < \infty. \tag{6}$$

We denoted this class by **B** in [29]. This class on $[0, \infty)$ is called *gcmed* (= generalized convolutions of mixtures of exponential distributions) in [5] et al. The class **CME** (= Convolutions of Mixtures of Exponential distributions) is closed under convolution and weak convergence ([29]). We define **CEME** by the class of convolutions of **CE** distributions and **ME** distributions. We denoted this class by **CME** in [29]. It seems that the name **CME** is more suitable to the above class in Definition 6.1. We say that a distribution μ belongs to class **CG** (Convolutions of gamma distributions) if it belongs to the class **CME** and the Q-measure Q is absolutely continuous and the density is nonincreasing on $(-\infty, 0)$ and nondecreasing on $(0, \infty)$. This class is called *ggc* in [5] et al. The class **CG** coincides with the weak closure of finite convolutions of gamma distributions ([5]). By the nondecreasingness of the density of the Q-measure of **CG** distribution, we see that every **CG** distribution is selfdecomposable. It is easy to see the inclusions.

$$\begin{array}{ccccccc} & & \mathbf{L} & & & & \\ & & \cup & & & & \\ \mathbf{CE} & \subset & \mathbf{CG} & \subset & \mathbf{CEME} & \subset \mathbf{CME} \subset & \mathbf{I}\,. \\ & & & & \cup & & \\ & & & & \mathbf{ME} & & \end{array}$$

Example 6.1. Every 1-dimensional stable distribution belongs to **CG**. Lévy measure of an α-stable ($0 < \alpha < 2$) distribution is written as follows:

$$\ell(x) = \begin{cases} c_+ x^{-\alpha-1} & = c_+ \Gamma(\alpha+1)^{-1} \int_0^\infty u^\alpha e^{-xu} du \quad for\ x > 0, \\ c_- |x|^{-\alpha-1} & = c_- \Gamma(\alpha+1)^{-1} \int_{-\infty}^0 |u|^\alpha e^{-xu} du \quad for\ x < 0 \end{cases}$$

where $c_+, c_- \geq 0$ and $c_+ + c_- > 0$.

The **CG** distribution on $[0, \infty)$ (= $\mathbf{CG}_+$ distribution) has a remarkable representation resemble to (3).

Theorem 6.1. ([5] *p.49*) *Let $\mu \in \mathbf{CG}_+$ with Laplace transform*

$$\mathcal{L}\mu(s) = \exp[\int_0^\infty (\frac{1}{u+s} - \frac{1}{u}) q(u) du]$$

where $q \geq 0$ is nondecreasing, $0 < \lim_{u \to \infty} q(u) = \alpha < \infty$ and satisfies $\int_0^1 u^{-1} q(u) du < \infty$, then the density of μ is represented as

$$f(x) = x^{\alpha-1} h(x)$$

where h is completely monotone.

Proof. For the proof of this theorem, the following fact is essential: Let X and Y be random variables with gamma distributions with parameters (α_1, β) and (α_2, β), then $(\frac{X}{X+Y}, \frac{Y}{X+Y})$ and $X + Y$ are independent. If a **CG**-distributed random variable X is a sum of n independent gamma distributed random variables, then it can be represented as $X = c_1 X_1 + c_2 X_2 + \cdots + c_n X_n$, where $X_1, X_2, \ldots, X_n$ are independent and scale parameter 1 gamma distributed random variables and $c_1, c_2, \ldots, c_n > 0$. Let $Y = \sum_{k=1}^n X_k$. Then Y is $(\alpha, 1)$-gamma distributed and $Y, X_1/Y, X_2/Y, \ldots, X_n/Y$ are independent. Hence, $Y, c_1 X_1/Y, c_2 X_2/Y, \ldots, c_n X_n/Y$ are also independent. This shows that X/Y and Y are independent. Hence X is a mixture of gamma distributions with shape parameter α. Taking a limit, we obtain the theorem. □

For **CME** on $[0, \infty)$ (= $\mathbf{CME}_+$) distribution, we have a similar but slightly weaker result.

Lemma 6.1. *Let μ_1 and μ_2 be mixtures of gamma distribution with shape parameters α and β, respectively. Then the convolution $\mu_1 * \mu_2$ is a mixture of gamma distributions with shape parameter $\alpha + \beta$.*

Proof. It is enough to notice that, for $x, \alpha, \beta, \gamma, \delta > 0$,

$$\int_0^x e^{-\gamma x}(x-y)^{\alpha-1}y^{-\beta-1}e^{-(\delta-\gamma)y}dy$$
$$= x^{\alpha+\beta-1}\int_0^1 u^{\beta-1}(1-u)^{\alpha-1}e^{-(\delta u+(1-u)\gamma)x}du.$$ □

Theorem 6.2. ([5]) *Let* $\mu \in \mathbf{CME}_+$ *with Laplace transform*

$$\mathcal{L}\mu(s) = \exp[\int_0^\infty (\frac{1}{u+s} - \frac{1}{u})q(u)du]$$

where $q \geq 0$, $0 < \operatorname{ess\,sup}_{u>0} q(u) = \alpha < \infty$ *and satisfies* $\int_0^1 u^{-1}q(u)du < \infty$, *then the density* f *of* μ *is represented as*

$$f(x) = x^{n-1}\int_0^\infty \lambda e^{-\lambda x}G(d\lambda) \tag{7}$$

where n *is the integer satisfying* $n-1 < \alpha \leq n$ *and* G *is a measure on* $(0,\infty)$ *satisfying*

$$\int_0^\infty \lambda^{1-n}G(\lambda) \leq \Gamma(n)^{-1}. \tag{8}$$

Moreover,

$$\int_0^\infty \lambda e^{-\lambda x}G(d\lambda) \leq x^{-n}n^n e^{-n}\Gamma(n)^{-1}. \tag{9}$$

Proof. The probability measure μ is represented as a convolution of n $\mathbf{ME}_+$ distributions. Hence by Lemma 6.1, we have (7). Since

$$1 \geq \int_0^\infty f(x)dx$$
$$= \int_0^\infty \left(\int_0^\infty x^{n-1}e^{-\lambda x}dx\right)\lambda G(d\lambda)$$
$$= \Gamma(n)\int_0^\infty \lambda^{1-n}G(d\lambda),$$

we have (8). Since $\lambda^n e^{-\lambda x} \leq x^{-n}n^n e^{-n}$ for $x > 0$ and $\lambda > 0$, we have (9) by (8). □

In the above theorem, n can not be replaced by α. A counter example is seen in [5]. A fact similar to Theorem 3.3 holds for $\mathbf{CME}_+$ distributions as follows.

Theorem 6.3. *Let* $\mu \in \mathbf{CME}_+$ *with Laplace transform*

$$\mathcal{L}\mu(s) = \exp[\int_0^\infty (\frac{1}{u+s} - \frac{1}{u})q(u)du]$$

where $0 \leq q(u) \leq \alpha$ *for* $\alpha \geq 1$ *and for all large* $u > 0$, $\lim_{u\to\infty} q(u) = \alpha$, *and satisfies* $\int_0^1 u^{-1}q(u)du < \infty$. *Let* f *be the density of* μ. *Then* $x^{1-\alpha}f(x)$ *is bounded if and only if*

$$\int_1^\infty \frac{1}{u}(\alpha - q(u))du < \infty.$$

If $q(u) \leq \alpha$ *u-a.e., then*

$$\lim_{x\downarrow 0} x^{1-\alpha}f(x) = \Gamma(\alpha)^{-1}\exp[\int_0^\infty \frac{\alpha u - (u+1)q(u)}{u(u+1)}du]. \tag{10}$$

Especially, if $\mu \in \mathbf{GC}_+$, *then the right-hand side of the above equality is represented as*

$$\Gamma(\alpha)^{-1}\exp[\int_0^\infty (\log u)q(du)]. \tag{11}$$

Proof. Let $q_1(u) = q(u)\wedge\alpha$ and let μ_1 be a **CME** distribution with Q-measure $q_1(u)du$. Since $\alpha > 0$, μ_1 has a density g. The quantity $s^\alpha \mathcal{L}\mu_1(s)$ is written as

$$\begin{aligned} s^\alpha \mathcal{L}\mu_1(s) &= \exp\left[\int_0^\infty (\frac{1}{u+s} - \frac{1}{u})q_1(u)du + \int_1^s \frac{\alpha}{u}du\right] \\ &= -\int_0^1 \frac{s}{u(u+s)}q_1(u)du + \int_1^s \frac{1}{u}(\alpha - q_1(u))du \\ &\quad + \int_1^s \frac{1}{u+s}q_1(u)du - \int_s^\infty \frac{s}{u(u+s)}q_1(u)du. \end{aligned}$$

The first and the second terms tend to

$$-\int_0^1 \frac{1}{u}q_1(u)du > -\infty \text{ and } \int_1^\infty \frac{1}{u}(\alpha - q_1(u))du \leq \infty,$$

respectively, by monotone convergence theorem. The third term minus fourth term is written as

$$\int_0^1 \frac{1}{u+1}q_1(su)du - \int_1^\infty \frac{1}{u(u+1)}q_1(su)du,$$

which converges to

$$\int_0^1 \frac{\alpha}{u+1}du - \int_1^\infty \frac{\alpha}{u(u+1)}du,$$

as $s \to \infty$ by the dominated convergence theorem. By Theorem 6.2, $g(x) = O(x^{-1})$ as $x \to \infty$. Hence $x^{1-\alpha}g(x)$ is bounded for large x. Note that $\lim_{s\to\infty} s^{\alpha}\mathcal{L}\mu_1(s)$ exists allowing infinity. Hence, $\int_1^\infty \frac{1}{u}(\alpha - q_1(u))du < \infty$ if and only if

$$\lim_{s\to\infty} s^{\alpha}\mathcal{L}\mu_1(s) = \exp\{\int_0^\infty \frac{\alpha u - (u+1)q_1(u)}{u(u+1)}du\} < \infty.$$

Assume that $\int_1^\infty \frac{1}{u}(\alpha - q(u))du < \infty$, equivalently, $\int_1^\infty \frac{1}{u}(\alpha - q_1(u))du < \infty$. By Tauberian theorem, we have $\mu_1([0,x]) \sim \frac{C}{\alpha}x^{\alpha}$ as $x \downarrow 0$ where C is the right hand side of (10). We have, by L'hospital's rule, $g(x) \sim Cx^{\alpha-1}$ as $x \downarrow 0$. This shows (10). We also have that $x^{1-\alpha}g(x)$ is bounded. The **CME** distribution μ_2 with Q-measure $q(u) - q_1(u)$ is written as $p\delta_0(dx) + (1-p)h(x)x$. Since $\alpha \geq 1$, we have

$$\begin{aligned} x^{1-\alpha}f(x) &= px^{1-\alpha}g(x) + (1-p)\int_0^x x^{1-\alpha}g(x-y)h(y)dy \\ &\leq px^{1-\alpha}g(x) + (1-p)\int_0^x (x-y)^{1-\alpha}g(x-y)h(y)dy \\ &\leq \sup_x x^{1-\alpha}g(x) < \infty. \end{aligned}$$

Conversely, assume that $x^{1-\alpha}f(x)$ is bounded. Since μ_2 has a point mass at the origin, $x^{1-\alpha}g(x)$ is bounded near the origin. Hence $x^{1-\alpha}g(x)$ is bounded. Then $s^{\alpha}\mathcal{L}\mu_1(s)$ is bounded. Hence we have $\int_1^\infty \frac{1}{u}(\alpha - q_1(u))du < \infty$, equivalently, $\int_1^\infty \frac{1}{u}(\alpha - q(u))du < \infty$.

(11) is straightforward. □

7. Unimodality

Definition 7.1. A probability measure μ on $\mathbb{R}$ is said to be unimodal if there is $a \in \mathbb{R}$ such that the distribution function of μ is convex on $(-\infty, a)$ and concave on (a, ∞).

Definition 7.2. A probability measure μ on $\mathbb{R}$ is said to be strongly unimodal if for every unimodal distribution ρ, the convolution $\mu * \rho$ is again unimodal.

Theorem 7.1. ([10]) *A probability measure μ on $\mathbb{R}$ is strongly unimodal if and only if it is absolutely continuous with logarithmic concave density* (**PF**$_2$*- density*).

Theorem 7.2. *Every* **CEME** *distribution is unimodal.*

Proof. **ME**-distributions are unimodal with mode 0. **CE**-distributions are strongly unimodal. Hence **CEME**-distributions are unimodal. □

It is easy to see that every stable distribution is selfdecomposable. Every selfdecomposable distribution is unimodal ([25]). Hence every stable distribution is unimodal. The proof of unimodality of selfdecomposable distributions is not simple. But, since we know that stable distributions are **CG**-distributions, the proof of unimodality of stable distributions is quite simple as follows.

Theorem 7.3. ([26]) *Every 1-dimensional stable distribution is unimodal.*

Proof. 1-dimensional stable distributions belongs to **CG** and **CG** is a subclass of **CEME**. Hence 1-dimensional stable distributions are unimodal by Theorem 7.2. □

8. Boundedness of transition densities of $\mathbf{CME}_+$ processes

Let $\{X(t)\}$ be a Lévy process. If the distribution of $X(1)$ is a $\mathbf{CME}_+$ distribution, then we call $\{X(t)\}$ a $\mathbf{CME}_+$ process. Assume that, in this section, the distribution of $X(1)$ is absolutely continuous. Namely, $\int_0^\infty \frac{1}{u} Q(du) = \infty$ for the Q-measure of the distribution of $X(1)$. We consider in this section, under what condition the transition density of $\{X(t)\}$ is bounded in the space variable x and the supremum in x tends to 0 as time variable t tends to ∞. The following result is shown in [24].

Theorem 8.1. *Let $\{X(t)\}$ be a $\mathbf{CME}_+$ process with transition density $p(t,x)$. Then for any $0 < t_0 < t_1$ and $x_0 > 0$,*

$$\sup_{t \in [t_0,t_1], x \in [x_0,\infty)} p(t,x) < \infty.$$

1. Let $0 < a < b$ and let

$$q(x) = \begin{cases} 1 & \text{for } x \in [a,b], \\ 0 & \text{for } x \in [a,b]^c. \end{cases}$$

Let $\mu \in \mathbf{CME}_+$ with Q-measure $q(u)du$. Then we have $\mu([0,x]) = p + q(1 - e^{-ax})$ for $x > 0$ where $p = \frac{a}{b}$ and $q = 1 - p$. Hence

$$\mu^{n*}(dx) = p^n \delta_0(dx) + p_n(x)dx$$

where

$$p_n(x) = \sum_{k=1}^{n} \binom{n}{k} p^{n-k} q^k \frac{a^k x^{k-1} e^{-ax}}{(k-1)!}.$$

Let $f_k(x) = \frac{a^k x^{k-1} e^{-ax}}{(k-1)!}$. Then, by Stirling formula $\Gamma(z) \sim \sqrt{2\pi} z^{z-1/2} e^{-z}, (z \to \infty)$,

$$\sup_x f_k(x) = \frac{a(k-1)^{k-1} e^{-(k-1)}}{(k-1)!} = O(k^{-1/2}) \text{ as } k \to \infty.$$

The quantity $\sup_x p_n(x)$ can be regarded as an integral of $\sup_x f_k(x)$ with respect to a binomial distribution $\{\binom{n}{k} p^{n-k} q^k\}_{k=0}^n$. The binomial distribution tends to $\delta_\infty(dx)$ as $n \to \infty$. Hence

$$\lim_{n\to\infty} \sup_x p_n(x) = 0.$$

2. Let $Q(dx) = \delta_a(dx)$ $(a > 0)$ and let μ be a **CME**$_+$- distribution corresponding to Q. Then

$$\mu^{t*}(dx) = e^{-t/a}\delta_0(dx) + p_t(x)dx$$

where

$$p_t(x) = \sum_{n=1}^{\infty} e^{-t/a} \frac{(t/a)^n}{n!} \frac{a^n x^{n-1} e^{-ax}}{(n-1)!}.$$

In this case, we also have

$$\sup_x p_t(x) \to 0 \text{ as } t \to \infty$$

Theorem 8.2. *Let* $\mu \in$ **CME**$_+$. *Let*

$$q_1(x) = \begin{cases} 1 & \text{for } x \in [a,b], \\ 0 & \text{for } x \in [a,b]^c \quad a.s. \end{cases}$$

$0 \le q_2(x) \le 1$ *a.s.*, $\int_1^\infty \frac{1}{x} q_2(x)dx = \infty$ *and* $\int_1^\infty \frac{1}{x}(1 - q_2(x))dx < \infty$ *where* $0 < a < b$. *If the Q-measure Q of* μ *is given by* $Q(dx) = (q_1(x) + q_2(x))dx + Q_3(dx)$, *then*

$$\lim_{t\to\infty} \sup_{0\le x<\infty} \frac{d\mu^{t*}}{dx}(x) = 0.$$

Proof. Let $1 \le n \le t < n+1$. Then Q-measure of μ^{t*} is writen as

$$nq_1(x)dx + q_2(x)dx + Q_4(dx)$$

where

$$Q_4(dx) = (t-n)q_1(x)dx + (t-1)q_2(x)dx + tQ_3(dx).$$

Probability measure corresponding to q_2 is absolutely continuous and the density h is bounded by Theorem3.3. Hence the probability measure corresponding to $q_2(x)dx + Q_4(dx)$ has a density $g(t,x)$ and it satisfies that

$$g(t,x) = \int h(x-y)\mu_2(dy) \le \sup_x h(x) < \infty$$

where μ_2 is a probability measure corresponding to $Q_4(dx)$. Let μ_1^{n*} be a probability measure corresponding to $nq_1(x)dx$. Then for $p = \frac{a}{b}$,

$$\mu_1^{n*}(dx) = p^n \delta_0(dx) + p_n(x)dx,$$

and

$$\sup_x p_n(x) \to 0 \text{ as } n \to \infty,$$

where $p_n(x) = \sum_{k=1}^n \binom{n}{k} p^{n-k} q^k \frac{a^k x^{k-1} e^{-ax}}{(k-1)!}$. Density $f(t,x)$ of μ^{t*} satisfies

$$\begin{aligned} f(t,x) &\le p^n g(t,x) + \sup_x p_n(x) \\ &\le p^n \sup_x h(x) + \sup_x p_n(x) \\ &\to 0 \quad \text{as } n \to \infty. \end{aligned}$$

□

Theorem 8.3. *Let* $\mu \in \mathbf{CME}_+$. *Let* $0 \le q_2(x) \le 1$ *a.s.,*

$$\int_1^\infty \frac{1}{x} q_2(x)dx = \infty \text{ and } \int_1^\infty \frac{1}{x}(1 - q_2(x))dx < \infty.$$

If the Q-measure of μ *is given by* $Q(dx) = \delta_a(dx) + q_2(x)dx + Q_3(dx)$ *where* $a > 0$, *then*

$$\lim_{t\to\infty} \sup_{0 \le x < \infty} \frac{d\mu^{t*}}{dx}(x) = 0.$$

Proof. The proof is the same as the proof of above theorem using 2 instead of 1. □

Remark 8.1. If q_2 does not satisfy $\int_1^\infty \frac{1}{u}(1 - q_2(u))du < \infty$, then there is an example for which the density of $\mu^{t*}(dx)$ is not bounded for all $t > 0$. The following Example 8.1 exhibits such an example.

Example 8.1. Let $\mu \in \mathbf{ME}_+$ with G-measure G. Assume that $G(\{\infty\}) = 0$ and $\int_0^x uG(du) \sim x(\log x)^{-2}$ as $x \to \infty$. Then μ is absolutely continuous with respect to Lebesgue measure and μ has a density

$$f(x) := \int_0^\infty ue^{-ux}G(du) \sim x^{-1}(\log \frac{1}{x})^{-2} \text{ as } x \downarrow 0.$$

Hence $\int_1^\infty \frac{1}{u}(1 - q(u))du < \infty$ by Theorem 3.3. Integration by parts yields that

$$F(x) := \int_0^x f(u)du \sim (-\log x)^{-1} \text{ as } x \downarrow 0.$$

By Abelian-Tauberian theorem for Laplace transform, this asymptotic relation is equivalent to that

$$\int e^{-sx}F(dx) \sim (-\log \frac{1}{s})^{-1} \; as \; s \to \infty.$$

Then,

$$\left(\int_0^\infty e^{-sx}F(dx)\right)^t \sim (\log s)^{-t} \; as \; s \to \infty. \tag{12}$$

This is equivalent to that

$$F^{t*}(x) \sim (-\log x)^{-t} \; as \; x \downarrow 0, \tag{13}$$

by Abelian-Tauberian theorem. Hence $f^{t*}(x)$ is unbounded in $x > 0$ for each $t > 0$. If $f^{t*}(x)$ is monotone for all small x, then $f^{t*}(x) \sim \frac{1}{x}t(-\log x)^{-t-1}$ as $x \downarrow 0$. Let

$$\exp[\int_0^\infty (\frac{1}{u+s} - \frac{1}{u})q(u)du]$$

be the Laplace transform of μ. Then

$$ts \int_0^\infty \frac{1}{u+s}\frac{q(u)}{u}du \sim t \log\log s \; as \; s \to \infty$$

by (12). By Abelian-Tauberian theorem for Stieltjes transform ([4]), this asymptotic relation is equivalent to that

$$t\int_0^x \frac{q(u)}{u}du \sim t\log\log x \; as \; x \to \infty.$$

If $\frac{q(u)}{u}$ is monotone for all large u, then $q(u) \sim \frac{1}{\log u}$ as $u \to \infty$. Conversely, assume that

$$\int_0^x \frac{q(u)}{u}du \sim \log\log x \; as \; x \to \infty.$$

Then we have (13) by Abelian-Tauberian theorems for Stieltjes transform and Laplace transform.

9. 1-dimensional generalized diffusion processes

Let $\{B(t)\}$ be a one-dimensional Brownian motion and let $\ell(t,x)$ be its local time. We denote by $\mathcal{M}$ the class of right continuous nondecreasing function m on $[-\infty,\infty]$ to $[-\infty,\infty]$ with $m(\pm\infty) = \pm\infty$ and $m(0-) = 0$. For $m \in \mathcal{M}$, we define $\ell_j = \ell_j(m)$ by

$$(-1)^j\ell_j = \inf\{(-1)^j x > 0 : (-1)^j m(x) = \infty\}$$

for $j = 1, 2$ and we define a measure $m(dx)$ on $[-\infty, \infty]$ by

$$\begin{aligned} m(dx) \quad &= dm(x) \; on \quad (\ell_1, \ell_2), \\ m([\ell_1, \ell_2]^c) &= 0, \\ m(\{\ell_j\}) \quad &= \infty \qquad for \quad j = 1, 2. \end{aligned}$$

Here $[\ell_1, \ell_2]^c$ is the complement of $[\ell_1, \ell_2]$. Let

$$\phi(t) = \int_{\mathbb{R}} \ell(t, x) m(dx).$$

Define a stochastic process $\{X(t), \zeta\}$ by $X(t) = B(\phi^{-1}(t))$ and the life time

$$\begin{aligned} \zeta &= \inf\{t > 0 : X(t) = \ell_1 \; or \; \ell_2\} \; if \quad \{\,\} \neq \emptyset, \\ &= \infty \qquad\qquad\qquad\qquad\qquad otherwise. \end{aligned}$$

This process is a strong Markov process with state space $E_m = (\text{supp}\, m)|_{(\ell_1, \ell_2)}$ and is called the generalized diffusion process corresponding to the function m (see [12]). The measure restricted to (ℓ_1, ℓ_2) is called the speed measure of the process $\{X(t)\}$. For $y \in E_m$, we define the hitting time of y by

$$\begin{aligned} \tau_y &= \inf\{t > 0 : X(t) = y\} \; if \quad \{\,\} \neq \emptyset, \\ &= \infty \qquad\qquad\qquad\qquad otherwise. \end{aligned}$$

If $|\ell_j| < \infty$ and $\ell_j \in \bar{E}_m$, where $\bar{E}_m$ is the closure of E_m in $\mathbb{R}$, then we define τ_{ℓ_j} by y by $\lim_{y \to \tau_{\ell_j}} \tau_y$ for $j = 1, 2$, respectively. We denote by $\tilde{E}_m$ the set with $\ell_j (j = 1, 2)$ adjoined to E_m whenever $|\ell_j| < \infty$ and $\ell_j \in \bar{E}_m$. If $P_x(\tau_y < \infty) > 0$ for x in E_m and y in $\tilde{E}_m$, we define

$$\mu_{xy}(dt) = P_x(\tau_y \in dt | \tau_y < \infty).$$

We denote by

$$\mathbf{H}_{gd} = \{\mu_{xy}(dy) : x \in E_m, y \in \tilde{E}_m, x \neq y, m \in \mathcal{M}\},$$

the class of conditionnal hitting time distributions of generalized diffusion processes.

In [27] (Theorem 1), the following characterization is obtained.

Proposition 9.1. *In order that a probability measure μ on $\mathbb{R}_+$ belongs to $\mathbf{H}_{gd}$, it is necessary and sufficient that there are a* $\mathbf{CE}_+$ *distribution μ_1 with Laplace transform $\mathcal{L}\mu_1(s) = \prod \frac{a_i}{a_i + s}$ and an* $\mathbf{ME}_+$ *distribution μ_2 with $\mu_2(\{0\}) = 0$ such that*

$$\mu = \mu_1 * \mu_2$$

and $\{a_i\}$ is either empty or a strictly increasing (finite or infinite) sequence and the spectral measure σ of $(s\mathcal{L}\mu_2(s))^{-1}$ has a positive point mass at a_i for each i.

This proposition shows that $\mathbf{H}_{gd}$ is a proper subclass of $\mathbf{CEME}_+$. We can restate this result in terms of Q-measure using Theorem 3.5 in Section 3.

Theorem 9.1. ([28]) *In order that a probability measure μ on $\mathbb{R}_+$ belongs to $\mathbf{H}_{gd}$, it is necessary and sufficient that its Laplace transform is represented as*

$$\mathcal{L}\mu(s) = \exp \int_0^\infty (\frac{1}{u+s} - \frac{1}{u})(q_1(u) + q_2(u))du,$$

where q_1 and q_2 satisfy the following conditions:
1. (a) $q_1 = 0$ or (b) q_1 is a non-decreasing step function with step size 1, $q_1(0) = 0$ and jump points $\{a_j\}$ of q_1 satisfy $\sum_j a_j^{-1} < \infty$ (hence, 0 is not a jump point).
2. $0 \le q_2(u) \le 1$, $\int_0^1 \frac{1}{u} q_2(u)du < \infty$, $\int_1^\infty \frac{1}{u} q_2(u)du = \infty$.
3. In case (b) in 1, q_2 satisfies

$$\int_{a_j-\epsilon_j}^{a_j+\epsilon_j} \frac{1}{u-a_j}(q_2(u) - 1_{(0,a_j)}(u))du < \infty$$

for $\{\epsilon_j\}$ with $a_j > \epsilon_j > 0$.

Proof. Apply Theorem 3.5 to the reciprocal of the expression of $s\mathcal{L}\mu(s)$ in Lemma 3.1. □

This theorem shows that every stable distribution μ with Laplace transform

$$\mathcal{L}\mu(s) = \exp \int_0^\infty (\frac{1}{u+s} - \frac{1}{u})cu^\alpha du$$

belongs to $\mathbf{H}_{gd}$. Here, $c > 0$ and $0 < \alpha < 1$. $\frac{1}{2}$-stable ($\alpha = \frac{1}{2}$) distribution is the hitting time distribution of Brownian motion. It is not known what kind of generalized diffusions correspond to other stable distributions. Gamma distribution does not belong to $\mathbf{H}_{gd}$ if and only if the shape parameter α is greater than 1.

10. Lévy processes appearing in mathematical finance

Recently, various types of Lévy processes, namely VG (Variance Gamma), NIG (Normal Inverse Gaussian), GIG (Generalized Inverse Gaussian), GH (Generalized Hyperbolic) processes often appear in Mathematical Finance literature as a model of stock price. Also, stationary processes of Ornstein Uhlenbeck type are used as volatility processes in stochastic volatility models (refer [3] and references therein). The class of stationary distributions of the statinary one-dimensional processes of Ornstein Uhlenbeck type coincides with the class of selfdecomposable distributions on

$\mathbb{R}$ ([18]). We show that the above classes (VG, NIG, GIG, GH) of processes (or distributions) belongs to the class **CG** or the class of selfdecomposable distributions.

Let $p(x;\rho,\delta)$ be the density of the positive $\rho/2$-stable distribution with Laplace transform

$$\int_0^\infty e^{-sx}p(x;\rho,\delta)dx = e^{-\delta(2s)^{\rho/2}}.$$

Boyarchenko and Levendorskii ([7]) called a probability measure on $[0,\infty)$ with density

$$p(x;\rho,\delta,\gamma) = e^{\delta\gamma^\rho}p(x;\rho,\delta)e^{-\frac{1}{2}\gamma^2 x},\ \gamma > 0$$

a Tempered Stable distribution and denoted $TS(\rho,\delta,\gamma)$ ([7]). Its Laplace transform is written as $e^{\Phi(s)}$, where

$$\Phi(s) = \delta[\gamma^\rho - (\gamma^2 + 2s)^{\rho/2}]$$

and the Lévy density is given by

$$\ell(x) = \frac{\delta\rho 2^{\rho/2-1}}{\Gamma(1-\rho/2)}x^{-\rho/2-1}e^{-\frac{1}{2}\gamma^2 x}.$$

They ([6]) called an infinitely divisible distribution μ on $\mathbb{R}$ a KoBoL distribution of order $\rho < 2$ if it is infinitely divisible with the Lévy measure

$$\nu(dx) = |x|^{-\rho-1}\{c_- e^{\lambda_- x}1_{\{x<0\}}(x) + c_+ e^{-\lambda_+ x}1_{\{x>0\}}(x)\}dx$$

where $c_\pm > 0$ and $\lambda_\pm > 0$. This shows that the Tempered Stable distribution is a one sided KoBoL distribution of order $\rho/2$. KoBoL distribution is a **CG** distribution with the density of its Q-measure

$$q(u) = \{\Gamma(\rho+1)\}^{-1}\Big\{c_-|u+\lambda_-|^\rho 1_{\{u<-\lambda_-\}}(u) + c_+(u-\lambda_+)^\rho 1_{\{u>\lambda_+\}}(u)\Big\}.$$

KoBoL distribution with $\rho = 0$ is called VG distribution. The use of VG distribution in finance is proposed by Madan and Seneta [13].

In [15], KoBoL distribution is called tilted stable distribution and the name "tempered stable" is used for an infinitely divisible distribution with Lévy measure

$$|x|^{-\alpha-1}\Big(1_{(-\infty,0)}(x)\int_{-\infty}^0 e^{-xu}q(u)du + 1_{(0,\infty)}(x)\int_0^\infty e^{-xu}q(u)du\Big)dx,$$

where $\alpha \in (0,2)$ and the measure $Q(du) := q(u)du$ on $\mathbb{R}\backslash\{0\}$ satisfies (6). The meaning of "tilted" is explained in [3].

We denote by K_λ the modified Bessel function of the third kind with index λ. Let $Y_t = \beta t + B_t$ where $\{B_t\}$ is a Brownian motion. Let $\{Z_t\}$ be a subordinator generated by $TS(\rho, \delta, \gamma)$-distribution independent of $\{Y_t\}$. Then the characteristic function of the subordination $X_t = Y_{Z_t}$ is of the form $e^{t\phi(z)}$, where $\phi(z) = \Phi(\frac{1}{2}z^2 - i\beta z)$. It is rewritten as

$$\phi(z) = \delta[(\alpha^2 - \beta^2)^{\rho/2} - (\alpha^2 - (\beta + iz)^2)^{\rho/2}] \tag{14}$$

where $\alpha = \sqrt{\gamma^2 + \beta^2}$. Since the 1-dimensional distributions of a subordinated process of a Brownian motion with drift by a selfdecomposable subordinator is selfdecomposable ([17]), the distribution of X_t is selfdecomposable for each $t > 0$. Adding $i\mu z$ to (14) and then letting $\rho = 1$, we get the NIG distribution with characteristic function

$$\exp\Big(i\mu z + \delta[(\alpha^2 - \beta^2)^{1/2} - (\alpha^2 - (\beta + iz)^2)^{1/2}]\Big).$$

The transition density is given by

$$p(t, x) = \frac{\alpha}{\pi} \exp[t(\delta\sqrt{\alpha^2 - \beta^2} - \beta\mu) + \beta x] \frac{K_1(\alpha\delta\langle(x/t - \mu)/\delta\rangle)}{\langle(x/t - \mu)/\delta\rangle}$$

where $\langle y \rangle = (1 + |y|^2)^{1/2}$.

A probability measure on $(0, \infty)$ is said to be a GIG distribution if it has a density

$$\frac{(\gamma/\delta)^\lambda}{2K_\lambda(\gamma\delta)} x^{\lambda-1} \exp\Big(-\frac{1}{2}(\delta^2/x + \gamma^2 x)\Big)$$

where the parameters satisfy

$$\begin{aligned} &\delta > 0, \gamma \geq 0 \text{ for } \lambda < 0, \\ &\delta > 0, \gamma > 0 \text{ for } \lambda = 0, \\ &\delta \geq 0, \gamma > 0 \text{ for } \lambda > 0. \end{aligned}$$

Halgreen [9] and, Shanbhag and Sreehari [20] showed that GIG distribution is a **CG** distribution.

A probability measure with characteristic function

$$e^{i\mu z}\Big(\frac{\alpha^2 - \beta^2}{\alpha^2 - (\beta + iz)^2}\Big)^{\lambda/2} \frac{K_\lambda(\delta\sqrt{\alpha^2 - (\beta + iz)^2})}{K_\lambda(\delta\sqrt{alpha^2 - \beta^2})}$$

is called GH distribution. It is absolutely continuous with respect to Lebesgue measure and the density is represented by $K_{\lambda-1/2}$. GH distributions are obtained by the subordination of Brownian motion with drift by GIG subordinator. They ([9], [20]) proved that GH distributions are selfdecomposable by showing that the one dimensional distributions of a subordinated process of Brownian motion with drift by **CG** subordinator are selfdecomposable. Sato's result ([17]) is its extension. We remark that NIG distribution is a GH distribution with $\lambda = -\frac{1}{2}$.

References

1. N. Aronszajn and W. F. Donoghue Jr., On exponential representations of analytic functions in the upper half-plane with positive imaginary part, J. Analyse Math., 5 (1956), 321-388.
2. N. Aronszajn and W. F. Donoghue Jr., A supplement to the paper on exponential representations of analytic functions in the upper half-plane with positive imaginary part, J. Analyse Math., 12 (1964), 113-127.
3. O. E. Barndorff-Nielsen and N. Shephard, Modelling by Lévy processes for financial econometrics, in Lévy Processes Theory and Applications, Birkhäuser (2001), 283-318.
4. N. H. Bingham, C. M. Goldie and J. L. Teugels, "*Regular Variation*", Cambridge University Press (1987), Cambridge.
5. L. Bondesson, Generalized gamma convolutions and related classes of distributions and densities, Lecture Notes in Statistics, 76 (1992) Springer-Verlag, New York.
6. S. Boyarchenko and S. Levendorskii, Perpetual American options under Lévy processes, SIAM J. Cotrol Optim. 40 (2002), 1663-1696.
7. S. Boyarchenko and S. Levendorskii, "*Non-Gaussian Merton-Black-Scholes Theory*", Advanced series of statistical science & applied probability Vol. 9, World Scientific (2002), New Jersey-London-Singapore-Hong Kong.
8. W. F. Donoghue, Jr., "*Monotone matrix functions and analytic continuation*", Springer 1974, Berlin Heidelberg New York.
9. C. Halgreen, Self-decomposability of the generalized inverse Gaussian and hyperbolic distributions, Z. Wahrsch. Verw. Gebiete 47 (1979) 13-17.
10. I. A. Ibragimov, On the composition of unimodal distributions, Theor. Probability Appl. 1 (1956) 255-260.
11. S. Karlin, "*Total Positivity*", Vol. 1, Stanford Univ. (1968), Stanford.
12. S. Kotani and S. Watanabe, Krein's spectral theory of strings and generalized diffusion processes, Functional Analysis in Markov Processes (M. Fukushima, ed.), Lecture Notes in Mathematics, 923 (1982), 235-259, Springer, Berlin Heidelberg New York.
13. D. B. Madan and E. Seneta, The VG model for share market returns, J. Business 63 (1990), 511-524.
14. E. Mammen, J. S. Marron and N. I. Fisher, Some asymptotics for multimodality tests based on kernel density estimates. Probab. Theory Relat. Fields 91 (1992), 115-132.
15. J. Rosínski, Tempered stable processes, Mini-proceedings : 2nd MaPhysto Conference on Lévy Processes Theory and Applications (2002), Aarhus University.
16. B. Roynette et M. Yor, Couples de Wald indéfiniment divisibles Examples liés

à la fonction gamma d'Euler et à la fonction zeta de Riemann, To appear in Ann. Inst. Fourier.

17. K. Sato, Subordination and selfdecomposability, Statistics & Probability Letters 54 (2001) 317-324.
18. K. Sato and M. Yamazato, Stationary processes of Ornstein-Uhlenbeck type, Lecture Notes in Math., 1021 Springer (1983) 541-551.
19. I. J. Schoenberg, On Pólya frequency functions I. The totally positive functions and their Laplace transforms, Journal d'Analyse Math., 1 (1951), 331-374.
20. D. N. Shanbhag and M. Sreehari, An extension of Goldie's result and further results in infinite divisibility, Z. Wahrsch. Verw. Gebiete 47 (1979) 19-25.
21. F. W. Steutel, *"Preservation of infinite divisibility under mixing and related topics"*, Mathematical Center Tracts 33 (1970), Matematisch Centrum, Amsterdam.
22. E. C. Titchmarsh, *"The zeta-function of Riemann"*, Hafner (1972), New York.
23. N. Tsilevich, A. Vershik and M. Yor, Distinguished properties of the gamma process, and related topics, Prépublication du Laboratoire de Probabilités et Modèles Aléatoires No. 575 (2000).
24. S. Watanabe, K. Yano and K. Yano, A density formula for the law of time spent on the positive side of one-dimensional diffusion processes, preprint.
25. M. Yamazato, Unimodality of infinitely divisible distribution functions of class L. Ann. Probability 6 (1978), 523-531.
26. M. Yamazato, On strongly unimodal infinitely divisible distributions, Ann. Probability 10 (1982), 589-601.
27. M. Yamazato, Hitting time distributions of single points for 1-dimensional generalized diffusion processes, Nagoya Math. J. 119 (1990), 143-172.
28. M. Yamazato, Characterization of the class of hitting time distributions of 1-dimensional generalized diffusion processes, Proc. 6th Japan-USSR Symposium on Probability Theory and Mathematical Statistics, (1992) 422-428, World Scientific, Singapore-New Jersey-London-Hong Kong.
29. M. Yamazato, On subclasses of infinitely divisible distributions on R related to hitting time distributions of 1-dimensional generalized diffusion processes, Nagoya Math. J. 127 (1992), 175-200.

On Stochastic Differential Equations Driven by Symmetric Stable Processes of Index α

Hiroya Hashimoto[1], Takahiro Tsuchiya[2] and Toshio Yamada[3]

[1]Sanwa Kagaku Kenkyusho Co.,LTD,
[2 3] Department of Mathematical Sciences, Ritsumeikan University

1. Introduction

In the first part of the present paper, famous Tanaka's equation is discussed in the case of symmetric stable processes. Then some important uniqueness results in one-dimensional case will be reviewed. The second part is devoted to obtain some results concerning comparison problems using Lamperti's method. Marcus integral plays an essential role to formulate comparison theorems. In the last part, a sufficient condition which guarantees the pathwise uniqueness in d-dimensional case, will be proposed.

2. On uniqueness problems: One dimensional case

We consider following stochastic differential equations driven by a symmetric stable process of index α.

$$dX_t = \sigma(X_{t-})\,dZ_t \tag{1}$$

$$dX_t = \sigma(t, X_{t-})\,dZ_t \tag{2}$$

where Z_t is a symmetric stable process of index α of which characteristic function is given by

$$E[e^{t\xi Z_t}] = e^{-|\xi|^\alpha t}, \quad 1 < \alpha < 2$$

As is well known, famous Tanaka's example shows that the weak uniqueness does not imply the pathwise uniqueness in the case of SDE driven by a Brownian motion. We will mention that Tanaka type argument is still applicable to the case of the equation with respect to a symmetric stable process of index α.

Theorem 2.1. *Consider the equation*

$$X_t = \int_0^t \operatorname{sgn}(X_{s-})\,dZ_s, \tag{3}$$

where

$$\operatorname{sgn}(x) = \begin{cases} 1, & \text{if } x > 0 \\ -1, & \text{if } x \leq 0 \end{cases}$$

Then, the weak uniqueness holds for solutions to (3), but the pathwise uniqueness fails.

To prove Theorem 2.1, the following theorem by Rosiński and Woyczyński [11] plays essential roles.

Theorem 2.2. *Let F be an* $\mathcal{F}_t := \sigma(Z_s; s \leq t)$ *-adapted process such that*

$$E(\int_0^T | F_s(\omega) |^\alpha \, ds) < \infty,$$

and the random time $\tau(u) := \int_0^u | F_t |^\alpha \, dt$ *satisfies that* $\tau(u) \to \infty, a.s.,$ *when* $u \to \infty$.

Consider the inverse of τ *and* $\mathcal{G}_t$*:*

$$\tau^{-1}(t) = \inf\{u : \tau(u) > t\} \quad \mathcal{G}_t = \mathcal{F}_{\tau^{-1}(t)}$$

Then the time changed stochastic integral

$$\tilde{Z}_t = \int_0^{\tau^{-1}(t)} F_s \, dZ_s \tag{4}$$

is a $\mathcal{G}_t$*adapted symmetric stable process of index* α*. x We have also,*

$$\int_0^t F_s \, dZ_s = \tilde{Z}_{\tau(t)}, \quad a.s. \tag{5}$$

[Proof of the Theorem 2.1] Let X_t be a solution to (3). Then, we observe that

$$\tau(u) := \int_0^u | \operatorname{sgn}(X_{t-}) |^\alpha \, dt = u \to \infty, as \ u \to \infty.$$

Then by the Theorem 2.2, X_t is a symmetric stable process of index α with respect to $\mathcal{F}_{\tau^{-1}(t)} = \mathcal{F}_t$. So, any solution to (3) has the same law. Then the solution to the equation (3) is unique in the weak sense.

We will observe that the weak existence of a solution to (3). Let X_t be a symmetric stable process of index α with $X_0 = 0$.

Then,

$$Z_t = \int_0^t \operatorname{sgn}(X_{s-}) \, dX_s$$

is also a symmetric stable of index α and X_t satisfies

$$X_t = \int_0^t \mathrm{sgn}(X_{s-})\mathrm{sgn}(X_{s-})\, dX_s = \int_0^t \mathrm{sgn}(X_{s-})\, dZ_s$$

This means the existence of a solution to (3) in the weak sense.

Now, let X_t be a solution to the equation (3):

$$X_t = \int_0^t \mathrm{sgn}(X_{s-})\, dZ_s.$$

Then $(-X_t)$ satisfies

$$-X_t = \int_0^t \mathrm{sgn}(-X_{s-})\, dZ_s$$

So, $(-X_t)$ is also a solution to (3). The pathwise uniqueness fails for the solution to (3). □

When is the solution to (1) or (2) is pathwise unique? If the coefficients are Lipschitz continuous, the Picard iteration method works very well and it proves the pathwise uniqueness for the solution to (1) or (2).

In one dimensional case much weaker conditions suffice for uniqueness for SDE's with respect to one dimensional Brownian motion. For example (see [12]), a sufficient condition for pathwise uniqueness is that $\int_{0+} \rho^{-2}(u)\, du = \infty$ where ρ is the modulus of continuity:

$$| \sigma(x) - \sigma(y) | \leq \rho(| x - y |).$$

In view of the above result, one would hope that analogous weaker conditions would suffice for the pathwise uniqueness to the solution to (1) or (2).

The following condition is due to Komatsu [7]. (See also Bass [2]).

Theorem 2.3. *Suppose that*

$$| \sigma(x) - \sigma(y) |^{\alpha} \leq \rho(| x - y |), \quad \forall x, \forall y \in R^1,$$

where the modulus of continuity ρ *is a continuous increasing function with* $\rho(0) = 0$ *such that*

$$\int_0^{\epsilon} \rho^{-1}(x)\, dx = \infty, \quad \forall \epsilon > 0.$$

Then, for all $x_0 \in R^1$ *the solution to the equation*

$$dX_t = \sigma(X_{t-})\, dZ_t, \quad X_0 = x_0,$$

is pathwise unique.

Concerning to the solution to the equation (2) of which coefficient is time inhomogeneous, following Nagumo type modification to the condition in the above theorem is proposed in [6] .

Theorem 2.4. *Suppose that*

$$|\sigma(t,x)-\sigma(t,y)|^{\alpha}\leq \frac{\rho(|x-y|)}{h(t)} \quad \forall x, \forall y \in R^1, \ \ \forall t \in [0,\infty),$$

where the function ρ *is continuous increasing with* $\rho(0)=0$ *such that*

$$\int_0^{\epsilon} \rho^{-1}(x)\,dx = \infty, \quad \forall \epsilon > 0,$$

and where the function $h(t)$ *is non-negative continuous such that*

$$\int_0^{t} h^{-1}(s)\,ds < \infty, \quad \forall t \in [0,\infty).$$

Then for all $x_0 \in R^1$, *the solution to the equation*

$$dX_t = \sigma(t, X_{t-})\,dZ_t, \quad X_0 = x_0$$

is pathwise unique.

Remark 2.1. The conditions given in the Theorem 2.2 and Theorem 2.3 are sharp and best possible in some sense.

3. Lamperti's method

Given $b(x)$ satisfying the Lipschitz condition:

$$|b(x)-b(y)|\leq M\,|x-y| \quad \forall x, \forall y \in R^1,$$

the equation driven by an one dimensional Brownian motion:

$$X_t = x_0 + B_t + \int_0^t b(X_s)\,ds \tag{6}$$

can be solved very simply using the following inequalities

$$D_n(t) := \max_{s\leq t} |X_s^{(n+1)} - X_s^{(n)}| \leq \int_0^t |b(X_s^{(n)}) - b(X_s^{(n-1)})|\,ds$$
$$\leq M \int_0^t D_{n-1}(s)\,ds,$$

where $X_t^{(0)} = x_0$ and $X_t^{(n)} = x_0 + B_t + \int_0^t b(X_s^{(n-1)})\,ds$.

Now we make a change of scale $x \to y = f(x)$ with $f \in C^2(R^1)$. Let Y_t be $Y_t = f(X_t)$. Then Ito's formula implies

(7) $$Y_t = Y_0 + \int_0^t \tilde{\sigma}(Y_s)\,ds + \int_0^t \tilde{b}(Y_s)\,ds$$

where,

(a) $\tilde{\sigma}(f) = f'$

(b) $\tilde{b}(f) = f'b + f''/2$.

Lamperti's idea is to construct the solution to the equation (7) by solving (**a**) and (**b**) for f and b.

Standard Picard's iteration procedure applies to a wider class of coefficients, but Lamperti's method is simpler, because it uses neither the martingale inequality nor Borel-Cantelli lemma. (See Lamperti [9] and also McKean [10].)

Just as in the same way as in one dimensional Brownian motion case, the following equation concerning a symmetric stable process of index α can be solved:

(8) $$X_t = X_0 + Z_t + \int_0^t b(X_s)\,ds.$$

Make a change of scale $Y_t = f(X_t)$ with increasing $f \in C^3(R^1)$.

Then in the language of Marcus integral Y_t satisfies

(9) $$Y_t = Y_0 + \int_0^t f'(X_{s-})\diamond dX_s,$$

or

(10) $$Y_t = Y_0 + \int_0^t f'(f^{-1}(Y_{s-}))\diamond dZ_s + \int_0^t f'(f^{-1}(Y_{s-}))\,b(f^{-1}(Y_{s-}))\,ds,$$

where $\diamond$ means the Marcus integral.

Remark 3.1. For the precise definition of Marcus integral, see Kurtz-Pardoux-Protter [8] and also Applebaum [1].

In the case where

$$X_t = X_0 + Z_t + \int_0^t b(X_s)\,ds,$$

Marcus integral:

$$\int_0^t \varphi(X_{s-})\diamond dX_s = \int_0^t \varphi(X_{s-})\diamond dZ_s + \int_0^t \varphi(X_{s-})\, b(X_s)\, ds$$

stands for

$$\begin{aligned}&\int_0^t \varphi(X_{s-})\, dZ_s + \int_0^t \varphi(X_{s-})\, b(X_s)\, ds \\ &+ \sum_{0\le s\le t} \{\Phi(X_{s-} + \Delta Z_s) - \Phi(X_{s-}) - \Delta Z_s \varphi(X_{s-})\},\end{aligned}$$

where $\Phi' = \varphi$.

4. Some comparison results

In the present section Z_t stands for a symmetric stable process of index α.

In view of comparison results for solutions to SDE's concerning a one dimensional Brownian motion, one would hope that that solutions to SDE's driven by a symmetric stable process of index α:

$$dX_t^{(i)} = \sigma(X_{t-})\, dZ_i + b_i(X_{t-})\, dt, \quad X_0^{(i)} = x_i, \ \ i = 1, 2,$$

enjoy comparison properties under the conditions such that

$$x_1 \le x_2, \ \ b_1 \le b_2$$

and σ satisfies the same condition as in Theorem 2.3.

But, unfortunately the next example would suggest that in our case, comparison problems for solutions would take much more complicated aspects.

Example 4.1. Consider the following two equations:

$$X_t^{(i)} = x_i + \int_0^t X_{s-}^{(i)}\, dZ_s, \quad i = 1, 2. \tag{11}$$

Then the solutions to (11) are given by Doléan-Dade as follows;

$$X_t^{(i)} = x_i \exp(Z_t)[\prod_{s\le t}(1 + \Delta Z_s)\exp(-\Delta Z_s)]. \tag{12}$$

Consider the following series of random times $\tau_n, n = 0, 1, 2, \ldots,$ such that

$$\tau_0 = 0, \ \ \tau_1 = \inf\{s; \Delta Z_s < -1\}, \cdots \tau_n = \inf\{s; s > \tau_{n-1}, \Delta Z_s < -1\}.$$

Under the condition $x_1 < x_2$, we observe that

$$X_t^{(1)} < X_t^{(2)} : \tau_{2n} \le t < \tau_{2n+1}, n = 0,1,2,\ldots$$
$$X_t^{(1)} > X_t^{(2)} : \tau_{2n+1} \le t < \tau_{2n+2}, n = 0,1,2,\ldots.$$

In the followings of the section, we will apply Lamperti's method to obtain some comparison results.

Consider the following SDE's:

$$X_t^{(i)} = X_0^{(i)} + Z_t + \int_0^t b_i(X_s)\,ds, \ \ i = 1,2 \tag{13}$$

where b_i satisfies the Lipschitz condition such that

$$| b_i(x) - b_i(y) | \le K | x - y |, \ \ \forall x, \forall y \in R^1$$

As we have seen in the section 3, solutions to (13) can be constructed by simple way.

We have two comparison lemmas.

Lemma 4.1. *(Weak comparison lemma)*

Suppose that for the equations (13)

(i) $X_0^{(1)} \le X_0^{(2)}$ *a.s.,*

and

(ii) $b_1(x) \le b_2(x), \ \ \forall x \in R^1$.

Then,

$$\mathbf{P}(X_t^{(1)} \le X_t^{(2)}, \ \ t \ge 0) = 1$$

holds.

Lemma 4.2. *(Strong comparison lemma)*

Suppose that for the equations (13)

(i) $X_0^{(1)} < X_0^{(2)}$ *a.s.,*

and

(ii) $b_1(x) < b_2(x), \ \ \forall x \in R^1$.

Then,

$$\mathbf{P}(X_t^{(1)} < X_t^{(2)}, \ \ t \ge 0) = 1$$

holds.

Let $f \in C^3(R^1)$ with $f' > 0$.

Make a change of scale $Y_t^{(i)} = f(X_t^{(i)})$. Then $Y_t^{(i)}$ satisfies following Marcus equation:

$$
\begin{aligned}
Y_t^{(i)} &= Y_0^{(i)} + \int_0^t f'(f^{-1}(Y_{s-}^{(i)}))\diamond dZ_s + \int_0^t f'(f^{-1}(Y_{s-}^{(i)}))b_i(f^{-1}(Y_{s-}^{(i)}))\,ds \\
&= \int_0^t \tilde{\sigma}(Y_{s-}^{(i)})\diamond dZ_s + \int_0^t \tilde{b}_i(Y_{s-}^{(i)})\,ds, \qquad (14)
\end{aligned}
$$

with

(a) $\tilde{\sigma}(f) = f'$ and

(b) $\tilde{b}_i(f) = f'b_i$

Suppose that the coefficients $\tilde{\sigma}$ and $\tilde{b}_i$ are given such that the equations (**a**) and (**b**) can be solved for $f \in C^3(R^1)$ $f' > 0$ and b_i.

Then we obtain following comparison results for solutions to Marcus equations (14).

Theorem 4.1. *(Comparison theorem for Marcus equations)*
Suppose that

$$\tilde{b}_1(x) \leq \tilde{b}_2(x),\ \forall x \in R^1,\ \textit{and}\ Y_0^{(1)} \leq Y_0^{(2)} a.s.,$$

Then

$$\mathbf{P}(Y_t^{(1)} \leq Y_t^{(2)}; t \in [0,\infty)) = 1$$

holds.

Theorem 4.2. *(Strong Comparison theorem for Marcus equations)*
Suppose that

$$\tilde{b}_1(x) < \tilde{b}_2(x),\ \forall x \in R^1,\ \textit{and}\ Y_0^{(1)} < Y_0^{(2)} a.s.,$$

Then

$$\mathbf{P}(Y_t^{(1)} < Y_t^{(2)}; t \in [0,\infty)) = 1$$

holds.

Remark 4.1. Lamperti's method is applicable to a wider class of Marcus equations driven by semimartingales.

5. Pathwise uniqueness: d-dimensional case

Let $\mathbf{Z}_t$ be a d-dimensional symmetric stable process of index α, $(1 < \alpha < 2)$:

$$E[\exp((\xi, \mathbf{Z}_t))] = \exp(-t \mid \xi \mid^{\alpha})$$

Consider the following stochastic differential equation driven by $\mathbf{Z}_t$:

$$\mathbf{Y}_t = \mathbf{Y}_0 + \int_0^t \sigma(\mathbf{Y}_{s-})\, d\mathbf{Z}_s, \tag{15}$$

or

$$Y_t^{(i)} = Y_0^{(i)} + \sum_{k=1}^{d} \int_0^t \sigma_{ik}(\mathbf{Y}_{s-})\, dZ_s^{(k)}, \quad i = 1, \dots, d. \tag{16}$$

Assumption 5.1.

Assume that the coefficient matrix $\sigma = [\sigma_{ik}]$ *satisfies*

$$\sigma_{ik}(\mathbf{x}) = \delta_{ik}\sigma(\mathbf{x}), \quad \mathbf{x} \in R^d$$

Under the Assumption 5.1, we have the following theorem.

Theorem 5.1.

Suppose that the continuous function ρ *defined on* $[0, \infty)$ *with* $\rho(0) = 0$ *is increasing and such that:*

$$G(u) := \rho^{\alpha}(u^{\frac{1}{\alpha-1}})u^{\frac{-1}{\alpha-1}}$$

is concave and

$$\int_0^{\epsilon} \frac{du}{G(u)} = \infty, \forall \epsilon > 0.$$

Let ρ *be the modulus of continuity:*

$$\mid \sigma(\mathbf{x}) - \sigma(\mathbf{y}) \mid \leq \rho(\mid \mathbf{x} - \mathbf{y} \mid), \quad \forall \mathbf{x}, \forall \mathbf{y} \in R^d.$$

Then for every $\mathbf{y}_0 \in R^d$, *the solution to*

$$Y_t^{(i)} = Y_0^{(i)} + \int_0^t \sigma(\mathbf{Y}_{s-})\, dZ_s^{(i)}, \quad i = 1, \dots, d. \quad \mathbf{Y}_0 = \mathbf{y}_0$$

is pathwise unique.

Remark 5.1. For examples, the functions such that

$$\rho(u) = u, \quad \rho(u) = u((\alpha - 1)\log \frac{1}{u})^{\frac{1}{\alpha}},$$
$$\rho(u) = u((\alpha - 1)\log \frac{1}{u})^{\frac{1}{\alpha}}(\log(\alpha - 1)\log \frac{1}{u})^{\frac{1}{\alpha}}, \dots,$$

satisfy the conditions imposed on ρ in the theorem 5.2.

Concerning the equation driven by a d-dimensional Brownian motion:

$$X_t^{(i)} = X_0^{(i)} + \int_0^t \sigma(\mathbf{X}_s)\, dB_s^{(i)}, \quad i = 1, \ldots, d,$$

the functions such that

$$\rho(u) = u, \quad \rho(u) = u(\log \frac{1}{u})^{\frac{1}{2}},$$
$$\rho(u) = u(\log \frac{1}{u})^{\frac{1}{2}}(\log \log \frac{1}{u})^{\frac{1}{2}}, \ldots,$$

imply the pathwise uniqueness. (See [13].)

References

1. D. Applebaum. *Lévy Processes and Stochastic Calculus*. Cambridge Univ. Press (2004).
2. R. F. Bass. Stochastic differential equations driven by symmetric stable processes. *Séminaire de Probabilités* XXXVI (2003), 302–313.
3. R. F. Bass, K. Burdzy and Z. Q. Chen. Stochastic differential equations driven by stable processes for which pathwise uniqueness fails, *Stoch. Proc. their Appl.* vol. 111 (2004), 1–15.
4. R. F. Bass. Stochastic differential equations with jumps. *Probability Surveys*. vol. 1 (2004), 1–19.
5. C. Doléans-Dade. Quelques applications de la formule de changement de variable pour les semimartingales. *Z. Wahr.* vol. 16 (1970), 181–194.
6. H. Hashimoto. On stochastic differential equations driven by symmetric stable processes—uniqueness and comparison problems (in Japanese), Master thesis (2004), Ritsumeikan University.
7. T. Komatsu. On the pathwise uniqueness of solutions of one-dimensional stochastic differential equations of jump type. *Proc. Japan Acad.*, vol. 58, Ser. A, no. 8 (1982), 353–356.
8. T. G. Kurtz, É.Pardoux and P. Protter. Storatonovich stochastic differential equations driven by general semimartingales. *Ann. Inst. H. Poicaré.* vol. 31, no. 2 (1995), 351–377.
9. J. Lamperti. A simple construction of certain diffusion processes. *Jour. Math. Kyoto Univ.*, vol. 4 (1964), 161–170.
10. H. P. McKean, Jr. *Stochastic Integrals*. Academic Press (1969).
11. J. Rosiński and W. A. Woyczyński. On Ito stochastic integration with respect to p-stable motion: Inner clock, integrability of sample paths,double and multiple integrals. *Ann. Prob.*, vol. 14 (1986), 271–286.

12. T. Tsuchiya, On the pathwise uniqueness of solutions of stochastic differential equations driven by multi-dimensional symmetric α stable class. submitted.
13. T. Yamada and S. Watanabe. On the uniqueness of solutions of stochastic differential equations. *Jour. Math. Kyoto Univ.*, vol. 11 (1971), 155–167.
14. S. Watanabe and T. Yamada. On the uniqueness of solutions of stochastic differential equations II. ibid., vol. 11 (1971), 553–563.

Martingale Representation Theorem and Chaos Expansion

Shinzo Watanabe

Ritsumeikan University

1. Introduction

Given a filtration $\mathbf{F} = \{\mathcal{F}_t\}$ (i.e. an increasing family of sub σ-fields of events), a martingale representation theorem is concerned with a representation of $\mathbf{F}$-martingales as stochastic integrals by *basic martingales*. In the case of *Brownian filtration*, of which we shall discuss in Section 2, a result is well-known as *Itô's representation theorem* which states that every square-integrable martingale can be represented as a stochastic integral by the path of Brownian motion. This theorem was first found by Itô ([I]) as a corollary of his theory on Wiener chaos expansion of L^2-Wiener functionals and plays an important role in the problem of financial markets. In this respect, we would quote the following remark by Daniel Stroock in page 180 of [S]: "*In fact, it (Itô's representation theorem) shares with Itô's formula responsibility for the widespread misconception in the financial community that Itô is an economist.*"

In this introduction, we will review such a theorem and discuss its idea in the simplest case of a random walk. Let $\{\xi_k\}_{k=1,2,\ldots}$ be a *coin tossing* sequence, i.e., i.i.d. sequence with $P(\xi_k = 1) = P(\xi_k = -1) = \frac{1}{2}$. It is also called an i.i.d. sequence of *random signs*. If we set

$$X_n = \begin{cases} 0, & n = 0 \\ \xi_1 + \cdots + \xi_n, & n = 1,2,\ldots \end{cases}$$

$X = (X_n)_{n=0,1,\ldots}$ is a simple random walk on $\mathbf{Z}$ starting from the origin. Since $\xi_k = X_k - X_{k-1}, k = 1,2,\ldots, X = (X_n)$ and $\{\xi_k\}$ generate the same filtration $\mathbf{F} = \{\mathcal{F}_n\}_{n=0,1,\ldots}$, where

$$\mathcal{F}_0 = \{\emptyset, \Omega\}, \quad \mathcal{F}_n = \sigma[X_1, \cdots, X_n] = \sigma[\xi_1, \cdots, \xi_n], \; n = 1,2,\ldots.$$

In the following, we take and fix $N \in \mathbf{Z}_{++}(:= \{ n \in \mathbf{Z} \mid n > 0 \})$ and consider the time up to N; N is called the *maturity* in financial problems. As usual, a family $Y = (Y_n)_{0 \le n \le N}$ of random variables is called an $\mathbf{F}$-*adapted process* if Y_n is $\mathcal{F}_n$-measurable for every $0 \le n \le N$, and a family $\Phi = (\Phi_k)_{1 \le k \le N}$ of random variables is called an **F**-*predictable process* if Φ_k is $\mathcal{F}_{k-1}$-measurable for every $1 \le k \le N$.

Given an **F**-predictable process $\Phi = (\Phi_k)_{1 \le k \le N}$ and a constant M_0, we define an **F**-adapted process $M = (M_n)_{0 \le n \le N}$ by

$$M_n = \begin{cases} M_0, & n = 0 \\ M_0 + \sum_{k=1}^n \Phi_k(X_k - X_{k-1}), & 1 \le n \le N \end{cases}$$

and call it the *martingale transform of* X, more precisely, $[M_0, \Phi]$-martingale transform of X. We denote this as $M = M_0 + \Phi \cdot X$, for short. It is immediately seen that M is an **F**-martingale. The martingale representation theorem for the filtration **F** states, conversely, that every **F**-martingale can be represented by a martingale transform of X. Namely,

Theorem 1.1. *Every* **F**-*martingale* $M = (M_n)$ *can be represented in the form*

$$M_n = M_0 + \sum_{k=1}^{n} \Phi_k(X_k - X_{k-1}), \quad n = 1, 2, \cdots, N$$

by the constant M_0 *and an* **F**-*predictable process* $\Phi = (\Phi_k)_{1 \le k \le N}$. *Here,* Φ_k *is determined uniquely by*

$$\Phi_k = (M_k - M_{k-1}) \cdot \xi_k, \quad k = 1, \cdots .N.$$

Note that $M_0 = E(M_N)$ *by the martingale property.*

Here is a simple proof: Since $M_k - M_{k-1}$ is $\mathcal{F}_k$-measurable, there exists a function $d_k(i_1, \cdots, i_k)$ on the product space $\{-1, 1\}^k$ such that

$$M_k - M_{k-1} = d_k(\xi_1, \cdots, \xi_k).$$

By the martingale property,

$$E(M_k - M_{k-1} | \mathcal{F}_{k-1}) = \frac{1}{2} d_k(\xi_1, \cdots, \xi_{k-1}, 1) + \frac{1}{2} d_k(\xi_1, \cdots, \xi_{k-1}, -1) = 0$$

and hence, we have $d_k(\xi_1, \cdots, \xi_{k-1}, 1) = -d_k(\xi_1, \cdots, \xi_{k-1}, -1)$. This implies, by setting $\Phi_k = d_k(\xi_1, \cdots, \xi_{k-1}, 1)$, that $d_k(\xi_1, \cdots, \xi_k) = \Phi_k \cdot \xi_k$. Then,

$$M_n = M_0 + \sum_{k=1}^{n} (M_k - M_{k-1}) = M_0 + \sum_{k=1}^{N} \Phi_k \cdot \xi_k.$$

Obviously $\Phi = (\Phi_k)_{1 \le k \le n}$ is **F**-predictable and $(M_k - M_{k-1}) \cdot \xi_k = \Phi_k \cdot \xi_k^2 = \Phi_k$. This completes the proof.

We will give still two different proofs; it may look too much exaggerated, as the proof given above is so simple. However, the ideas of the following proofs can be applied to the case of continuous time: indeed, they provide us with two different prototypes of the proof in continuous time, as we shall see in Section 2.

Proof A. A real $\mathcal{F}_N$-measurable random variable Y has a representation as $Y = f(\xi_1, \cdots, \xi_N)$ by a function $f(i_1, \cdots, i_N)$ on the product space $\{-1,1\}^N$, and the totality of such Y forms a 2^N-dimensional (real) Hilbert space ($\cong \mathbf{R}^{2^N}$) with the L^2-inner product $(Y, Z) = E(YZ)$. A orthonormal basis (ONB) $\{\Xi_S\}$ is given, indexed by a subset $S \subset \{1, 2, \cdots, N\}$, as

$$\Xi_S := \prod_{i \in S} \xi_i, \quad S \subset \{1,2,\cdots,N\}, \quad \text{with the convention} \quad \Xi_\emptyset \equiv 1.$$

Take any Y and consider the orthogonal expansion of $Y - E(Y)$ by the orthonormal system $\{\Xi_S; S \neq \emptyset\}$. Writing each Ξ_S as $\Xi_{S\setminus\{k\}} \cdot \xi_k$ where k is the largest element in S, we immediately have the following expansion

$$Y - E(Y) = \sum_{k=1}^{N} \Phi_k \cdot \xi_k \quad \text{where} \quad \Phi_k \text{ is } \mathcal{F}_{k-1} - \text{measurable.}$$

Define an **F**-martingale Z by a martingale transform $Z = E(Y) + \Phi \cdot X$. Then we have $Y = Z_N$. Hence, $E(Y|\mathcal{F}_n) = Z_n$. Noting that every **F**-martingale (M_n) is given in the form $M_n = E(Y|\mathcal{F}_n)$ for some Y, we can now complete the proof. □

Proof B. Given an **F**-martingale $M = (M_n)$, define an **F**-predictable process $\Phi = (\Phi_k)$ by setting

$$\Phi_k = E[(M_k - M_{k-1}) \cdot \xi_k | \mathcal{F}_{k-1}], \quad k = 1, \cdots, N. \tag{1.1}$$

Define an **F**-martingale $\widetilde{M} = (\widetilde{M}_n)$ by the $[M_0, \Phi]$-martingale transform of X. Then we have

$$\Phi_k = (\widetilde{M}_k - \widetilde{M}_{k-1}) \cdot \xi_k = E[\widetilde{M}_k - \widetilde{M}_{k-1}) \cdot \xi_k | \mathcal{F}_{k-1}]. \tag{1.2}$$

Therefore, for an **F**-martingale $L = M - \widetilde{M}$, we see by (1.1) and (1.2) that

$$E[(L_k - L_{k-1}) \cdot \xi_k | \mathcal{F}_{k-1}] = E[(L_k - L_{k-1}) \cdot (X_k - X_{k-1}) | \mathcal{F}_{k-1}] = 0, \ k = 1, \ldots, N.$$

By applying the next lemma, we can conclude that $L = 0$, implying $M = \widetilde{M}$, which completes the proof. □

Lemma 1.1. *If an **F**-martingale $L = (L_n)$ satisfies that*

$$L_0 = 0 \quad \text{and} \quad E[(L_k - L_{k-1}) \cdot (X_k - X_{k-1}) | \mathcal{F}_{k-1}] = 0, \ k = 1, \ldots, N, \tag{1.3}$$

then, $L = 0$.

Proof. Every $\mathcal{F}_N$-measurable random variable is bounded and hence, we can take a constant $c > 0$ such that $\max |L_N| \le 1/(2c)$. Then, noting $E(L_N) = E(L_0) = 0$, $1 + cL_N \ge \frac{1}{2}$ and $E(1 + cL_N) = 1$.

Define a new probability Q on $(\Omega, \mathcal{F}_N)$ by setting

$$Q(A) = E[(1 + cL_N) \cdot 1_A], \quad A \in \mathcal{F}_N.$$

Then $X = (X_n)$ is an $\mathbf{F}$-martingale under Q. Indeed, we have

$$\begin{aligned}
&E^Q[(X_n - X_{n-1})|\mathcal{F}_{n-1}] \\
&= (1 + cL_{n-1})^{-1} E[(1 + cL_N)(X_n - X_{n-1})|\mathcal{F}_{n-1}] \\
&= (1 + cL_{n-1})^{-1} E[(1 + cL_n)(X_n - X_{n-1})|\mathcal{F}_{n-1}] \\
&= c(1 + cL_{n-1})^{-1} E[L_n(X_n - X_{n-1})|\mathcal{F}_{n-1}] \\
&= c(1 + cL_{n-1})^{-1} E\left[\sum_{k=1}^{n} (L_k - L_{k-1})(X_n - X_{n-1})|\mathcal{F}_{n-1}\right] \\
&= c(1 + cL_{n-1})^{-1} \sum_{k=1}^{n} E[(L_k - L_{k-1})(X_n - X_{n-1})|\mathcal{F}_{n-1}] = 0,
\end{aligned}$$

because $E[(L_n - L_{n-1})(X_n - X_{n-1})|\mathcal{F}_{n-1}] = 0$ by (1.3), and $E[(L_k - L_{k-1})(X_n - X_{n-1})|\mathcal{F}_{n-1}] = (L_k - L_{k-1})E[(X_n - X_{n-1})|\mathcal{F}_{n-1}] = 0$ when $k < n$. Since $X_n - X_{n-1}$ takes values 1 or -1 only, this implies that

$$Q(X_n - X_{n-1} = 1|\mathcal{F}_{n-1}) = Q(X_n - X_{n-1} = -1|\mathcal{F}_{n-1}) = \frac{1}{2},$$

that is, $X = (X_n)$ is also a symmetric random walk under Q. We have thus deduced that $P = Q$ on $(\Omega, \mathcal{F}_N)$, which implies that $L_N = 0$ and hence $L_n = E[L_N|\mathcal{F}_n] = 0$ for all $0 \le n \le N$. □

Remark 1.1. Generally for $\mathbf{F}$-martingales L and M, we define an $\mathbf{F}$-predictable process $\langle L, M \rangle = (\langle L, M \rangle_n)$ by

$$\langle L, M \rangle_n = \sum_{k=1}^{n} E[(L_k - L_{k-1})(M_k - M_{k-1})|\mathcal{F}_{k-1}], \quad n = 1, \cdots, N.$$

We say that two martingales are *orthogonal* if $\langle L, M \rangle = 0$. Now, Lemma 1.1 can be stated as follows: An $\mathbf{F}$-martingales L which is orthogonal to the basic martingale $X = (X_n)$ must be a constant.

Remark 1.2. The notion of martingale transforms has been introduced by D. L. Burkholder ([B]). In continuous time, a corresponding notion is that of stochastic integrals.

Example 1.1. We consider the case where above random signs $\{\xi_k\}$ are replaced by more complicated i.i.d. random variables.

For a positive integer $L \geq 3$, let $E = \{a_1, \cdots, a_L\}$ where $a_1, \cdots, a_L$ are different real numbers. Let ξ be an E-valued random variable such that $P(\xi = a_l) > 0$ for all l, and $E(\xi) = 0$. We denote $\sigma^2 = E(\xi^2)$.

We consider an i.i.d. sequence $\xi_1. \cdots, \xi_N$ such that $\xi_k \stackrel{d}{=} \xi, k = 1, \ldots, N$, and define the process $X = (X_n)_{0 \leq n \leq N}$ and $\mathbf{F} = \{\mathcal{F}_n\}_{0 \leq n \leq N}$ in the same way as above so that $\mathbf{F}$ is the natural filtration of X. So, as above, X is an $\mathbf{F}$-martingale which generates the filtration $\mathbf{F}$. However, Theorem 1.1 does not hold in this case, that is, there exists an $\mathbf{F}$-martingale which cannot be given as a martingale transform by X. We now obtain a martingale representation theorem in this case.

The set of all real functions on E forms a real vector space of dimension L which is a real Hilbert space with respect to the L^2-inner product by the law μ of ξ. Let $f_l(a)$, $l = 0, 1, \ldots, L-1$, be functions on E which form an orthonormal basis in this Hilbert space. We choose, as we may, $f_0(a) \equiv 1$, and $f_1(a_l) = a_l/\sigma, l = 1, \ldots, L$. Then, random variables $Y_l = f_l(\xi)$, $l = 0, \ldots, L-1$, are orthonormal in $L^2(\Omega, P)$ and $\sigma Y_1 = \xi$. Set, for each $l = 1, \ldots, L-1$, an $\mathbf{F}$-adapted stochastic process $Y^{(l)} = (Y_n^{(l)})_{0 \leq n \leq N}$ by

$$Y_n^{(l)} = \begin{cases} 0, & n = 0 \\ \sum_{k=1}^n f_l(\xi_k), & n = 1, \ldots, N \end{cases}.$$

Then, $Y^{(l)}$ are all $\mathbf{F}$-martingales and, obviously, $\sigma Y^{(1)} = X$. It is easy to see that $\langle Y^{(l)}, Y^{(l')} \rangle_n = \delta_{l,l'} n, 0 \leq n \leq N$ for every $l, l' = 1, \ldots, L-1$.

We can show by the same idea as above that every $\mathbf{F}$-martingale $M = (M_n)$ with $M_0 = 0$ has a representation as a sum of martingale transforms by these martingales in which $\mathbf{F}$-predictable integrands $\Phi^{(l)} = (\Phi_k^{(l)})_{1 \leq k \leq N}$, $l = 1, \ldots, L-1$, are uniquely determined:

$$M_n = \sum_{l=1}^{L-1} \left\{ \sum_{k=1}^{n} \Phi_k^{(l)} \cdot (Y_k^{(l)} - Y_{k-1}^{(l)}) \right\}, \quad n = 0, \ldots, N.$$

Thus, we see that the martingale representation by a single basic martingale X is impossible.

In this example, we consider $S_n = e^{\alpha X_n + \beta n}$, $(\alpha, \beta$; constants), as a stock price in a financial market. We choose constants so that $(S_n)_{0 \leq n \leq N}$ is an $\mathbf{F}$-martingale. Then the above fact can be interpreted that the market is *incomplete*; (S_n) is an $\mathbf{F}$-martingale under P on $(\Omega, \mathcal{F}_N)$ and it generates $\mathbf{F}$. And there are different equivalent probabilities on $(\Omega, \mathcal{F}_N)$ under which (S_n) is still an $\mathbf{F}$-martingale. If $L = 2$, then, essentially by Theorem 1.1, the market is *complete*; an equivalent martingale measure is unique, which is a key in Proof B.

2. The Case of Continuous Time

In continuous time, we have several technical difficulties in defining corresponding notions such as predictability, martingale transforms which are called in this case as stochastic integrals, and so on. We would summarize here some of basic notions, cf. e.g., [IW], [KK], [P], [RW], [RY].

2.1 Fundamental notions

The time parameter set $\mathbf{T}$ is usually the interval $[0, \infty)$; we often take it to be an finite interval $[0, T]$ for some $T > 0$. A filtration $\mathbf{F} = \{\mathcal{F}_t\}_{t \in \mathbf{T}}$ is an increasing family of sub σ-fields of events. We assume the *usual condition* unless otherwise stated:

(i) The probability space $(\Omega, \mathcal{F}, P)$ is complete and every $\mathcal{F}_t$ contains all P-null sets.

(ii) $\mathbf{F}$ is right-continuous, that is, $\mathcal{F}_{t+}(:= \cap_{\epsilon>0}\mathcal{F}_{t+\epsilon}) = \mathcal{F}_t$ for every $t \in \mathbf{T}$.

We introduce the following notations.

$$\begin{aligned}
\mathcal{M}_2 &= \{\, M = (M_t);\ \text{a square-integrable } \mathbf{F}-\text{martingale},\ M_0 = 0 \,\} \\
\mathcal{M}_{2,loc} &= \{\, M = (M_t);\ \text{a locally square-integrable } \mathbf{F}-\text{martingale},\ M_0 = 0 \,\} \\
\mathcal{M}_2^c &= \{\, M \in \mathcal{M}_2 \mid t \mapsto M_t \text{ is continuous, a.s.} \,\} \\
\mathcal{M}_{2,loc}^c &= \{\, M \in \mathcal{M}_{2,loc} \mid t \mapsto M_t \text{ is continuous, a.s.} \,\}
\end{aligned}$$

A stochastic process $X = (X_t(\omega))$ is a function of $(t, \omega) \in \mathbf{T}\times\Omega$. It is called $\mathbf{F}$-*adapted* if X_t is $\mathcal{F}_t$-measurable for every $t \in \mathbf{T}$. We identify two processes $X = (X_t(\omega))$ and $Y = (Y_t(\omega))$ and write $X = Y$, if $P(\{\omega | \exists t \in \mathbf{T},\ X_t(\omega) \neq Y_t(\omega)\}) = 0$. A real process $\Phi = (\Phi_t(\omega))$ is called a *simple* $\mathbf{F}$-*predictable process* if it is given, by a sequence $0 = s_0 < s_1 < \cdots < s_n$ of times and bounded real $\mathcal{F}_{s_i}$-measurable random variables f_i, as

$$\Phi_t(\omega) = \textstyle\sum_{i=0}^{n} f_i(\omega) 1_{(s_{i-1}, s_i]}(t). \tag{2.1}$$

The smallest σ-field $\mathcal{P}$ on $\mathbf{T} \times \Omega$ with respect to which all simple $\mathbf{F}$-predictable processes are measurable, is called the *predictable σ-field with respect the filtration* $\mathbf{F}$, or simply, $\mathbf{F}$-*predictable σ-field.* A process $\Phi = (\Phi_t(\omega))$ is called $\mathbf{F}$-*predictable* if it is $\mathcal{P}$-measurable. Every left-continuous $\mathbf{F}$-adapted process is $\mathbf{F}$-predictable.

A real process $A = (A_t(\omega))$ for which $t \mapsto A_t$ is right-continuous and increasing (in the wide sense) is called an *increasing process*, and a real process which can be represented as a difference of two increasing processes is called a *process of bounded variation*. The following fact holds: Given $M, N \in \mathcal{M}_{2,loc}$, there exists a unique **F**-predictable process of bounded variation $\langle M, N\rangle = (\langle M, N\rangle_t)$ such that the process $(M_tN_t - \langle M, N\rangle_t)$ is an **F**-local martingale. We denote $\langle M, M\rangle$ simply by $\langle M\rangle$. $\langle M\rangle$ is an **F**-predictable increasing process. Furthermore, if, at least, one of M and N is continuous, then $\langle M, N\rangle$ is a continuous process.

Let the samplewise total variation of $s \in [0, t] \mapsto \langle M, N\rangle_s$ be denoted by $\|\langle M, N\rangle\|_t$. Then it holds that $\|\langle M, N\rangle\|_t \leq \sqrt{\langle M\rangle_t \cdot \langle N\rangle_t}$, a.s. Hence, if $M, N \in \mathcal{M}_2$, then $E[\|\langle M, N\rangle\|_t] \leq E[M_t^2]^{1/2}E[N_t^2]^{1/2} < \infty$ for every $t \in \mathbf{T}$.

Given $M \in \mathcal{M}_{2,loc}$, we denote by $\mathcal{L}_{2,loc}(M)$ ($\mathcal{L}_2(M)$) the totality of real **F**-predictable processes $\Phi = (\Phi_t)$ such that $\int_0^t |\Phi_s|^2 d\langle M\rangle_s < \infty$, a.s. for every $t \in \mathbf{T}$, (resp. $E\left[\int_0^t |\Phi_s|^2 d\langle M\rangle_s\right] < \infty$ for every $t \in \mathbf{T}$).[1] Then, for given $M \in \mathcal{M}_{2,loc}$ and $\Phi \in \mathcal{L}_{2,loc}(M)$, there exists a unique $N \in \mathcal{M}_{2,loc}$ such that the following holds for any $L \in \mathcal{M}_{2,loc}$:

$$\langle N, L\rangle_t = \int_0^t \Phi_s d\langle M, L\rangle_s \text{ a.s.} \quad \text{for every } t \in \mathbf{T}.$$

We call this $N = (N_t)$ the *stochastic integral of Φ by M* and denote it by $N_t = \int_0^t \Phi_s dM_s$, or simply, by $N = \int \Phi dM$. $\int \Phi dM \in \mathcal{M}_{2,loc}^c$ if $M \in \mathcal{M}_{2,loc}^c$, and $\int \Phi dM \in \mathcal{M}_2$ if $\Phi \in \mathcal{L}_2(M)$.

If Φ is given by a simple **F**-predictable process as given by (2.1), then $\Phi \in \mathcal{L}_{2,loc}(M)$ for every $M \in \mathcal{M}_{2,loc}$ and $\int \Phi dM$ is given by the following Riemann sum:

$$\int_0^t \Phi_s dM_s = \begin{cases} 0, & t = 0 \\ \sum_{i=0}^{k-1} f_i(M_{s_{i+1}} - M_{s_i}) + f_k(M_t - M_{s_k}), & s_k < t \leq s_{k+1} \end{cases}.$$

When a right-continuous process $X = (X_t)$ is such that $M^X \in \mathcal{M}_{2,loc}$ where M^X is defined by $M_t^X = X_t - X_0$, then the stochastic integral $\int \Phi dM^X$ is also denoted by $\int \Phi dX$. We call it also the stochastic integral of Φ by X.

When an **F**-predictable process $\Phi = (\Phi_t)$ is given by $\Phi_t = !_{[\sigma \geq t]}$ where $\sigma \in \mathbf{T} \cup \{\infty\}$ is an **F**-stopping time, the process $(X_0 + \int_0^t \Phi_s dX_s)_{t \in \mathbf{T}}$ coincides with the *stopped process $X^\sigma = (X_{t\wedge\sigma})$ of X by σ*.

To realize the idea of Proof B above in the continuous time case, we introduce the following notion. We consider the case of $\mathcal{M}_2$, the case of $\mathcal{M}_{2,loc}$ can be discussed similarly by an obvious modification. For $M_1, M_2 \in \mathcal{M}_2$, we say that *$M_1$ and M_2 are orthogonal*, and denote this by $M_1 \perp M_2$,

[1] Two $\Phi, \Psi \in \mathcal{L}_{2,loc}(M)$ are identified and denoted by $\Phi = \Psi$ if $\int_0^t |\Phi_s - \Psi_s|^2 d\langle M\rangle_s = 0$, a.s. for every t.

if $\langle M_1, M_2\rangle = 0$. If $M_1 \perp M_2$ and $\Phi_i \in \mathcal{L}_2(M_i)$, $i = 1,2$, then it holds that $\int \Phi_1 dM_1 \perp \int \Phi_1 dM_2$. We have the following orthogonal decomposition result (cf. [KW], [KK], [P]).

Prop 2.1. Let $M_1, \ldots, M_n \in \mathcal{M}_2$ and $M_i \perp M_j$ if $i \neq j$. Then, for every $M \in \mathcal{M}_2$, there exists unique[2] $\Phi_i \in \mathcal{L}(M_i)$, $i = 1, \ldots, n$, and $N \in \mathcal{M}_2$ such that

$$M = \sum_{i=1}^{n} \int \Phi_i dM_i + N \quad \text{and} \quad N \perp M_i, \ i = 1, \ldots, n. \tag{2.2}$$

Φ_i is an **F**-predictable process satisfying $\langle M, M_i\rangle_t = \int_0^t (\Phi_i)_s d\langle M_i\rangle_s$, $i = 1, \ldots, n$.

2.2 The case of Brownian filtrations

Let $W = (W_t)$ be the d-dimensional standard Brownian motion (Wiener process) such that $W_0 = 0$. Let $\mathbf{F}^W = \{\mathcal{F}_t\}$ be the natural filtration, that is, $\mathcal{F}_t$ is the σ-field generated by $W_s, 0 \leq s \leq t$ (and P-null sets). Then it is known that it satisfies the usual condition. $\mathbf{F}^W$ is called the (d-dimensional) *Brownian filtration*.

Denoting $W_t = (W_t^1, \cdots, W_t^d)$ by its components, $W^i = (W_t^i) \in \mathcal{M}_2^c(\mathbf{F}^W)$, $i = 1, \ldots, d$. This system of martingales satisfies

$$\langle W_i, W_j\rangle_t \equiv \delta_{i,j} t, \quad \text{so that, in particular, } \langle W_i\rangle_t \equiv t, \quad i, j = 1, \ldots, d. \tag{2.3}$$

Hence,

$$\mathcal{L}_{2,loc}(W^i)(:= \mathcal{L}_{2,loc}) = \{\Phi = (\Phi_t); \mathbf{F}^W\text{-predictable and } \int_0^t \Phi_s^2 ds < \infty, \text{ a.s. } \forall t\}$$

and

$$\mathcal{L}_2(W^i)(:= \mathcal{L}_2) = \{\Phi = (\Phi_t); \mathbf{F}^W\text{-predictable and } E\left[\int_0^t \Phi_s^2 ds\right] < \infty, \text{ a.s. } \forall t\}.$$

For $\Phi = \left(\Phi_t = (\Phi_t^1 \ldots, \Phi_t^d)\right) \in (\mathcal{L}_{2,loc})^d$, the stochastic integral $\int \Phi dW \in \mathcal{M}_{2,loc}^c(\mathbf{F}^W)$ is defined by

$$\int_0^t \Phi_s dW_s = \sum_{i=1}^{d} \int_0^t \Phi_s^i dW_s^i.$$

The martingale representation theorem for the Brownian filtration $\mathbf{F}^W$ asserts that every martingale in $\mathcal{M}_{2,loc}(\mathbf{F}^W)$ can be represented by such a

---

[2] cf. Footnote 1.

stochastic integral. This results implies, in particular, that every martingale with respect to the Brownian filtration is necessarily continuous. Namely, we have

Theorem 2.1. (i) *For every* $M \in \mathcal{M}_{2,loc}(\mathbf{F}^W)$, $\left(\mathcal{M}_2(\mathbf{F}^W)\right)$, *there exist unique* $\Phi^i \in \mathcal{L}_{2,loc}$ *(resp.* $\mathcal{L}_2$*)* , $i = 1, \ldots, d$, *such that*

$$(2.4) \qquad M = \int \Phi dW = \textstyle\sum_{i=1}^{d} \int \Phi^i dW^i.$$

In particular, it holds that

$$(2.5) \qquad \mathcal{M}_{2,loc}(\mathbf{F}^W) = \mathcal{M}^c_{2,loc}(\mathbf{F}^W) \quad \textit{and} \quad \mathcal{M}_2(\mathbf{F}^W) = \mathcal{M}^c_2(\mathbf{F}^W).$$

(ii) *Every* $\mathbf{F}^W$*-locally square integrable martingale* $X = (X_t)$ *can be represented in the form*

$$(2.6) \qquad X_t = X_0 + \int_0^t \Phi_s dW_s$$

by some $\Phi \in (\mathcal{L}_{2,loc})^d$ *and* X_0 *is a constant. In particular,* X *is continuous.*
(iii) *Every* $\mathbf{F}^W$*-local martingale* $X = (X_t)$ *is continuous so that it is locally square integrable. Hence the result (ii) can be applied to obtain the representation (2.6).*

We give two proofs (which correspond to Proofs A and B of Section 1) in the case of $M \in \mathcal{M}_2(\mathbf{F}^W)$. The case of $M \in \mathcal{M}_{2.loc}(\mathbf{F}^W)$ can be treated by a standard localization argument. (ii) is obviously deduced from (i), and (iii) can be shown by approximating X by bounded martingales which are necessarily continuous by (ii). Such an approximation is easily obtained by applying a standard maximal inequality for martingales.
Proof A. This is based on the following Itô representation theorem of which we mentioned in Introduction.

Theorem 2.2. *Let* $T > 0$ *be arbitrary but fixed. Let* $Y \in L^2(\mathcal{F}_T^W)$*. Then there exists a unique* $\Phi = (\Phi^i) \in (\mathcal{L}_2)^d$ *such that*

$$(2.7) \qquad Y = E[Y] + \int_0^T \Phi_s dW_s.$$

Proof. The proof is based on the well-known Wiener chaos expansion (the expansion by multiple Wiener-Itô integrals) as will be given by Theorem 2.4 below.

$$(2.8) \qquad Y = E[Y] + \textstyle\sum_{n=1}^{\infty} \int \cdots \int_{0<t_1<\ldots<t_n<T} f_n(t_1, \cdots, t_n) dW(t_1) \cdots dW(t_n).$$

Here $f_n \in L^2\left(\Delta_T^{(n)} \to (\mathbf{R}^d)^{\otimes n}\right)$, $\Delta_T^{(n)} := \{(t_1, \cdots, t_n) | 0 < t_1 < \ldots < t_n < T\}$, and we use the following notational convention; for $A = (A^{i_1, \cdots, i_n}) \in (\mathbf{R}^d)^{\otimes n}$ and

$a_1 = (a_1^i) \in \mathbf{R}^d, ..., a_n = (a_n^i) \in \mathbf{R}^d$, $Aa_1 \cdots a_n$ is a real number defined by

$$Aa_1 \cdots a_n = \sum_{i_1=1}^{d} \cdots \sum_{i_n=1}^{d} A^{i_1,\cdots,i_n} a_1^{i_1} \cdots a_n^{i_n}.$$

Setting $\left(\widetilde{f_{n-1}^{i,t}}(t_1, \cdots, t_{n-1})\right)^{i_1,\cdots,i_{n-1}} = (f_n(t_1, \cdots, t_{n-1}, t))^{i_1,\cdots,i_{n-1},i}$, define

$$\Phi_t^i = f_1^i(t) + \sum_{n=2}^{\infty} \int \cdots \int_{0<t_1<\ldots<t_{n-1}<t} \widetilde{f_{n-1}^{i,t}}(t_1, \cdots, t_{n-1}) dW(t_1) \cdots dW(t_{n-1}).$$

Then $\Phi^i = (\Phi_t^i)$ has a version as an $\mathbf{F}$-predictable process so that $\Phi = (\Phi^i) \in (\mathcal{L}_2)^d$ and (2.7) follows from (2.8). □

The proof of Theorem 2.1 (i) follows at one from Theorem 2.2.

Proof B. By Prop. 2.1, it is enough to show that every $N \in \mathcal{M}_2(\mathbf{F}^W)$ which is orthogonal to every W^i, $i = 1, \ldots, d$ must be 0. When $N_T \in L^\infty(\mathcal{F}_T^W)$, we can choose $c > 0$ such that $1 + cN_T$ is bounded from above and below by positive constants, so that a probability Q on $(\Omega, \mathcal{F}_T^W)$ can be defined which is equivalent to P with density $1+cN_T$. We deduce that any stochastic integral $\int \Phi dW$ for $\Phi \in \mathcal{L}_2$ is in $\mathcal{M}_2(\mathbf{F}^W)$ with respect to Q. In particular, W_t^i and $W_t^i W_t^j - \delta_{i,j} t$ are all Q-martingales. This implies, by Lévy's martingale characterization of Wiener process, that $W = (W^i)$ is also a d-dimensional Wiener process with respect to Q, that is, $P = Q$ on $(\Omega, \mathcal{F}_T^W)$. Hence, $1 + cN_T = 1$, that is, $N_T = 0$, a.s. Then $N_t = E(N_T|\mathcal{F}_t) = 0$, a.s. for $0 \leq t \leq T$.

In the general case of unbounded N, the proof can be reduced to the bounded case by approximation (cf. e.g. [IW], p. 82, [RY], p. 210). □

2.3 The Clark-Bismut-Ocone Formula and Wiener chaos expansion.

The Clark-Bismut-Ocone Formula is concerned with an expression of the integrand Φ in the Itô representation (2.7). In discussing this, we apply some functional analysis (Malliavin calculus) on Wiener process so that it is convenient to set up the Wiener process canonically: We take, as our basic probability space $(\Omega, \mathcal{F})$, the path space $\mathbf{W}_0(\mathbf{R^d}) := \{ w;\ [0, T] \ni t \mapsto w(t) \in \mathbf{R}^d$, continuous, $w(0) = 0 \}$, which is a Banach space with the usual maximum norm, endowed with the d-dimensional Winer measure P^W on the σ-field $\mathcal{F}$ of P^W-measurable sets. The natural filtration $\mathbf{F}^W = \{\mathcal{F}_t^W\}$ is defined as usual, so that $\mathcal{F} = \mathcal{F}_T^W$. $x(t, w) := w(t)$, $w \in \mathbf{W}_0(\mathbf{R^d})$, is the canonical realization of d-dimensional Wiener process.

Let $H \subset \mathbf{W}_0(\mathbf{R^d})$ be the Cameron-Martin subspace;

$$H = \left\{ h \in \mathbf{W}_0(\mathbf{R^d}) \mid h(t) = \int_0^t h'(s) ds,\ h' \in L^2\left([0, T] \to \mathbf{R}^d\right) \right\},\ \|h\|_H = \|h'\|_{L^2}.$$

A P^W-measurable function on $\mathbf{W}_0(\mathbf{R^d})$ is called a Wiener functional. The Malliavin calculus is a differential and integral calculus for Wiener functionals (cf. e.g. [IW], [M]). Typical differential operators are, the gradient operator or Gross-Malliavin-Shigekawa operator D which sends a real Wiener functional to an H-valued Wiener functional, its dual operator or Skorohod operator D^* and the Ornstein-Uhlenbeck operator $L = -D^*D$. D is defined formally, for a Wiener functional $F = (F(w))$, by

$$(DF, h)_H = \lim_{\epsilon \to 0}(F(w + \epsilon h) - F(w))/\epsilon, \quad h \in H.$$

For a real separable Hilbert space E, we denote by $L^p(E), 1 \le p < \infty$, the usual L^p-space of E-valued Wiener functionals. In the Malliavin calculus, a family of Sobolev-type spaces $\mathbf{D}_k^p(E), 1 \le p < \infty, k = 0, 1, \cdots$, is introduced; roughly, $F \in \mathbf{D}_k^p(E)$ if $F \in L^p(E)$ and $D^kF \in L^p(E \otimes H^{\otimes k})$. When $E = \mathbf{R}$, $L^p(E)$ and $\mathbf{D}_k^p(E)$ are denoted simply by L^p and $\mathbf{D}_k^p$, respectively.

Let $F \in \mathbf{D}_1^2$. Then $DF \in L^2(H)$ and we denote $D_tF = (DF)'(t)$, that is, $\langle DF, h\rangle_H = \int_0^T (D_tF, h'(t))_{\mathbf{R}^d} dt$, for every $h \in H$.

The Clark-Bismut-Ocone formula may be stated as follows:

Theorem 2.3. *Let $F \in \mathbf{D}_1^2 (\subset L^2)$. Then we can define a version Φ_t of $E(D_tF|\mathcal{F}_t^W)$ suitably such that $\Phi = (\Phi_t) \in (\mathcal{L}_2)^d$ and it is the integrand of the Itô representation (2.7) for F. Thus, we may write formally*

$$F = E[F] + \int_0^T E(D_tF|\mathcal{F}_t^W)dW_t. \tag{2.9}$$

A proof based on the Wiener chaos expansion will be given below.

Example 2.1. Let $d = 1$. For the canonical representation of one-dimensional Wiener process $w(t) = x(t, w)$, we define $m(t) = m(t, w) := \max_{0 \le s \le t} w(s)$.

Let $f(x, y)$ be a (piecewise) smooth function on $\mathbf{R}^2$ with bounded first order derivatives $\frac{\partial f}{\partial x}$ and $\frac{\partial f}{\partial y}$. We take $F(w) = f(w(T), m(T))$. Then $\Phi_t = E(D_tF|\mathcal{F}_t^W)$, $0 < t < T$, is given by the following formula; letting $r(t) = r(t, w) := m(t) - w(t)$,

$$\Phi_t(w) = \int_{-\infty}^{r(t)} dx \sqrt{\frac{1}{2\pi(T-t)}} \left(\exp\left\{ -\frac{x^2}{2(T-t)} \right\} - \exp\left\{ -\frac{[2r(t)-x]^2}{2(T-t)} \right\} \right)$$
$$\times \frac{\partial f}{\partial x}(x + w(t), m(t))$$
$$+ \int_{-\infty}^{\infty} dx \int_{r(t)\vee x}^{\infty} dy \sqrt{\frac{2}{\pi(T-t)^3}} \exp\left\{ -\frac{[2y-x]^2}{2(T-t)} \right\} \cdot (2y - x)$$
$$\times \left(\frac{\partial f}{\partial x} + \frac{\partial f}{\partial y} \right)(x + w(t), y + w(t))$$
$$= \int_{-\infty}^{r(t)} dx \sqrt{\frac{1}{2\pi(T-t)}} \left(\exp\left\{ -\frac{x^2}{2(T-t)} \right\} - \exp\left\{ -\frac{[2r(t)-x]^2}{2(T-t)} \right\} \right)$$
$$\times \frac{\partial f}{\partial x}(x + w(t), m(t))$$
$$+ \int_{-\infty}^{r(t)} dx \int_{r(t)}^{\infty} dy \sqrt{\frac{2}{\pi(T-t)^3}} \exp\left\{ -\frac{[2y-x]^2}{2(T-t)} \right\} \cdot (2y - x)$$
$$\times \left(\frac{\partial f}{\partial x} + \frac{\partial f}{\partial y} \right)(x + w(t), y + w(t))$$
$$+ \int_{r(t)}^{\infty} dx \int_{x}^{\infty} dy \sqrt{\frac{2}{\pi(T-t)^3}} \exp\left\{ -\frac{[2y-x]^2}{2(T-t)} \right\} \cdot (2y - x)$$
$$\times \left(\frac{\partial f}{\partial x} + \frac{\partial f}{\partial y} \right)(x + w(t), y + w(t)).$$

In particular, if $f(x, y) = g(x)$, then

$$\Phi_t = \int_{-\infty}^{\infty} dx \sqrt{\frac{1}{2\pi(T-t)}} \exp\left\{ -\frac{x^2}{2(T-t)} \right\} g'(x + w(t))$$

and, if $f(x, y) = g(y)$, then

$$\Phi_t = 2 \int_{r(t)}^{\infty} dy \sqrt{\frac{1}{2\pi(T-t)}} \exp\left\{ -\frac{y^2}{2(T-t)} \right\} g'(y + w(t)).$$

A proof can be deduced from the following facts: If $F(w) = m(T)$, then $F \in \mathbf{D}_1^2$ and $D_t = 1_{[0,\tau]}(t)$, where τ is a unique random time in $[0, T]$ such that $w(\tau) = m(T)$. The condition $\tau > t$ is equivalent to the condition that $m(T) > m(t)$. If, for each $0 < t < T$, a continuous path w_t^+ is defined by $w_t^+(s) = w(t + s) - w(t), 0 \leq s \leq T - t$, then $w(T) = w(t) + x(T - t, w_t^+)$, and

$$m(T) = m(t)1_{[m(t)=m(T)]} + (w(t) + m(T-t, w_t^+))\,1_{[m(t)<m(T)]}$$
$$= m(t)1_{[m(T-t,w_t^+)<r(t)]} + (w(t) + m(T-t, w_t^+))\,1_{[m(T-t,w_t^+)>r(t)]}.$$

We use the independence of $\mathcal{F}_t^W$ and w_t^+, and also the following well-known formula for the joint distribution of $w(t)$ and $m(t)$, cf. [IM],

$$P(w(t) \in dx, m(t) \in dy) = 1_{[x<y,y>0]} \cdot \sqrt{\frac{2}{\pi t^3}} \exp\left\{-\frac{[2y-x]^2}{2t}\right\}(2y-x)dxdy.$$

The next example is concerned with a stochastic differential equation (SDE). Consider the following SDE for a diffusion process $X = (X_t^\alpha)$ on $\mathbf{R}^m$ with the initial point $x = (x^\alpha)$: [3]

(2.10) $\quad dX_t^\alpha = \sigma_i^\alpha(X_t)dw^i(t) + b^\alpha(X_t)dt, \quad X_0^\alpha = x^\alpha, \quad \alpha = 1, \ldots, m.$

Here, coefficients σ_i^α and b^α are assumed to be smooth with bounded first order derivatives. Then a solution $X^x = (X_t^x(w))$ exists, unique in pathwise, so that X_t^x defines an $\mathcal{F}_t^W$-measurable Wiener functional with values in $\mathbf{R}^m$. Let P_t be the transition operator: $P_t f(x) = E[f(X_t^x)]$, acting on $\widehat{C}(\mathbf{R}^m)$. Then, if $f \in \widehat{C}^2(\mathbf{R}^m)$, $P_t f \in \widehat{C}^2(\mathbf{R}^m)$, and $(t, x) \mapsto P_t f(x)$ is C^1 in t and C^2 in x. Here, $\widehat{C}(\mathbf{R}^m) = \{f;$ real, continuous on $\mathbf{R}^m$ having limits as $|x| \to \infty\}$, and $\widehat{C}^k(\mathbf{R}^m) = \{f \in \widehat{C}(\mathbf{R}^m);$ derivatives up to the order k are in $\widehat{C}(\mathbf{R}^m)\}$.

Then, $u(t, x) := P_t f(x)$, $f \in \widehat{C}^2(\mathbf{R}^m)$, satisfies the heat equation:

(2.11) $\quad \frac{\partial u}{\partial t} = Lu, \quad u|_{t=0} = f,$

where L is a second order differential operator given by

(2.12) $\quad L = \frac{1}{2}a^{\alpha\beta}(x)\frac{\partial^2}{\partial x^\alpha \partial x^\beta} + b^\alpha(x)\frac{\partial}{\partial x^\alpha} \quad \text{with} \quad a^{\alpha\beta}(x) = \sum_{i=1}^d \sigma_i^\alpha(x)\sigma_i^\beta(x).$

Example 2.2. Consider the case when $F \in L^2$ is given by $F(w) = f(X_T^x)$ where $f \in \widehat{C}^2(\mathbf{R}^m)$. Then the Itô representation (2.7) for F is given by

(2.13) $\quad F(w) = P_T f(x) + \int_0^T \frac{\partial P_{T-t} f}{\partial x^\alpha}(X_t^x)\sigma_i^\alpha(X_t^x)dw^i(t).$

It is well-known that $F \in \mathbf{D}_1^2$ and, by (2.13), we see that the Clark-Bismut-Ocone density $E\left((D_tF)^i|\mathcal{F}_t\right)$ is given by

$$E\left((D_tF)^i|\mathcal{F}_t\right) = \frac{\partial P_{T-t}f}{\partial x^\alpha}(X_t^x)\sigma_i^\alpha(X_t^x), \quad i = 1, \ldots, d.$$

This follows at once from the Itô formula: If $u(t, x) := P_t f(x)$, then

$$\begin{aligned} &u(T-t, X_t^x) - u(T, x) \\ &= \int_0^t \frac{\partial u}{\partial x^a}(T-s, X_s^x)\sigma_i^\alpha(X_s^x)dw^i(s) + \int_0^t \left(-\frac{\partial u}{\partial t} + Lu\right)(T-s, X_s^x)ds. \end{aligned}$$

[3] We omit the summation sign by following the usual convention.

Noting that u satisfies (2.11), (2.13) follows readily by letting $t \to T$.

Wiener chaos expansion. Consider the space $L^2 = L^2(W_0(\mathbf{R}^d), \mathcal{F}_T^W, P^W)$ of square-integrable Wiener functionals. Then we have its Wiener chaos decomposition into sum of mutually orthogonal subspaces $C_T^{(n)}, n = 0, 1, \dots$:

$$L^2 = \bigoplus \textstyle\sum_{n=0}^{\infty} C_T^{(n)}, \quad C_T^{(0)} \cong \mathbf{R}. \tag{2.14}$$

$C_T^{(0)}$ is one-dimensional, consisting of all constant functionals. $C_T^{(n)}$ for $n \geq 1$ is usually described by Itô's multiple Wiener integrals (cf. [I-1]) and, as is remarked in this paper, a multiple Wiener integral may be defined by an iteration of Itô's stochastic integrals as follows:

Let, as above, $\Delta_T^{(n)} := \{(t_1, \cdots, t_n) | 0 < t_1 < \ldots < t_n < T\} \subset [0, T]^n$ and $L^2(\Delta_T^{(n)})$ be the usual L^2-space of real functions f on $\Delta_T^{(n)}$ such that

$$\|f\|^2_{L^2(\Delta_T^{(n)})} := \int \cdots \int_{\Delta_T^{(n)}} |f(t_1, \cdots, t_n)|^2 dt_1 \cdots dt_n < \infty.$$

Let $f = (f^{i_1, \cdots, i_n}) \in L^2\left(\Delta_T^{(n)} \to (\mathbf{R}^d)^{\otimes n}\right)$, so that $f^{i_1, \cdots, i_n} = (f^{i_1, \cdots, i_n}(t_1, \cdots, t_n)) \in L^2(\Delta_T^{(n)})$ for every $(i_1, \cdots, i_n) \in \{1, \cdots, d\}^n$. We define $I^{(n)}(f) \in L^2$ by the following iterated Itô stochastic integrals:[4]

$$I^{(n)}(f)(w) = \int_0^T \left\{ \int_0^{t_n} \left\{ \cdots \left\{ \int_0^{t_2} f^{i_1, \cdots, i_n}(t_1, \cdots, t_n) dw^{i_1}(t_1) \right\} \cdots \right\} dw^{i_{n-1}}(t_{n-1}) \right\} dw^{i_n}(t_n).$$

As in (2.8), the right-hand side is denoted simply by

$$\int \cdots \int_{\Delta^{(n)}} f(t_1, \cdots, t_n) dw(t_1) \cdots dw(t_n).$$

It holds that

$$\|I^{(n)}(f)\|_{L^2} = \|f\|_{L^2(\Delta_T^{(n)} \to (\mathbf{R}^d)^{\otimes n})}. \tag{2.15}$$

Theorem 2.4. *For each n, the subspace $C_T^{(n)}$ is isometrically isomorphic to $L^2\left(\Delta_T^{(n)} \to (\mathbf{R}^d)^{\otimes n}\right)$ by the correspondence $I_n(f) \rightleftharpoons f$, so that every $F \in L^2$ has*

---

[4] We omit the summation sign for $i_1, \cdots, i_n$.

an orthogonal expansion by uniquely determined $f_n \in L^2\left(\Delta_T^{(n)} \to (\mathbf{R}^d)^{\otimes n}\right)$, $n = 1, 2, \ldots,$ *:*

(2.16) $\quad F = E(F) + \sum_{n=1}^{\infty} \int \cdots \int_{\Delta_T^{(n)}} f_n(t_1, \cdots, t_n) dw(t_1) \cdots dw(t_n).$

We have by (2.15) that

(2.17) $\quad \|F\|_{L^2}^2 = E(F)^2 + \sum_{n=1}^{\infty} \|f_n\|_{L^2(\Delta_T^{(n)} \to (\mathbf{R}^d)^{\otimes n})}^2.$

Example 2.3. (Veretennikov-Krylov expansion [VK]) We consider the SDE (2.10) and consider the same $F(w) = f(X_T^x)$ as in Example 2.2. Let the transition operator P_t, $t \in [0, T]$, be defined as in Example 2.2 by $P_t f(x) = E[f(X_t^x)]$, and define the operator Q_t^i, $t \in [0, T]$, $i = 1, \cdots, d$, by

$$Q_t^i f(x) = \sigma_\alpha^i(x) \frac{\partial P_t f}{\partial x^\alpha}(x), \quad f \in \widehat{C}^2(\mathbf{R}^m).$$

Take $f \in \widehat{C}^2(\mathbf{R}^m)$. Then a smoothness of $P_t f$ is guaranteed as in Example 2.2: If σ_α^i is nondegenate so that the operator L is elliptic, we may take any $f \in \widehat{C}(\mathbf{R}^m)$. Then the Wiener chaos expansion of $F(w) = f(X_T^x)$, for a fixed $x \in \mathbf{R}^m$, is given by the following *Veretennikov-Krylov formula*: $E(F) = P_T f(x)$ and, for $n = 1, 2, \ldots,$

$$f_n^{i_1, \cdots, i_n}(t_1, \cdots, t_n) = P_{t_1}\left(Q_{t_2 - t_1}^{i_1}\left(\cdots\left(Q_{T - t_n}^{i_n} f\right)\cdots\right)\right)(x).$$

A proof can be given by applying the following formula successively: If $g \in \widehat{C}^2(\mathbf{R}^m)$, then

$$g(X_s^x) - P_s g(x) = \sum_{i=1}^{d} \int_0^s Q_{s-t}^i g(X_t^x) dw^i(t) \quad \text{for every } 0 < s < T.$$

This formula is obtained from the Itô formula applied to $u(t, X^x(t))$, $t \in [0, s]$, where $u(t, x) = P_{s-t} g(x)$, for each fixed $0 < s < T$.

Proof of the Clark-Bismut-Ocone formula (Theorem 2.3). It is enough to show this when F is in the n-th order Wiener chaos, i.e. $F \in C_T^{(n)}$, for $n \geq 1$. So we assume that $F = \int \cdots \int_{\Delta_T^{(n)}} f(t_1, \cdots, t_n) dw(t_1) \cdots dw(t_n)$, $f \in L^2\left(\Delta_T^{(n)} \to (\mathbf{R}^d)^{\otimes n}\right)$. Then, for each $0 < s < T$ and $i = 1, \ldots, d$, $(DF_s)^i$ in the $(n-1)$-th order Wiener chaos given by

$$(DF_s)^i = \begin{cases} f^i(s), & \text{when } n = 1 \\ \int \cdots \int_{\Delta_T^{(n-1)}} \widetilde{f^{s,i}}(t_1, \cdots, t_{n-1}) dw(t_1) \cdots dw(t_{n-1}), & \text{when } n > 1 \end{cases},$$

where, if $n > 1$, $\widetilde{f^{s,i}} \in L^2\left(\Delta_T^{(n-1)} \to (\mathbf{R}^d)^{\otimes(n-1)}\right)$ is defined by

$$\begin{aligned}&\left(\widetilde{f^{s,i}}\right)^{(i_1,\cdots,i_{n-1})}(t_1,\cdots,t_{n-1})\\&=\sum_{k=1}^{n} 1_{[t_{k-1},t_k]}(s)\cdot f^{(i_1,\cdots,i_{k-1},i,i_{k+1},\cdots,i_{n-1})}(t_1,\cdots,t_{k-1},s,t_k,\cdots,t_{n-1})\end{aligned}$$

with the conventions $t_0 = 0$ and $t_n = T$. Then,

$$\begin{aligned}&E\left[(D_sF)^i|\mathcal{F}_s^W\right]\\&=\int\cdots\int_{\Delta_s^{(n-1)}}\widetilde{f^{s,i}}(t_1,\cdots,t_{n-1})dw(t_1)\cdots dw(t_{n-1})\\&=\int\cdots\int_{\Delta_s^{(n-1)}} f^{(i_1,\cdots,i_{n-1},i)}(t_1,\cdots,t_{n-1},s)dw^{i_1}(t_1)\cdots dw^{i_{n-1}}(t_{n-1}),\end{aligned}$$

because, if $(t_1,\cdots,t_{n-1}) \in \Delta_s^{(n-1)}$, that is, if $0 < t_1 < \cdots < t_{n-1} < s$, then

$$\left(\widetilde{f^{s,i}}\right)^{(i_1,\cdots,i_{n-1})}(t_1,\cdots,t_{n-1}) = f^{(i_1,\cdots,i_{n-1},i)}(t_1,\cdots,t_{n-1},s).$$

We can now conclude that

$$\int_0^T E\left[(D_sF)^i|\mathcal{F}_s^W\right]dw^i(s) = \int_0^T E\left[D_sF|\mathcal{F}_s^W\right]dw(s) = F. \qquad \square$$

2.4 The case of one-dimensional diffusion processes.

Let $= -\infty \le l < r \le \infty$ be given and fixed. A most general regular diffusion process $X = (X(t))$ on the interval $I = [l, r]$ is determined by a *canonical scale* $s(x)$, a *speed measure* $dm(x)$ and, when the boundary l or r is regular, a *Feller boundary condition* on the boundary: $s(x)$ is a strictly increasing continuous function on $I^o = (l, r)$, $dm(x)$ is an everywhere positive Radon measure on $I^o = (l, r)$, and l (r), is regular if $-\infty < s(l+) + m((l, c])$, (resp. $s(r-) + m[c, r) < \infty$). Here (and below), we fix a point c such that $l < c < r$. The local generator of X is given by a generalized second-order differential operator $\frac{d}{dm}\frac{d}{ds}$, (cf. [IM], [I]). By changing the coordinate by the scale function, we may and do assume from now that $s(x) \equiv x$, unless otherwise stated.

If $X(t)$ starting at a point in I^o cannot hit a boundary in a finite time, we may delete the boundary from the state space I. For simplicity, we assume that *every boundary point is a trap*[5]. That is, we assume $X(t) = X(m_\partial)$ for all

[5] cf. Remark 2.1 below for the case of more general boundary conditions.

$t \geq m_\partial$ when $m_\partial < \infty$, where we set $m_a = \inf\{ t \mid X(t) = a \}$ for $a \in I$ and $m_\partial = m_l \wedge m_r$.

It is well-known that, if $X(0) = a$, $a \in I^o$, $X(t)$ is obtained from one-dimensional Brownian motion by a time change: Let $W(t)$ be a one-dimensional Wiener process with $W(0) = 0$, $B^a(t) = a + W(t)$ and $l(t,x) := \lim_{\epsilon\downarrow 0}(4\epsilon)^{-1} \int_0^t 1_{(x-\epsilon,x+\epsilon)}(B^a(s))ds$ be the local time of $B^a = (B^a(t))$. Define an increasing process $A(t)$ by

$$A(t) = \begin{cases} \int_{I^o} l(t,x)dm(x), & \text{if} \quad t < m_\partial^{B^a} \\ \infty, & \text{if} \quad t \geq m_\partial^{B^a} \end{cases},$$

where $m_\partial^{B^a}$ is the hitting time to the boundaries l and r for $B^a = (B^a(t))$ defined similarly as above. Let $t \mapsto \tau_t$ be the right-continuous inverse of $t \mapsto A(t)$. Then, $X(t) = B^a(\tau_t)$.

We fix $a \in I^o$. Let $\mathbf{F}^X = \{\mathcal{F}_t^X\}$ be the natural filtration of $X = (X(t))$ with $X(0) = a$. From the above time change expression, we see that $X = (X(t))$ is a local martingale.

Theorem 2.5. *Every $M = (M(t)) \in \mathcal{M}_{2,loc}(\mathbf{F}^X)$ has the following representation*

$$M(t) = \int_0^t \Phi(s)dX(s), \quad \Phi = (\Phi(t)) \in \mathcal{L}_{2,loc}(X).$$

Proof. Let $C(I)$ be the space of all real continuous functions on $I = [l,r]$; I is compact even in the case of unbounded boundaries. The resolvent operator G_λ, $\lambda > 0$, on $C(I)$ is defined, as usual, by $G_\lambda f(b) = E\left[\int_0^\infty e^{-\lambda t} f(X^b(t))dt\right]$, where $X^b = (X^b(t))$ denote the diffusion starting at $b \in I$. The generator L of the diffusion is defined (independently of λ) by $L = \lambda - G_\lambda^{-1}$ with domain $\mathcal{D}(L) = G_\lambda(C(I))$. Then, for $u \in \mathcal{D}(L)$, $M^u(t) = u(X^b(t)) - u(b) - \int_0^t Lu(X^b(s))ds \in \mathcal{M}_2(\mathbf{F}^{X^b})$, so that the semimartingale decomposition of $u(X^b(t))$ is given by

$$u(X^b(t)) = u(b) + M^u(t) + \int_0^t Lu(X^b(s))ds. \tag{2.18}$$

On the other hand, $u \in \mathcal{D}(L)$ is absolutely continuous on I^o and the Radon-Nikodym derivative u' is a function of bounded variation. Therefore, we can apply an extension of Itô's formula (Itô-Tanaka formula, cf. e.g. [RW], [RY]) to obtain

$$u\left(B^b(t \wedge m_\partial^{B^b})\right) - u(b) = \int_0^{t\wedge m_\partial^{B^b}} u'(B^b(s))dB^b(s) + \int_{I^o} l(t \wedge m_\partial^{B^b}, x)du'(x).$$

Then, by the time change $t \mapsto \tau_t$,

$$u(X^b(t)) - u(b) = \int_0^{\tau_t} u'(B^b(s))dB^b(s) + \int_{I^o} l(\tau_t, x)du'(x),$$

and we have (cf. [IW], p. 102) that $\int_0^{\tau_t} u'(B^b(s))dB^b(s) = \int_0^t u'(X^b(s))dX^b(s)$. Comparing this with the semimartingale decomposition (2.18), we readily see that $M^u(t) = \int_0^t u'(X^b(s))dX^b(s)$, which coincides with $\int_0^t u'(X(s))dX(s)$ when we take $b = a$.

We have shown in [KW] that $\{M^u; u \in \mathcal{D}(L)\}$ generates the space $\mathcal{M}_2(\mathbf{F}^{X^b})$; any $M \in \mathcal{M}_2(\mathbf{F}^{X^b})$, which is orthogonal to every M^u, $u \in \mathcal{D}(L)$, must be zero. Theorem follows at once from this. □

We would ask the following question: When is the local martingale $X = X(t)$ with $X(0) = a \in I^o$ a true martingale? In order to answer this, we first introduce the well-known classification of boundary points. Define

$$\sigma(l) = \int\int_{l<y<x<c} dm(x)dy, \quad \mu(l) = \int\int_{l<y<x<c} dxdm(y)$$

$$\sigma(r) = \int\int_{c<x<y<r} dm(x)dy, \quad \mu(r) = \int\int_{c<x<y<r} dxdm(y).$$

In the following, we state the classification in the case of the boundary l only; the case of the boundary r can be stated similarly by replacing $\sigma(l)$ and $\mu(l)$ with $\sigma(r)$ and $\mu(r)$, respectively.

Definition 2.1. The boundary l is called

(i) *regular* if $\sigma(l) < \infty$ and $\mu(l) < \infty$,

(ii) *exit* if $\sigma(l) < \infty$ and $\mu(l) = \infty$,

(iii) *entrance* if $\sigma(l) = \infty$ and $\mu(l) < \infty$,

(iv) *natural* if $\sigma(l) = \infty$ and $\mu(l) = \infty$.

The following criteria are sometimes more practical:

- l is regular if and only if $-\infty < l$ and $m((l, c]) < \infty$.
- l is exit if and only if $m((l, c]) = \infty$ and $\int_l^c m((x, c])dx < \infty$.
- l is entrance if and only if $l = -\infty$ and $\int_l^c |x|dm(x) < \infty$.

The following result is due to S. Kotani ([K]):

Theorem 2.6. $X = (X(t))$ *(with* $X_0 = a \in I^o$*) is a martingale if and only if the both boundaries* l *and* r *are* not *entrance. In other words, an existence of entrance boundary, and only this, can destroy the true martingale character of the local martingale* X*.*

Before giving a proof, we discuss some examples.

Example 2.4. (Two-dimensional Bessel diffusion process.) A two-dimensional Bessel diffusion process, that is, the radial motion of a two-dimensional Brownian motion, is a diffusion $X = (X(t))$ on $(0, \infty)$ with the generator $L = \frac{1}{2}\left(\frac{d^2}{d^2x} + \frac{1}{x}\frac{d}{dx}\right)$. A canonical scale is given by $s(x) = \log x$, so that, choosing this as the coordinate, we are considering a diffusion $Y = (Y(t))$ on $(-\infty, \infty)$ with the canonical scale $s(y) \equiv y$ and the speed measure $dm(y) = 2e^{2y}dy$, i.e. the generator given by $L = \frac{1}{2}e^{-2y}\frac{d^2}{dy^2}$. By the above criterion, the boundary $-\infty$ is entrance and the boundary ∞ is natural. Then, by the theorem, the local martingale $Y(t)(= \log(X(t)))$ is not a true martingale. This is a little surprising, as the all moments of $Y(t)$ are finite.

Example 2.5. (Positive diffusions.) Motivated by a problem in mathematical finance, Delbaen and Shirakawa (cf. [DS]) studied the following diffusion process $X = (X(t))$ on $[0, \infty)$ which is given by a solution of the following SDE:

$$dX_t = \sigma(X_t)dW_t, \quad X_0 = 1.$$

The coefficient $\sigma(x)$ is assumed to be 0 when $x \leq 0$, and is strictly positive and continuous for $x > 0$. Then a solution exists, unique in the law sense, until it reaches the origin and, we assume that it is stopped as soon as the origin is reached in a finite time.

This is equivalent to considering a diffusion process $X = (X(t))$, $X(0) = 1$, on $[0, \infty)$, that is, $l = 0$ and $r = \infty$, with the scale function $s(x) \equiv x$ and the speed measure $dm(x) = \frac{2}{\sigma(x)^2}dx$. The origin is assumed to be a trap when it can be reached in a finite time. According to the above definition and criteria, the origin 0 is regular if $\int_0^1 \frac{1}{\sigma(x)^2}dx < \infty$, exit if $\int_0^1 \frac{1}{\sigma(x)^2}dx = \infty$ but $\int_0^1 \frac{x}{\sigma(x)^2}dx < \infty$, and natural if $\int_0^1 \frac{x}{\sigma(x)^2}dx = \infty$. It cannot be entrance. As for the boundary point ∞, it is either entrance or natural, and it is so according as $\int_1^\infty \frac{x}{\sigma(x)^2}dx < \infty$ or $\int_1^\infty \frac{x}{\sigma(x)^2}dx = \infty$.

It is known (cf. [IM], [IW, p. 450]) that X reaches the origin in a finite time a.s. if the origin is either regular or exit, equivalently, if $\int_0^1 \frac{x}{\sigma(x)^2}dx < \infty$, and X cannot reach the origin in a finite time, if the origin is natural, equivalently, if $\int_0^1 \frac{x}{\sigma(x)^2}dx = \infty$.

Also, by the theorem above, we can conclude that $X(t)$ is a true martingale if and only if ∞ is natural, equivalently, if $\int_1^\infty \frac{x}{\sigma(x)^2}dx = \infty$.

In [DS], these results are rediscovered by a method based on the Ray-Knight Theorem for Brownian local times.

Proof of Theorem 2.6. First, we note that $X = (X(t))$ is a martingale if and only if $X^b(t)$ is integrable and $E[X^b(t)] = b$ for any $t > 0$ and $b \in I^o$. This is a simple consequence of the Markov property of the diffusion. By taking the Laplace transform, these conditions can be equivalently stated in terms of the resolvent kernel: Let, as above, $G_\lambda f(x) = \int_0^\infty e^{-\lambda t} E\left[f(X^x(t))\right] dt, f \in C(I)$. Then, $G_\lambda f(x) = \int_I G_\lambda(x, dy) f(y)$ by the resolvent kernel $G_\lambda(x, dy)$. $X = (X(t))$ is a martingale if and only $\int_I G_\lambda(x, dy)|y| < \infty$ and $\int_I G_\lambda(x, dy)y = \frac{x}{\lambda}$, for every $\lambda > 0$ and $x \in I^o$.

Let $u_1(x)$ be a positive and increasing solution of the homogeneous equation $\lambda u - Lu = 0$ on I^o, where $L = \frac{d}{dm}\frac{d}{dx}$ is the local generator of X. *When l is regular, we impose the condition that* $u(l+) = 0$. Then u_1 exists and it is unique up to a multiplicative constant.

Similarly, let $u_2(x)$ be a positive and decreasing solution of the homogeneous equation $\lambda u - Lu = 0$ on I^o with the condition $u(r-) = 0$ when r is regular. Then u_2 exists and it is unique up to a multiplicative constant.

The following property of functions u_1 and u_2 is a key in the proof, cf. [IM], [I]; a nice and detailed proof can be found in the latter reference. We state it for the increasing solution u_1 and its derivative u_1': For the decreasing solution u_2 and $-u_2'$, it can be stated similarly by exchanging the role of boundaries l and r.

- $u_1(l+) > 0$ if and only if l is entrance; so, if l is regular, exit or natural, then $u_1(l+) = 0$.

- $u_1(r-) < \infty$ if and only if r is regular or exit; so, if r is entrance or natural, $u_1(r-) = \infty$.

- $u_1'(l+) > 0$ if and only if l is regular or exit; so, if l is entrance or natural, $u_1'(l+) = 0$.

- $u_1'(r-) < \infty$ if and only if r is regular or entrance; so, if r is exit or natural, $u_1'(r-) = \infty$.

Define

$$\phi_1(x) = \begin{cases} \frac{u_1(x)}{u_1(r-)}, & \text{if } r \text{ is regular or exit,} \\ 0, & \text{if } r \text{ is entrance or natural,} \end{cases}$$

and similarly

$$\phi_2(x) = \begin{cases} \frac{u_2(x)}{u_2(l+)}, & \text{if } l \text{ is regular or exit,} \\ 0, & \text{if } l \text{ is entrance or natural.} \end{cases}$$

We define the λ-order Green function $g_\lambda(x, y),\ x, y \in I^o,$ by

$$g_\lambda(x, y) = \begin{cases} \frac{u_1(x)u_2(y)}{h(\lambda)}, & \text{if } x \leq y, \\ \frac{u_1(y)u_2(x)}{h(\lambda)}, & \text{if } y \leq x, \end{cases}$$

where $h(\lambda) = u_1'(x)u_2(y) - u_1(x)u_2'(x)$ is the Wronskian of u_1 and u_2 (which is independent of x). Then the resolvent kernel $G_\lambda(x, dy),\ x \in I^o,\ dy \subset I,$ is given by

$$G_\lambda(x, dy) = 1_{I^o}(y) g_\lambda(x, y) dm(y) + \frac{\phi_1(x)}{\lambda} \cdot \delta_r(dy) + \frac{\phi_2(x)}{\lambda} \cdot \delta_l(dy).$$

Note that, for a fixed $x \in I^o$, the measure $G_\lambda(x, dy)$ has its full measure $1/\lambda$ on $I \cap (-\infty, \infty)$.

Now we can show that $\int_I G_\lambda(x, dy)|y| < \infty$, and the following relation:holds:

(2.19) $$\int_I G_\lambda(x, dy) y = \frac{x}{\lambda} + \frac{u_1(l+)}{\lambda h(\lambda)} \cdot u_2(x) - \frac{u_2(r-)}{\lambda h(\lambda)} \cdot u_1(x).$$

For this, we need the following lemma:

Lemma 2.1. *(1)*

$$\int_{(l,x]} y u_1(y) dm(y)$$
$$= \begin{cases} x \int_{(l,x]} u_1(y) dm(y) - \frac{u_1(x)}{\lambda} + \frac{u_1'(l+)(x-l)}{\lambda}, & \text{if } l \text{ is regular or exit,} \\ x \int_{(l,x]} u_1(y) dm(y) - \frac{u_1(x)}{\lambda} + \frac{u_1(l+)}{\lambda}, & \text{if } l \text{ is entrance or natural.} \end{cases}$$

(2)

$$\int_{(x,r)} y u_2(y) dm(y)$$
$$= \begin{cases} x \int_{(x,r)} u_2(y) dm(y) + \frac{u_2(x)}{\lambda} + \frac{u_2'(r-)(r-x)}{\lambda}, & \text{if } r \text{ is regular or exit,} \\ x \int_{(x,r)} u_2(y) dm(y) + \frac{u_2(y)}{\lambda} - \frac{u_2(r-)}{\lambda}, & \text{if } r \text{ is entrance or natural.} \end{cases}$$

Note also that

$$u_1'(l+) = \frac{h(\lambda)}{u_2(l+)} \quad \text{if } l \text{ is regular or exit,}$$

and

$$u_2'(r-) = -\frac{h(\lambda)}{u_1(r-)} \quad \text{if } r \text{ is regular or exit.}$$

Proof of Lemma 2.1. Suppose that l is regular or exit. Then $u_1(l+) = 0$ and we have

$$u_1(x) = \int_l^x u_1'(z)dz = \int_l^x dz\left[u_1'(l+) + \int_{(l,z]} du_1'(y)\right].$$

Since u_1 is a solution of $\lambda u - Lu = 0$, we have $du_1'(y) = \lambda u_1(y)dm(y)$. Hence,

$$\begin{aligned} u_1(x) &= u_1'(l+)(x-l) + \lambda \int\int_{l<y<z<x} dz\, u_1(y)dm(y) \\ &= u_1'(l+)(x-l) + \lambda \int_{(l,x]} (x-y)u_1(y)dm(y) \\ &= u_1'(l+)(x-l) + \lambda x \int_{(l,x]} u_1(y)dm(y) - \lambda \int_{(l,x]} yu_1(y)dm(y). \end{aligned}$$

From this, we obtain

$$\int_{(l,x]} yu_1(y)dm(y) = x\int_{(l,x]} u_1(y)dm(y) - \frac{u_1(x)}{\lambda} + \frac{u_1'(l+)(x-l)}{\lambda}.$$

The proof in the case when l is entrance or natural is similar: We note, in this case, that $u(_1(l+) \geq 0$, $(u(_1(l+) > 0$ if and only if l is entrance), and $u_1'(l+) = 0$ always.

The proof in the case of the right boundary point r is the same if the role of l and r are exchanged. □

We have

$$\int_{I^o} g_\lambda(x,y)|y|dm(y) = \frac{u_1(x)}{h(\lambda)}\int_{(x,r)} |y|u_2(y)dm(y) + \frac{u_2(x)}{h(\lambda)}\int_{(l,x]} |y|u_1(y)dm(y)$$

and, by noting $\int_{I^o} g(x,y)dm(y) \leq 1/\lambda$ and using Lemma 2.1, we can estimate this as $O(1+|x|)$.

We have also,

$$\int_{I^o} g_\lambda(x,y)ydm(y) = \frac{u_1(x)}{h(\lambda)}\int_{(x,r)} yu_2(y)dm(y) + \frac{u_2(x)}{h(\lambda)}\int_{(l,x]} yu_1(y)dm(y).$$

Substituting the expressions in Lemma 2.1 to the integrals in the right-hand side, we can obtain (2.19).

From (2.19), we see that $G_\lambda(x,dy)y = x/\lambda$, $x \in I^o$, if and only if $u_1(l+)u_2(x) - u_2(r-)u_1(x) = 0$ on I^o. We can easily deduce that this holds if and only if $u_1(l+) = u_2(r-) = 0$. This holds if and only if neither l nor r is entrance. This completes the proof of Theorem 2.6. □

Remark 2.1. When l or r is regular, we assumed it to be a trap for our diffusion. Of course, there is a variety of possible boundary behaviors and $X(t)$ is generally not a local martingale, any more.

References

DS. F. Delbaen and H. Shirakawa, No arbitrage condition for positive diffusion price processes, *Asia-Pacific Financial Markets* **9** (2002), 159–168.

IW. N. Ikeda and S. Watanabe, *Stochastic Differential Equations and Diffusion Processes*, Second Edition, North-Holland/Kodansha, Amsterdam/Tokyo, 1988.

I-1. K. Itô, Multiple Wiener Integral, *J. Math. Soc. Japan*, **3** (1951), 157–169.

I-2. K. Itô, *Kakuritu Katei II* (Stochastic Processes, II), Iwanami Shoten, Tokyo, 1957 (in Japanese); English Translation by Yuji Ito, Yale University, 1961.

IM. K. Itô and H. P. McKean, Jr., *Diffusion Processes and Their Sample Paths*, Springer, Berlin, 1965, Second Printing 1974, in *Classics in Mathematics*, 1996.

KK. G. Kallianpur and R. L. Karandikar, *Introduction to Option Pricing Theory*, Birkhäuser, Boston/Basel/Berlin, 2000.

K. S. Kotani, *On a condition that one-dimensional diffusion processes are martingales*, 2003.

KW. H. Kunita and S. Watanabe, On square integrable martingales, *Nagoya Math. J.* **30** (1967), 209–245.

M. P. Malliavin, *Stochastic Analysis*, Springer, Berlin, 1997.

P. P. Protter, *Stochastic Integration and Differential Equations, A New Approach*, Springer Verlag, Berlin/Heidelberg/New York, 1990.

RW. L. C. G. Rogers and D. Williams, *Diffusion, Markov Processes, and Martingales, Vol. 2, Itô Calculus*, John Wiley & Sons, Chichester/New York/Brisbane/Toronto/Singapore, 1987

S. D. W. Stroock, *Markov Processes from K. Itô's Perspective*, *Annals of Mathematical Studies* **155** (2003), Princeton University Press, Princeton/Oxford.

VK. A. Ju. Veretennikov and N. V. Krylov, On explicit formulas for solutions of stochastic differential equations, *Math. USSR Sbornik* **29** (1976), 239–256.

TINNITUS

New Therapeutic Tools
and Techniques

TINNITUS
New Therapeutic Tools and Techniques

Editors

Mehdi Abouzari
Hamid R Djalilian
University of California, Irvine, USA

World Scientific

NEW JERSEY • LONDON • SINGAPORE • BEIJING • SHANGHAI • TAIPEI • CHENNAI

Published by

World Scientific Publishing Co. Pte. Ltd.
5 Toh Tuck Link, Singapore 596224
USA office: 27 Warren Street, Suite 401-402, Hackensack, NJ 07601
UK office: 57 Shelton Street, Covent Garden, London WC2H 9HE

Library of Congress Control Number: 2025006463

British Library Cataloguing-in-Publication Data
A catalogue record for this book is available from the British Library.

TINNITUS
New Therapeutic Tools and Techniques

ISBN 978-981-98-0733-8 (hardcover)
ISBN 978-981-98-0734-5 (ebook for institutions)
ISBN 978-981-98-0735-2 (ebook for individuals)

For any available supplementary material, please visit
https://www.worldscientific.com/worldscibooks/10.1142/14162#t=suppl

Desk Editors: Soundararajan Raghuraman/Julio Hong/Steven Patt

Typeset by Stallion Press
Email: enquiries@stallionpress.com

To my girls, Elham and Niki.

Mehdi

To my parents and siblings, for your unwavering love, sacrifices, and the values you instilled in me.

To my wife, my partner in every journey, whose support and belief make every dream possible. And to my children, the brightest lights in my life, who remind me daily of the beauty of wonder and hope.

Last, but not least, to my many mentors, whose guidance, wisdom, and encouragement not only shaped my path but also fostered my love for research and the pursuit of knowledge. I wanted to especially thank Drs. Walter Hall, Sam Levine, Hamed Sajjadi, and the late Michael Paparella who were pivotal in my education.

Many years of research on tinnitus allowed me to make the discoveries that form the basis of this book, and none would be possible without all of you. This book is for all of you, with my deepest gratitude and love.

Hamid

https://doi.org/10.1142/9789819807345_fmatter

Foreword

Tinnitus is a "phantom" phenomenon that is very real for those who experience it and unfortunately can be disturbing and even disabling or overwhelming for a sizeable subset. Since we view tinnitus as a disease and not as a symptom, patients and healthcare providers alike can become frustrated at the lack of a single or small number of effective treatments. However, a mind shift to viewing tinnitus as a symptom that may result from a number of causes and may be worsened or improved by a number of other factors allows all of us who are in the tinnitus workup and treatment pathway to approach it in a reasonable, systematic fashion, and achieve good levels of success.

Drs. Mehdi Abouzari and Hamid Djalilian have taken on this huge subject and, because of their perspective and experience, have divided it up into sections that enable all of us to experience the necessary mind shift and begin thinking of tinnitus not as a single, large, confusing problem that is difficult to approach but as a symptom with different causes and exacerbating and alleviating factors that can be approached in a shared decision-making style with the patient and their family. This book covers the current thinking exceptionally well.

Starting from the KISS concept of "keep it simple, stupid", the chapters highlight dietary and lifestyle modifications, cognitive behavioral techniques, sound and music therapy, and internet-based self-help, which is accessible to all. Hearing aid effectiveness is well known and is described in detail. A good majority of tinnitus sufferers will benefit from understanding what is conveyed in this chapter. From there, this book goes on to discuss pharmacological, complementary or alternative medicine, and a new intervention, botulinum toxin. Neurostimulation can be delivered in a number of ways; this is described well and should be understood to this degree of detail by all tinnitus care providers so that counseling can be as thorough as possible. Finally, this book concludes with chapters devoted to very difficult issues in tinnitus — hyperacusis, misophonia, pulsatile tinnitus, and childhood tinnitus, before tackling the future of safe, effective stem cell therapy.

I congratulate Drs. Abouzari and Djalilian on their clinical and research work leading up to this book, and on convening the right subjects and authors to make this work an excellent reference for all. We are now moving from "you'll have to learn to live with it" to "let's make sense of what's going on and target our treatments together". This is a giant leap forward and deserves our highest praise.

Sujana S. Chandrasekhar, MD
Partner, ENT & Allergy Associates, LLP
Past President, American Academy of Otolaryngology-Head and Neck Surgery
Past President, American Otological Society
Consulting Editor, Otolaryngologic Clinics of North America
Co-Host and Co-Executive Producer, She's On Call

https://doi.org/10.1142/9789819807345_fmatter

Preface

Tinnitus is a condition that affects millions of people worldwide, with 50 million people experiencing tinnitus in the US alone. For many, it can significantly impact their quality of life, interfering with daily activities and causing distress. As our understanding of this complex condition evolves, so too do the therapeutic approaches and tools available to manage it. Recent advancements in tinnitus research have led to a paradigm shift in our understanding of this complex condition. This book embraces a fresh perspective that recognizes tinnitus as a brain disorder influenced by multiple systems, necessitating a comprehensive approach to care that accounts for the intricate connection between the senses, brain, body, and mind. This new understanding challenges traditional notions of tinnitus, expanding beyond its roots in hearing loss to embrace other systemic and multifactorial nature. By incorporating this broader understanding of tinnitus as a complex, multifaceted condition influenced by brain hypersensitivity, this book aims to provide a more comprehensive and effective approach to tinnitus management.

This book brings together insights from leading researchers, clinicians, and experts in the field, offering readers an up-to-date exploration of novel approaches to tinnitus management. Whether you are a healthcare professional seeking to expand your knowledge of tinnitus management, an advanced student or researcher exploring new avenues for investigation, in industry seeking new business opportunities, or an individual living with tinnitus looking for information on the latest treatment options, this book offers valuable insights and practical information. As we continue to make strides in our understanding and treatment of tinnitus, it is our hope that this book will serve as a valuable resource, contributing to improved outcomes and quality of life for those affected by this challenging condition.

Finally, we would like to hereby acknowledge the expertise of all contributors, for generously devoting their time and considerable effort in preparing their respective chapters. We would also like to express our gratitude to World Scientific for providing us the opportunity to publish this book. Last but not least, we would like to thank Dr. Karen Tawk for her immense editorial assistance to this book.

Mehdi Abouzari
Hamid R. Djalilian

https://doi.org/10.1142/9789819807345_fmatter

About the Editors

Mehdi Abouzari, MD, PhD is a translational physician-scientist, an assistant professor of otolaryngology, and the director of clinical research of otology and neurotology at the University of California, Irvine. He is an associate fellow of the American Neurotology Society, associate editor of the *Frontiers in Audiology and Otology*, and editorial board member of *Research in Vestibular Science*. He has authored 110 plus peer-reviewed papers and 8 book chapters and presented 90 abstracts at national and international conferences with over 2000 citations to date. His most impactful work has been a group of studies showing that a number of ear disorders represent migraine phenomena. He has successfully led several clinical trials on the development of technology-based platforms and devices that address hearing loss and tinnitus. He is also the director of the ENTelligence Lab (https://entelligencelab.com/) which focuses on the application of artificial intelligence and machine learning in otolaryngology and hearing sciences.

Hamid R. Djalilian, MD is a professor of otolaryngology, neurosurgery, and biomedical engineering and the director of otology and neurotology at the University of California, Irvine. He completed his residency at the University of Minnesota and his fellowship at the Minnesota Ear, Head and Neck Clinic. He has published two books, over 200 peer-reviewed papers, and 150 invited papers and chapters, and lectured more than 250 times at regional, national, and international meetings. In addition to a busy clinical practice, his clinical research interests are the effects of migraine on the ear, developing novel treatments for tinnitus, medical devices, and improvement of surgical outcomes. He has developed several medical devices that are undergoing clinical trials and has several granted and pending patents related to these devices and tinnitus treatment. His basic science lab is studying gene therapy for restoration of hearing, mitochondrial preservation, and function enhancement for prevention and treatment of hearing loss. He was the founding president of the Migraine in Otolaryngology Society (miosociety.org) and is the chief medical advisor to NeuroMed Care (neuromedcare.com), a telemedicine clinic to treat tinnitus using medical treatments developed by his laboratory for tinnitus.

https://doi.org/10.1142/9789819807345_fmatter

Contents

https://doi.org/10.1142/9789819807345_0001

Chapter 1

Dietary and Lifestyle Modifications in Tinnitus Treatment

Ella J. Lee[*,†], Najva Mazhari[*,†], Ryka Vahidi[†], Avissa Khoshsar[†], Madeleine Chao[†], Allison Oh[†], and Hamid R. Djalilian[†,‡,§,¶]

[†]*Department of Otolaryngology-Head and Neck Surgery, University of California, Irvine, CA, USA*

[‡]*Department of Biomedical Engineering, University of California, Irvine, CA, USA*

[§]*Department of Neurosurgery, University of California, Irvine, CA, USA*

[¶]*hdjalili@hs.uci.edu*

Abstract

Tinnitus, the perception of noise without external stimuli, is a prevalent and sometimes debilitating condition impacting sleep, attention, mood, and quality of life. Currently, the full pathophysiology of tinnitus remains unclear. What has been recently understood is that processes related to migraine play a significant role in the loudness of tinnitus. Given that these processes are influenced by diet and lifestyle factors (stress, sleep, overstimulation), modifying diet and lifestyle offers a promising strategy

[*]These two authors contributed equally to this manuscript.

for the management of tinnitus. Establishing and maintaining regular meal schedules and adequate hydration play a pivotal role in managing tinnitus. Several food items identified as migraine triggers also exacerbate tinnitus symptoms through similar mechanisms in some patients. These include foods containing caffeine, tyramine, aspartame, phenylethylamine, monosodium glutamate, nitrates/nitrites, sulfites, and some forms of alcohol. Conversely, certain supplements have demonstrated the potential to support inner ear health. Magnesium, vitamin B_2 (riboflavin), and coenzyme Q10 are among the most notable. Adopting good sleep hygiene practices and stress management techniques are also crucial elements for general health and brain function. Poor sleep quality and lack of sleep have been linked to aggravation of tinnitus. Therefore, maintaining a regular sleep schedule, controlling sleep apnea, and using guided meditation before sleep can significantly improve sleep quality and help prevent or alleviate tinnitus. Additionally, stress management techniques, such as regular physical activity, mindfulness, and meditation, have shown promise in reducing both the frequency and intensity of loud tinnitus episodes. Integrating dietary adjustments, supplements, and lifestyle modifications represents a holistic approach to managing tinnitus without the need for invasive treatments. These strategies serve as a valuable initial step in tinnitus management, offering meaningful relief and fostering a sense of control over this challenging condition.

Introduction

Tinnitus is a common condition affecting over 740 million people globally to date [1]. It is characterized by the perception of noise, such as buzzing, rustling, or whistling, in the absence of external acoustic stimuli. Tinnitus can affect individuals of all ages, although it is more common among older adults [1]. It can present substantial challenges to those experiencing it due to its widespread occurrence, compromising various aspects of life including sleep, mood, and overall well-being [2].

While the precise pathophysiology remains unclear, various mechanisms have been suggested, including genetic, iatrogenic, neurological, and vascular pathways [2]. Prominent theories regarding its origin, such as the cortical theory and the dorsal cochlear nucleus theory, propose that tinnitus may arise from damage to the vestibulocochlear nerve and the trigeminal nerve, both of which relay signals to the dorsal cochlear nucleus [2].

Tinnitus and several other vestibulocochlear disorders are closely related to the pathogenesis of migraine headaches. Migraine is believed to result from the activation of peripheral and central trigeminovascular neurons, which in turn, release neuropeptides and cytokines such as CGRP, substance P, neurokinin A, and pituitary adenylate cyclase-activating polypeptide [4]. This cascade of events induces vasodilation, plasma extravasation, and mast cell degranulation, ultimately establishing a chronic neuroinflammatory state within the meninges and causing headaches [3]. These processes can also lead to alteration in the inner ear vasculature, which is directly innervated by the trigeminal system [5]. For instance, during a migraine attack, changes in cochlear blood flow can result in temporary or permanent hearing decline. Additionally, activation of the trigeminal nerve may amplify central sensitivity. As a result, individuals may experience heightened awareness or perception of tinnitus, and exacerbation of symptoms [6]. Although a common notion is that migraine always accompanies headaches, some patients with migraine never develop a headache. Atypical migraine, such as ocular migraine, vestibular migraine, and cochlear migraine, may not feature the typical unilateral, pulsating headaches, but can manifest with other symptoms like increased tinnitus loudness, vertigo, hyperacusis, aural fullness, and otalgia [6–9]. The unconventional presentations often pose challenges for diagnosis using standard criteria [10], suggesting that some patients experiencing tinnitus may have undiagnosed atypical migraine. Indeed, tinnitus occurs frequently across various types of migraine, including vestibular migraine and migraine with aura [11, 12]. Furthermore, there is a significant relationship between tinnitus and headache laterality, implying potentially shared pathophysiological mechanisms [13]. From these associations, one can deduce that migraine therapies can also be effective in managing tinnitus symptoms.

Currently, there are no medications approved by the US Food and Drug Administration (FDA) for the treatment of tinnitus [14]. Still, various classes of medications, including anesthetics, antiarrhythmics, antidepressants, antihistamines, calcium channel blockers, and vasodilators, are often used off-label to treat patients with tinnitus. However, the efficacy of these as a singular modality of treatment remains uncertain [14, 15]. Therefore, managing tinnitus often involves non-pharmacological interventions including counseling and psychoeducation, cognitive behavioral therapy (CBT), and tinnitus retraining therapy (e.g., sound therapy, hearing aids) [16]. Emerging research suggests that dietary and lifestyle factors may play a major role in either exacerbating or mitigating tinnitus symptoms [17], much like their impact on migraine. The American Headache Society recommends that the management of migraine attacks involve patient education about appropriate lifestyle modifications [18]. This entails identifying and minimizing exposure to migraine triggers such as psychological and physical stress, variations in sleep pattern (e.g., too much or too little sleep, interrupted sleep), dietary patterns (e.g., skipping meals, certain foods, and dehydration), hormonal changes, and exposure to intense stimuli (e.g., bright light, intense sounds, and intense motion) [18–22]. Given the parallels between the two conditions, a comparable approach is warranted for managing tinnitus. Implementing a comprehensive lifestyle adjustment, dietary modification, and the incorporation of supplements would not only be instrumental for improving tinnitus outcomes but also empower individuals to take proactive steps toward improving their conditions and overall well-being. In our clinic, we commonly use lifestyle and dietary changes as a first step in treatment, prior to initiating medical therapy. We have found that combining medical therapy with lifestyle and dietary changes significantly reduces the loudness of tinnitus.

Dietary Modifications

Regular eating habits and dietary factors

A simple yet effective approach to dietary modification is maintaining a regular meal schedule. This involves eating meals at consistent times daily and avoiding hunger and intermittent fasting. Previous studies have linked migraine to hypothalamus activity, which regulates hunger and satiety by responding to various hormones [23]. Alteration of its activity, such as due to fasting, is believed to be responsible for migraine premonitory symptoms like hunger, thirst, fatigue, and desire to micturate [24]. Consequently, skipping meals is often cited as a common trigger in these patients [21]. Moreover, hypothalamic activity is also implicated in cochlear physiology [25], as evidenced by its communication with the vestibular system [26], highlighting its broader impact on sensory functions. Fasting can also induce hypoglycemia, often discussed as an inciting factor for migraine headaches. Interestingly, hypoglycemia has been observed as a trigger in a subset of other vestibulocochlear disorders such as sudden sensorineural hearing loss [27] and Ménière's disease, both of which are likely migraine variants [9]. This connection offers an additional explanation for why an irregular meal schedule might provoke migraine headaches and related conditions like tinnitus.

Adequate hydration is an essential aspect of overall health and well-being. Its importance has also been implicated in inner ear health, although the exact mechanisms are not fully understood [28]. Research indicates that reduced water intake and low body water percentage correlate with greater tinnitus-related distress and tinnitus severity [22, 29]. To maintain optimal hydration, it is recommended to consume at least 2 liters (≥67 oz.) of water daily, with increased intake during exercise or exposure to hot weather.

Dietary fat intake and obesity are potential risk factors for tinnitus. Elevated levels of free fatty acids can alter the inner ear's pH balance, leading to vascular injuries in the cochlea [22, 30]. According to one study, obese individuals were found to be 2.14 times more likely to have tinnitus compared to those with normal weight [31]. This observation has been attributed to greater oxidative response and generation of free radicals in obese individuals, which is believed to promote damage in the vestibulocochlear system [22, 32]. This correlation may also be influenced by the higher prevalence of obstructive sleep apnea in obese patients. Therefore, moderating fat intake and reducing body mass index (BMI) through dietary changes, physical activity, or a combination of thereof can aid improve tinnitus outcomes.

Food triggers

Studies indicate a notable overlap between dietary triggers for tinnitus and migraine, suggesting that avoiding certain food items can help reduce the severity, frequency, or duration of symptoms in both conditions. While research linking specific food triggers to tinnitus is ongoing, it is hopeful that tinnitus patients may benefit similarly from identifying and avoiding sensitive foods [3, 22]. Common triggers in migraine patients, such as caffeine, tyramine (found in fermented foods such as cheese, wine, yeast, etc.), aspartame (artificial sweeteners), monosodium glutamate (soy sauce, frozen or canned foods, pickled foods, salad dressings, etc.), nitrates (cured meats and hot dogs), and phenylethylamine (chocolates and cacao-containing products), have also shown sensitivity in some tinnitus patients.

Identifying food triggers can be challenging due to their variable nature and the delayed onset of symptoms, which may manifest hours after ingestion. Keeping a comprehensive food and symptom diary (e.g., migraine diary) over a few weeks can assist in pinpointing specific triggers for each individual. This approach facilitates personalized dietary interventions tailored to individual preferences and needs [33]. In our clinic, we encourage patients to begin by avoiding daily consumption of foods containing potentially offending molecules. If a patient experiences an increase in tinnitus levels, we advise them to assess their diet and other factors such as dehydration, hunger, stress, sleep, and overstimulation in the 6 hours prior to the spike in the tinnitus to identify the trigger.

Caffeine

Caffeine is a substance commonly found in coffee, tea, and soda, and is also a component of certain medications including Fioricet, Esgic, Anacin, and Excedrin [20]. It is a central nervous system stimulant that belongs to the methylxanthines class [34]. The primary action of caffeine is to block adenosine receptors, thereby modulating neurotransmitter release and neuronal activities [35].

Caffeine is frequently discussed in relation to auditory and vestibular systems, particularly regarding its effect on tinnitus. While many patients with tinnitus identify caffeine as a trigger, a few studies have found that caffeine consumption may have no effect or even a protective effect against tinnitus. For example, a meta-analysis of 384 studies investigating lifestyle risk factors found mixed results regarding the relationship between caffeine intake and tinnitus and was unable to draw a definitive conclusion [36]. Within this meta-analysis was a large-scale prospective cohort study that found a significant inverse association between caffeine intake and tinnitus incidence in women [37]. Similarly, both pseudo- and true randomized placebo-controlled trials found no relationship between caffeine consumption and tinnitus severity, concluding caffeine abstinence as an ineffective strategy to alleviate tinnitus [34, 38]. On the other hand, in a cross-sectional study of 258 adults, it was found that nearly 6% of subjects reported worsening tinnitus after drinking coffee or tea, while only 0.4% reported feeling better afterward [39]. An important consideration regarding caffeine is its impact on sleep quality. Caffeine consumption is associated with prolonged sleep onset latency, shorter sleep duration, and poor perceived sleep quality [40]. Decreased sleep quality and duration are correlated with increased tinnitus severity, as will be discussed in subsequent sections.

Caffeine has been recognized as a common trigger of migraine attacks as well as migraine-associated vertigo attacks [41]. Consequently, eliminating caffeine from one's diet can decrease the frequency, onset, and severity of migraine attacks [41]. Also, caffeine overuse can escalate episodic migraine into chronic migraine [42]. It is important to note that most studies indicate chronic or high-dose caffeine consumption (≥ 400 mg/day) as a trigger, rather than low-to-moderate amounts [43]. Paradoxically, an abrupt withdrawal from caffeine also induces migraine [42]. Given the varied research outcomes, individuals with tinnitus should be cautious with their caffeine consumption. It is recommended to gradually taper caffeine intake over several weeks and avoid excessive intake to minimize the risk of triggering migraine and worsening auditory symptoms [42]. While not every patient benefits from caffeine elimination, it is difficult to predict which patients will experience benefits before they stop consuming it. In our clinic, we have found caffeine to be the worst of the dietary triggers. Patients are encouraged to avoid not only caffeinated beverages but also decaffeinated coffee, which still contains trace amounts of caffeine. We have seen many patients whose primary dietary trigger for loud tinnitus has been caffeine.

Tyramine

Tyramine is a monoamine derived from the amino acid tyrosine. Tyramine is formed from the breakdown of protein during aging, processing, or fermentation. It is found in fermented foods such as aged cheese, sour cream, and beer, as well as processed protein (e.g., protein shakes or protein bars), and overly-ripened fruits [44]. In cheese, tyramine levels increase with the aging process, making certain varieties like cheddar, mozzarella, and parmesan rich in this compound. On the other hand, cheeses made from pasteurized milk, such as cottage cheese, ricotta, and American cheese, contain little or no tyramine and are considered better options [3]. Cured, smoked, or processed proteins, including sausages and bacon, are rich in tyramine and can trigger migraine and tinnitus attacks [3, 45]. Tyramine acts as an indirect sympathomimetic by causing the release of neurotransmitters and vasoactive substances. It is through this mechanism that tyramine is thought to provoke headaches [46]. Moreover, tyramine is metabolized by the enzyme monoamine oxidase (MAO). A byproduct of this reaction is hydrogen peroxide, which can cause damage to the mitochondrial DNA. It has also been suggested that this oxidative stress generated by tyramine contributes to the pathogenesis of migraine [32], and therefore, to tinnitus loudness [9].

Aspartame

Aspartame (NutraSweet) is an artificial sweetener often used as a substitute for sucrose and is 180 to 200 times sweeter than sucrose [47]. Evidence suggests that aspartame can trigger migraine headaches in those who consume moderate to high doses (900 to 3000 mg/d) over time [20]. A single 12-oz can of diet soda (e.g., Diet Coke) provides approximately 185 mg of aspartame [48], and drinking several cans of diet soda daily could potentially exceed the threshold associated with migraine triggers. Although the incidence may be low, some individuals may be genetically predisposed to aspartame toxicity, as illustrated in a case report involving a pair of siblings [49]. The report showed that cochlear symptoms persisted for at least 15 months after discontinuing aspartame use, highlighting the possible long-lasting effects of aspartame on the cochlea. When experiencing dizziness, vertigo, tinnitus, or fluctuating hearing loss, it may be beneficial to monitor and consider discontinuing aspartame consumption [45, 49]. Sucrose and high-fructose corn syrup pose no known risks [3]. Other artificial sweeteners such as Saccharin (Sweet'n Low) and Sucralose (Splenda) are considered better alternatives to table sugar, however, contrary findings have been reported [50, 51]. We generally recommend discontinuing aspartame (primarily in diet sodas) and reducing sugar intake overall. Some patients report an increase in tinnitus after sugar consumption, possibly due to compensatory hypoglycemia post-insulin action or a neuroinflammatory mechanism.

Phenylethylamine

Phenylethylamine is a derivative of phenylalanine found in cocoa beans and its products [20]. While it is most concentrated in chocolate, trace amounts can also be present in other fermented foods such as cheese and wine [52]. There are currently very few studies directly exploring the relationship between phenylethylamine and tinnitus. However, it is often grouped with other bioamines like histamine and tyramine as being responsible for dietary-induced migraine [32]. Patients with migraine have a reduced ability to deaminate all monoamines, particularly phenylethylamine [53]. Phenylethylamine has been found to cause vasodilation and increase cerebral blood flow, thus elevating intracranial pressure [53]. Studies also suggest that other components, such as flavonoids in chocolate and wine, might be more responsible for migraine headaches [32].

Monosodium glutamate

Glutamate is an amino acid naturally present in food, either in free form, bound to proteins, or in the form of salt [54]. It is often added to the diet as monosodium glutamate (MSG), a widely used food additive and flavor enhancer, particularly in Asian foods/sauces and packaged fast foods [55]. MSG is associated with the "Chinese restaurant syndrome", characterized by symptoms such as headaches, tinnitus, facial pressure, and throbbing of the head and ears [56]. For instance, a case report documented a man who experienced transient vertigo after ingesting spicy Chinese sausages containing MSG [57]. The authors attributed this reaction to MSG's interaction with glutamate receptor subunit δ1, which is highly expressed in the inner hair cells of the organ of Corti. Other studies suggest that this effect may result from high concentrations of glutamate causing vasoconstriction, leading to vascular-like headaches in susceptible individuals [58]. Research has investigated increased endogenous glutamate release as a potential contributor to migraine as well as tinnitus and its intensity [25, 59], but no studies have extensively evaluated oral MSG as a direct risk factor for tinnitus. However, knowledge accumulated so far indicates that MSG consumption could be detrimental to the auditory system [60, 61]. Similarly, it has been indicated as a dietary trigger for migraine and other inner ear conditions, such as Ménière's disease [28, 62, 63].

Nitrates and nitrites

Nitrate is present in vegetables and fruit, but it is found in particularly high levels in processed meats. When ingested, nitrites are reduced by facultative anaerobic bacteria of the oral cavity to yield nitrite [64]. Nitrites are also consumed in the form of sodium nitrite, a commonly used food preservative. Nitrate- and nitrite-containing compounds have been identified as triggers of headache. The vasodilation by nitric oxide (NO), the product of the nitrate-nitrite-NO pathway, is thought to be responsible for this reaction [65]. This mechanism may be similarly implicated in tinnitus. An animal model of tinnitus has revealed that increased NO production in the ventral cochlear nucleus contributes to the onset and persistence of tinnitus [66]. It is noteworthy that NO also participates in the NO/ONOO^{-} cycle, where its activity can induce tinnitus and contribute to other inner ear symptoms such as hearing loss, vertigo, and dizziness [67]. Being mindful of nitrate and nitrite intake from foods and additives is worthwhile for individuals to assess their potential impact on tinnitus symptoms. Even low-nitrate or "nitrate-free" processed meats contain high levels of tyramine. Therefore, in our clinic, we advise patients to avoid consuming processed meats.

Sulfites

Sulfite is an inorganic salt with antioxidant and preservative properties [68]. It is commonly used as a food additive, found in a variety of dried and frozen foods, jams, and sauces. Sulfites occur naturally in wine during the fermentation process and are also commonly added to wine to help the preservation of its taste and color. While most people tolerate sulfites well, some individuals can have sensitivities, possibly due to sulfite oxidase deficiency. For these individuals, consumption of sulfite-containing food will lead to the accumulation of sulfites in the serum. The buildup can inhibit several enzymes, affecting the metabolism in the nucleus accumbens, and possibly contribute to tinnitus [68]. Therefore, individuals with tinnitus should exercise caution when consuming foods containing preservatives or additives, which are sources of both nitrites and sulfites. At the very least, patients should consider them as possible triggers. Although "sulfite-free" wines are available, which lack added sulfite, they still contain the natural sulfites and tyramine produced from the fermentation process. Therefore, we recommend against wine consumption for our tinnitus patients.

Alcohol

Alcohol consumption is a common modifiable risk factor for many health conditions. It is also identified as a dietary trigger in over 20% of migraine patients [69]. Like caffeine, its effect on tinnitus is ambiguous. For instance, a meta-analysis of 25 observational studies found negative associations indicating that high alcohol consumption might have preventative effects [70]. In contrast, a pre-post study suggested that while alcohol might help individuals cope with tinnitus transiently, it negatively impacts brain regions responsible for managing tinnitus-related distress [71]. Lastly, a cross-sectional study of 2,952 Italian adults found a bimodal association between alcohol consumption and the risk of tinnitus: moderate drinkers had a significantly lower risk of tinnitus, while high alcohol drinkers showed an increased risk compared to abstainers [31]. However, studies comparing health outcomes between drinkers and non-drinkers might be misleading, as the non-drinker group may include individuals who quit drinking due to poor health [72].

On the contrary, alcohol consumption is associated with the risk of hearing loss [73], and tinnitus frequently occurs secondary to hearing loss of any cause [74]. Moreover, alcoholic beverages often contain other biogenic amines known to cause migraine, which can in turn cause or intensify tinnitus. In our clinic, we recommend that patients with tinnitus avoid consuming wine or beer. Based on our experience, hard (distilled) alcohol does not typically worsen tinnitus, as the primary triggering compounds like tyramine are often removed during the distillation process. However, since alcohol has a diuretic effect, we recommend patients consume extra water to prevent dehydration. As a general guideline, for every 30 mL of hard alcohol consumed, it is recommended to drink an additional 150 mL of water.

Supplementations

Incorporating beneficial nutrients into a diet can be a helpful means to support inner ear health. Supplementary magnesium, vitamin B_2, and coenzyme Q10 have shown promise in various studies for their potential to improve outcomes in migraine and tinnitus. With established safety profiles and minimal side effects, utilizing these supplements can offer valuable support alongside dietary adjustments.

Magnesium

Magnesium is an essential intracellular cation that acts as a cofactor for more than 350 enzymes [75]. Its involvement in the pathogenesis of migraine has been well-documented. Magnesium plays an important role in the prevention of central sensitization, a process known to underlie the mechanisms of migraine and thought to play a major role in the loudness of tinnitus. In the postsynaptic neurons of the spinal dorsal horn, extracellular magnesium blocks the *N*-methyl-*D*-aspartate (NMDA) receptor in a voltage-dependent manner, preventing calcium ions from entering cells and transmitting pain signals [76]. Consequently, low serum magnesium can reduce the pain threshold and lead to central sensitization [77]. Studies have shown that inadequate consumption of magnesium is associated with migraine in adults [78]. Additionally, magnesium has proven effective both as a prophylactic and an acute treatment for migraine [79]. Current data points to a plausible link between tinnitus and magnesium, similar to that observed with migraine. For example, a small case-control study found significantly lower magnesium levels in patients experiencing severe bilateral subjective tinnitus compared to the control group [80]. Furthermore, a single-arm pre-post study demonstrated a notable reduction in tinnitus severity after 532 mg daily oral magnesium supplementation [81]. These preliminary findings, coupled with tinnitus' close association with migraine, suggest that individuals suffering from tinnitus can also benefit from magnesium supplementation.

Vitamin B_2 (Riboflavin)

Vitamin B_2, also known as riboflavin, serves as a cofactor for flavin adenine dinucleotide (FAD) and flavin mononucleotide (FMN). It is an essential component for energy metabolism, cellular antioxidant potential, and metabolic interactions with other micronutrients such as iron, vitamin B_6 (pyridoxine), and vitamin B_9 (folate) [82]. Vitamin B_2 is often recommended as a supplementation for migraine sufferers because it can effectively reduce the number of migraine days, duration, and frequency, without serious side effects [83, 84]. Additionally, it can also help improve migraine-related cochlear and vestibular disorders [85]. Previous reports have shown that reduced intake of vitamin B_2 is an independent nutritional factor associated with increased tinnitus and tinnitus-related annoyance [29].

Coenzyme Q10 (CoQ10)

Coenzyme Q10 (CoQ10 or ubiquinone) is a fat-soluble, vitamin-like quinone synthesized naturally within the human body [86]. It functions as a mobile electron carrier within mitochondria and as an antioxidant. Clinically, it has been investigated for its therapeutic benefits in cardiac, neurological, oncological, and immunological conditions [87]. Cochlear damage is linked to lipid oxidation, protein oxidation, and DNA damage, processes in which CoQ10 plays a role [87]. Consequently, CoQ10 has been examined for its prospective effects on vestibular and cochlear symptoms. Earlier studies have shown its effectiveness in supporting the recovery of auditory hair cells damaged by hypoxia and protecting against noise-induced hearing loss [88]. Furthermore, a nonrandomized clinical trial showed that supplementation may be particularly beneficial for patients with low CoQ10 levels in reducing tinnitus severity [89]. Patients taking statin, a cholesterol-lowering medication, have a decreased concentration of CoQ10 in muscle and serum [90]. Therefore, it is recommended that these patients consider supplementing with CoQ10. While further randomized clinical trials are needed to substantiate these tentative findings, its potential as a therapeutic supplement for migraine and its low side effect profile suggest promising applications for managing tinnitus symptoms.

Sleep Modifications

Sleep is essential for maintaining homeostasis and well-being, closely linked to circadian rhythms and hormone regulations [91]. Consequently, sleep disturbances and disorders are some of the well-recognized physiologic triggers of many health conditions, including vestibulocochlear disorders [3, 92]. An animal study has demonstrated that repeated episodes of sleep deprivation aggravate noise-induced cochlea injury, highlighting its impact on conditions like tinnitus and sudden hearing loss [91]. Several cross-sectional studies have also indicated a significant association between lack of sleep and increased tinnitus intensity [39, 93, 94]. Individuals with ongoing tinnitus and chronic tinnitus are more likely to have sleep disturbances and unhealthy sleep behaviors, such as short sleep durations, late chronotypes, and sleeplessness, than those without tinnitus [95, 96].

This relationship between sleep and tinnitus is likely bidirectional, with tinnitus symptoms possibly impacting sleep quality and duration, and insufficient sleep potentially exacerbating tinnitus symptoms. In some patients, having an adequate amount of sleep helps alleviate tinnitus symptoms [39]. Therefore, breaking this cycle by addressing sleep problems is an important step in initiating tinnitus management.

Proper sleep hygiene is essential for improving sleep quality. This involves maintaining a regular sleep routine, achieving 7 to 9 hours of sleep, and adopting a contemplative practice [40]. Waking up and going to bed at the same time daily, including weekends, can help establish a consistent sleep schedule. Light exposure stimulates the brain by reducing melatonin secretion, therefore it is advisable to minimize exposure to ambient light and light from electronic devices before bedtime [40]. Additionally, practicing yoga or meditation can benefit both tinnitus management and sleep quality [97]. It is also important to treat any underlying sleep conditions, such as sleep apnea and insomnia due to other causes, to optimize sleep health [3]. In our clinic, we frequently consult sleep medicine specialists to assist in managing patients with tinnitus. Circadian rhythm disorders are common among migraine patients, and given migraine's impact on tinnitus loudness, we advocate for sleep therapy in the form of cognitive behavioral therapy for insomnia (CBTi). We advise patients to obtain 15 minutes of outdoor light exposure without sunglasses in the first 2 hours after waking up and to eat breakfast within 30 minutes of awakening. Furthermore, we have a low threshold for conducting a sleep study to evaluate for sleep apnea in patients with anatomical abnormalities suggestive of sleep apnea (e.g., large tongue, low palate, large neck, significant nasal obstruction), snoring, or morning somnolence upon awakening.

Stress Modifications

Stress is intricately linked to tinnitus through the activation of the hypothalamus-pituitary-adrenal (HPA) axis. Tinnitus patients, particularly those with chronic tinnitus, exhibit elevated cortisol levels due to prolonged stress. This persistent elevation can lead to HPA axis insufficiency, similar to what is observed in chronic fatigue syndrome, posttraumatic stress disorder, and burnout syndrome [98]. These findings are supported by a cross-sectional study of 180 adults with tinnitus, which revealed a notable prevalence of stress symptoms (65%) and a positive correlation between scores on the Tinnitus Handicap Inventory (THI) and Lipp's Stress Symptoms Inventory for Adults (LSSI) [99]. The heightened stress response in individuals with tinnitus is further linked to increased rates of depression, anxiety, and somatic symptom disorders [100]. While the causality between psychological distress and tinnitus is still under investigation, the correlation is evident. Drawing from our clinical experience, we observe that stress is a major trigger of migraines, which in turn increases tinnitus loudness. Therefore, addressing underlying stress through medication or psychotherapy is essential in treating patients with tinnitus, especially those who identify stress as a primary trigger.

Effective stress management strategies, are therefore, a crucial element in tinnitus treatment, particularly when implemented early to prevent neuroplastic changes and chronification. Relaxation techniques such as meditation, yoga, and deep breathing exercises, much like their benefits for sleep, are also effective in reducing stress [17]. Daily physical activity and exercise have been shown to reduce the risk of tinnitus and associated distress [17, 101]. Furthermore, exercise has been observed to enhance inner ear function in animal models, potentially counteracting age-related balance loss by promoting proprioceptive and vestibular connectivity in motor neurons [102]. Establishing a consistent exercise routine can be challenging; starting with 5–7 minutes daily and gradually increasing by a few minutes each week is recommended, with a goal of reaching 20–30 minutes, three to five times a week.

Conclusions

- Tinnitus is a common condition affecting millions of people worldwide, impacting daily life with sleep disturbances, concentration issues, and mood changes.
- Migraine and tinnitus frequently co-occur, suggesting shared underlying mechanisms involving neuroinflammation and vascular changes.
- While standard diagnostic measures and treatments for tinnitus remain limited, dietary and lifestyle modifications traditionally employed in migraine therapy have been used effectively in our clinic.
- Dietary modifications include maintaining regular meal schedules and ensuring adequate hydration.
- Identifying and eliminating food triggers such as caffeine, tyramine, aspartame, phenylethylamine, monosodium glutamate, nitrates/nitrites, sulfites, and some forms of alcohol through a food diary can improve tinnitus outcomes.
- If the patient experiences an increase in the level of tinnitus, they are encouraged to assess their diet and other factors (dehydration, hunger, stress, sleep, overstimulation) in the 6 hours preceding the spike in tinnitus to identify the trigger.

- Supplementation with magnesium, vitamin B_2, and coenzyme Q10 shows potential in preventing and alleviating tinnitus symptoms, paralleling their benefits in migraine treatment.
- Proper sleep hygiene practices and avoiding both sleep deprivation and excessive sleep are important for both general health and inner ear function.
- Adopting effective stress management and coping strategies, such as relaxation techniques (e.g., meditation, yoga) and regular physical activity, can help prevent neuroplastic changes that exacerbate tinnitus symptoms.
- The findings underscore the importance of educating individuals on lifestyle adjustments to effectively prevent and manage tinnitus in clinical practice.

References

1. Jarach CM, Lugo A, Scala M, *et al.* Global prevalence and incidence of tinnitus. *Journal of the American Medical Association Neurology*. **79**(9):888–900 (2022).
2. Tang D, Tran Y, Lewis JR, *et al.* Associations between intake of dietary flavonoids and the 10-year incidence of tinnitus in older adults. *European Journal of Nutrition*. **61**(4):1957–1964 (2022).
3. Umemoto KK, Tawk K, Mazhari N, Abouzari M, and Djalilian HR. Management of migraine-associated vestibulocochlear disorders. *Audiology Research*. **13**(4):528–545 (2023).
4. Frederiksen SD, Bekker-Nielsen Dunbar M, Snoer AH, Deen M, and Edvinsson L. Serotonin and neuropeptides in blood from episodic and chronic migraine and cluster headache patients in case-control and case-crossover settings: A systematic review and meta-analysis. *Headache: The Journal of Head and Face Pain*. **60**(6):1132–1164 (2020).
5. Vass Z, Shore SE, Nuttall AL, and Miller JM. Direct evidence of trigeminal innervation of the cochlear blood vessels. *Neuroscience*. **84**(2):559–567 (1998).
6. Lee A, Abouzari M, Akbarpour M, Risbud A, Lin HW, and Djalilian HR. A proposed association between subjective nonpulsatile tinnitus and migraine. *World Journal of Otorhinolaryngology — Head & Neck Surgery*. **9**(2):107–114 (2022).
7. Abouzari M, Tan D, Sarna B, *et al.* Efficacy of multi-modal migraine prophylaxis therapy on hyperacusis patients. *Ann Otol Rhinol Laryngol*. **129**(5):421–427 (2020).

8. Risbud A, Muhonen EG, Tsutsumi K, Martin EC, Abouzari M, and Djalilian HR. Migraine features in patients with isolated aural fullness and proposal for a new diagnosis. *Otol Neurotol.* **42**(10):1580–1584 (2021).
9. Benjamin T, Gillard D, Abouzari M, Djalilian HR, and Sharon JD. Vestibular and auditory manifestations of migraine. *Current Opinion in Neurology.* **35**(1):84–89 (2022).
10. Francis MV. Atypical migraine in clinical practice: Are we missing it? *Annals of Indian Academy of Neurology.* **25**(3):347–349 (2022).
11. García A, Madrigal J, and Castillo-Bustamante M. Vestibular Migraine and tinnitus: A challenging narrative. *Cureus: Journal of Medical Science.* **13**(6):e15998 (2021).
12. Guichard E, Montagni I, Tzourio C, and Kurth T. Association between headaches and tinnitus in young adults: Cross-sectional study. *Headache.* **56**(6):987–994 (2016).
13. Langguth B, Hund V, Busch V, *et al.* Tinnitus and headache. *BioMed Research International.* **2015**:797416 (2015).
14. Kim SH, Kim D, Lee JM, Lee SK, Kang HJ, and Yeo SG. Review of pharmacotherapy for tinnitus. *Healthcare (Basel).* **9**(6):779 (2021).
15. Lee HY, and Jung DJ. Recent updates on tinnitus management. *Journal of Audiology & Otology.* **27**(4):181–192 (2023).
16. Han BI, Lee HW, Ryu S, and Kim JS. Tinnitus update. *Journal of Clinical Neurology.* **17**(1):1–10 (2021).
17. Wadhwa S, Jain S, and Patil N. The role of diet and lifestyle in the tinnitus management: A comprehensive review. *Cureus Journal of Medical Science.* **16**(4):e59344 (2024).
18. Turner LC, Molgaard CA, Gardner CH, Rothrock JF, and Stang PE. Migraine trigger factors in non-clinical Mexican-American population in San Diego County: Implications for etiology. *Cephalalgia.* **15**(6):523–530 (1995).
19. Ailani J, Burch RC, and Robbins MS. Board of directors of the American headache society. The American headache society consensus statement: Update on integrating new migraine treatments into clinical practice. *Headache.* **61**(7):1021–1039 (2021).
20. Sun-Edelstein C, and Mauskop A. Foods and supplements in the management of migraine headaches. *Clinical Journal of Pain.* **25**(5):446–452 (2009).
21. Kelman L. The triggers or precipitants of the acute migraine attack. *Cephalalgia.* **27**(5):394–402 (2007).
22. Özbey-Yücel Ü, Uçar A, Aydoğan Z, Tokgoz-Yilmaz S, and Beton S. The effects of dietary and physical activity interventions on tinnitus symptoms: An RCT. *Auris Nasus Larynx.* **50**(1):40–47 (2023).

23. Roger C, Lasbleiz A, Guye M, Dutour A, Gaborit B, and Ranjeva JP. The role of the human hypothalamus in food intake networks: An MRI perspective. *Frontiers in Nutrition.* **8**:760914 (2022).
24. Alstadhaug KB. Migraine and the hypothalamus. *Cephalalgia.* **29**(8):809–817 (2009).
25. Szczepek AJ, and Mazurek B. Neurobiology of stress-induced tinnitus. *Current Topics in Behavioral Neurosciences.* **51**:327–347 (2021).
26. McKeown J, McGeoch PD, and Grieve DJ. The influence of vestibular stimulation on metabolism and body composition. *Diabetic Medicine.* **37**(1):20–28 (2020).
27. Parkin JL, and Tice R. Hypoglycemia and fluctuating hearing loss. The Annals of Otology Rhinology and Laryngology. **79**(5):992–997 (1970).
28. Oğuz E, Cebeci A, and Geçici CR. The relationship between nutrition and Ménière's disease. *Auris Nasus Larynx.* **48**(5):803–808 (2021).
29. Lee DY, and Kim YH. Relationship between diet and tinnitus: Korea national health and nutrition examination survey. *Clinical and Experimental Otorhinolaryngology.* **11**(3):158–165 (2018).
30. Dawes P, Cruickshanks KJ, Marsden A, Moore DR, and Munro KJ. Relationship between diet, tinnitus, and hearing difficulties. *Ear and Hearing.* **41**(2):289–299 (2020).
31. Gallus S, Lugo A, Garavello W, *et al.* Prevalence and determinants of tinnitus in the Italian adult population. *Neuroepidemiology.* **45**(1):12–19 (2015).
32. Borkum JM. Migraine triggers and oxidative stress: A narrative review and synthesis. *Headache.* **56**(1):12–35 (2016).
33. Baos V, Ester F, Castellanos A, *et al.* Use of a structured migraine diary improves patient and physician communication about migraine disability and treatment outcomes. *International Journal of Clinical Practice.* **59**(3):281–286 (2005).
34. Ledesma ALL, Leite Rodrigues D, Monteiro de Castro Silva I, Oliveira CA, and Bahmad F. The effect of caffeine on tinnitus: Randomized triple-blind placebo-controlled clinical trial. *PLoS One.* **16**(9):e0256275 (2021).
35. Ghahraman MA, Farahani S, and Tavanai E. A comprehensive review of the effects of caffeine on the auditory and vestibular systems. *Nutritional Neuroscience.* **25**(10):2181–2194 (2022).
36. Biswas R, Lugo A, Genitsaridi E, *et al.* Modifiable lifestyle-related risk factors for tinnitus in the general population: An overview of smoking, alcohol, body mass index and caffeine intake. *Progress in Brain Research.* **263**:1–24 (2021).
37. Glicksman JT, Curhan SG, and Curhan GC. A prospective study of caffeine intake and risk of incident tinnitus. *The American Journal of Medicine.* **127**(8):739–743 (2014).

38. Claire LS, Stothart G, McKenna L, and Rogers PJ. Caffeine abstinence: An ineffective and potentially distressing tinnitus therapy. *International Journal of Audiology*. **49**(1):24–29 (2010).
39. Pan T, Tyler RS, Ji H, Coelho C, and Gogel SA. Differences among patients that make their tinnitus worse or better. *American journal of Audiology*. **24**(4):469–476 (2015).
40. Baranwal N, Yu PK, and Siegel NS. Sleep physiology, pathophysiology, and sleep hygiene. *Progress in Cardiovascular Diseases*. 77:59–69 (2023).
41. Hindiyeh NA, Zhang N, Farrar M, Banerjee P, Lombard L, and Aurora SK. The role of diet and nutrition in migraine triggers and treatment: A systematic literature review. *Headache*. **60**(7):1300–1316 (2020).
42. Nowaczewska M, Wiciński M, and Kaźmierczak W. The ambiguous role of caffeine in migraine headache: From trigger to treatment. *Nutrients*. **12**(8):2259 (2020).
43. Zhang L, Yin J, Li J, Sun H, Liu Y, and Yang J. Association between dietary caffeine intake and severe headache or migraine in US adults. *Scientific Report*. **13**:10220 (2023).
44. Stratton JE, Hutkins RW, and Taylor SL. Biogenic amines in cheese and other fermented foods: A review. *The Journal of Food Protection*. **54**(6):460–470 (1991).
45. Tomanic M, Belojevic G, Jovanovic A, Vasiljevic N, Davidovic D, and Maksimovic K. Dietary factors and tinnitus among adolescents. *Nutrients*. **12**(11):3291 (2020).
46. Ghose K, and Carroll JD. Mechanism of tyramine-induced migraine: Similarity with dopamine and interactions with disulfiram and propranolol in migraine patients. *Neuropsychobiology*. **12**(2–3):122–126 (1984).
47. Czarnecka K, Pilarz A, Rogut A, *et al.* Aspartame-true or false? Narrative review of safety analysis of general use in products. *Journal of Nutrients*. **13**(6):1957 (2021).
48. Mackey SA, and Berlin CM. Effect of dietary aspartame on plasma concentrations of phenylalanine and tyrosine in normal and homozygous phenylketonuric patients. *Clinical Pediatrics (Phila)*. **31**(7):394–399 (1992).
49. Pisarik P, and Kai D. Vestibulocochlear toxicity in a pair of siblings 15 years apart secondary to aspartame: Two case reports. *Cases Journal*. **2**:9237 (2009).
50. Patel RM, Sarma R, and Grimsley E. Popular sweetner sucralose as a migraine trigger. *Headache: The Journal of Head and Face Pain*. **46**(8):1303–1304 (2006).
51. Martin VT, and Vij B. Diet and Headache: Part 1. *Headache: The Journal of Head and Face Pain*. **56**(9):1543–1552 (2016).
52. Borah A, Paul R, Mazumder MK, and Bhattacharjee N. Contribution of β-phenethylamine, a component of chocolate and wine, to dopaminergic

neurodegeneration: Implications for the pathogenesis of Parkinson's disease. *Neuroscience Bulletin*. **29**(5):655–660 (2013).

53. McCulloch J, and Harper AM. Phenylethylamine and cerebral blood flow. Possible involvement of phenylethylamine in migraine. *Neurology*. **27**(9):817–821 (1977).
54. Loï C, and Cynober L. Glutamate: A safe nutrient, not just a simple additive. *Annals of Nutrition & Metabolism*. **78**(3):133–146 (2022).
55. Zaeem Z, Zhou L, and Dilli E. Headaches: A review of the role of dietary factors. *Current Neurology and Neuroscience Reports*. **16**(11):101 (2016).
56. Mazzoli M. Complementary tinnitus therapies. In Møller AR, Langguth B, De Ridder D, Kleinjung T, (Eds.) *Textbook of Tinnitus*. Springer; pp. 733–747 (2011).
57. Leussink VI, Hartung HP, Stüve O, and Kieseier BC. Vestibular hypofunction after monosodium glutamate ingestion: Broadening the spectrum of 'Chinese restaurant syndrome.' *Journal of Neurology*. **263**(5):1027–1028 (2016).
58. Merritt JE, and Williams PB. Vasospasm contributes to monosodium glutamate-induced headache. *Headache*. **30**(9):575–580 (1990).
59. Cacace AT, Hu J, Romero S, Xuan Y, Burkard RF, and Tyler RS. Glutamate is down-regulated and tinnitus loudness-levels decreased following rTMS over auditory cortex of the left hemisphere: A prospective randomized single-blinded sham-controlled cross-over study. *Hearing Research*. **358**:59–73 (2018).
60. Altaher W, Alhelo H, Chosky D, and Kulesza RJ. Neonatal exposure to monosodium glutamate results in impaired auditory brainstem structure and function. *Hearing Research*. **405**:108243 (2021).
61. Guven SG, Ersoy O, Topuz RD, Bulut E, Kizilay G, and Uzun C. Does oral Monosodium Glutamate have a Cochleotoxic effect? An experimental study. *Audiol Neurootol*. **27**(2):109–121 (2022).
62. Kalatharan V, and Al-Karagholi MAM. Targeting peripheral N-Methyl-D-Aspartate receptor (NMDAR): A novel strategy for the treatment of migraine. *Journal of Clinical Medicine*. **12**(6):2156 (2023).
63. Maria SA, Claudia C, Pamela G, Andrea C, and Roberto A. Medical therapy in Ménière's disease. *Audiological Medicine*. **10**(4):171–177 (2012).
64. Hord NG, Tang Y, and Bryan NS. Food sources of nitrates and nitrites: The physiologic context for potential health benefits. *The American Journal of Clinical Nutrition*. **90**(1):1–10 (2009).
65. Gonzalez A, Hyde E, Sangwan N, Gilbert JA, Viirre E, and Knight R. Migraines are correlated with higher levels of nitrate-, nitrite-, and nitric oxide-reducing oral microbes in the American gut project cohort. *mSystems*. **1**(5) (2016).

66. Hockley A, Berger JI, Palmer AR, and Wallace MN. Nitric oxide increases gain in the ventral cochlear nucleus of guinea pigs with tinnitus. *European Journal Neuroscience*. **52**(9):4057–4080 (2020).
67. Pall ML, and Bedient SA. The NO/ONOO- cycle as the etiological mechanism of tinnitus. *International Tinnitus Journal*. **13**(2):99–104 (2007).
68. Kronberg MT. Sulfite intolerance: A cause of tinnitus? *Bioscience Hypotheses*. **1**(4):185–188 (2008).
69. Pellegrino ABW, Davis-Martin RE, Houle TT, Turner DP, and Smitherman TA. Perceived triggers of primary headache disorders: A meta-analysis. *Cephalalgia*. **38**(6):1188–1198 (2018).
70. Biswas R, Genitsaridi E, Trpchevska N, *et al.* Low evidence for tinnitus risk factors: A systematic review and meta-analysis. *The Journal of the Association for Research in Otolaryngology*. **24**(1):81–94 (2023).
71. Vanneste S, and De Ridder D. The use of alcohol as a moderator for tinnitus-related distress. *Brain Topography*. **25**(1):97–105 (2012).
72. Dawes P, Newall J, Stockdale D, and Baguley DM. Natural history of tinnitus in adults: A cross-sectional and longitudinal analysis. *British Medical Journal Open*. **10**(12):e041290 (2020).
73. Chen HL, Tan CT, Wu CC, and Liu TC. Effects of diet and lifestyle on audio-vestibular dysfunction in the elderly: A literature review. *Journal of Nutrients*. **14**(22):4720 (2022).
74. Shapiro SB, Noij KS, Naples JG, and Samy RN. Hearing loss and tinnitus. *Medical Clinics of North America*. **105**(5):799–811 (2021).
75. Dolati S, Rikhtegar R, Mehdizadeh A, and Yousefi M. The role of magnesium in pathophysiology and migraine treatment. *Biological Trace Element Research*. **196**(2):375–383 (2020).
76. Mayer ML, Westbrook GL, and Guthrie PB. Voltage-dependent block by Mg2+ of NMDA responses in spinal cord neurones. *Nature*. **309**(5965):261–263 (1984).
77. Shin HJ, Na HS, and Do SH. Magnesium and pain. *Journal of Nutrients*. **12**(8):2184 (2020).
78. Slavin M, Li H, Khatri M, and Frankenfeld C. Dietary magnesium and migraine in adults: A cross-sectional analysis of the national health and nutrition examination survey 2001–2004. *Headache: The Journal of Head and Face Pain*. **61**(2):276–286 (2021).
79. Chiu HY, Yeh TH, Huang YC, and Chen PY. Effects of intravenous and oral magnesium on reducing migraine: A meta-analysis of randomized controlled trials. *Pain Physician*. **19**(1):E97–E112.
80. Uluyol S, Kılıçaslan S, and Yağız Ö. Relationship between serum magnesium level and subjective tinnitus. *Kulak Burun Bogaz Ihtis Derg*. **26**(4):225–227 (2016).

81. Cevette MJ, Barrs DM, Patel A, *et al.* Phase 2 study examining magnesium-dependent tinnitus. *International Tinnitus Journal.* **16**(2):168–173 (2011).
82. McNulty H, Pentieva K, and Ward M. Causes and clinical sequelae of riboflavin deficiency. *Annual Review of Nutrition.* **43**:101–122 (2023).
83. Chen YS, Lee HF, Tsai CH, *et al.* Effect of Vitamin B2 supplementation on migraine prophylaxis: A systematic review and meta-analysis. *Nutritional Neuroscience.* **25**(9):1801–1812 (2022).
84. Namazi N, Heshmati J, and Tarighat-Esfanjani A. Supplementation with Riboflavin (Vitamin B2) for migraine prophylaxis in adults and children: A review. *International Journal for Vitamin and Nutrition Research.* **85**(1–2): 79–87 (2015).
85. Abouzari M, Abiri A, and Djalilian HR. Successful treatment of a child with definite meniere's disease with the migraine regimen. *American Journal of Otolaryngology.* **40**(3):440–442 (2019).
86. Arenas-Jal M, Suñé-Negre JM, and García-Montoya E. Coenzyme Q10 supplementation: Efficacy, safety, and formulation challenges. *Comprehensive Reviews in Food Science and Food Safety.* **19**(2):574–594 (2020).
87. Ahn Jh, Yoo Mh, Lee Hj., *et al.* Coenzyme Q10 in combination with steroid therapy for treatment of sudden sensorineural hearing loss: A controlled prospective study. *Clinical Otolaryngology.* **35**(6):486–489 (2010).
88. Sugahara K, Hirose Y, Mikuriya T, *et al.* Coenzyme Q10 protects hair cells against aminoglycoside. *PLoS One.* **9**(9):e108280 (2014).
89. Khan M, Gross J, Haupt H, *et al.* A pilot clinical trial of the effects of coenzyme Q10 on chronic tinnitus aurium. *Otolaryngology–Head and Neck Surgery.* **136**(1):72–77 (2007).
90. Qu H, Guo M, Chai H, Wang WT, Gao ZY, and Shi DZ. Effects of Coenzyme Q10 on Statin-Induced Myopathy: An updated meta-analysis of randomized controlled trials. *The Journal of the American Heart Association.* **7**(19):e009835 (2018).
91. Li P, Bing D, Wang S, *et al.* Sleep deprivation modifies noise-induced cochlear injury related to the stress hormone and autophagy in female mice. *Frontiers Neuroscience.* **13**:1297 (2019).
92. Tiseo C, Vacca A, Felbush A, *et al.* Migraine and sleep disorders: A systematic review. *Journal of Headache and Pain.* **21**(1):126 (2020).
93. Stouffer JL, and Tyler RS. Characterization of tinnitus by tinnitus patients. *Journal of Speech Language and Hearing Research.* **55**(3):439–453 (1990).
94. Bhatt JM, Bhattacharyya N, and Lin HW. Relationships between tinnitus and the prevalence of anxiety and depression. *Laryngoscope.* **127**(2):466–469 (2017).
95. Peng J, Dong Y, Luo Y, *et al.* The relationship between sleep traits and tinnitus in UK biobank: A population-based cohort study. *Ear and Hearing.* **44**(1):53–60 (2023).

96. Inagaki Y, Suzuki N, Oishi N, Goto F, Shinden S, and Ogawa K. Personality and sleep evaluation of patients with tinnitus in Japan. *Psychiatric Quarterly*. **92**(1):249–257 (2021).
97. Gunjawate DR, and Ravi R. Effect of yoga and meditation on tinnitus: A systematic review. *The Journal of Laryngology and Otology*. **135**(4):284–287 (2021).
98. Mazurek B, Haupt H, Olze H, and Szczepek AJ. Stress and tinnitus — From bedside to bench and back. *Frontiers in Systems Neuroscience*. **6**:47 (2012).
99. Ciminelli P, Machado S, Palmeira M, *et al.* Tinnitus: The sound of stress? *Clinical Practice & Epidemiology in Mental Health*. **14**:264–269 (2018).
100. Hackenberg B, Döge J, O'Brien K, *et al.* Tinnitus and its relation to depression, anxiety, and stress — A population-based cohort study. *Journal of Clinical Medicine*. **12**(3):1169 (2023).
101. Chen S, Yang X, Jiang Y, *et al.* Associations between physical activity, tinnitus, and tinnitus severity. *Ear and Hearing*. **44**(3):619–626 (2023).
102. Battilana F, Steurer S, Rizzi G, Delgado AC, Tan KR, and Handschin C. Exercise-linked improvement in age-associated loss of balance is associated with increased vestibular input to motor neurons. *Aging Cell*. **19**(12):e13274 (2020).

https://doi.org/10.1142/9789819807345_0002

Chapter 2

Cognitive Behavioral Therapy for Tinnitus Rehabilitation

Karen Tawk[*,†], Ghaidaa S. Khlaifat[*,†], Joshua K. Kim[†], and Hamid R. Djalilian[†,‡,§,¶]

[†]*Department of Otolaryngology-Head and Neck Surgery, University of California, Irvine, CA, USA*

[‡]*Department of Biomedical Engineering, University of California, Irvine, CA, USA*

[§]*Department of Neurosurgery, University of California, Irvine, CA, USA*

[¶]*hdjalili@hs.uci.edu*

Abstract

Tinnitus is a widespread condition affecting millions of people worldwide, with some experiencing a detrimental impact on their quality of life. Although the U.S. Food and Drug Administration (FDA) has not approved any pharmaceutical agent for tinnitus treatment, the American, European, and Japanese clinical guidelines strongly recommend cognitive behavioral therapy (CBT) as a preferred option for managing chronic and bothersome tinnitus over alternative modalities. CBT is a form of psychotherapy that was successful in changing how people

---

[*]These authors contributed equally to this manuscript.

think (cognitive) and how they react (behavior) to tinnitus. The "first" and "second" waves of Cognitive Behavioral Therapy (CBT) focused on behavioral principles and cognitive restructuring respectively, while the "third" wave represented a shift toward embracing mindfulness and contextual factors, incorporating approaches such as mindfulness-based CBT and acceptance and commitment therapy. Mindfulness-based CBT helped reduce tinnitus severity and associated distress when compared to relaxation techniques alone. Similarly, acceptance and commitment therapy demonstrated effectiveness in decreasing tinnitus severity when compared to tinnitus rehabilitation training or waitlist control. Another new advance is the CBT delivery method. Audiologist-delivered CBT is a new promising option to expand CBT accessibility for patients with tinnitus. In addition, social media and telehealth have become vital weapons in managing tinnitus. For instance, internet-based CBT and smartphone-based tinnitus intervention could offer a valuable alternative for providing care for patients who are reluctant to seek clinic-based care.

Introduction

Tinnitus, derived from the Latin verb "tinnire" meaning "to ring", denotes the perception of a phantom sound without any external stimulus [1]. This phenomenon, described as ringing, buzzing, or hissing, finds historical mention in the Egyptian Papyrus Ebers (1550 B.C.) as well as the Babylonian Talmud (500 C.E.) [2, 3]. In the latter, tinnitus was regarded as Titus's curse, symbolized by a gnat buzzing in the brain, responsive to sound, and subsequent habituation [3]. It is a widespread condition, affecting millions of people [4, 5]. Epidemiologic studies report its prevalence between 10% and 15% in the United States and various nations [1, 6]. In specific populations, such as veterans, its prevalence even exceeds that of chronic diseases such as back pain, knee problems, and diabetes [7, 8]. While some patients are relatively not bothered by their tinnitus, others experience a significant disruption of their daily life activities. Adrian and El Refaie reported that among adults with tinnitus, 2.8% were moderately annoyed by their tinnitus, 1.6% were severely annoyed, and 0.5% were not able to lead a normal life [9]. In these patients, the detrimental impact of tinnitus on quality of life was well-established and extended across all five subsections (mobility, self-care, usual activities performance, pain/discomfort, and anxiety/depression) of validated health-related quality of life surveys [6, 10]. Therefore, untreated tinnitus may lead to symptoms of depression, anxiety, impaired sleep, impaired hearing, and hopelessness and result in poorer health-related quality of life. Additionally, the healthcare cost for managing tinnitus poses a considerable burden on a nation's healthcare system [11–13]. This strain is further exacerbated by the lack of effective treatment strategies, contributing to long-term disability payments reaching $1.2 billion per year in the US Veteran population in 2012 [14].

Tinnitus management is classified into three categories: (1) treatments targeting tinnitus perception, including hearing devices, sound therapy, and medical treatment, (2) approaches targeting the reaction to tinnitus such as cognitive behavioral therapy (CBT) and mindfulness techniques, and (3) combined approaches modulating both tinnitus perception and reactions to it such as tinnitus retraining therapy [15]. While the U.S. Food and Drug Administration has not approved any pharmaceutical agent for tinnitus treatment [16], the clinical guidelines set by the American Academy of Otolaryngology-Head and Neck Surgery, as well as newer European and Japanese clinical guidelines, strongly recommend CBT as a preferred option for managing chronic and bothersome tinnitus over alternative modalities [17–19]. However, the advantages are mainly limited to addressing the distress associated with tinnitus [17–19].

As most of us are aware, CBT stands out as one of the extensively researched methodologies in the realm of tinnitus management, and its application is widely recognized as a leading evidence-based practice within the field, garnering significant acceptance in tinnitus management and rehabilitation. In this chapter, we summarize the latest advances and data on CBT as an option for tinnitus management.

The Power of CBT: Transforming Thoughts for a Better Life

CBT is an active approach that focuses on the recognition of maladaptive cognition (cognitive) and the modification of negative thoughts along with effective behaviors (behavioral) [20]. This involves providing psycho-education, using positive mental imagery, practicing relaxation techniques, cognitive restructuring, exposure, problem-solving, mindfulness, and attention exercises [21]. In stressful situations, certain individuals may experience feelings of pessimism and a sense of being unable to address issues effectively. CBT promotes the cultivation of more balanced thinking patterns to enhance stress-coping mechanisms. It also increases adaptive behaviors and reduces emotional distress, teaching individuals to adopt a problem-solving strategy [22].

Two of the earliest versions of CBT emerged during the 1950s, with Albert Ellis introducing Rational Emotive Behavior Therapy as one of the initial forms of CBT [23]. Later, in the 1960s, Aaron T. Beck developed cognitive therapy as another version of this therapy approach [24]. The "ABC Technique of Irrational Beliefs" as described by Ellis, has been widely adopted as the main framework in CBT. The ABC model was established throughout his work on rational emotive behavior therapy. According to Ellis, individuals hold distinct beliefs regarding the world and themselves that guide their lives and reactions. He proposed that some of these belief systems are "irrational" or maladaptive, triggering people to engage in unhealthy or undesirable behaviors. Ellis's theory suggests that the ABC model may be implemented to identify irrational beliefs. By paying close attention to a situation (without bias) and examining how a person internally reacts, individuals can identify recurring patterns of thoughts or beliefs. Once an irrational belief has been identified, one can examine the behavior or feelings that resulted from it [23]. Cognitive interventions were first introduced into therapeutic strategies in the 1970s by Beck [25]. According to Ellis' theory, the A-B-C model (antecedents, beliefs, and consequences) suggests that the activating event (A) does not directly cause the emotional and behavioral consequences (C). Instead, it is the beliefs (B), encompassing interpretations and assumptions concerning the event, that significantly define the emotional and behavioral consequences (C). This combined approach assists patients in recognizing and altering maladaptive behaviors to improve their mental health and well-being (Figure 1).

CBT, a form of psychotherapy, serves as a treatment for a broad spectrum of psychiatric disorders, from anxiety and depression to the distress associated with tinnitus [26]. Exploring the effects of CBT, a search of the Cochrane Library database [27] revealed that numerous medical conditions have been the subject of randomized controlled trials (RCTs). These conditions included psychosomatic illnesses like chronic fatigue syndrome and irritable bowel syndrome, alongside psychiatric disorders such as anxiety, depression, and post-traumatic stress disorder, as well as sociobehavioral issues like substance abuse and smoking. CBT became a feasible treatment option for subjective tinnitus in the mid-1980s [28].

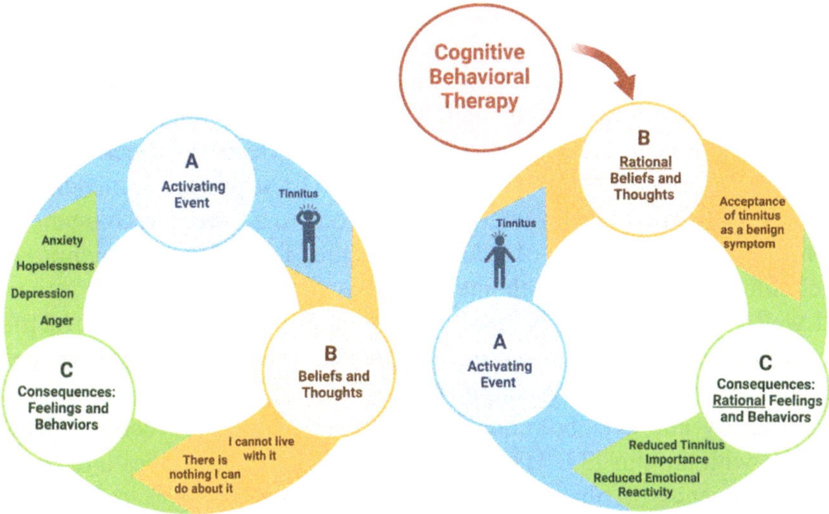

Figure 1. The impact of CBT on interrupting the negative pattern of thoughts in tinnitus patients

Subsequently, numerous clinical trials and several reviews have assessed the overall effectiveness of CBT in managing tinnitus [29–37]. These reviews provided initial empirical evidence supporting the effectiveness of CBT in alleviating tinnitus-related distress. The primary goal of CBT is not to reduce the acoustic features of tinnitus, such as its loudness or pitch, but rather to support people in managing particular challenges in their daily lives by fostering more positive and realistic thoughts about their condition [38]. In this way, CBT aims to reduce the distress and handicap caused by tinnitus.

According to Woods and Theodoroff [39], each session focuses on a specific topic and may vary between patients because therapy is tailored to each individual. Despite this customization, CBT consistently incorporates goal-oriented problem-solving techniques, the analysis and modification of dysfunctional thoughts, and the adjustment of negative behavior patterns [39]. This approach aims to facilitate habituation, where patients become accustomed to their tinnitus to the extent that it no longer feels unpleasant or threatening. Patients in distress due to tinnitus may need varying amounts of CBT. Like most therapeutic approaches, CBT is adaptable to meet an individual's needs. Some patients might benefit from a few sessions that include education and self-help techniques, while others may require more extensive sessions and possibly medication [39]. The full habituation process may take up to 18 months, although some patients adapt more quickly [40]. The ultimate goal is to equip patients with the necessary skills to manage their recovery autonomously.

The American Academy of Otolaryngology-Head and Neck Surgery clinical practice guidelines on tinnitus gave CBT a strong recommendation, making it the only tinnitus intervention to receive a "strong recommendations" [17]. In the absence of a definitive "cure" for any specific symptom, the ultimate role of any health intervention is to enhance the patient's overall well-being. From this perspective, any improvement in quality of life or reduction in depression scores in individuals with tinnitus can be considered an effective intervention. Indeed, research has shown that CBT significantly improves both quality of life and depression scores in individuals with tinnitus [33, 41]. These findings align with studies indicating the effectiveness of this intervention in treating depression and anxiety [42, 43], which are conditions frequently associated with tinnitus [44].

Research on the effects of CBT on tinnitus-related outcomes has begun to accumulate, yielding promising results. Martinez-Devesa's systematic review in 2010 evaluated the therapeutic impact of CBT on tinnitus-related distress by examining eight trials with a total of 468 participants. The review indicated that while CBT is not intended to directly alleviate the perceived loudness of tinnitus, it effectively targeted the behavioral, cognitive, and physiological responses. Therefore, when compared to waiting-list controls, CBT was found to be more effective in improving depression and quality of life [33]. Similarly, Hesser's 2011 meta-analysis [34] sought to provide an updated and detailed analysis of controlled trials examining the impact of CBT on tinnitus-related distress. Across 15 studies involving 1091 participants, the analysis revealed significant positive outcomes of CBT. It was found to diminish not only tinnitus-related symptoms but also associated anxiety and depression symptoms, with persistent improvements noted during follow-up assessments. In comparison to both waitlist and active control conditions, CBT demonstrated superiority [34]. In a more recent Cochrane review [35], 28 primarily European studies were analyzed, encompassing 2733 individuals who suffered from chronic tinnitus (lasting a minimum of three months) and aged between 43 and 70 years. The duration of CBT varied from 3 to 22 weeks and was primarily provided online or in hospital settings. The review found that CBT had the potential to alleviate the impact of tinnitus on quality of life when compared to no intervention/waitlist control, audiological treatment, tinnitus retraining therapy, and other active controls. However, there was no apparent difference in anxiety and depression outcomes, which was attributed to the lack of long-term follow-up [35]. Despite the evident beneficial effects of CBT for tinnitus in these reviews, its influence on crucial factors such as tinnitus loudness continues to be inadequately documented, and evidence regarding the long-term outcomes is still lacking.

Recent Advances

In exploring the evolution of CBT, Hayes used the metaphorical categorization of waves [45]. The "first wave" originating in the 1950s was characterized by a focus on neo-behavioral principles targeting directly problematic behaviors. Progressing to the "second wave" in the late 1960s, CBT was characterized by cognitive restructuring in addition to behavioral and emotional therapies [46]. The second wave prioritized cognitive changes as the first step toward behavioral and emotional changes. The emergence of the "third wave" therapies marked a shift by placing less emphasis on pathological and focusing more on contextual aspects. As such, the new generation of CBT fosters a broader scope of skills such as mindfulness and acceptance [46–48], including various approaches such as mindfulness-based CBT (MBCT) and acceptance and commitment therapy (ACT) [49]. However, it is important to note that this terminology has sparked debate, with Hofmann *et al.* arguing that the distinctions between these generations are minimal [50].

Fast forward to the advent of COVID-19 in late 2019, which brought a remarkable impact on the world of medicine and related fields. In response, social media and telehealth have become vital weapons in managing tinnitus. These technology-driven interactive solutions have been instrumental in offering both physical and emotional support to patients suffering from tinnitus during the pandemic and beyond.

Mindfulness-Based Cognitive Behavioral Therapy

MBCT involves giving attention to the present moment without passing judgment. By encouraging acceptance, resilience, and mental well-being, MBCT can help people learn to acquire adaptive coping techniques that will lessen the negative impact of their illnesses on their quality of life [51]. Unfortunately, the early suggestion to "just learn to live with it, because there is no cure" for tinnitus patients frequently results in iatrogenic distress. As a result, one starts to feel both helpless and accountable for controlling the noise. By fostering mindfulness, patients can perceive tinnitus-related emotions, thoughts, and physical sensations without becoming overwhelmed by them, thereby disengaging from maladaptive cognitive patterns and reducing emotional reactivity [52]. Research has shown that MBCT is superior to intensive relaxation therapy in reducing the severity of chronic tinnitus on the Tinnitus Questionnaire [53]. It also lessens disabilities and psychological discomfort [52, 53]. Philippot *et al.* [54] conducted a randomized clinical trial comparing the effectiveness of MBCT to relaxation training on patients suffering from significant distress due to their tinnitus. They found that MBCT maintained or enhanced the benefits of psychoeducation in reducing negative emotions, rumination, and psychological challenges associated with tinnitus at a later follow-up. However, these effects were diminished in the relaxation group [54]. This study suggested some potential for MBCT in managing tinnitus, although they only offered 6 sessions (*vs.* typical 8-session MBCT protocol) [49]. Furthermore, several studies demonstrated the efficacy of MBCT in reducing tinnitus severity by examining functional MRI (fMRI) data [55, 56]. For instance, Zimmerman *et al.* recorded significant changes in the resting state connectivity involving the default mode network, cingulo-opercular network, and amygdala after 8-week MBCT intervention [56]. In addition, they found that patterns of resting state connectivity linking the amygdala, dorsal attention network, and frontoparietal network closely correlated with the severity of tinnitus-related disability. Importantly, reduced connectivity between the amygdala and parietal regions, alongside the negative correlation with Tinnitus Functional Index (TFI) was suggestive of a brain imaging marker of successful tinnitus treatment with MBCT. Additionally, they observed a trend of reduced functional segregation which predicted a decrease in TFI throughout the MBCT program. Therefore, the skills cultivated in MBCT can present promising avenues for tinnitus-related distress, and the use of brain imaging, particularly resting state functional connectivity analyses, constitutes an important tool for tinnitus intervention effectiveness.

Acceptance and commitment therapy (ACT)

In 1999, Hayes, Strosahl, and Wilson published the pioneering book "Acceptance and Commitment Therapy: An Experiential Approach to

Behavior Change", which outlined the theoretical principles and practical applications of ACT, a variant of psychotherapy that aids individuals in accepting their emotions and thoughts rather than seeking to change or control them [57]. Since then, ACT has continued to gain popularity and recognition in the psychotherapy and psychology fields. One perspective is to conceptualize ACT as a second-generation CBT method. However, unlike traditional CBT, which aims to alleviate symptoms, ACT emphasizes enhancing psychological well-being [58]. In the context of tinnitus, ACT is focused on encouraging tinnitus patients to accept their thoughts as normal human experiences while simultaneously committing to behaviors aligned with their values and goals, rather than attempting to change or eliminate those undesirable thoughts and feelings [57]. Both ACT and CBT have demonstrated effectiveness in reducing tinnitus-related distress in comparative studies. These effects persisted through a one-year follow-up, indicating the potential of ACT procedures as an alternative to traditional CBT techniques for managing tinnitus [59]. On the other hand, Westin *et al.* conducted a randomized controlled trial comparing the effects of ACT and Tinnitus Retraining Therapy (TRT) on tinnitus. According to the study's findings, ACT was a more effective treatment for tinnitus than TRT or being on a waitlist. Notably, 54.5% of the patients experienced statistically significant improvement in their condition, with the effects sustained for 6 and 8 months [60].

Recent updates on the management of tinnitus in 2023 highlighted the success of counseling-based approaches such as CBT, mindfulness, ACT, and TRT. These methods are recommended as priority treatments based on their proven efficacy [61].

Audiologist-delivered cognitive behavioral therapy

In the United States, a survey revealed that approximately half of tinnitus sufferers had not consulted a physician regarding their condition. Despite the efficiency of CBT in coping with annoying tinnitus, consultations with physicians primarily focused on medical therapy, which has less evidence of effectiveness. Medical therapy was the most frequently suggested option, whereas CBT, which has more evidence of effectiveness, was the least commonly offered in clinical settings [6]. This situation is partly attributed to a lack of experts qualified to provide CBT for tinnitus, high treatment costs, patients' fear of stigma, and limited accessibility issues [62–65]. One suggested solution to these challenges would be to further educate and recruit psychologists. However, until psychologists can fulfill this demand, one potential alternative is to train audiologists. The development of audiologist-guided CBT represents a significant advancement. Traditionally, psychological therapies for patients with tinnitus-related distress were exclusively delivered by psychologists. However, the Department of Health in the United Kingdom [66] recommended that in cases where psychologists were not available, audiologists should extend their responsibilities to encompass providing psychological treatment using CBT or other appropriate counseling techniques. In fact, prompt intervention for controlling tinnitus and/or hyperacusis is crucial, as 12.7% of individuals who sought professional help for these conditions experienced thoughts of self-harm or suicide, with such incidents associated with the severity of their tinnitus and hyperacusis-handicap [67]. These findings highlight the necessity of training more audiologists to advance the delivery of psychological therapies.

Research indicated that the level of training and experience of eHealth therapists delivering guided online interventions had no discernible impact on therapeutic success [68]. Incorporating audiologist-delivered CBT into the management of tinnitus is becoming more popular [69, 70].The presence of a trained audiologist providing weekly feedback and on-demand help resulted in a decrease in the severity of tinnitus when compared to a control group [70]. Furthermore, one year following the intervention, patients reported maintained improvements in depression, anxiety, insomnia, and tinnitus distress [71]. Taylor *et al.* sought to develop a manualized, low-intensity psychological intervention that was feasible for audiologists to administer, cost-effective, and supported by evidence [69]. Their goal was to assess how well audiologists could implement a "psychologically informed" manual for managing tinnitus. Despite being acceptable by patients and audiologists alike, the intervention's feasibility for testing in a multi-site efficacy trial was nevertheless hindered by participant recruitment and retention issues [69]. The study revealed that even experienced audiologists require more thorough training to proficiently use the psychological treatment components. The participating audiologists stated that their ability to provide the treatment was hampered by the inadequate training they received, which lacked role-play and feedback from a supervisor. As recommended by the involved audiologists, creating reusable learning materials — like video recordings of role-play scenarios in which an audiologist applies a set of psychological techniques from the manual — may be an efficient way for more productive management [69].

To assess patients' feedback on audiologist-delivered CBT for tinnitus and/or hyperacusis; 31 patients underwent CBT by an audiologist at a clinic in the United Kingdom [72]. Patients expressed significant effectiveness of the therapy (median response of 8 out of 10) and acknowledged that the treatment had a major positive impact on their capability to cope with their diseases (median response of 9 out of 10). Additionally, they showed high levels of acceptance of receiving CBT from audiologists, with a median response of 10 out of 10 [72]. Discovering large effect sizes in the treatment outcomes, particularly in Tinnitus Handicap Inventory (THI) and Generalized Anxiety Disorder (GAD) scores, as well as other scores, indicates the power of audiologist-delivered CBT in addressing tinnitus and hyperacusis from the patients' viewpoints [72]. In a similar study, conducted retrospectively, 68 patients received CBT from an audiologist. All measures, such as scores from the THI, the Visual Analog Scale for various aspects of tinnitus severity, the Hyperacusis Questionnaire, and the Insomnia Severity Index, demonstrated significant improvements. Patients showed noteworthy improvements, including those who did not finish CBT. This implies that audiologists-delivered CBT can successfully reduce self-reported symptoms of insomnia, hyperacusis, and tinnitus [73]. Researchers have focused on the lasting effects of an audiologist-guided, Internet-based CBT for tinnitus distress in the US [74]. Of the 200 invited participants, 132 completed a follow-up questionnaire after one year, resulting in a 66% response rate. During the therapy, an audiologist guided the participants by tracking their progress, providing feedback on completed worksheets, offering new modules, and answering their questions. The primary focus was on changes in tinnitus distress measured by the TFI from the beginning of the study. Secondary measures, including hearing disability, tinnitus-related thoughts, hyperacusis, insomnia, anxiety, depression, and overall quality of life were also evaluated. The study concluded that the improvements observed immediately after the intervention were sustained even a year later [74].

Internet-based cognitive behavioral therapy

Internet interventions are increasingly popular for delivering accessible and affordable healthcare and for promoting self-management and engagement. In addition to traditional face-to-face administration of CBT by a clinician to individuals or groups, CBT can be administered remotely over the Internet, often in the form of self-help or self-directed tinnitus treatment [75]. Certain Internet CBT (ICBT) programs involve guidance through regular interaction with a clinician using a support messaging system, such as email or telephone. Consequently, these programs are more accurately categorized as self-directed rather than self-help programs. Through this approach, the clinician monitors the patient's progress, provides feedback on assigned tasks, addresses inquiries, and encourages the patient to complete the program. Conversely, interventions lacking clinician involvement are typically classified as self-help.

The content of the ICBT intervention was derived from a CBT self-help program originally developed for a Swedish audience [76]. The first RCT that utilized these materials in an online setting produced promising and encouraging results [77]. Evidence from the literature demonstrated that engaging in ICBT diminished tinnitus distress, insomnia, depression, and anxiety and improved quality of life [77–80]. ICBT effectively targets altered cognitive and behavioral processes by identifying negative thoughts and emotions linked to tinnitus, fostering new habits, and reducing unhelpful feelings and thoughts about tinnitus. These strategies are key to better-controlling tinnitus and reducing the associated annoyance and anxiety. In an RCT of internet-based versus in-person CBT, both modalities were found equally effective in reducing tinnitus-related distress, as assessed by the TFI [81]. Furthermore, the positive effects of ICBT on tinnitus distress persisted even at one-year post-intervention, with a large effect size [74].

Considering alternative delivery methods is important while considering patient preferences, accessibility, and costs. Additionally, cultural differences and disparities should be carefully addressed, especially in the context of internet-delivered interventions, as cultural acceptance can significantly influence treatment efficacy [82].

Smartphone-based CBT

Mobile applications are rapidly becoming an integral part of our daily routines, offering a wide range of features from entertainment and networking to educational endeavors. This pattern has extended into healthcare, particularly in the awareness and management of conditions like tinnitus. As a result, the market for smartphone applications offers a diverse range of applications specifically designed for CBT aimed at addressing tinnitus symptoms. These apps are developed to assist users in masking, controlling, mitigating, or managing tinnitus symptoms through smartphone-delivered CBT, sound therapy, and symptom tracking. In the realm of tinnitus management via smartphone apps, Mehdi *et al.* (2020) [83] conducted a systematic review that categorizes these apps into six main categories: CBT, tinnitus relief, hearing testing, hearing enhancement, hearing protection, and smartphone-based electroencephalogram EEG applications. Each category targeted different aspects of tinnitus management. For instance, the smartphone app TrackYourTinnitus tracks fluctuations in tinnitus symptoms over time, aiding patients to better understand and manage their condition by identifying factors contributing to symptom swings [84]. The mobile app "Tinnitus Coach" [85] is designed for tinnitus management, assisting patients in cultivating coping skills. The feedback from participants regarding the app's features was generally positive, with many expressing a willingness to purchase it due to its effectiveness in promoting relaxation and distraction from tinnitus [85]. Other applications, such as Diapason for Tinnitus, Beltone Tinnitus Calmer, and ResoundRelief, include elements specifically designed for CBT [83]. One key aspect of CBT within these apps involved reframing negative thoughts associated with tinnitus into more neutral or positive ones. Tinnitus-related negative thoughts included feeling helpless, fears of tinnitus ruining one's life, distress caused by loud tinnitus, having a miserable day due to tinnitus, or feeling alone and unsupported by others. Another important component of CBT in these apps was encouraging users to engage in pleasurable activities. Users are prompted to identify and participate in enjoyable activities such as spending time with friends, acquiring unfamiliar skills, or playing music. Users may receive weekly reminders to engage in these activities, based on the understanding that indulging in enjoyable endeavors can significantly improve quality of life, even for those living with tinnitus [86]. In 2021, Abouzari and colleagues [87] developed a novel smartphone app for tinnitus patients, combining tailored sound therapy with CBT (xtinnitus.com and BeyondTinnitus.com run through the Beyond Tinnitus Sound App). Participants with persistent tinnitus demonstrated a significant decrease in THI scores following a pilot trial evaluating the app's potential efficacy. Generalized Anxiety Disorder-7 (GAD-7) and Perceived Stress Scale surveys were administered in the study, noting that while improvements were observed, they did not reach statistical significance. This outcome may be attributed to the broad nature of these two assessments and the short duration of the trial [87]. A larger randomized clinical trial by the same group with 92 patients (45 control, 47 treatment) showed reduction in TFI, GAD,

Patient Health Questionnaire-9 (PHQ-9), and Pittsburgh Sleep Quality Index (PSQI). These patients were treated with the ICBT and internet-based sound therapy with no provider intervention/visits.

The rapid expansion of smartphone applications brings challenges for users, healthcare providers, and researchers alike in discerning and recommending high-quality apps [88]. Information regarding app quality is limited, often relying on star ratings on the app store, while app reviews are typically subjective and may originate from unreliable sources [89]. A survey investigating the usage of mobile apps for managing tinnitus [86] revealed various reasons behind the reluctance to use such apps among respondents, with the most common factor being a lack of awareness about available apps. Other reasons included not being comfortable with technology, an inability to find a suitable app, or not owning a mobile device or tablet. Among the motivations for app exploration were desired for sleep improvement, tinnitus masking, and healthcare professional recommendations. Additionally, when choosing an app, ease of use, reliability, and cost were prioritized, followed by recommendations from medical professionals or fellow individuals experiencing tinnitus [86].

Given the potential benefits of app-based tinnitus services, apps must be continually evaluated for the validity and usefulness of their content. App-based tinnitus intervention could offer a valuable alternative for providing care for patients who are reluctant to seek clinic-based care. Transitioning to telehealth delivery methods may help overcome hurdles to in-person care.

Conclusion

- CBT is the only strongly recommended option for managing chronic and bothersome tinnitus over alternative modalities as noted by the American Academy of Otolaryngology-Head and Neck Surgery Clinical Practice Guidelines.
- The aim of CBT is not to "cure" tinnitus but rather to support people in managing their condition by fostering more positive and realistic thoughts with the aim of increasing their quality of life.
- While the "first" and "second" waves of CBT focused on behavioral principles and cognitive restructuring respectively, the "third" wave represented a shift toward embracing mindfulness and contextual factors, incorporating approaches such as MBCT and ACT.

- Skills developed through MBCT helped reduce chronic tinnitus severity and related distress, showing advantages over intensive relaxation techniques.
- Similarly, ACT has been effective in decreasing tinnitus severity when compared to TRT or waitlist control groups.
- There is a lack of accessible CBT intervention for tinnitus. In response, creative approaches to delivering CBT have been developed. These included audiologist-delivered CBT, as well as CBT delivered through internet-based platforms and smartphone applications, expanding accessibility to tinnitus management.
- To date, research has demonstrated the effectiveness of audiologist-delivered CBT, highlighting the need for additional training to expand the scope of practice and confidence of audiologists while delivering CBT.
- Considering patient preferences, accessibility, and costs, it is crucial to explore alternative delivery methods for tinnitus management. ICBT emerges as the most searched online therapy, demonstrating efficacy comparable to face-to-face CBT.
- App-based tinnitus intervention could offer a valuable alternative for providing care for patients who are reluctant to seek clinic-based care. However, ongoing research is warranted to ensure the validity of their content.

References

1. Baguley D, McFerran D, and Hall D. Tinnitus. *The Lancet.* **382**:1600–1607 (2013).
2. Dietrich S. Earliest historic reference of 'tinnitus' is controversial. *The Journal of Laryngology and Otology.* **118**:487–488 (2004).
3. Dan B. Titus's tinnitus. *Journal of the History of the Neurosciences.* **14**:210–213 (2005).
4. Shargorodsky J, Curhan GC, and Farwell WR. Prevalence and characteristics of tinnitus among US adults. *American Journal of Medicine.* **123**:711–718 (2010).
5. Mahboubi H, Oliaei S, Kiumehr S, Dwabe S, and Djalilian HR. The prevalence and characteristics of tinnitus in the youth population of the United States. *The Laryngoscope.* **123**:2001–2008 (2013).
6. Bhatt JM, Lin HW, and Bhattacharyya N. Tinnitus epidemiology: Prevalence, severity, exposures and treatment patterns in The United States. *JAMA Otolaryngology — Head & Neck Surgery.* **142**:959–965 (2016).

7. Lewis MS, *et al.* The influence of tinnitus and hearing loss on the functional status of military service members and veterans. *International Journal of Audiology.* **62**:44–52 (2023).
8. Department of Veterans Affairs and Department of Defense. Clinical Practice Guideline for the Management of Tinnitus. Updated 2024. https://www.healthquality.va.gov/guidelines/CD/tinnitus/VADoD-CPG-Tinnitus-Full-CPG-2024_Final_508.pdf.
9. Adrian D, and El Refaie A. The Epidemiology of Tinnitus. In Tyler R, (Ed.) *The Handbook of Tinnitus,* pp. 1–23 (Singular, 2000).
10. Joo Y-H, Han K, and Park KH. Association of hearing loss and tinnitus with health-related quality of life: The Korea national health and nutrition examination survey. *PLoS One.* **10**:e0131247 (2015).
11. Piccirillo JF, Rodebaugh TL, and Lenze EJ. Tinnitus. *Journal of the American Medical Association.* **323**:1497–1498 (2020).
12. Maes IHL, Cima RFF, Vlaeyen JW, Anteunis LJC, and Joore MA. Tinnitus: A cost study. *Ear and Hearing.* **34**:508–514 (2013).
13. Stockdale D. *et al.* An economic evaluation of the healthcare cost of tinnitus management in the UK. *BMC Health Service Research.* **17**:577 (2017).
14. Goldstein E. *et al.* Cost of care for subjective tinnitus in relation to patient satisfaction. *Otolaryngology–Head and Neck Surgery: Official Journal of American Academy of Otolaryngology-Head and Neck Surgery.* **152**:518–523 (2015).
15. Beukes E, Andersson G, Manchaiah V, and Kaldo V. *Cognitive Behavioral Therapy for Tinnitus.* Plural Publishing. (2021).
16. Elgoyhen AB, and Langguth B. Pharmacological approaches to the treatment of tinnitus. *Drug Discovery Today.* **15**:300–305 (2010).
17. Tunkel DE, *et al.* Clinical practice guideline: Tinnitus. *Otolaryngology — Head and Neck Surgery: Official Journal of American Academy of Otolaryngology-Head and Neck Surgery.* **151**:S1–S40 (2014).
18. Ogawa K. *et al.* Clinical practice guidelines for diagnosis and treatment of chronic tinnitus in Japan. *Auris Nasus Larynx*. **47**:1–6 (2020).
19. Cima RFF. *et al.* A multidisciplinary European guideline for tinnitus: Diagnostics, assessment, and treatment. *HNO.* **67**:10–42 (2019).
20. Tang D, Li H, and Chen L. Advances in understanding, diagnosis, and treatment of tinnitus. *Advances in Experimental Medicine and Biology*, **1130**:109–128 (2019).
21. Wenzel A. Basic strategies of cognitive behavioral therapy. *Psychiatric Clinics of North America.* **40**:597–609 (2017).
22. Hofmann SG, Asnaani A, Vonk IJJ, Sawyer AT, and Fang A. The efficacy of cognitive behavioral therapy: A review of meta-analyses. *Cognitive Therapy and Research.* **36**:427–440 (2012).
23. Ellis A. Rational psychotherapy and individual psychology. *The Journal of Individual Psychology*. **13**:38 (1957).

24. Beck AT. *Depression: Clinical, Experimental, and Theoretical Aspects.* New York (1967).
25. Beck AT. *Depression: Causes and Treatment.* University of Pennsylvania Press, Philadelphia, (1972).
26. Nakao M, Shirotsuki K, and Sugaya N. Cognitive–behavioral therapy for management of mental health and stress-related disorders: Recent advances in techniques and technologies. *BioPsychoSocial Medicine.* **15**:16 (2021).
27. Cochrane database of systematic reviews: All issues. Cochrane Library. https://www.cochranelibrary.com/cdsr/table-of-contents/2024/3.
28. Scott B, Lindberg P, Lyttkens L, and Melin L. Psychological treatment of tinnitus an experimental group study. *The International Journal of Audiology.* **14**:223–230 (1985).
29. Robinson SK. *et al.* A randomized controlled trial of cognitive-behavior therapy for tinnitus. *International Tinnitus Journal.* **14**:119–126 (2008).
30. Schmidt CJ, Kerns RD, Finkel S, Michaelides E, and Henry JA. Cognitive behavioral therapy for veterans with tinnitus. *Federal Practitioner.* **35**:36–46 (2018).
31. Andersson G, and Lyttkens L. A meta-analytic review of psychological treatments for tinnitus. *British Journal of Audiology.* **33**:201–210 (1999).
32. Andersson G, Melin L, Hägnebo C, Scott B, and Lindberg P. A review of psychological treatment approaches for patients suffering from tinnitus. *Annals of Behavioral Medicine.* **17**:357–366 (1995).
33. Martinez-Devesa P, Perera R, Theodoulou M, and Waddell A. Cognitive behavioural therapy for tinnitus. *Cochrane Database of Systematic Reviews.* Issue 9. Art. No.: CD005233, (2010).
34. Hesser H, Weise C, Westin VZ, and Andersson G. A systematic review and meta-analysis of randomized controlled trials of cognitive–behavioral therapy for tinnitus distress. *Clinical Psychology Review.* **31**:545–553 (2011).
35. Fuller T, *et al.* Cognitive behavioural therapy for tinnitus. *Cochrane Database of Systematic Reviews.* **1**:CD012614 (2020).
36. Cima RFF, Andersson G, Schmidt CJ, and Henry JA. Cognitive-behavioral treatments for tinnitus: A review of the literature. *Journal of the American Academy of Audiology.* **25**:29–61 (2014).
37. Aazh H, Landgrebe M, Danesh AA, and Moore BC. Cognitive behavioral therapy for alleviating the distress caused by tinnitus, Hyperacusis and Misophonia: Current perspectives. *Psychology Research and Behavior Management.* **12**:991–1002 (2019).
38. Jun HJ, and Park MK. Cognitive behavioral therapy for tinnitus: Evidence and efficacy. *Korean Journal of Audiology & Otology.* **17**:101–104 (2013).
39. Woods JW, and Theodoroff SM. Cognitive behavioral therapy for tinnitus. *Audiology Today.* **31**:22–28 (2019).

40. Burle NL, de O, Resende LM, de Alves LM, Garcia VS, and Mancini PC. Effect of tinnitus habituation therapy on auditory abilities. *International Archives of Otorhinolaryngology.* **25**:e18–e26 (2021).
41. Grewal R, Spielmann PM, Jones SEM, and Hussain SSM. Clinical efficacy of tinnitus retraining therapy and cognitive behavioural therapy in the treatment of subjective tinnitus: A systematic review. *The Journal of Laryngology and Otology*, **128**:1028–1033 (2014).
42. Luo C, *et al.* A comparison of electronically-delivered and face to face cognitive behavioural therapies in depressive disorders: A systematic review and meta-analysis. *EClinicalMedicine* **24**:100442 (2020).
43. Natsky AN. *et al.* Economic evaluation of cognitive behavioural therapy for insomnia (CBT-I) for improving health outcomes in adult populations: A systematic review. *Sleep Medicine Reviews.* **54**:101351 (2020).
44. Henry JL, and Wilson PH. Coping with tinnitus: Two studies of psychological and audiological characteristics of patients with high and low tinnitus-related distress. *International Tinnitus Journal.* **1**:85–92 (1995).
45. Hayes SC. Acceptance and commitment therapy, relational frame theory, and the third wave of behavioral and cognitive therapies. *Behavior Therapy.* **35**:639–665 (2004).
46. Hayes SC, and Hofmann SG. The third wave of cognitive behavioral therapy and the rise of process-based care. *World Psychiatry.* **16**:245–246 (2017).
47. Kabat-Zinn J. Mindfulness-based interventions in context: Past, present, and future. *Clinical Psychology: Science and Practice.* **10**:144–156 (2003).
48. Feliu-Soler A, *et al.* Economic impact of third-wave cognitive behavioral therapies: A systematic review and quality assessment of economic evaluations in randomized controlled trials. *Behaviour Research and Therapy.* **49**:124–147 (2018).
49. Segal Z, Williams M, and Teasdale J. *Mindfulness-Based Cognitive Therapy for Depression, Second Edition.* Guilford Publications, (2018).
50. Hofmann SG, Sawyer AT, and Fang A. The empirical status of the "New Wave" of cognitive behavioral therapy. *Psychiatric Clinics of North America.* **33**:701–710 (2010).
51. Anderson N, and Ozakinci G. Effectiveness of psychological interventions to improve quality of life in people with long-term conditions: Rapid systematic review of randomised controlled trials. *BMC Psychology.* **6**:11 (2018).
52. Marks E, Smith P, and McKenna L. Living with tinnitus and the health care journey: An interpretative phenomenological analysis. *British Journal of Health Psychology.* **24**:250–264 (2019).
53. McKenna L, Marks EM, Hallsworth CA, and Schaette R. Mindfulness-based cognitive therapy as a treatment for chronic tinnitus: A randomized controlled trial. *Psychotherapy and Psychosomatics.* **86**:351–361 (2017).

54. Philippot P, Nef F, Clauw L, de Romrée M, and Segal Z. A randomized controlled trial of mindfulness-based cognitive therapy for treating tinnitus. *Clinical Psychology & Psychotherapy.* **19**:411–419 (2012).
55. Husain FT, *et al.* Assessing mindfulness-based cognitive therapy intervention for tinnitus using behavioural measures and structural MRI: A pilot study. *International Journal of Audiology.* **58**:889–901 (2019).
56. Zimmerman B, *et al.* Functional brain changes during mindfulness-based cognitive therapy associated with tinnitus severity. *Frontiers in Neuroscience,* **13**:747 (2019).
57. Hayes SC, Strosahl KD, Strosahl K, and Wilson KG. *Acceptance and Commitment Therapy: An Experiential Approach to Behavior Change.* Guilford Publications, (2003).
58. Hayes SC, Levin ME, Plumb-Vilardaga J, Villatte JL, and Pistorello J. Acceptance and commitment therapy and contextual behavioral science: Examining the progress of a distinctive model of behavioral and cognitive therapy. *Behavior Therapy.* **44**:180–198 (2013).
59. Hesser H, *et al.* A randomized controlled trial of internet-delivered cognitive behavior therapy and acceptance and commitment therapy in the treatment of tinnitus. *Journal of Consulting and Clinical Psychology.* **80**:649–661 (2012).
60. Westin VZ, *et al.* Acceptance and commitment therapy versus tinnitus retraining therapy in the treatment of tinnitus: A randomised controlled trial. *Behaviour Research and Therapy,* **49**:737–747 (2011).
61. Lee HY, and Jung DJ. Recent updates on tinnitus management. *Journal of Audiology & Otology.* **27**, 181–192 (2023).
62. Titov N. Internet-delivered psychotherapy for depression in adults. *Current Opinion in Psychiatry.* **24**:18–23 (2011).
63. Barney LJ, Griffiths KM, Christensen H, and Jorm AF. Exploring the nature of stigmatising beliefs about depression and help-seeking: Implications for reducing stigma. *BMC Public Health.* **9**:61 (2009).
64. Schröder J, *et al.* Development of a questionnaire measuring Attitudes towards psychological online interventions-the APOI. *Journal of Affective Disorders.* **187**:136–141 (2015).
65. Mohr DC, *et al.* Barriers to psychotherapy among depressed and nondepressed primary care patients. *Annals of Behavioral Medicine.* **32**:254–258 (2006).
66. Health D. of. *Provision of Services for Adults with Tinnitus. A Good Practice Guide.* Central Office of Information London (2009).
67. Aazh H, and Moore BCJ. Thoughts about suicide and self-harm in patients with tinnitus and hyperacusis. *Journal of the American Academy of Audiology.* **29**:255–261 (2018).
68. Baumeister H, Reichler L, Munzinger M, and Lin J. The impact of guidance on Internet-based mental health interventions — A systematic review. *Internet Interventions.* **1**:205–215 (2014).

69. Taylor JA, *et al.* The TinMan study: Feasibility trial of a psychologically informed, audiologist-delivered, manualised intervention for tinnitus. *International Journal of Audiology.* **59**:905–914 (2020).
70. Beukes EW, Baguley DM, Allen PM, Manchaiah V, and Andersson G. Audiologist-guided internet-based cognitive behavior therapy for adults with tinnitus in the United Kingdom: A randomized controlled trial. *Ear and Hearing.* **39**:423–433 (2018).
71. Beukes EW, Allen PM, Manchaiah V, Baguley DM, and Andersson G. Internet-based intervention for tinnitus: Outcome of a single-group open trial. *Journal of the American Academy of Audiology.* **28**:340–351 (2017).
72. Aazh H, Bryant C, and Moore BCJ. Patients' perspectives about the acceptability and effectiveness of audiologist-delivered cognitive behavioral therapy for tinnitus and/or hyperacusis rehabilitation. *American Journal of Audiology*. **28**:973–985 (2019).
73. Aazh H, and Moore BCJ. Effectiveness of audiologist-delivered cognitive behavioral therapy for tinnitus and hyperacusis rehabilitation: Outcomes for patients treated in routine practice. *American Journal of Audiology,* **27**:547–558 (2018).
74. Beukes EW, Andersson G, and Manchaiah V. Long-term efficacy of audiologist-guided internet-based cognitive behaviour therapy for tinnitus in the United States: A repeated-measures design. *Internet Interventions.* **30**:431–447 (2022).
75. Kaldo V, *et al.* Internet versus group cognitive-behavioral treatment of distress associated with tinnitus: A randomized controlled trial. *Behaviour Research and Therapy.* **39**:348–359 (2008).
76. Andersson G, and Kaldo V. Internet-based cognitive behavioral therapy for tinnitus. *Journal of Clinical Psychology*, **60**:171–178 (2004).
77. Andersson G, Strömgren T, Ström L, and Lyttkens L. Randomized controlled trial of internet-based cognitive behavior therapy for distress associated with tinnitus. *Psychosomatic Medicine.* **64**:810–816 (2002).
78. Beukes EW, Manchaiah V, Allen PM, Baguley DM, and Andersson G. Internet-based interventions for adults with hearing loss, tinnitus, and vestibular disorders: A systematic review and meta-analysis. *Trends and Hearing.* **23**:2331216519851749 (2019).
79. Weise C, Kleinstäuber M. and Andersson, G. Internet-delivered cognitive-behavior therapy for tinnitus: A randomized controlled trial. *Psychosomatic Medicine.* **78**:501–510 (2016).
80. Kaldo-Sandström V, Larsen HC, and Andersson G. Internet-based cognitive-behavioral self-help treatment of tinnitus: Clinical effectiveness and predictors of outcome. *American Journal of Audiology.* **13**:185–192 (2004).
81. Beukes EW, Andersson G, Allen PM, Manchaiah V, and Baguley DM. Effectiveness of guided internet-based cognitive behavioral therapy vs

dace-to-face clinical care for treatment of tinnitus: A randomized clinical trial. *JAMA Otolaryngology — Head & Neck Surgery.* **144**:1126–1133 (2018).

82. Rochlen AB, Zack JS, and Speyer C. Online therapy: Review of relevant definitions, debates, and current empirical support. *Journal of Clinical Psychology.* **60**:269–283 (2004).
83. Mehdi M, *et al.* Smartphone apps in the context of tinnitus: Systematic review. *Sensors* **20**:1725 (2020).
84. Schlee W, *et al.* Measuring the moment-to-moment variability of tinnitus: The track your tinnitus smart phone app. *Frontiers in Aging Neuroscience,* **8**:294 (2016).
85. Henry, J. A. *et al.* Development and field testing of a smartphone 'App' for tinnitus management. *International Journal of Audiology*, **56**:784–792 (2017).
86. Sereda M, Smith S, Newton K, and Stockdale D. Mobile apps for management of tinnitus: Users' survey, quality assessment, and content analysis. *JMIR mHealth and uHealth.* **7**:e10353 (2019).
87. Abouzari M, *et al.* Adapting personal therapies using a mobile application for tinnitus rehabilitation: A preliminary study. *Annals of Otology Rhinology and Laryngology.* **130**:571–577 (2021).
88. Cummings E, Borycki EM, and Roehrer E. Issues and considerations for healthcare consumers using mobile applications. *Studies in Health Technology and Informatics.* **183**:227–231 (2013).
89. Kuehnhausen M, and Frost VS. Trusting smartphone apps? To install or not to install, that is the question. In *2013 IEEE International Multi-Disciplinary Conference on Cognitive Methods in Situation Awareness and Decision Support (CogSIMA)* pp. 30–37 (2013).

https://doi.org/10.1142/9789819807345_0003

Chapter 3

Sound and Music Therapy for Tinnitus Rehabilitation

Brooke Reid

Department of Audiology, UCSF,

San Francisco, CA, USA

brooke.reid@ucsf.edu

Abstract

Tinnitus, characterized by the perception of sound in the absence of an external source, poses a significant challenge to millions worldwide due to its elusive etiology and impact on quality of life. Over the years, sound and music therapy have emerged as promising options for tinnitus management and rehabilitation. This chapter highlights current evidence on the efficacy of sound and music-based interventions in alleviating tinnitus-related distress and improving subjective well-being. The review begins by explaining the neurophysiological mechanisms underlying tinnitus perception and explores how sound therapy can help modulate these processes to reduce symptom severity. Various sound therapy approaches, including sound masking, sound enrichment, and sound therapy delivered through specialized devices, are examined in terms of their efficacy and feasibility in clinical settings. Moreover, the potential synergistic effects of combining sound therapy with cognitive-behavioral techniques are explored to optimize treatment outcomes. Furthermore, the therapeutic role of music in tinnitus rehabilitation is investigated, highlighting its unique ability to engage emotional, cognitive, and sensorimotor processes. Music-based interventions are evaluated for their capacity to promote habituation to tinnitus sensations and enhance coping mechanisms. Lastly, this chapter addresses practical considerations for implementing sound and music therapy protocols, including treatment duration, patient adherence, and individualized treatment planning. Indeed, sound and music therapy represent promising avenues for tinnitus rehabilitation, offering a holistic approach that addresses both the sensory and emotional components of tinnitus perception. By integrating insights from neuroscience, psychology, and sound therapy, these interventions hold the potential to transform the landscape of tinnitus management and improve the quality of life for individuals living with this prevalent condition.

Sound and Music Therapy for Tinnitus Rehabilitation

Tinnitus is a condition where an individual perceives phantom sounds that are not objectively present in their environment. The perceptions reported by individuals take the shape of many sounds with ringing, hissing, buzzing or static-like noise most often reported. The level of discomfort experienced by an individual can vary greatly, ranging from being not noticeable to completely intolerable. Studies have shown that as many as 45% of tinnitus patients also suffer from reduced loudness tolerance, commonly known as hyperacusis [1, 2]. Factors such as depression, stress, and anxiety can also contribute to the level of disability reported by the individual [3]. These observations can be explained by the widely accepted theory about the anatomical basis of tinnitus and what keeps it persistent. Tinnitus is primarily understood as a perception within a spectrum and is often a result of changes in neural networks, both auditory and non-auditory, following damage to the cochlea. Homeostatic mechanisms that aim to compensate for hearing loss may alter the balance of neurotransmitters, leading to the emergence of chronic tinnitus and related symptoms in many individuals. Whilst most people are not bothered by tinnitus, others may experience negative emotions when auditory network dysfunction is coupled with limbic-gating dysfunction, providing significance to the otherwise meaningless perception of tinnitus [4].

Studies show that while many individuals perceive their tinnitus as confined to one or both ears, it originates from a neurological disorder caused by the hyperactive synchronous activity of tonotopic neurons in the primary auditory cortex [3]. Unfortunately, for many that suffer from the condition, what causes this neuronal hypersensitivity is mostly unknown. Some research has likened it as a form of phantom limb syndrome where the peripheral signal — mainly involving cranial nerve V or VIII — is reduced or inhibited causing the neurons to become overly sensitive as they attempt to elicit a conditioned neuronal response [5].

Studies estimate that over 90% of those with tinnitus have a measurable hearing loss. However, only 50% of those with hearing loss will develop tinnitus [6]. Hearing loss, temporary or permanent, is not the only condition associated with tinnitus as other illnesses including temporomandibular joint disorder (TMJ), vitamin deficiencies, ototoxic medications and depression/anxiety have also been reported as being associated with the disorder [12]. No matter what the cause of an individual's tinnitus may be, appropriate management can improve patient outcomes.

Indeed, numerous management approaches for tinnitus are available. Common examples include tinnitus retraining therapy (TRT), tinnitus activities treatment (TAT), cognitive behavioral therapy (CBT) and mindfulness-based therapy [3]. There are also several proprietary devices available including Lenire, Neuromonics and Levo which will be discussed later in this chapter. While all the different therapies listed above have different approaches to tinnitus management, they are all ultimately aimed towards the same goal of desensitization and emotional relief from the bothersome stimuli. It is important to note that although pitch and intensity matching of tinnitus may be utilized in some therapeutic approaches, they are not a reliable means of measuring the degree of distress an individual may experience due to their tinnitus perception. There does not seem to be an association between the psychophysical assessments of tinnitus, such as pitch and loudness, and the distress it causes as measured through standardized patient outcome measures [3, 12].

Researchers at the University of Iowa developed TAT [13]. The therapy is heavily influenced by cognitive-behavioral strategies and concentrates on four key areas, including hearing and communication, thoughts and emotions, concentration, and sleep. Partial Masking Sound Therapy is usually applied in TAT, which involves setting a noise or music at the lowest level possible to offer relief. A structured and comprehensive counseling approach is provided by using a picture-based method to encourage patient engagement. To facilitate progress, homework and activities are assigned to the patient to ensure they have a thorough understanding of their goals for tinnitus habituation.

The University of Iowa team also developed two questionnaires that are available in a growing number of languages to determine what situations are most affected by tinnitus, while also used to measure habituation success throughout management The Iowa Tinnitus Handicap Questionnaire was created and subjected to psychometric testing by Francis Kuk and colleagues in 1990. This questionnaire evaluates three aspects: the physical, emotional, and social impacts of tinnitus, the patient's hearing ability, and their personal perception of the condition. These questionnaires can be a helpful tool for a provider in any specialty who is trying to help refer a patient to the right specialty clinic for their tinnitus management [14].

In contrast, TRT aims to address the unconscious, automatic processes that perpetuate the response to and perception of tinnitus. The therapy involves educational counseling on the neurophysiological model of tinnitus and sound therapy, with the goal of achieving habituation to the perception of and reaction to the tinnitus percept [15, 16].

While numerous approaches to managing tinnitus have been developed, this chapter will concentrate on sound therapy and its connection to TRT which integrates both sound therapy and stress management techniques. While there is no widely used medication, surgical intervention or cure currently approved as a viable treatment for tinnitus [17], research has shown that TRT can be taught to successfully manage the bothersome level of tinnitus through a process called habituation [18]. This process involves using sound therapy to gradually desensitize the brain to the tinnitus sound, helping the brain to perceive it as less important over time. The goal being that the tinnitus is no longer brought to an individual's forefront of attention.

Although sound enrichment and music therapy are crucial components of the treatment, it is essential to recognize that these approaches alone may not be adequate for achieving successful habituation. What drives the bothersome level of tinnitus is not the anatomical hypersensitivity of the auditory neurons themselves. It is estimated that over 50 million people in the United States have some form of tinnitus, but only about 10–20% of those are bothered enough to seek medical guidance and treatment [7]. What is the difference between individuals who have tinnitus but are not sufficiently bothered by it to seek medical guidance and those who are bothered enough to seek assistance? While there are many theories and much research being done to answer this question, there is increasing evidence that an atypical form of migraine could possibly be the cause of the enhanced attention to the tinnitus percept. Anxiety/stress and poor sleep significantly exacerbate this migraine process. At the initial onset of tinnitus, patients often have increased anxiety and poor sleep which perpetuates the migraine process which causes the tinnitus to be perceived at a louder level causing a feedback loop [8–11].

Our brains must filter through so much information throughout the day that it is impossible for us to focus our attention on everything that we experience. Indeed, our brain has a mechanism that determines the most significant information and prioritizes bringing it to our attention. This decision maker is known as the limbic system and houses many important systems including those for memory, decision making and most importantly our flight or fight response. Individuals develop bothersome tinnitus when their limbic systems activate the flight or fight response as a defense to the stimuli [18]. Our brain cannot habituate to a stimulus that is causing it stress and activating this fight or flight response. It is critical to remove the emotional hold that tinnitus may have on an individual for TRT and sound therapy to be successful. Mindfulness therapy and cognitive-behavioral therapy are two such stress management approaches guided by psychologists for those suffering from tinnitus [19]. Those that are not as bothered by their tinnitus may benefit from a less regimented form of stress management including short duration meditation, deep breathing exercises, and several mobile apps for relaxation.

Sound and music therapy for tinnitus consists of several key elements that should be followed, However, there is no specific type of music or ambient noise that has been proven to be more effective than others in achieving long-term habituation. It is important that an individual with tinnitus find a sound that is soothing to them and something that they genuinely enjoy listening to [20]. Sound therapy will not be helpful if the person listening to it is just as annoyed by the sound itself as they are with their tinnitus [21]. It is also important that the sound involves an act of passive listening rather than something that the individual is actively paying attention to. It should be highlighted that even though an individual may experience a temporary reduction in their tinnitus when watching TV or listening to an audiobook or podcast due to a shift in their attention, this does not qualify as engagement in sound therapy for habituation [22]. The limbic system still needs to perceive the tinnitus in some capacity along with the sound therapy to retrain its response to the stimuli [17]. The most crucial element of sound therapy to remember is that the goal of sound enrichment is for long-term habituation and not short-term complete masking. Where masking is defined as the process of using a sound stimulus to temporarily mask out the tinnitus at a louder volume level [23].

In order to achieve long-term habituation, one must begin by removing the emotional response triggered by the limbic system and using positive reinforcement to help the brain retrain its negative response to the tinnitus stimuli, using a method often referred to as operant conditioning [24]. An excellent illustration of this concept is Pavlov's dog study, which was published in 1897. In this study, Pavlov demonstrated that certain cortical reflexes are innate and do not necessitate any deliberate learning on the part of the individual. While this technique cannot be applied verbatim as tinnitus and its perception are constant and cannot physically be eliminated, a modified version of this theory can be applied to sound therapy and its relation to long-term habituation. Sound therapy aims to achieve more than just a temporary masking of tinnitus. Its objective is to assist in the process of retraining the brain's response to tinnitus, which will eventually reduce the emotional grip that the limbic system has on the tinnitus stimuli. By doing so, the brain can adequately compartmentalize tinnitus and prevent it from becoming the primary focus of attention. This results in a decrease in the frequency and intensity of tinnitus by maintaining a lower sensory signal over an extended period [25].

Sound therapy has been an effective method of tinnitus management by offering both immediate temporary relief and long-term habituation [26]. Three primary types of sound therapy exist: the first involves the stimulation of a broad range of frequencies, while the second allows for the subjective pitch-matched frequency of tinnitus to be removed from the broad band frequency range, known as notched therapy [16]. The third increases the amount of sound provided around the tinnitus frequency utilized by beyondtinnitus.com and Neuromonics [27, 28]. Music therapy can fit into all three categories and individuals can make it as simplistic or as complex as they want. Some prefer home-based sound therapy, which involves getting access to sound enrichment via desktop noise generators, mobile apps, sleep headbands, or sound pillows. Others utilize in-ear noise generators or hearing aids for sound therapy which will be discussed in the upcoming chapter on hearing aids for tinnitus.

Generalized sound therapy does not consider an individual's ability to pitch match their tinnitus. Some tinnitus sufferers can confidently identify the pitch and intensity level of their phantom sound, while others describe a broader range of frequencies perceived that cannot be so easily narrowed into one specific range. These individuals with more of a broadband tinnitus percept typically prefer generalized sound therapy or passive music as a form of sound enrichment to be used at times when their tinnitus is more bothersome [29]. There are several mobile apps that can provide single or layered soundscapes. For instance, one can listen to the sound of the ocean in isolation or overlay the sounds of rain and thunder for a more intricate listening encounter. Some individuals prefer even more simplistic soundscapes by choosing to listen to a broad range of frequencies dependent on what intensity level is added at the low, mid and high range. Examples of this are white noise which provides equal intensity for all frequencies, pink noise offering greater intensity at the low and mid frequency range with reduced intensity for high frequencies, and brown noise with even more of an intensity at the low frequency range [30]. Sound therapy can be accessed in the sound field or by using ear level devices with Bluetooth or wired streaming capability.

Sound and music therapy can also be utilized for notch therapy, which is a form of sound therapy where the subjective pitched-matched frequency of tinnitus can be notched out of the broadband frequency range. Another form of sound and music therapy is harmonic sound therapy which has increased sound around the tinnitus pitch and subharmonics of the tinnitus pitch. For this form of sound enrichment to be successful, providers must rely on a patient's ability to subjectively pitch match their tinnitus successfully [31]. In the pitch matching procedure with the audiologist, two pitches are presented to the patient above their threshold, and the patient is instructed to choose which pitch is closer to their tinnitus pitch. The procedure involves bracketing back and forth around the perceived pitch. On the other hand, in the loudness match procedure, the pitch-matched sound is played in increasing levels until the patient reports that the external sound matches the loudness of their tinnitus perception [7]. Alternatively, patients can also subjectively pitch match their tinnitus via a specialized mobile app.

Notch therapy is based on a theory called lateral habituation, where the focus should be on de-energizing a cluster of hyperactive neurons in the auditory cortex instead of providing low level stimulation throughout the tonotopic regions through generalized sound therapy [33]. For a neuron to become active and transmit an electrical impulse for encoding, the surrounding neuronal regions must be deactivated to allow for a more precise signal to be sent. This technique is applied in notched therapy, where the frequency that the tinnitus was pitch matched to is removed to target the specific region and deactivate it while stimulating the frequency region surrounding it. This approach is particularly useful for individuals with true tonal tinnitus [3]. For example, if an individual pitch matches their tinnitus at 4,000 Hz, then the 4,000 Hz frequency will be removed from the sound or music therapy. By directly stimulating all frequencies around 4,000 Hz with a broad frequency range, it will target the synchronous cluster of spontaneous firing neurons in that specific 4,000 Hz region. By doing so, the therapy promotes long term habituation via neuroplasticity by constantly deactivating that frequency range. In harmonic sound therapy, the tinnitus pitch and selected subharmonics are stimulated. The theory behind harmonic sound therapy is that tinnitus is due to peripheral deafferentation and by stimulating more at the tinnitus frequency, more peripheral input will lead to reduced central spontaneous activity and thus tinnitus. Regardless of the method used, true habituation of tinnitus using any form of sound therapy, if achieved, often takes between 6–18 months [33].

Neuroplasticity refers to the brain's ability to change its structure and function in response to experiences and environmental stimuli. This process involves the formation of new neural connections, the strengthening of existing ones, and the pruning of unused or unnecessary connections. Neuroplasticity can occur naturally throughout life and is involved in processes such as learning and memory formation. It can also be harnessed for therapeutic purposes, such as in rehabilitation after brain injury or stroke [35]. In the context of tinnitus, neuroplasticity can be harnessed to modify the activity of the nervous system over time to reduce the perception of tinnitus [36].

Neuroplasticity devices, mostly accessed with ear level devices via noise generator or Bluetooth streaming, also typically focus on delivering either generalized or notched sound therapy [34]. While sound therapy can be used as needed, neuroplasticity devices often recommend a specific wear time of approximately 1–4 hours per day [37]. One common type of sound therapy using ear-level neuroplasticity devices is the layering of multiple sounds. An illustration of this is the incorporation of low-level white noise in the background of an individual's music therapy. As the individual becomes more habituated to the therapy, the white noise is gradually reduced until it is no longer necessary. This process continues until the music therapy itself is gradually decreased and eliminated as well.

The primary neuroplasticity device utilized for this type of music therapy is Neuromonics, which originated in Australia more than two decades ago and was tested in four clinical trials involving over 200 tinnitus patients. Around 80 to 90% of individuals reported a significant reduction in their tinnitus symptoms [38, 39]. Over a six-month timeframe, patients gradually became less aware and less affected by their tinnitus, even when they were not undergoing treatment. Additionally, many participants reported experiencing benefits during the initial stages of treatment, such as relief from tinnitus and improved sleep [39]. With numerous options offered to individuals via Bluetooth streaming for sound therapy, the physical device for Neuromonics is no longer available. However, it is now offered through a mobile app. The patient simply needs an access code from a tinnitus specialist who has been trained by the company.

Neuromonics is an acoustic-based therapy that includes counseling, monitoring, support, and education from a tinnitus management audiologist. Its goal is to retrain the neural pathways responsible for tinnitus by using specific patterns of acoustic stimuli, as described in a study by Davis and colleagues. It is a more structured program for individuals who need defined guidelines and routine follow up with a tinnitus specialist to monitor progress while the specialist adjusts the sound generator. Neuromonics is not to be mistaken for a simplistic tinnitus masker, and instead is understood to provide positive engagement of the limbic system for tinnitus desensitization [39]. The goal of any form of sound or music therapy is to reduce the perception of tinnitus, not its loudness level; which can be contradictory to some of the research we are about to discuss. But keep in mind that if tinnitus is no longer brought to the forefront of attention, and therefore no longer perceived by an individual, the loudness level no longer matters.

The BeyondTinnitus.com program is a web-based sound and music therapy that allows patients from areas without access to a tinnitus care provider to receive therapy or to be used in conjunction with provider care. The Elinava Technologies Tinnitus Rehabilitation Program has now completed 3 studies on their proprietary therapy. One such study on this system involved 18 subjects who were randomized to receive the BeyondTinnitus.com sound/music vs. white noise therapy. Tinnitus loudness, annoyance, and minimum masking level (MML) were measured after 3 months. After a 1-month washout period, the groups switched therapies for another 3 months, and measurements were repeated. Results showed that customized sound therapy reduced mean loudness by 23%, Tinnitus Handicap Inventory score by 25%, anxiety by 22%, and MML by 23%, while white noise therapy showed no significant changes. In another study a total of 30 patients enrolled in this study consisting of 20 treatment and 10 control patients [40]. The treatment group was treated with a web-based cognitive behavioral therapy (xtinnitus.com) in addition to the sound/music therapy using a mobile app. After 8 weeks, the treatment group showed a significantly greater improvement in Tinnitus Handicap Inventory scores, with improvements 3.3 times greater than the control group.

Similar to BeyondTinnitus.com, AudioNotch.com is fully automated software that allows one to create their own Notched Sound Therapy and listen to it after subjective tinnitus pitch matching. Through their website you can upload your favorite music or use their sounds to create custom sound therapy tailored for the individual. A study completed in 2011 developed and evaluated a 5-day intensive (6 hours/day) tailor-made notched music training (TMNMT) for chronic, tonal tinnitus. The study assessed TMNMT efficacy and its persistence over four weeks in two patient groups with low (≤8,000 Hz, $N = 10$) or high (>8,000 Hz, $N = 10$) tinnitus frequencies. Results showed that TMNMT significantly reduced tinnitus loudness, distress, and auditory cortex activity in patients with ≤8,000 Hz frequencies but not in those with >8,000 Hz frequencies. These effects were not persistent, suggesting the need for longer-term training [41]. The original German team that pioneered TMNMT enrolled 100 users in a randomized controlled trial to evaluate the efficacy of this same treatment 5 years later [42]. This was completed in parallel group design and performed in a double-blinded manner. After a three-month duration, tinnitus loudness was found to be decreased in the treatment group. The reduction in loudness was observed to be small but persisted one month after the cessation of therapy.

Neuromodulation and neuroplasticity are related but distinct concepts in neuroscience, and both have shown promise in the treatment of tinnitus [43, 44]. Neuromodulation involves the use of external means to modify the activity of neurons in the nervous system, such as electrical, chemical, or magnetic stimulation, to achieve therapeutic effects. On the other hand, neuroplasticity is an internal process that occurs naturally in response to experiences and can be harnessed for therapeutic purposes by modifying the activity of the nervous system over time [43, 44].

Both approaches have shown promise in the treatment of tinnitus, and future research may explore the potential of combining these approaches for more effective outcomes.

Most recently, Neuromod received FDA approval in March 2023 for a bimodal neuromodulation device called Lenire. This device targets the neural pathways by stimulating multiple cranial nerves for habituation. As previously discussed, the vestibulocochlear nerve (CN VIII) may not be the only cranial nerve beneficial to target for tinnitus management solely with sound therapy. Lenire uses bimodal stimulation to target both the vestibulocochlear nerve (CN VIII) and trigeminal nerve (CN V) by providing sound therapy via headphones along with mild electrical pulses to the tongue to be used 60 minutes per day (one 60-minute session or two 30-minute sessions) as part of a monitored treatment plan by the audiologist. Both devices are paired to a controller that is used by the patient with baseline sound therapy and electrical pulse stimuli determined by their audiologic and tinnitus profile [45].

Lenire has published two clinical trials with over 500 participants. The first study used their proprietary PS1 setting consisting of synchronized sound and tongue stimulation. A pure tone (ranging approximately from 500 to 8000 Hz, with repetition period of about 80 ms) was presented at the same time that a train of pulses was presented to a specific location on the tongue. Participant progress was monitored using the tinnitus functional index (TFI) and tinnitus handicap inventory (THI), which are two widely used and validated questionnaires. This study revealed a significant improvement in tinnitus perception after the first 6 weeks with a plateau in improvement during the second 6 weeks [47]. The second study aimed to correct this plateau by changing the parameter settings after 6 weeks of use to promote an even greater increase in neuroplasticity and overcome treatment habituation effects as seen in the first study [46]. The second 6 weeks used PS6 which consisted of only pure tone bursts paired with spatially randomized tongue stimulation with varying interstimulus delays (700–800 ms) and without any background noise. The Tent-A2 study revealed 95% improvement after just 12 weeks of use with 91% of participants maintaining improvement when re-evaluated at 12 months [46], though when patients were asked if they subjectively felt better, the number was far lower.

Of note, additional clinical trials on Lenire should be considered as the current trials lacked a proper control group providing active treatment to all participants. While the Tent-A2 trial did test sound therapy alone compared to sound therapy and tongue stimulation they still chose to use proprietary sound therapy instead of generic background noise. This could lead some to argue the tongue stimulation may not have a role in the improvement seen and that the perceived improvement may be partially or entirely due to a placebo effect or the sound therapy portion and the daily relaxation alone.

Another FDA approved neuromodulation device for tinnitus management is the Otoharmonics Levo system. Levo is unique to other neuromodulation devices as it is used by a patient while they sleep. The creation of this device was based on the understanding that the auditory system remains active and processes information during sleep. Therefore, introducing sound therapy during this time may be more effective as there is less competing information, allowing the brain to focus on neural habituation. The device is an app compatible with Apple devices with proprietary software that the patient connects to via customized earbuds that are more comfortable for sleep. The patient will then utilize the device as they sleep nightly for at least 3 months. Its use is accompanied by follow up with the tinnitus specialist and personalized programming of the device. An independent study completed by the International Tinnitus Journal in 2010 evaluated 11 patients who were stimulated with Levo sound therapy during sleep for a period of 6 months [48]. The stimulus used for each individual was programmed to mimic the pitch and intensity level that was subjectively measured as a match to their tinnitus perception. While this was a small and uncontrolled study, it did show a statistically significant reduction of tinnitus for all individuals with most reporting improvement after just one week and all within 3 weeks [48].

Another study of 12 patients of the Levo device showed that after 3 months of continuous use of the device at bedtime, subjects reported a significant decrease in tinnitus perception that was measured by an average decrease in tinnitus intensity of 14.1 dB SPL or the equivalent of a 62% reduction based on pre measured intensity levels. A statistically significant improvement was also observed in all questionnaires with a group-level decrease up to 78% (TRQ), 65% (THI) and 77% (TFI) [49].

In this chapter, we explored the applications of sound and music therapy, ranging from simple home-based sound enrichment to sophisticated ear-level neuromodulation devices. The level of knowledge regarding the available treatment options for tinnitus may differ among healthcare providers of different specialties. The chosen method for treating tinnitus can greatly affect the individual patient. The providers knowledge and experience with different management strategies, as well as the ability to be adaptable and customize the approach to each patient, is crucial. Indeed, continuous learning and improvement around tinnitus management is essential to maximize patient outcomes.

Conclusion

- Sound and music therapy offer promising non-invasive approaches for managing tinnitus, leveraging the brain's plasticity to modulate perception and reduce distress.
- Various sound therapy techniques, including sound masking and enrichment, provide customizable solutions tailored to individual preferences and needs.
- Music-based interventions engage multiple cognitive and emotional processes to promote habituation and coping.
- The combination of sound therapy with cognitive-behavioral techniques enhances treatment efficacy by addressing both the sensory and psychological aspects of tinnitus.
- Practical considerations, including treatment duration, patient adherence, and individualized planning, are crucial for optimizing therapeutic outcomes.
- Challenges remain in standardizing protocols and measuring treatment outcomes due to the heterogeneous nature of tinnitus and subjective variability in symptom perception.
- Public awareness and education initiatives are needed to destigmatize tinnitus, promote early intervention, and encourage individuals to seek appropriate care from a tinnitus specialist.
- By optimizing the benefits of sound and music therapy, providers can empower individuals living with tinnitus to reclaim control over their auditor experiences and improve their overall quality of life.

References

1. Jastreboff PJ. Tinnitus habituation therapy (THT) and tinnitus retraining therapy (TRT). In *Tinnitus Handbook*. pp. 99–120 (2000).
2. Aazh H, and Allott R. Cognitive behavioural therapy (CBT) for tinnitus management: A qualitative review of patient experiences and treatment outcomes. *Hearing, Balance and Communication*. **16**(3):181–191 (2018).

3. Langguth B, *et al.* Tinnitus: Causes and clinical management. *The Lancet Neurology*. **12**(9):920–930 (2013).
4. Ryan D, and Bauer CA. Neuroscience of tinnitus. *Neuroimaging Clinics of North America*. **26**(2):187–196 (2016).
5. Rauschecker JP, and Leaver AM. Tinnitus: A neuroscience perspective. *Current Opinion in Neurology*. **23**(5):413–417 (2010).
6. Tyler RS. The psychoacoustical measurement of tinnitus. In Møller M, (Ed.) *Textbook of Tinnitus*. pp. 83–90 (2012).
7. Spankovich C. 20Q: Tinnitus — Developing a practical management protocol. *AudiologyOnline, Article*. 25780 (2019).
8. Langguth B, *et al.* The role of migraine in the perception of tinnitus. In *Progress in Brain Research*, Elsevier, Amsterdam, Netherlands, 2007.
9. Stolzberg D, and Salvi RJ. Tinnitus and its relationship to migraine. In *Migraine and Other Headache Disorders*, Springer International Publishing, Cham, Switzerland, 2018.
10. Jorgensen S, and Klokker M. Sleep disturbance and the onset of the perceived intensity of tinnitus. *Ear and Hearing*. **33**(2):179–180 (2012).
11. Ostermann K, and Tochtermann U. Migraine, stress and the temporal relationship to exacerbation of chronic tinnitus. *Zeitschrift für Allgemeinmedizin*. **81**(11):454–457 (2005).
12. Shore SE, *et al.* Maladaptive plasticity in tinnitus — Triggers, mechanisms and treatment. *Nature Reviews Neurology*. **12**(3):150–160 (2016).
13. Tyler RS, *et al.* Tinnitus retraining therapy: Mixing point and total masking are equally effective. *Ear and Hearing*. **28**(5_suppl):1S–7S (2007).
14. Kuk FK, *et al.* The psychometric properties of a tinnitus handicap questionnaire. *Ear and Hearing*. **11**(6):434–445 (1990).
15. Jastreboff PJ, and Hazell JW. A neurophysiological approach to tinnitus: Clinical implications. *British Journal of Audiology*. **27**(1):7–17 (1993).
16. Jastreboff PJ, and Jastreboff MM. Tinnitus retraining therapy (TRT) as a method for treatment of tinnitus and hyperacusis patients. *Journal of the American Academy of Audiology*. **11**(3):162–177 (2000).
17. Baguley DM, *et al.* Tinnitus. *The Lancet*, **382**(9904):1600–1607 (2013).
18. Jastreboff PJ, and Jastreboff MM. Tinnitus retraining therapy: A different view on tinnitus. *ORL*. **68**(1):23–29 (2006).
19. Hesser H, *et al.* The effect of mindfulness-based cognitive therapy in patients with tinnitus and co-morbid depressive symptoms: A randomized controlled trial. *Journal of Psychosomatic Research*. **70**(4):377–384 (2011).
20. Cima RF, *et al.* Tinnitus: From pathophysiology to treatment. *Journal of International Advanced Otology*. **15**(2):159–166 (2019).
21. De Ridder D, *et al.* Phantom percepts: Tinnitus and pain as persisting aversive memory networks. *Proceedings of the National Academy of Sciences*. **108**(20):8075–8080 (2011).

22. Aazh H, and Moore BC. Factors associated with depression in patients with tinnitus and hyperacusis. *American Journal of Audiology*. **27**(3S):424–432 (2018).
23. Hobson J, *et al.* Sound therapy (masking) in the management of tinnitus in adults. *Cochrane Database of Systematic Reviews.* **11**, CD006371 (2012).
24. Tyler RS, and Baker LJ. Difficulties experienced by tinnitus sufferers. *Journal of Speech and Hearing Disorders*. **48**(2):150–154 (1983).
25. Davis PB, and Wilde RA. Treatment of tinnitus with a customized acoustic neural stimulus: A controlled clinical study. *Ear and Hearing.* **21**(4):239–254 (2000).
26. Schlee W, *et al.* Phantom sounds after cochlear implantation reflect central neural plasticity. *Neural Plasticity*. **2011**:1–10 (2011).
27. Mahboubi H, Haidar YM, Kiumehr S, Ziai K, Djalilian HR. Customized versus noncustomized sound therapy for treatment of tinnitus: A randomized crossover clinical trial. *Ann Otol Rhinol Laryngol*. **126**(10):681–687 (2017).
28. Davis PB, Paki B, and Hanley PJ. Neuromonics tinnitus treatment: Third clinical trial. *Ear and Hearing.* **28**(2):242–59 (2007).
29. Tunkel DE, *et al.* Clinical practice guideline: Tinnitus. *Otolaryngology–Head and Neck Surgery.* **151**(2_suppl):S1–S40 (2014).
30. Liu W-H, and Wu C-W. A study of noise generation and its characteristic in digital audio systems. *IEEE Transactions on Consumer Electronics*. **51**(3):868–876 (2005).
31. Searchfield GD, *et al.* An evaluation of the effectiveness of notch therapy for tinnitus management. *International Journal of Audiology*. **49**(8):554–563 (2010).
32. Baguley DM. Mechanisms of tinnitus. *British Medical Bulletin.* **97**(1):73–100 (2010).
33. Seydell-Greenwald A, and Leaver AM. Tinnitus: Neuroscience, diagnosis, and treatment. *Annual Review of Neuroscience*. **44**:101–118 (2021).
34. Tyler RS. Neurophysiological models, psychological models, and treatments for tinnitus. In Werner JS, Spetzler RF, and Bucholz EL, (Eds.) *Neuromodulation: Comprehensive Textbook of Principles, Technologies, and Therapies*, pp. 147–160. Springer (2012).
35. Merzenich MM, *et al.* Brain plasticity-based therapeutics. *Frontiers in Human Neuroscience*. **5**(8):385 (2014).
36. Schaette R, and McAlpine D. Tinnitus with a normal audiogram: Physiological evidence for hidden hearing loss and computational model. *The Journal of Neuroscience*. **31**(10):13452–13457 (2011).
37. Searchfield GD, and Kaur M. Refining tinnitus retraining therapy for prolonged tonal tinnitus. *International Journal of Audiology.* **51**(1):3–7 (2012).

38. Davis PB, *et al.* Clinical trial findings of a neurophysiologically-based tinnitus rehabilitation technique using tinnitus desensitization music. In Patuzzi R, (Ed.) *Seventh International Tinnitus Seminar.* pp. 74–77. Fremantle, Australia: University of Western Australia (2002a).
39. Davis PB, Wilde RA, and Steed L. Neurophysiologically-based tinnitus rehabilitation using tinnitus desensitization music. In Patuzzi R, (Ed.) *Seventh International Tinnitus Seminar*. Fremantle, Australia: University of Western Australia (2002b).
40. Abouzari M, *et al.* Adapting personal therapies using a mobile application for tinnitus rehabilitation: A preliminary study. *Annals of Otology Rhinology and Laryngology.* **130**(6):571–577 (2021).
41. Teismann H, *et al. Short and Intense Tailor-Made Notched Music Training against Tinnitus: The Tinnitus Frequency Matters*. (2011).
42. Stein A, *et al.* Clinical trial on tonal tinnitus with tailor-made notched music training. *BMC Neurology*, **16**:55 (2016).
43. Perera T, *et al.* Neuromodulation therapies for tinnitus. *Handbook of Clinical Neurology.* **146**:427–440 (2017).
44. Kaźmierczak M, and Janczewski G. Neuromodulation and neuroplasticity in tinnitus treatment: Current state and future directions. *Journal of Clinical Medicine*. **9**(9):2936 (2020).
45. Sullivan MD, and Thayer S. A new neuromodulation device for tinnitus management. *The Hearing Journal*. **75**(12):6–8 (2022).
46. Conlon B, *et al.* Different bimodal neuromodulation settings reduce tinnitus symptoms in a large randomized trial. *Scientific Report*, **12**(1):10845 (2022).
47. Conlon B, *et al.* Bimodal neuromodulation combining sound and tongue stimulation reduces tinnitus symptoms in a large randomized clinical study. *Science Translational Medicine*. **13**(602):eabf2212 (2020).
48. Pedemonte M, *et al.* Tinnitus treatment with sound stimulation during Sleep. *International Tinnitus Journal Study.* **16**(1):37–43 (2010).
49. Drexler D, *et al*. (2016). Impact of reduction of tinnitus intensity on patients' quality of life. *International Journal of Audiology*, **55**(1):11–19.

https://doi.org/10.1142/9789819807345_0004

Chapter 4

Internet-Based Self-Help Treatment for Tinnitus

Ali Kouhi* **and Alireza Sharifi**

Department of Otolaryngology-Head and Neck Surgery, Amir-Alam Hospital, Tehran University of Medical Sciences, Tehran, Iran

Otorhinolaryngology Research Center, Tehran University of Medical Sciences, Tehran, Iran

**a-kouhi@tums.ac.ir*

Abstract

Recently, internet-based approaches for hearing-related services have presented considerable reliability and efficacy in comparison with traditional in-clinic methods. In this chapter, all aspects of internet-based self-help treatment for tinnitus patients are discussed. Internet-based therapies can cover different aspects of tinnitus management such as online counseling, internet Cognitive Behavioral Therapy, peer support, and sound therapy. However, these modules need guidance during treatment courses including supports regarding patients' questions, outlining the content of modules, providing feedback, monitoring weekly scores, and monitoring progress. It is shown that internet-based methods can have comparable efficacy and outcomes to in-person methods, and they can also reduce the general costs per patient per year. It is worth mentioning that the characteristics or occurrence of side effects following internet-based treatment for tinnitus have not yet been known clearly. The need of internet-based methods for tinnitus patients was highlighted during the COVID-19 pandemic. Tinnitus patients often report a more noticeable tinnitus in the quiet, and the absence of masking sounds in daily life due to the quarantine may have made the tinnitus perception more noticeable. Additionally, the increased anxiety and stress levels are potential risk factors for worsening tinnitus. All of these factors during the pandemic have presented new insights about the priority of the internet-based self-help treatment for tinnitus patients. The next generation of internet-based self-help treatments will probably enroll Artificial Intelligence to combine multimodal sets. In recent years, the use of AI in various fields of medicine have surged and considering its function and mode of action, different aspects of tinnitus management may be covered by AI in future.

What Is Internet Based Treatment

The internet has played an important role in developing the health-related information transfer and care in recent decade, especially after worldwide use of smart phones that made internet available in everybody pocket. In audiology, platforms are designed for self-management programs, rehabilitation, and research. There is vast evidence supporting that internet-based interventions can improve the psychological distress, disease control, and health status [1]. Internet-based approaches give the opportunity to help considerable proportion of older patients who might not have convenient access to in person hearing-related services. To date, applications are available to help tinnitus psychological therapy, hearing counseling, cochlear implant programming, and hearing aid verification and fitting. These internet-based modules are shown to be comparable to traditional in-clinic methods regarding effectiveness and reliability. These interventions may include interactive homework, audio, video, text, and asynchronous clinician contact [2].

To maximize the efficacy, it is crucial to understand the underlying mechanisms of these internet-based self-help interventions. In order to optimize this intervention for patients these factors should be considered: understanding the user's interactions process, components of interventions, predicting factors related to the outcome, issues that can affect the acceptance and motivation of patients, adverse events of each intervention, and cost-effectiveness [3]. In this chapter, all aspects of internet-based self-help treatment for tinnitus patients are discussed.

How It Is Done

Content of applications

Internet-based self-help therapies can cover different aspects of tinnitus management such as: online counseling, internet Cognitive Behavioral Therapy (iCBT), peer support and sound therapy. Psychological interventions can effectively reduce tinnitus distress. iCBT has the highest evidenced support and addresses unhelpful emotional reactions and thought patterns caused by tinnitus. It typically includes text-based modules to work through. Different types of apps cover iCBT to manage tinnitus, including self-help, chatbots, Acceptance Commitment Therapy (ACT), and visual copes [4].

Other aspects of tinnitus management are covered by applications, too; Examples are, sound therapy, tinnitus masking, tinnitus habituation, acoustic neuromodulation, Zen sounds, notched therapy, and distraction of tinnitus. Tinnitus habituation is a process that helps patients finding unconscious ways to psychologically and emotionally cope with tinnitus. Notched sound online therapy consists of customizing sound by omitting band of reported tinnitus pitch from the sound. The patient chooses the sound signal they want to be notched, then the online process applies the notch [5].

The most prevalent function of sound therapy is tinnitus masking. The most common approach in iCBT strategies is self-help type. Evidence is lacking on other approaches and future studies will help defining their role in management of patients.

Self-help devices

Recent years smartphone apps and internet-based platforms have obtained significant attention because of advantages such as being wearable, possibility of reporting objective and subjective data in any time of a day [6]. They provide a platform that can be developed with new devices at negligible costs. To date, different modalities for the treatment of tinnitus exist, but all of them can be categorized as two types of treatment:

Sound-related therapy, CBT [7]. Here we briefly review the most popular devices:

1. *Tinnitus relief using sound therapy*

Most of these apps that are published on markets suggest sound therapies using various techniques like notched sound, acoustic neuromodulation, or amplitude modulation, or tinnitus masking. A list of most popular apps in this category is summarized in Table 1. Some pictures from various parts of Headspace application, as a sample of applications of sound therapy, are depicted in Figure 1 [8].

Table 1. Most popular apps based on sound therapy in order to tinnitus therapy.

				Rating (out of 5)		Pricing	
No.	**App Name**	**Platform**	**Description**	**Android**	**iOS**	**Android**	**iOS**
1	Audio Notch	Android, iOS	Customized notched sound therapy	2.8	1.8	F	I-A-P
2	H & T Sound Therapy	Android	Tinnitus masking	4.3	—	F	—
3	Headspace	Android, iOS	Sound stimulation	4.5	4.8	I-A-P	I-A-P
4	Kalmeda mynoise	Android, iOS	Tinnitus therapy	3.3	3.5	I-A-P	I-A-P
5	My Noise	Android, iOS	Tinnitus controlling	4.3	4.6	I-A-P	I-A-P
6	Oticon Tinnitus Sound	Android, iOS	Tinnitus controlling	3.1	2.8	F	F
7	Relax Melodies	Android, iOS	Sound stimulation	4.3	4.8	F	I-A-P
8	Relax Noise 3	Android	Tinnitus masking	4.2	—	F	—
9	Simply Noise	Android, iOS	Tinnitus controlling and management	3.3	4.3	F	I-A-P
10	Sleep Pillow	Android, iOS	Sound stimulation	3.6	4.4	F	I-A-P
11	Soothing Sounds Lite	Android, iOS	Sound stimulation	4.6	4.7	I-A-P	F
12	Starkey Relax	Android, iOS	Tinnitus masking	3.7	4.0	F	F
13	Stop Tinnitus	Android, iOS	Tinnitus masking	2.7	1.3	$	$

Table 1. (*Continued*)

No.	App Name	Platform	Description	Rating (out of 5) Android	Rating (out of 5) iOS	Pricing Android	Pricing iOS
14	Tinnitus Aid	iOS	Tinnitus management	—	4.6	—	I-A-P
15	Tinnitracks	Android, iOS	Tinnitus controlling and management	3.8	3.6	I-A-P	I-A-P
16	Tinnitus Help	Android, iOS	Tinnitus masking	3.0	4.4	$	$
17	Tinnitus Notch	Android	Customized notched sound therapy	2.7	—	I-A-P	—
18	Tinnitus Peace	Android	Tinnitus controlling	3.8	—	F	—
19	Tinnitus Play	iOS	Tinnitus masking	—	4.4	—	F
20	Tinnitus Relief	Android	Tinnitus controlling	4.2	—	I-A-P	—
21	Tinnitus Sound Therapy	Android	Tinnitus masking	3.9	—	F	—
22	Tinnitus Tailor	Android, iOS	Customized sound therapy	4.2	4.1	I-A-P	I-A-P
23	Tinnitus Therapy (Lite)	Android, iOS	Tinnitus masking	3.3	4.7	I-A-P	I-A-P
24	Tonal Tinnitus Therapy	Android	Tinnitus controlling and management	4.1	—	I-A-P	—
25	Track Your Tinnitus	Android, iOS	Tinnitus management	2.1	1.0	F	F
26	Whist	Android, iOS	Tinnitus controlling	4.2	3.7	$	$
27	White Noise (Lite)	Android, iOS	Tinnitus masking	4.6	4.8	F	F
28	Widex Zen	Android, iOS	Tinnitus masking	3.0	4.3	F	F
29	Tinnitus Balance App	Android, iOS	Tinnitus controlling	3.7	2.3	F	F

Notes: I-A-P: In-app Purchases/ F: Free/ $: Should pay money.

2. *CBT*

The smartphone application markets contain a wide variety of choices that are particularly designed for CBT. Most of them focus on managing anxiety and depression, which are the most common comorbidities related to

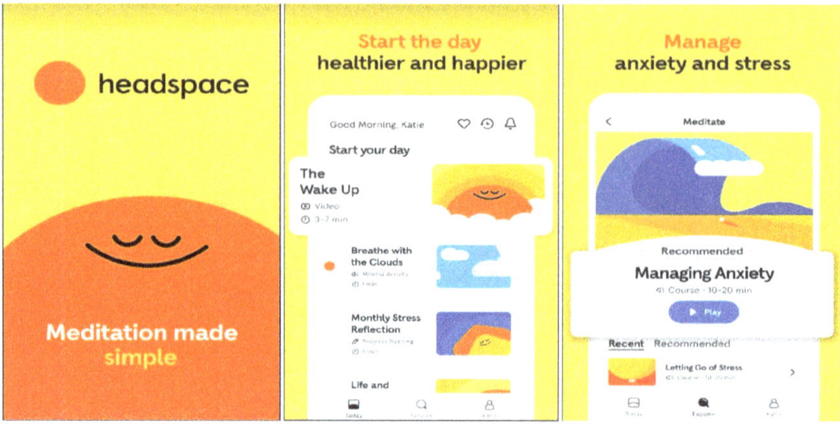

Figure 1. Different parts of Headspace application.

tinnitus. Some examples are summarized in Table 2 [5]. Some pictures from various parts of ReSound Relief application, as a sample of applications of CBT, are depicted in Figure 2 [8].

These apps are changing and ever-growing due to the quick development of new software and technology infra-structure on different platforms. Therefore, it is difficult to estimate the exact number of available tinnitus-related apps at any time, but recent studies have reported over 200 [9, 10]. A review of existing apps showed that there are five apps which are most common among patients based on their reported outcome and rating. They are as follows: (1) Tinnitus Therapy Lite, (2) my Noise, (3) Relax Melodies, (4) Oticon Tinnitus Sound, and (5) White Noise Free (11).

Another systematic review have qualified the apps with a more comprehensive tool, which is called Mobile Application Rating Scale(MARS) and evaluate 4 categories including information quality, aesthetics, functionality, and engagement. Four individuals rated the apps using MARS and the mean score was calculated and compared. Sanvello-Stress and Anxiety Help and Tinnitus Peace had the highest and the lowest mean score among all considered apps, respectively [12]. Modfit-Stress and Anxiety and Sanvello-Stress and Anxiety Help had the lowest and the highest mean score among CBT apps, respectively. Other apps with high scores were: (1) ReSound Relief, (2) Beltone Tinnitus Calmer, (3) Woebot — Your Self-Care Expert, and (4) Youper — Emotional Health.

Table 2. Most popular apps based on CBT in order to tinnitus therapy.

No.	App Name	Platform	Description	Rating (out of 5) Android	Rating (out of 5) iOS	Pricing Android	Pricing iOS
1	Beltone Tinnitus Calmer	Android, iOS	Tinnitus masking	4.5	4.8	I-A-P	F
2	CBT Companion	Android, iOS	Learn and practice CBT technique	4.5	4.6	I-A-P	I-A-P
3	Diapason for tinnitus	Android, iOS	Game-based digital therapy	2.9	3.9	I-A-P	I-A-P
4	MindShift CBT	Android, iOS	Anxiety controlling and management	3.7	4.2	F	F
5	Moodfit — Stress and Anxiety	Android	Stress & anxiety controlling and management	4.1	—	I-A-P	—
6	Quirk CBT	Android, iOS	Self-help CBT app	4.4	4.7	F	F
7	ReSound Relief	Android, iOS	Tinnitus masking	4.7	4.7	I-A-P	I-A-P
8	Sanvello-Stress and Anxiety Help	Android, iOS	Anxiety management	4.5	4.8	I-A-P	F
9	Stress and Anxiety Companion	Android, iOS	Stress & anxiety management	3.2	4.1	F	F
10	Tinnitus E-program	Web-based	Tinnitus controlling and management	—	—	—	—
11	Tinnitus Tunes	Web-based	Tinnitus controlling and management	—	—	—	—
12	What's Up? A Mental Health	Android, iOS	Stress & anxiety & depression management	3.7	4.4	I-A-P	I-A-P
13	Woebot — Your Self-Care Expert	Android, iOS	Stress & anxiety management	4.1	4.7	F	F
14	Wysa: Mental Health Therapy	Android, iOS	A chatbot offering CBT and DBT techniques	4.6	4.9	I-A-P	I-A-P
15	Youper — Emotional Health	Android, iOS	A chatbot based on CBT and ACT techniques, monitoring and tracking mood changes	4.0	4.8	I-A-P	I-A-P

Notes: Abbreviations: I-A-P: In-app Purchases/ F: Free/ $: Should pay money.

Figure 2. Different parts of ReSound Relief application.

Tinnitus Peace and Relax Melodies had the lowest and the highest mean score among sound therapy apps, respectively. Other apps with high scores were: (1) Widex Zen, (2) Tinnitus Therapy Lite, (3) Starkey Relax, and (4) Tinnitracks. In conclusion, only 7 apps out of 34 apps including Mindshift, Woebot, Wysa, Audio Notch, Simply Noise, ReSound Relief, and Tinnitus Therapy Lite were evidence-based. Likewise, some apps such as GGTI, Tinnitus Coach, IT-Mist, and iTerapi are recently developed and there have been validated by few studies [7, 12]. The majority of apps need certain clinical validation by more studies.

Internet based vs in-person cognitive behavioral therapy

Cognitive Behavioral Therapy (CBT) is the most effective option for the treatment of tinnitus associated distress to date [13]. In spite of many favorable outcomes, there is limited accessibility to CBT due to the shortage of well-trained clinicians [14]. Therefore, an internet-based CBT (iCBT) intervention would be attractive to improve access to the best evidence-based care [15]. ICBT was first described in Sweden and then adapted in United Kingdom population [16].

There are some important limitations for in-person tinnitus treatment. The health care burden for tinnitus in the United States is estimated to be about $660 per year per patient [17]. Clinics service is not available always due to the service and geographic restrictions [14, 18]. Finally, although there are various management approaches for tinnitus, level of evidence to support these options are not enough [19]. All of these negative aspects affect the face-to-face treatment approaches. There are some studies reporting no significant differences in terms of efficacy between iCBT and in-person group CBT (GCBT) [20, 21]. Additionally, the only clinical trial that compared the effectiveness of iCBT to individualized face to face treatment for managing tinnitus, showed equal effectiveness regarding decreasing tinnitus severity and most related difficulties [19]. Therefore, the iCBT can be used in place of individualized in-person treatment with equal efficacy , further research is needed to compare comprehensively other aspects such as cost-effectiveness and efficacy of iCBT.

Guided vs non guided use of self-help therapy

Like any other chronic problem, tinnitus needs guidance during treatment course including supports regarding patients' questions, outlining the content of modules, providing feedback, monitoring weekly scores, and monitoring progress [22, 23]. Although the efficacy of iCBT has been shown in several randomized controlled trials, role of audiologists support during the treatment has not yet been defined. This is because the findings of the previous studies are controversial. In traditional in-clinic therapy, the positive impact of patients' expectations, therapist confidence, and therapy outcome are better understood and performed. Some studies showed of the importance of such therapeutic support in internet-based treatments and clinicians support is assumed to play a key role in the iCBT's efficacy. However, there are evidence to the contrary. These mixed findings emphasize on the need for defining the specification and amount of therapeutic support needed to improve the efficacy of internet-based self-management approaches.

These supports usually needs to be individualized and each patient needs different type and amount of guidance. Some factors may play a role in providing support such as patients‘ symptoms, possible comorbidities, psychologic status, and patients' expectations [24]. Individuals with comorbidities such as anxiety, personality disorder, or depression might need more guidance. It is shown that patients with insomnia or social anxiety disorder need less support than patients with depression disorder.

Although evidence is lacking in this field, a recent RCT, compared the iCBT therapy with and without guidance, and reported that there is no significant higher improvement by guided iCBT than unguided iCBT. However, they showed cost-effective advantage for iCBT [25].

What Are the Results

Outcome & efficacy

Efficacy of iCBT is shown in clinical trials in different cultures and countries. However, the longevity of effect is not well understood. It is shown that the efficacy of iCBT on most tinnitus-related problems especially distress last for at least 1-year [26, 27], and it seems to be the same in different populations. Research on in person CBT and increasing its effectiveness duration would apply to iCBT too.

There are some factors that can predict and affect the outcome, and they can indirectly have impact on the duration of efficacy. Predicting factors are categorized as demographical, clinical, and treatment-related variables. In terms of demographical issues, several studies highlighted that the educational level of participants is an important predicting factor for successful outcome. It may be due to higher ability to understand and follow instructions of the internet-based self-help intervention. Patients with decreased work-related function due to tinnitus have lower chance to achieve successful outcome. This factor was highlighted during the COVID-19 pandemic. This may be because work may provide some distraction from tinnitus. In addition, having history of previous treatment can worsen the outcome. It might be due to the ineffectiveness of alternative therapy which is not proven evidence-based. Some variables such as gender, age, positive expectations, the amount of time spent for treatment, quality of interactions with therapist, and compliance of patients with homework may affect the outcome, although good level of evidence is lacking.

In terms of clinical issues, one of predicting factors is the level of depression. Patients with high levels of depression show better outcomes. In contrast, patients with insomnia have lower success rate. Also, the participants who have more severe tinnitus had a better chance to improve. Patients who suffer from clicking tinnitus, lower-pitched, or musical presentations are less likely to achieve positive outcome. Coexistence of hearing loss with tinnitus may result in more optimal outcomes.

In terms of treatment-related issues, participants who wear at least one hearing aid are at better chance of iCBT success [28–31]. Future research should consider each of these factors to determine their strength in prediction of outcome.

Cost-benefit

Tinnitus can have high societal costs:

1. The direct costs of health care such as treatment session or consultation by clinicians and medication.
2. Additional costs such as travel costs or the expenses of over-the-counter medication for patient and/or his/her family.
3. Indirect costs due to decreased efficiency and absence at work [32, 33].

Mean cost for tinnitus per patient per year are calculated between 1544-3429 euro for direct costs, between 69-115 euro for a patient and/or his/her family, and between 2565-3702 euro for societal costs in Europe (32).

A meta-analysis in 2022 reported that internet-based CBT (iCBT) has comparable costs in comparison to other forms of CBT; It had lower costs than in-clinic type but similar effectiveness. Also, patients showed greater adoption with iCBT in clinical practice in comparison to the in-clinic type [34].

An internet-based self-help treatment can decrease the costs in all three aspects. If it would be effective and can improve the self-management skills, it could reduce the patient's need for face-to-face health care, and it leads to decrease the direct and additional costs. In addition, their quality of life would be improved which can increase the efficiency and work presence. It can also result in reducing the indirect costs. Therefore, an effective internet-based treatment can reduce the general costs per patient per year [35].

Acceptance and motivation

Acceptability attributes to the patients' willingness to use a service. It is very important that innovative internet-based interventions are developed which are likely to be used. Although the effectiveness of iCBT for tinnitus has been proven, some studies have reported that the high dropout rates and the patient's preference for in-person therapy are the main challenges for acceptance of internet-based self-help interventions [36].

Previous studies showed that generally the users of internet-based self-help approaches express positive views and experiences on this type of treatment. The patient-reported outcome emphasized the increase in knowledge about "How does the internet-based approach work?", "What are its benefits?", and "Explain the options of an approach" in accepting the internet-based self-help treatment by patients [37–39].

Some factors are crucial for patients to accept and choose an app. Being user friendly is the most important one, it might be very helpful to improve the patient's willingness and ability to use the app. Clear and easy to understand instructions makes the app more likable by the patients. Trustworthy source, reviews, and cost are also commonly mentioned as key factors. The source of recommendation plays a key role in accepting an app. Among the sources, medical professional and another person with tinnitus are the most commonly listed. It seems that the name of an app is not crucial for tinnitus patients [11].

There are studies showing that motivational factors have a crucial effect on the process and outcomes of the treatment. A successful psychotherapy must avoid pessimistic and unrealistic expectations, which lead to disappointment. Also, it is recommended for psychotherapeutic interventions to focus on motivational issues in order to integrate the new strategies into patients' everyday life. Therefore, having knowledge about the specific motivation and expectations of patients can be helpful in designing of strategies to encourage them to start and keep up with psychotherapy. There is lack of enough data about what encourages tinnitus patients to seek out internet-based self-help approaches [11].

To date, addressing the sleep problems is reported as the most common motivation factor to try an app by patients. Tinnitus masking ability, recommendation from a family member, hearing professional, or people reactions on the web are also listed as motivational parameters. Other factors include managing their tinnitus, coping with it, frustration and desperation, curiosity of patients to explore new therapies, and so on. A low number of patients try to use an app because of its convenience and portability when they travel. Moreover, streaming via hearing aids, various types of sounds in an app, and being free trial are other motivational factors, which are attributed to the specifications of the app.

It is beneficial to encourage individuals to stay motivated and have an active role during the treatment. In order to achieve this, some suggestions have been made:

1. Interventions should focus on patients' self-efficacy and resources.
2. Using particular training features such as treatment platform or adding the therapist feedback.
3. Emotional benefits such as comfort and relaxation when they use the internet-based intervention.
4. Treatment success plays an important role in maintaining motivation [11, 40].

Knowing the main motivational factors and patients' preferences to accept and choose an internet-based self-help treatment can help the clinicians design and suggest a more appropriate form of treatment.

Side effects

The characteristics or occurrence of side effects following internet-based treatment for tinnitus have not yet been known clearly. The type of side effects and the potential reasons for its happening may be revealed in future studies. One potential side effect may be decreased success rate resulting in diminished interest of patients to use this type of treatment [27]. A meta-analysis of clinical trials of internet-based CBT (iCBT) ($n = 2866$ participants) reported that 17.4% of cases in control group and 5.8% of intervention group showed a deterioration in final outcome after receiving iCBT. This finding highlighted the role of internet-based treatment for tinnitus. Also the clinical severity before treatment, being in a emotional relationship, being older, and having an academic degree were related to lower odds of worsening in interventional group [41].

Beukes *et al.* in 2018 demonstrated higher odds for female than males to involve with side effects. Among their participants, only 11% (n=11) of patients receiving iCBT presented side effects such as negative feeling, worsening of symptoms, prolongation of treatment, and emergence of new symptoms. In another study they evaluated unwanted events in different population again, but they reported one patient with prolongation of treatment. It is supposed that they reduced the underlying factors which were related to occurrence of side effects [27].

Knowledge of these side effects can be helpful in improving future of internet-based self-help approach for tinnitus. Minimizing the adverse events, like any other treatment modality, will improve the outcome and increase willingness to consider this type of treatment.

Reasons for use and non-use

Several internet-based applications have been developed to help in Tinnitus management. These applications have several possible advantages: (1) lowering the cost of care, (2) improving the quality of care, and (3) improving the accessibility of care [11]. However, there are some concerns regarding internet-based self-help interventions, including misuse or safety [42], risk, responsibility, and effectiveness. All of these issues can affect the use of this type of treatment and the attitudes of patients and audiologists towards internet-based approaches is important [11].

Earlier in this chapter, acceptability and motivating factors to use internet-based apps was explained. One study in 2019 on 643 patients investigate the main reasons for not using these apps. The most commonly reason was lack of awareness of such a treatment modality. The second most common reason was that patients were not comfortable with technology, generally. Probably, the age of patients has a key role in this. Another reason was that they could not find suitable application to be helpful for their specific situation. This is mainly due to the lack of knowledge or accessibility. Audiologists can help patients find the proper platform and application regarding their problem. Another etiology was deficient access to handheld devices with applications support. Other reasons included not willing to use technology, not interested in paying attention to tinnitus, exacerbation of tinnitus with application use, too many number of choices, and preferring personal contact [11]. Even in patients willing to use the application there was some challenges such as technical problems in software or interruption of application use by phone calls and notifications [36].

The internet-based self-help treatment for tinnitus has not yet been developed properly. Providing information and awareness may help patients find and use these applications more easily.

Special Considerations

Cultural and linguistic suitability in different cultures

There are studies suggesting that ethnicity, race, culture, linguistic adoption, and health related knowledge of patients affects the level of patient engagement with health [43–45]. For instance, the patient engagement levels of African-Americans and Hispanic are lower than Caucasians [46]. Additionally, individuals with limited English knowledge may have problem due to the cultural differences, language barriers, and less health-related non-English resources [47]. Likewise, interventions provided in patient's native language have twice efficacy than the one in English. Therefore, personalized and culturally-sensitive interventions are essential in order to have a sustainable and active involvement of patients in treatment process [48].

CBT has highest evidence in the literature regarding the effectiveness of reduction in the severity of the tinnitus. In order to increase accessibility and supply culturally-sensitive intervention, the non-English version of internet-based form of CBT (iCBT) has been produced. To date, the iCBT has been used in United Kingdom [43], Germany [49], Sweden [50], Spain [44], and United States [45], and its efficacy and feasibility have been evaluated. It is noteworthy that iCBT was first developed in

Sweden by Andersson *et al.* [50]. Although the iCBT shows acceptable efficacy and feasibility in above mentioned countries but it is not available in many countries. Further studies need to investigate the iCBT key role in treatment protocols globally. Also, translation of the resources to different languages is necessary to overcome the cultural differences and language barriers all over the world.

Internet-based therapy in cochlear implant patients

Cochlear Implants (CI) are shown to have potential benefit of decreased tinnitus. There are two hypotheses about this potential effect:

(1) Presentation of background sound on the CI contributes to the decreasing of tinnitus. An increased neural activity along the auditory system happens after CI which can increase the contrast between the background neural activity and auditory representation of tinnitus. this may lead to tinnitus awareness and audibility reduction.
(2) Increasing attention to the signal provided by CI can decrease distraction due to the tinnitus and provide the patient with some control in addition to a greater sense of well-being [51].

Some studies evaluated the effectiveness of electrical and acoustic stimulation in CI users to reduces tinnitus severity. Their findings showed that external sounds are effective in reducing tinnitus annoyance and loudness; however, a considerable variability in external sound preference in terms of intensity and characteristics was observed among CI users [52, 53]. There are a lot of sound therapy options for CI patients on the market [51].

Internet-based self-help approaches can be useful for these patients due to:

(1) They can be easily used anywhere and be streamed to the CI.
(2) They usually have a wide range of external sounds. Therefore, the CI users can use the most favorable one for themselves (51).

There are two studies which reported the efficacy and outcomes of internet-based self-help applications for tinnitus therapy in CI patients using external sounds. The first one in 2018 reported that the use of internet-based self-help application can provide relief for some CI users who suffer from tinnitus. Also, there is not any significant difference between in-person and internet-based therapy in CI users. In addition, there were some individual differences in the acceptability of each external sound which were solved by providing different sounds in the application [51]. Other Study in 2021 showed that the sound therapy using an internet-based application can be effective for majority of CI users who suffer from tinnitus. The most acceptable sound and the best level of sound for tinnitus masking should be recommended by audiologists according to tinnitus evaluation to achieve the highest level of relief [54]. It is noteworthy that both of these studies used the Resound Relief app as the internet-based self-help approach in CIs. Although these two studies may not be enough to provide sufficient evidence, but the results are very promising and future research will help define specific aspects of this treatment for CI users.

Role of COVID-19 pandemic

Coronavirus diseases 2019 (COVID-19) has a disrupted impact on different aspects of health of the community. It is considered as an important stressor for many diseases [55]. During this pandemic, quarantine conditions and various restrictions have increased the stress levels, and stress is a well know risk factor for perception and severity of tinnitus [56].

Tinnitus patients often report a more noticeable tinnitus in the quite [57]. Avoiding a quite environment has been recommended as an adjunctive measure to overcome this problem and increase symptom masking [58]. During the COVID-19 pandemic, the absence of masking sounds in daily life due to the quarantine may have made the tinnitus perception more noticeable. Additionally, the increased anxiety and stress levels are potential risk factors for worsening tinnitus [59, 60]. Aydogan *et al.* showed that during the pandemic tinnitus annoyance and loudness increased significantly and the perception of tinnitus was affected negatively [56]. All of these consequences during the pandemic should seek the audiologists to prioritize the internet-based interventions in tinnitus therapy in addition to these two important factors:

(1) Severe limitations of personal contacts and social activities result in reducing the in-person type of treatment approaches.
(2) Pandemic condition and its related underlying mechanism in worsening the tinnitus, increases the need for tinnitus therapy more than usual [56].

All of these factors during the pandemic have presented new insights about the priority of the internet-based self-help treatment for tinnitus patients.

Future of Treatment

In recent years, the interest in using and developing electronic health care for tinnitus patients has been increased. Different aspects of this tool include information for patient, easier communication with healthcare provider, assessment of tinnitus symptoms, internet-based self-help treatments, and smartphone apps. However, most of these interventions have not enough supporting data yet [61].

The first generation of internet-based treatment was mostly a self-help book as an information source. ICBT was developed as second generation of internet-based treatment for tinnitus. Then the notched music/sound developed as a third generation of online self-help therapy. To date, about 200 apps, which are designed according to the iCBT and sound therapy functions, are available for tinnitus treatment.

Recently, some games, which train auditory system to perceive distinctions among sounds and ignore the distractor ones, also have been developed. They maintain patients motivated more easily during the treatment period. Several studies reported that these games were effective and feasible.

Multimodal therapies can be effective in tinnitus patients as internet-based self-help treatment. But the evidence for this type of treatment is limited to animal models. Among multimodal therapies, perceptual training using computer, which combines visual and tactile stimuli with sound, can be used in human. Further studies should approve its efficacy though.

The next generation of internet-based self-help treatments will probably enroll Artificial Intelligence (AI) to combine multimodal sets. In recent years, the use of AI in various fields of medicine have surged and considering its function and mode of action, different aspects of tinnitus management may be covered by AI in future [5].

Conclusion

- Internet-based self-help therapies can cover different aspects of tinnitus management such as online counseling, internet Cognitive Behavioral Therapy (iCBT), peer support and sound therapy.

- It is shown that internet-based methods can have comparable efficacy and outcomes to in-person methods, and they can also reduce the general costs per patient per year.
- Different modalities for the treatment of tinnitus exist that can be categorized as two types of treatment: Sound-related therapy and CBT.
- There is limited accessibility to CBT, as the most effective technique to treat tinnitus related depression, due to the shortage of well-trained clinicians, and an iCBT intervention would be attractive to improve access to the best evidence-based care.
- Internet-based self-help approaches can be useful for CI patients, and there are a lot of sound therapy options for them on the market.
- Tinnitus needs guidance during treatment courses including supports regarding patients' questions, outlining the content of modules, providing feedback, monitoring weekly scores, and monitoring progress.
- The characteristics or occurrence of side effects following internet-based treatment for tinnitus have not yet been known clearly.
- The next generation of internet-based self-help treatments will probably enroll AI to combine multimodal sets.

References

1. Beatty L, LS. A systematic review of internet-based self-help therapeutic interventions to improve distress and disease-control among adults with chronic health conditions. *Clinical Psychology Review*. **33**(4):609–622 (2013).
2. Greenwell K, HD. Use and mediating effect of interactive design features in audiology rehabilitation and self-management internet-based interventions. *American Journal of Audiology*. **25**(3S):278–283 (2016).
3. Moore GF, Audrey S, Barker M, *et al.* Process evaluation of complex interventions: Medical Research Council guidance. *BMJ*. **19**:350 (2015).
4. Greenwell K, FD, and Hoare DJ. The application of intervention coding methodology to describe the tinnitus e-programme, an internet-delivered self-help intervention for tinnitus. *American Journal of Audiology*. **24**(3):311–315 (2015).
5. Searchfield GD, SP, Doborjeh Z, *et al.* A state-of-art review of digital technologies for the next generation of tinnitus therapeutics. *Frontiers in Digital Health*. **3**:97 (2021).
6. Mehdi M, Schwager D, Pryss R, *et al.* Towards automated smart mobile crowdsensing for tinnitus research. In *2019 IEEE 32nd International*

Symposium on Computer-Based Medical Systems (CBMS), IEEE: 75–80 (2019).
7. Mehdi M, Dode A, Pryss R, *et al.* Contemporary review of smartphone apps for tinnitus management and treatment. *Brain Sciences.* **10**(11):867 (2020).
8. Abouzari M, Goshtasbi K, Sarna B, Ghavami Y, Parker EM, Khosravi P, *et al.* Adapting personal therapies using a mobile application for tinnitus rehabilitation: A preliminary study. *Annals of Otology, Rhinology & Laryngology.* **130**(6):571–577 (2021).
9. Nagaraj MK, P. P. Internet/smartphone-based applications for the treatment of tinnitus: A systematic review. *European Archives of Oto-Rhino-Laryngology.* **277**:649–657 (2020).
10. Deshpande AK, ST. A comprehensive evaluation of tinnitus apps. *American Journal of Audiology.* **28**(3):605–616 (2019).
11. Sereda M, Smith S, Newton K, *et al.* Mobile apps for management of tinnitus: Users' survey, quality assessment, and content analysis. *JMIR mHealth and uHealth.* **7**(1):e10353 (2019).
12. Mehdi M, Stach M, Riha C, *et al.* Smartphone and mobile health apps for tinnitus: Systematic identification, analysis, and assessment. *JMIR mHealth and uHealth.* **8**(8): e21767 (2020).
13. Hesser H, Weise C, Westin VZ, *et al.* A systematic review and meta-analysis of randomized controlled trials of cognitive–behavioral therapy for tinnitus distress. *Clinical Psychology Review.* **31**(4):545–553 (2011).
14. Hoare DJ, Broomhead E, Stockdale D, *et al.* Equity and person-centeredness in provision of tinnitus services in UK National Health Service audiology departments. *European Journal of Person Centered Healthcare.* **3**(3):318–326 (2015).
15. Andersson G, Strömgren T, Ström L, *et al.* Randomized controlled trial of internet-based cognitive behavior therapy for distress associated with tinnitus. *Psychosomatic Medicine.* **64**(5):810–816 (2002).
16. Thornton L, Batterham PJ, Fassnacht DB, *et al.* Recruiting for health, medical or psychosocial research using Facebook: Systematic review. *Internet Interventions.* **4**:72–81 (2016).
17. Goldstein E, Ho C-X, Hanna R, *et al.* Cost of care for subjective tinnitus in relation to patient satisfaction. *Otolaryngology–Head and Neck Surgery.* **152**(3):518–523 (2015).
18. Gander PE, Hoare DJ, Collins L, *et al.* Tinnitus referral pathways within the National Health Service in England: A survey of their perceived effectiveness among audiology staff. *BMC Health Services Research.* **11**:1–11 (2011).
19. Beukes EW, Andersson G, Allen PM, *et al.* Effectiveness of guided internet-based cognitive behavioral therapy vs face-to-face clinical care for treatment of tinnitus: A randomized clinical trial. *JAMA Otolaryngology–Head and Neck Surgery.* **144**(12):1126–1133 (2018).

20. Kaldo V, Levin S, Widarsson J, *et al.* Internet versus group cognitive-behavioral treatment of distress associated with tinnitus: A randomized controlled trial. *Behavior Therapy.* **39**(4):348–359 (2008).
21. Jasper K, Weise C, Conrad I, *et al.* Internet-based guided self-help versus group cognitive behavioral therapy for chronic tinnitus: A randomized controlled trial. *Psychotherapy and Psychosomatics.* **83**(4):234–246 (2014).
22. Lambert MJ. Early response in psychotherapy: Further evidence for the importance of common factors rather than "placebo effects". *Journal of Clinical Psychology.* **61**(7):855–869 (2005).
23. Lambert MJ, and Ogles BM. The general efficacy of psychotherapy. In *Bergin and Garfield's Handbook of Psychotherapy and Behavior Change*, 5th ed. New York: John Wiley & Sons. pp. 139–193 (2004).
24. Andersson G, Carlbring P, Ljótsson B, and Hedman E. Guided internet-based CBT for common mental disorders. *Journal of Contemporary Psychotherapy.* **43**:223–233 (2013).
25. Rheker J, AG, and Weise C. The role of "on demand" therapist guidance vs. no support in the treatment of tinnitus via the internet: A randomized controlled trial. *Internet Interventions.* **2**(2):189–199 (2015).
26. Beukes EW, AG, and Manchaiah V. Long-term efficacy of audiologist-guided Internet-based cognitive behaviour therapy for tinnitus in the United States: A repeated-measures design. *Internet Interventions.* **30**:100583 (2022).
27. Beukes EW, Allen PM, Baguley DM, *et al.* Long-term efficacy of audiologist-guided internet-based cognitive behavior therapy for tinnitus. *American Journal of Audiology.* **27**(3S):431–447 (2018).
28. Rodrigo H, BE, Andersson G, and Manchaiah V. Internet-based cognitive–behavioural therapy for tinnitus: Secondary analysis to examine predictors of outcomes. *BMJ Open.* **11**(8):e049384 (2021).
29. Rodrigo H, BE, Andersson G, and Manchaiah V. Exploratory data mining techniques (decision tree models) for examining the impact of internet-based cognitive behavioral therapy for tinnitus: Machine learning approach. *Journal of Medical Internet Research.* **23**(11):e28999 (2021).
30. Beukes EW, CS, Ratinaud P, *et al.* Experiential characteristics among individuals with tinnitus seeking online psychological interventions: A cluster analysis. *Brain Sciences.* **12**(9):1221 (2022).
31. Rodrigo H, BE, Andersson G, and Manchaiah V. Predicting the outcomes of internet-based cognitive behavioral therapy for tinnitus: Applications of artificial neural network and support vector machine. *American Journal of Audiology.* **31**(4):1167–1177 (2022).
32. Trochidis I, Lugo A, Borroni E, *et al.* Systematic review on healthcare and societal costs of tinnitus. *International Journal of Environmental Research and Public Health.* **18**(13):6881 (2021).

33. Maes IH, Cima RF, Vlaeyen JW, Anteunis LJ, and Joore MA. Tinnitus: A cost study. *Ear and Hearing*. **34**(4):508–514 (2013).
34. Patel N, MA, McGinnity S, *et al.* Cost effectiveness of cognitive behavioral therapy for the treatment of subjective tinnitus in Australia. *Ear and Hearing*. **43**(2):507–518 (2022).
35. Demoen S, Jacquemin L, Timmermans A, *et al.* Cost-effectiveness of a smartphone application for tinnitus treatment (the CATT trial): A study protocol of a randomized controlled trial. *Trials*. **23**(1):435 (2022).
36. Lan T, Zhao F, and Xiong B. The acceptability and influencing factors of an Internet-based tinnitus multivariate integrated sound therapy for patients with tinnitus. *Ear, Nose & Throat Journal*. **99**(9):597–606 (2020).
37. Greenwell K, Sereda M, Coulson NS, *et al.* Understanding user reactions and interactions with an Internet-based intervention for tinnitus self-management: Mixed-methods evaluation. *American Journal of Audiology*. **28**(3):697–713 (2019).
38. Beukes EW, Manchaiah V, Valien TE, *et al.* Positive experiences related to living with tinnitus: A cross-sectional survey. *Clinical Otolaryngology*. **43**(2):489–495 (2018).
39. Aazh H, and Danesh AA. Internet-based cognitive behavioral therapy for tinnitus: Insights from health care professionals. *Hearing Journal*. **74**(2):20–21 (2021).
40. Heinrich S, Rozental A, Carlbring P, *et al.* Treating tinnitus distress via the Internet: A mixed methods approach of what makes patients seek help and stay motivated during Internet-based cognitive behavior therapy. *Internet Interventions*. **4**:120–130 (2016).
41. Rozental A, Magnusson K, Boettcher J, *et al.* For better or worse: An individual patient data meta-analysis of deterioration among participants receiving Internet-based cognitive behavior therapy. *Journal of Consulting and Clinical Psychology*. **85**(2):160–171 (2017).
42. Misra S, Lewis TL, and Aungst TD. Medical application use and the need for further research and assessment for clinical practice: Creation and integration of standards for best practice to alleviate poor application design. *JAMA Dermatology*. **149**(6):661–662 (2013).
43. Beukes EW, Vlaescu G, Manchaiah V, *et al.* Development and technical functionality of an Internet-based intervention for tinnitus in the UK. *Internet Interventions*. **6**:6–15 (2016).
44. Beukes EW, Munzo MF, Andersson G, *et al.* Internet-based cognitive behavioural therapy for tinnitus in Spanish: A global feasibility trial. *International Journal of Audiology*. **61**(8):632–641 (2022).
45. Beukes EW, Fagelson M, Aronson EP, *et al.* Readability following cultural and linguistic adaptations of an Internet-based intervention for tinnitus for use in the United States. *American Journal of Audiology*. **29**(2):97–109 (2020).

46. Cunningham PJ, Hibbard J, and Gibbons CB. Raising low 'patient activation' rates among Hispanic immigrants may equal expanded coverage in reducing access disparities. *Health Affairs*. **30**(10):1888–1894 (2011).
47. Schyve PM. Language differences as a barrier to quality and safety in health care: The Joint Commission perspective. *Journal of General Internal Medicine*. **22**(3):360–361 (2007).
48. Griner D, and Smith TB. Culturally adapted mental health intervention: A meta-analytic review. *Psychotherapy*. **43**(4):531–548 (2006).
49. Weise C, Kleinstäuber M, and Andersson G. Internet-delivered cognitive-behavior therapy for tinnitus: A randomized controlled trial. *Psychological Medicine*. **78**(4):501–510 (2016).
50. Andersson G, Kaldo V. Internet-based cognitive behavioral therapy for tinnitus. *Journal of Clinical Psychology*. **60**(2):171–178 (2004).
51. Tyler RS, Owen RL, Bridges J, *et al.* Tinnitus suppression in cochlear implant patients using a sound therapy app. *American Journal of Audiology*. **27**(3):316–323 (2018).
52. Vernon JA. Masking of tinnitus through a cochlear implant. *Journal of the American Academy of Audiology*. **11**(6):293–294 (2000).
53. Tyler RS, Keiner A, Walker K, *et al.* A series of case studies of tinnitus suppression with mixed background stimuli in a cochlear implant. *American Journal of Audiology*. **24**(3):398–410 (2015).
54. Perreau AE, Tyler RS, Frank V, *et al.* Use of a smartphone app for cochlear implant patients with tinnitus. *American Journal of Audiology*. **30**(3):676–687 (2021).
55. Remuzzi A, and Remuzzi G. COVID-19 and Italy: What next? *The Lancet*. **395**(10231):1225–1228 (2020).
56. Aydoğan Z, Satekin MÇ, Ocak E, *et al.* Effects of the coronavirus disease 2019 pandemic on subjective tinnitus perception. *Journal of Laryngology and Otology*. **136**(5):410–413 (2022).
57. Härter M, Maurischat C, Weske G, *et al.* Psychological stress and impaired quality of life in patients with tinnitus. *HNO*. **52**(2):125–131 (2004).
58. Gritsenko K, Caldwell W, Shaparin N, *et al.* Resolution of long standing tinnitus following radiofrequency ablation of C2-C3 medial branches — A case report. *Pain Physician*. **17**(1):E95 (2014).
59. Zenner H-P, Delb W, Kröner-Herwig B, *et al.* A multidisciplinary systematic review of the treatment for chronic idiopathic tinnitus. *European Archives of Oto-Rhino-Laryngology*. **274**:2079–2091 (2017).
60. Wilson PH, Henry J, Bowen M, *et al.* Tinnitus reaction questionnaire: Psychometric properties of a measure of distress associated with tinnitus. *Journal of Speech, Language, and Hearing Research*. **34**(1):197–201 (1991).
61. Kleinjung T, LB. Avenue for future tinnitus treatments. *Otolaryngologic Clinics of North America*. **53**(4):667–683 (2020).

https://doi.org/10.1142/9789819807345_0005

Chapter 5

Emerging Pharmacological Therapeutic Agents for Tinnitus

Alireza Zonnour[*], Mehdi Abouzari[†], and Hamid R. Djalilian[†,‡,§,¶]

[*]*Otorhinolaryngology Research Center, Department of Otolaryngology-Head and Neck Surgery, Amir-Alam Hospital, Tehran University of Medical Sciences, Tehran, Iran*

[†]*Department of Otolaryngology-Head and Neck Surgery, University of California, Irvine, CA, USA*

[‡]*Department of Biomedical Engineering, University of California, Irvine, CA, USA*

[§]*Department of Neurosurgery, University of California, Irvine, CA, USA*

[¶]*hdjalili@hs.uci.edu*

Abstract

Tinnitus, the perception of sound without external stimuli, is a significant health burden globally. Effective treatment remains elusive due to an incomplete understanding of its mechanisms, heterogeneous manifestations, and limited treatment options.

Current pharmacological options for acute tinnitus treatment follow sudden hearing loss guidelines, recommending systemic or intratympanic steroids. Existing medications for chronic tinnitus mostly focus on alleviating the intensity of symptoms including tricyclic antidepressants, benzodiazepines, and lidocaine, showing varied efficacy and safety. However, various guidelines recommend against currently available pharmacological interventions due to insufficient efficacy and potential side effects.

Ongoing studies on emerging pharmacological treatments target various pathways to treat tinnitus effectively. Migraine treatments address the migraine-tinnitus link, offering relief by targeting underlying migraine pathology. Inflammatory pathways represent another promising target, with TNF-α inhibitors showing promise in animal models. Potassium channel modulators, such as Kv3 K+ channel activators and BK channel openers, are being investigated for their potential in modulating neuronal hyperexcitability and administrating agents targeting those demonstrated promising results in animal studies. GABAergic drugs, including extrasynaptic GABAA receptor modulators, offer a strategy to reduce abnormal hyperactivity which needs further investigations. Glutamate receptor antagonists, like mGluR agonists and AMPA receptor antagonists, hold promise in suppressing tinnitus in animal models. Also, potential treatment targets such as the limbic system, lipid metabolism, and synaptic genes have gained interest in recent studies. Moreover, ongoing clinical trials investigate novel treatment options like migraine medications, Etanercept, Brexanolone, and Curcumin for tinnitus management.

Challenges include methodological flaws in trials, lack of standardization across studies, tinnitus heterogeneity, and limited pharmacological options. Overcoming these requires standardized criteria, longer follow-ups, and diversified trial designs. Despite challenges, advancements in neuroinflammation and genetics and brain-targeted interventions offer promising drug targets. Lastly, establishing personalized treatment approaches will be crucial for improving tinnitus care and outcomes.

Introduction

Tinnitus, the perception of sound without an external stimulus, poses a significant health burden globally, with its prevalence escalating alongside advancing age [1].

Despite extensive research efforts, an effective cure for tinnitus remains elusive, largely due to the incomplete understanding of its underlying mechanisms, the heterogeneous nature of tinnitus manifestations, the lack of sensitive outcome measures, and the difficulty in translating preclinical findings to clinical applications [2].

Pharmacotherapeutic strategies for tinnitus predominantly focus on either directly targeting the perception of tinnitus or alleviating associated comorbidities. While the former remains largely elusive, advancements in antidepressant medications hold promise in addressing secondary symptoms and enhancing patients' overall well-being. Conventional treatments, such as dietary modifications, cognitive behavioral therapy, and sound therapy, offer limited relief, with no medication currently approved by the Food and Drug Administration (FDA) specifically for tinnitus management. Patient dissatisfaction with current treatment modalities underscores the urgent need for innovative approaches [3, 4].

The limited pharmacological options available for tinnitus underscore the importance of expanding our understanding of its underlying mechanisms. Despite advancements in genetics and plasma biomarker research, definitive diagnostic tools and targeted therapies for tinnitus are lacking. By integrating insights from basic science with clinical practice, we can pave the way for more personalized and efficacious treatments for tinnitus sufferers [5].

In this chapter, we aim to review the latest advancements in pharmacological treatments for tinnitus. By exploring emerging therapeutic agents, we provide insights into potential breakthroughs in managing this challenging condition.

Existing Pharmacological Treatments

Acute tinnitus

When addressing acute tinnitus, treatment protocols closely follow those recommended for acute sudden hearing loss, as outlined in established guidelines. These guidelines advocate for the use of systemic or intratympanic steroid treatment, although the evidence supporting their efficacy remains moderate. Conversely, if acute tinnitus occurs in the absence of measurable hearing loss, the standard cortisone therapy typically prescribed for hearing loss may not be recommended [3].

The mechanisms of action of steroids in the management of acute tinnitus are thought to involve various pathways. They may work by suppressing the irritability or hypersensitivity of sensory cells within the inner ear, particularly in cases of cochlear hair cell damage. Additionally, steroids may reduce inflammation caused by immune-mediated or autoimmune dysfunction within the inner ear. Furthermore, they might enhance inner ear vascularity and exert direct effects on the neuroepithelium of the inner ear [6].

These diverse actions underscore the rationale for incorporating steroids into the treatment regimen for acute tinnitus, especially when it is associated with sudden hearing loss.

Chronic tinnitus

Although no medication has received FDA approval for tinnitus management, various drugs are employed in clinical practice, albeit with controversial efficacy and safety profiles. The primary focus of pharmacotherapy is to alleviate symptoms, reduce tinnitus intensity, and alleviate associated psychological distress, such as depression and anxiety [4, 7]. In this section, we provide a brief overview of the commonly prescribed medications utilized in clinical practice to address chronic tinnitus.

Among the pharmacotherapies investigated, tricyclic antidepressants, benzodiazepines, and lidocaine have shown promising results in trials. However, lidocaine's efficacy is limited due to its intravenous administration requirement, short half-life, and adverse side effects. While lidocaine has been explored intravenously or trans-tympanically, its beneficial effects on tinnitus suppression appear to be transient, with associated risks of increased tinnitus severity [5, 8]. Also, a recent network meta-analysis revealed significant reductions in tinnitus severity and response rates with pharmacological interventions targeting the brain (e.g., amitriptyline, acamprosate, gabapentin) and those with anti-inflammatory effects (e.g., steroids, melatonin) compared to placebo or waiting-list controls [9].

However, several trials testing various pharmacotherapies failed to demonstrate efficacy over placebos. Studies on trazodone and selective serotonin reuptake inhibitors yielded no significant improvements in tinnitus severity or quality of life compared to placebos. Additionally, due to limited evidence and concerns regarding their side effect profiles, benzodiazepines are not recommended for treating subjective tinnitus alone, while cannabinoids may exacerbate symptoms [5, 10]. Also, Cochrane meta-analyses have reported negative findings for Ginkgo biloba, anticonvulsants, and antidepressants in tinnitus management [11].

Furthermore, various guidelines, including the US, Swiss, German, NICE, European, and Japanese, do not recommend pharmacological interventions for tinnitus treatment due to insufficient efficacy data and potential side effects. Recommendations are based on systematic reviews and randomized trials, with evidence suggesting non-recommendation for betahistine, ginkgo, antidepressants, benzodiazepines, zinc, melatonin, cannabis, oxytocin, steroids, and gabapentin [3, 11]. Table 1 demonstrates the summary of existing pharmacological treatments for chronic tinnitus.

Table 1. Summary of Existing Pharmacological Treatments for Tinnitus [11–16]

Drugs	Dosage	Results
Acamprosate	333 mg TDS for 3 months*	Improvement
Alprazolam	0.25–0.5 mg/day for 1 week (Max dosage of 1 mg/day)*	Decreased loudness
Amitriptyline	50–100 mg/day for 6 weeks	Decreased intensity and subjective relief
Betahistine	48 mg/day for 3 months	Slight improvement in loudness and tinnitus handicap inventory
Cilostazol	200 mg/day for 4 weeks	Decreased Visual Analogue Scale score
Clonazepam	N/A	Decreased annoyance and intensity
Gabapentin	2400 mg/day for 20 weeks*	Decreased annoyance
Lidocaine	1–2 mg/kg of body weight Intravenously for 3–4 min	Complete or partial suppression
Melatonin	3 mg/day for 30 days	Decreased tinnitus intensity in patients with insomnia and sleep quality improvement
Misoprostol	Increasing dosage for 4 months	Decreased loudness
Neramexane	25–75 mg/day for 16 weeks	Decreased annoyance
Nortriptyline	50–150 mg/day for 6 weeks	Decreased loudness
Ondansetron	16 mg/day for 4 weeks	Improvements in tinnitus severity index score
Oxytocin	16 IU single dose	Decreased Clinical Global Impressions score

(*Continued*)

Table 1. (*Continued*)

Drugs	Dosage	Results
Pramipexole	Max dosage of 0.7 mg TDS for 4 weeks	Decreased annoyance
Sertraline	25–50 mg/day for 16 weeks	Decreased loudness and severity
Vitamin B12	Weekly dosage of 2500 μg for 6 weeks*	Improvements in vitamin B12 deficient patients
Atorvastatin, Baclofen, Carbamazepine, Cyclandelate, Diazepam, G. Biloba, Lamotrigine, Memantine, Paroxetine, Piribedil, Trimipramine, Vardenafil, Zinc		No significant difference

Note: *No significant difference at lower doses and with shorter duration of treatment.

Overview of Emerging Pharmacological Treatments and Potential Treatment Targets

Various mechanisms contribute to the development of tinnitus, encompassing maladaptive neuroplasticity, vascular dysfunction, oxidative stress, and genetic predisposition. Recent studies have also revealed significant hyperactivity in multiple brain regions among tinnitus patients without identifiable origins, underscoring the intricate neurobiological underpinnings of the condition. Additionally, individuals with tinnitus often experience concurrent conditions such as depression and anxiety, which can complicate both treatment and assessment of treatment efficacy [1, 12]. The heterogeneous nature of tinnitus necessitates a multifaceted approach, targeting diverse pathogenic mechanisms and actively managing associated comorbidities for optimal treatment outcomes.

In this section, we will delve into a detailed overview of emerging pharmacological treatments for tinnitus, exploring potential treatment targets and novel therapeutic approaches. By examining recent research findings, we aim to shed light on promising avenues for addressing the complex pathophysiology of tinnitus and improving treatment outcomes for affected individuals.

Migraine treatments and tinnitus

The close association between migraine headaches and various vestibulocochlear disorders, including tinnitus, has garnered significant attention in recent studies. Alongside migraine-associated vertigo, conditions such as vestibular migraine, persistent postural-perceptual dizziness, and benign paroxysmal positional vertigo are now recognized as manifestations of migraine within the inner ear [7].

Epidemiological investigations, including analyses of the National Health and Nutrition Examination Survey (NHANES) U.S. national database, have revealed compelling evidence linking migraine with subjective hearing loss (HL) and tinnitus, even after adjusting for confounders [13]. Studies have highlighted the dynamic relationship between tinnitus and migraine, with tinnitus severity often escalating during migraine episodes or presenting as a concurrent symptom that subsides post-headache. Notably, there exists a significant correlation between the laterality of headache and tinnitus, suggesting a shared pathophysiological mechanism [8, 13]. Additionally, the association between hyperacusis and chronic migraine, often referred to as "phonophobia" in migraine literature, underscores the interconnectedness of auditory symptoms in migraineurs [14]. Shared triggering factors between tinnitus and migraine further support their mechanistic link, including stress, hormonal changes, sleep disturbances, dietary factors, and intense sensory stimuli [4, 7].

These findings suggest a complex interplay between migraine and central auditory pathway dysregulation, offering insights into potential therapeutic strategies for managing both conditions.

The pathophysiological link between migraine and tinnitus stems from the broader concept of migraine as a state of brain sensitivity, rather than solely a headache disorder. This sensitivity, characterized by electrophysiological changes in the brain, can manifest in various forms, including ocular migraine, vestibular migraine, cochlear migraine, and abdominal migraine, among others [4]. Central to the understanding of migraine pathophysiology is the activation of peripheral and central trigeminovascular neurons. These neurons release neuropeptides and cytokines such as calcitonin gene-related peptide (CGRP), substance P, neurokinin A, and pituitary adenylate cyclase-activating polypeptide. The aberrant release of these molecules triggers vasodilation, plasma extravasation, and mast cell degranulation, culminating in chronic neuroinflammatory changes in the meninges. Importantly, the inner ear (vestibule and cochlea) is intricately connected to the trigeminal nervous system via neurovascular branches. Consequently, the activation of trigeminal meningeal nociceptors, inflammation of the trigeminal nerve, and vasodilation of the inner ear circulation contribute to migraine-related cochlear alterations [7, 8]. These neurovascular changes provide a plausible explanation for a myriad of cochleovestibular symptoms observed in migraine, including tinnitus.

In clinical practice, implementing a migraine treatment protocol has shown promising results in alleviating the intensity of loud tinnitus among affected individuals. This approach, termed "integrative neurosensory rehabilitation", incorporates comprehensive lifestyle modifications, dietary adjustments, and supplementation, alongside pharmacological interventions if necessary. Lifestyle changes include stress reduction, optimizing sleep hygiene, and dietary modifications such as increased hydration and avoidance of potential triggers such as caffeine and tyramine [4, 8]. Additionally, patients are often prescribed supplements such as vitamin B2 (Riboflavin) and magnesium oxide to augment migraine management efforts.

For patients who do not respond adequately to lifestyle modifications and supplements within six to eight weeks, pharmacotherapy may be initiated to increase the threshold for migraine triggers and manage underlying comorbidities. Antidepressants like nortriptyline are commonly prescribed, with careful titration to minimize side effects. Topiramate and verapamil may also be considered if initial treatments prove ineffective. Furthermore, individualized combinations of pharmacotherapies may be necessary for optimal symptom relief, taking into account patient comorbidities and tolerance to potential side effects [7] (Figure 1).

Although the persistent, subtle level of tinnitus remains unchanged by migraine treatment, as it originates from the foundational cochlear damage and central activation, it seems that tackling the underlying cause of migraine effectively alleviates loud and fluctuating tinnitus [8].

Considering the growing understanding of the link between tinnitus and migraine, traditional migraine treatments should be considered for tinnitus management as notable improvements were observed in tinnitus perception when addressing the underlying migraine pathology. While this treatment doesn't cure tinnitus, it notably reduces its impact. Future studies may explore shared dietary triggers and neuroanatomical connections between migraine, tinnitus, and psychiatric disorders, potentially advancing tinnitus management and improving the quality of life for sufferers of both conditions.

Inflammatory pathways and tinnitus

The pathophysiological link between inflammation and tinnitus has garnered increasing attention, particularly regarding the involvement of

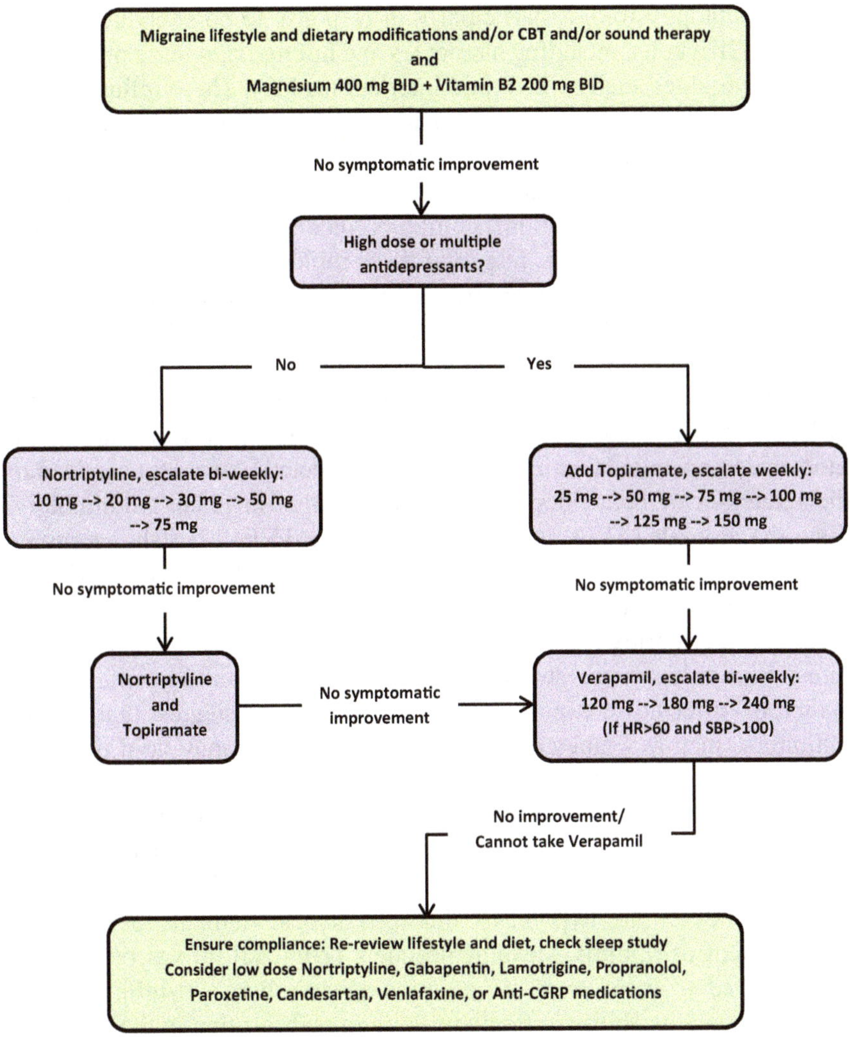

Figure 1. Proposed algorithm for treatment of tinnitus with migraine medications. BID, twice daily; CBT, cognitive behavioral therapy; HR, heart rate; SBP, systolic blood pressure; CGRP, calcitonin-gene related peptide.

TNF-α, IL-1, and microglia as potential treatment targets [15, 16]. Accumulating evidence suggests that inflammation plays a significant role in subjective tinnitus pathogenesis, with both noise exposure and salicylate administration triggering inflammation throughout the auditory pathway. Emerging research emphasizes the importance of glial cells in tinnitus pathophysiology, particularly in response to noise-induced hearing loss. Glial cells, including astrocytes and microglia, release proinflammatory cytokines such as TNF and interleukins [17]. These inflammatory mediators influence *N*-methyl-*D*-aspartate (NMDA) and γ-Aminobutyric acid type A (GABAA) receptors leading to altered neurotransmission and subsequent neuroplastic changes associated with chronic tinnitus [18–20]. While the precise role of inflammatory mediators in chronic tinnitus awaits further clarification, targeting these molecules holds promise as a potential effective treatment approach. Investigating the role of inflammation in auditory and limbic system plasticity may offer valuable insights into tinnitus pathogenesis and therapeutic strategies.

Studies have explored potential treatments targeting underlying inflammation to address tinnitus, with a focus on TNF-α inhibitors. Studies demonstrated that intraperitoneally injected Etanercept can lead to the reduction of behavioral evidence of tinnitus in mice. Additionally, dithiothalidomide (dTT), another drug targeting TNF-α, has shown promise. In experiments with mice, dTT injections prevented increases in TNF-α, IL-1β, IL-18, NLRP3, and microglial morphological changes in the primary auditory cortex (AC), while also mitigating behavioral signs of tinnitus. Similarly, in rat studies, dTT did alleviate subsequent tinnitus symptoms after noise exposure [15]. These findings suggest that TNF-α inhibitors, such as Etanercept and dithiothalidomide, may hold potential as treatments for tinnitus.

In addressing the underlying inflammation associated with tinnitus, intra-tympanic steroid injections, particularly using dexamethasone, have emerged as a potential treatment avenue. The theoretical basis for this approach lies in the hypothesis linking cochlear damage to the early development of tinnitus. Dexamethasone's beneficial effects on tinnitus are believed to stem from its immune-suppressive, anti-inflammatory properties, and its ability to facilitate sodium reabsorption in the inner ear via interaction with steroid receptors in the human temporal bone [21]. Moreover, the intra-tympanic administration route offers the advantage of achieving regionally high peri-lymphatic drug levels, minimizing the risk of systemic adverse events [21]. However, studies have indicated controversy about the effectiveness of intra-tympanic dexamethasone injection in chronic tinnitus treatment. Yet, there is potential for combination treatments involving intra-tympanic dexamethasone injection with other drugs such as oral melatonin to be effective as shown in the literature [6, 12, 22]. These findings highlight the potential of steroid-based interventions in managing tinnitus associated with inflammation, which needs further research.

Potassium channels and tinnitus

The next category of tinnitus drug development programs is potassium channel modulators, a field witnessing significant advancements in both pathophysiologic knowledge and research methodology. Recent years have unveiled an increasing number of potassium channels implicated in the neural mechanism of tinnitus, highlighting their potential as therapeutic targets.

Dysfunctions or downregulations of voltage-gated Kv3 K+ channels have been linked to neuronal network hyperexcitability, a precursor to tinnitus onset. Despite the Kv3.1 channel activator AUT00063's disappointing clinical trial results, it remains a promising therapeutic target. Similarly, Kv7.2/3 channels have emerged as potential targets, showing involvement in tinnitus mechanisms and warranting further clinical investigation [23, 24].

Moreover, preclinical studies have emphasized the significance of calcium-activated potassium (BK) channels in the cochlear neural mechanism of tinnitus. BK channels modulate pre- and post-synaptic excitability, potentially regulating tinnitus-associated activity. Studies administering BK channel openers like BMS-191011 have shown promise in suppressing tinnitus biomarkers, suggesting a potential therapeutic avenue [24, 25].

Furthermore, investigations into the role of KCNQ2/3 channels in the dorsal cochlear nucleus (DCN) have revealed insights into tinnitus generation. RL-81, a potent KCNQ2-3 opener, has demonstrated efficacy in preventing tinnitus development in animal models, presenting a promising lead candidate for further studies [26, 27].

Overall, the exploration of tinnitus-related potassium channels and corresponding drugs, along with potential drug combinations, holds immense potential for understanding potassium ion channel involvement in tinnitus and advancing pharmacological development in this area.

Gamma-aminobutyric acid (GABA) neurotransmitter and tinnitus

Studies have indicated that patients with tinnitus experience altered cerebral activity due to cerebral hyperactivity. This perception of severe tinnitus is associated with abnormal neurotransmitter secretion akin to pain perception in the somatosensory system, characterized by decreased GABAergic inhibition. Considering GABA as the principal inhibitory neurotransmitter in the central nervous system, augmenting GABA inhibitory activity has been proposed as a potential treatment avenue to mitigate tinnitus severity. Evidence from various studies, including neurochemical, anatomical, physiological, and pharmacological investigations, supports the GABAA hypothesis. Notably, data implicating extrasynaptic GABAA receptors in tinnitus control suggests the potential for extrasynaptic GABAA receptor modulators, offering a strategy to reduce abnormal hyperactivity and potentially alleviate tinnitus severity [12, 28, 29].

In a recent study, compelling evidence was provided regarding the efficacy of a small molecule NMDAR-PAM, M-8324, in increasing the spiking of GABAergic neurons and preventing noise exposure-induced tinnitus. Understanding the specificity, efficacy, limitations, and therapeutic applications of enhancing inhibition, particularly in the early stages of disease progression, requires additional research to fully comprehend its therapeutic potential [30].

Overall, the GABA neurotransmitter system represents a compelling therapeutic target for tinnitus management. Further research into the precise mechanisms underlying GABAergic dysfunction in tinnitus, as well as the development of novel GABAergic drugs, holds promise for improving treatment options for individuals affected by tinnitus.

Glutamate receptors and tinnitus

Numerous research studies have implicated excess glutamate as a primary trigger, leading to various pathologies in noise-exposed cochleae, particularly damage to postsynaptic afferent nerves mediated by α-amino-3-hydroxy-5-methyl-4-isoxazolepropionic acid (AMPA) receptor activity. Notably, after a sound stimulus, there is a temporary suppression of spontaneous firing in the auditory pathway, known as residual inhibition. This suppression involves metabotropic glutamate receptors (mGluRs), particularly in the inferior colliculus (IC). Also, Noise exposure has been shown to induce excitotoxic swelling at auditory nerve terminals, leading to synaptopathy, a process dependent on NMDA receptors [26, 31].

Antagonists of both mGluR and AMPA glutamate receptors have been targeted in tinnitus drug development efforts. For instance, Eglumegad (LY354740), a mGluR agonist, has demonstrated the ability to suppress spontaneous firing in the IC of mice with behavioral evidence of tinnitus, showing promise for potential clinical trials. Similarly, BGG492 (selurampanel), a competitive AMPA glutamate receptor antagonist, has shown significant reductions in tinnitus loudness and annoyance after administration, supporting further investigation into its clinical efficacy [26, 32].

Additionally, OTO-313, a sustained exposure formulation of the NMDA receptor antagonist gacyclidine, aims to block heightened activation of cochlear NMDA receptors implicated in tinnitus generation. Despite promising initial results, the superiority of OTO-313 over placebo was not confirmed in a larger Phase 2 study, emphasizing the need for careful consideration of study design in future investigations [26, 33].

Overall, targeting glutamate receptors holds promise as a potential pharmacological approach for tinnitus management. However, further research is needed.

Other potential treatment targets for tinnitus

In the pursuit of potential treatment targets for tinnitus, emerging avenues of research delve into various aspects of neural circuitry and biochemical pathways. Neuroimaging studies have illuminated the involvement of limbic system structures in tinnitus, suggesting altered synaptic plasticity as a potential mechanism. Endocannabinoids, particularly the CB1 and CB2 receptor subtypes, have garnered interest due to their modulatory effects on synaptic transmission within auditory pathways, especially concerning their impact on limbic and auditory areas. Similarly, dopamine, a key neurotransmitter in the basal ganglia, may influence tinnitus perception through dopaminergic pathways connecting the limbic system and cochlea. Understanding these neural mechanisms could provide insights into potential therapeutic interventions for tinnitus [16, 34].

Furthermore, recent explorations into novel plasma biomarkers have uncovered associations between lipid dysregulation and tinnitus pathogenesis. Compounds such as triglycerides and diglycerides have been linked positively to tinnitus, while others like cholesteryl esters and lysophosphatidylcholines show inverse associations. These findings underscore the potential role of lipid metabolism in tinnitus [5].

Moreover, genetic studies have shed light on rare synaptic genes implicated in severe tinnitus. These synaptic genes, involved in membrane trafficking and cytoskeletal protein binding, offer insights into the molecular underpinnings of severe tinnitus across diverse patient populations. Continued research into these genetic markers promises to deepen our understanding of tinnitus etiology and may pave the way for personalized treatment approaches [5].

Currently active clinical trials targeting tinnitus treatment:

Treatment of tinnitus with migraine medications (NCT04404439)

In this phase 4 clinical trial, the objective is to evaluate the efficacy of medications in reducing the impact of tinnitus on patients' daily lives over 8 wees. The trial comprises three arms: nortriptyline (7.5 mg) plus topiramate (10 mg), verapamil (30 mg) plus paroxetine (4 mg), and a placebo group. It is a double-blinded trial. Dosage adjustments may occur weekly if symptoms persist, with nortriptyline, topiramate, verapamil, and paroxetine doses potentially increasing by 7.5 mg, 10 mg, 30 mg, and 4 mg, respectively, per week (up to maximum doses). Symptom survey scores will be collected weekly, with pre- and post-treatment assessments conducted.

SPI-1005 for the treatment of meniere's disease (STOPMD-3) (NCT04677972)

The ongoing phase 3 randomized, double-blind, placebo-controlled, multicenter clinical trial (RCT) assesses SPI-1005 in adults with Meniere's disease and recent active symptoms. Developed by Sound Pharmaceuticals, SPI-1005 targets tinnitus associated with Meniere's disease by enhancing glutathione peroxidase (GPx) activity. GPx scavenges reactive oxygen species (ROS), protecting against cochlear damage and hearing loss induced by noise exposure. SPI-1005 aims to prevent outer hair cell loss and reduce hearing threshold shifts following noise exposure. This study aims to evaluate SPI-1005's efficacy and safety in this patient population.

Investigation of the NMDA antagonist ketamine as a treatment for tinnitus (*NCT03336398*)

Research indicates that tinnitus involves maladaptive plasticity, affecting glutamatergic and GABA signaling. Ketamine, an anesthetic, modulates these pathways, showing efficacy in depression and chronic pain treatment. This phase 2 trial explores ketamine's potential for tinnitus associated with hearing loss, assessing its impact on depression, anxiety, and GABA levels in the auditory cortex using MRS imaging.

TNF-α treatment of blast-induced tinnitus (*NCT04066348*)

In this phase 2 trial, the primary objectives are to assess whether Etanercept can significantly alleviate tinnitus distress measured by the Tinnitus Functional Index (TFI) and improve hearing. Secondary objectives include determining if Etanercept sustains therapeutic effects over time and reduces tinnitus loudness measured by visual numeric scale (VNS) rating. Participants will be randomly assigned to receive either 2×25 mg Etanercept injections or placebo injections.

The efficacy of incobotulinum toxin a injections for treatment of tinnitus: a randomized controlled trial (*NCT05650645*)

This phase 2 trial aims to evaluate the efficacy of Xeomin (incobotulinumtoxinA) injections into the auricular muscles for alleviating tinnitus, as assessed by the Tinnitus Handicap Inventory (THI) questionnaire completed by patients.

An open-label study evaluating brexanolone in adults with tinnitus (*NCT05645432*)

The main goal of this phase 2 trial is to assess the safety and tolerability of brexanolone in individuals experiencing tinnitus, through a single 6-hour continuous intravenous infusion. Brexanolone, also known as Zulresso, is an antidepressant consisting of allopregnanolone and sulfobutylether-beta-cyclodextrin, administered intravenously. Originally intended for postpartum depression treatment in adults, it is now being explored for tinnitus, among other conditions. The drug utilizes allopregnanolone, a progesterone metabolite acting on GABAA receptors. Administered intravenously, it employs a PANAM chemistry platform focusing on positive allosteric modulation (PAM) of GABAA receptors, offering a non-benzodiazepine approach.

Effect of combined antioxidant therapy on oxidative stress markers and inflammatory cytokines in patients with tinnitus (NCT05646693)

The primary aim of this study is to assess the impact of combined antioxidant therapy with Adepsique® (amitriptyline, perphenazine, and diazepam) on individuals suffering from chronic tinnitus symptoms, with a focus on inflammatory cytokine and oxidative stress levels in the bloodstream. The study will involve 58 participants, divided into two groups randomly assigned to receive either antioxidant pills or a placebo, to be taken once daily for 3 months. Throughout the trial, the clinical manifestations of tinnitus, as well as levels of inflammatory cytokines and oxidative stress markers will be monitored.

Effect of highly bioavailable curcumin on subjective tinnitus (NCT04800107)

Participants experiencing severe, subjective tinnitus related to sensorineural hearing loss will initially undergo TFI and THI assessments and will be randomly assigned to two groups. The treatment group will receive capsules containing a blend of 500 mg curcumin-phosphatidylcholine and 250 mg Boswellia-phosphatidylcholine, aimed at enhancing curcumin absorption. Meanwhile, individuals in the placebo group will receive a placebo compound. After 30 days, participants from both groups will undergo TFI and THI assessments again to evaluate treatment outcomes.

Action of intra-auricular topical lidocaine on tinnitus (NCT05711641)

This placebo-controlled trial aims to evaluate intraauricular lidocaine's efficacy in reducing tinnitus intensity compared to a placebo (distilled water) when applied topically in the external auditory canal. Both lidocaine and placebo will be administered to each participant at 15-day intervals, with the application order randomized. Tinnitus intensity will be measured before and after lidocaine or placebo application using visual analog scales and acuphenometry.

Effects of herbal sleep formula on patients with insomnia and tinnitus (NCT05212298)

This project aims to enroll non-dialysis adults experiencing both sleep and tinnitus disorders. A total of 120 adults will be randomly assigned to three groups. The control group will receive a placebo treatment. The first intervention group will receive a low-dose herbal compound, totaling 557 mg of herbal compound capsules. The second intervention group will receive a high-dose herbal compound, totaling 1114 mg of herbal compound capsules. Participants will complete health, sleep, and tinnitus questionnaires to assess the effects of the interventions.

Challenges, Limitations, and Prospects for the Future

Addressing the challenges and limitations in tinnitus research and treatment is paramount to advancing our understanding and improving patient outcomes. Methodological flaws plague trials RCTs in this field, ranging from inconsistent inclusion criteria to variable outcome measures and interventions. Moreover, the lack of standardization across studies hampers data interpretation and hinders progress. Overcoming these challenges requires strict criteria for participant selection and outcome assessment, as well as the recognition of tinnitus heterogeneity and the need for tailored interventions. Longer-term follow-up in trials is essential to gauge treatment efficacy accurately and address the complexities of tinnitus management comprehensively. Additionally, the limited pharmacological options available underscore the urgency of expanding research efforts to uncover new therapeutic targets and mechanisms underlying tinnitus.

Despite the existing challenges, the future of tinnitus treatment holds promise with advancements in research methodologies and a deeper understanding of its pathophysiology. Diversified trial designs that acknowledge tinnitus's multifaceted nature offer opportunities for targeted interventions tailored to specific subgroups of patients. Preclinical studies investigating neuroinflammation and genetic influences pave the way for the discovery of novel drug targets, potentially revolutionizing tinnitus management. Furthermore, exploring brain-targeted interventions to address abnormal brain hyperactivity, oxidative stress, and underlying migraine associated with tinnitus presents exciting prospects for future randomized controlled trials. As research progresses, establishing meaningful criteria for subtyping tinnitus and identifying personalized treatment approaches will be crucial for improving patient care and outcomes.

Conclusion

- Tinnitus management remains challenging due to its complex mechanisms and limited treatment options.
- Current pharmacological treatments primarily focus on alleviating symptoms and associated comorbidities rather than directly targeting tinnitus perception.
- Emerging pharmacological approaches, such as migraine treatments, inflammation modulation, potassium channel targeting agents, and GABAergic medications, offer promising avenues for managing tinnitus.
- Ongoing clinical trials investigating various treatments demonstrate commitment to improving treatment outcomes, with a focus on personalized approaches and novel interventions.
- Challenges in tinnitus research, include methodological inconsistencies, lack of standardization across studies, tinnitus heterogeneity, and limited pharmacological options.
- Prospects for tinnitus treatment include diversified trial designs, targeted interventions based on tinnitus subtyping, and advancements in understanding the underlying pathophysiology, offering hope for improved patient care.

References

1. Liu D, Hu Y, Wang D, Han H, Wang Y, Wang X, Zhou Z, Ma X, and Dong Y. Herbal medicines in the treatment of tinnitus: An updated review. *Frontiers in Pharmacology*. **13**:1037528 (2023).
2. Cederroth CR, Dyhrfjeld-Johnsen J, and Langguth B. An update: Emerging drugs for tinnitus. *Expert Opinion on Emerging Drugs*. **23**(4):251–260 (2018).
3. Mazurek B, Hesse G, Sattel H, *et al*. S3 guideline: Chronic tinnitus: German Society for Otorhinolaryngology, Head and Neck Surgery E. V. (DGHNO-KHC). *HNO*. **70**:795–827 (2022).
4. Abouzari M, and Djalilian HR. Tinnitus is modulated by migraine. *The Hearing Journal*. **76**(10):27, 28, 30, 32 (October 2023).
5. Park KW, Kullar P, Malhotra C, and Stankovic KM. Current and emerging therapies for chronic subjective tinnitus. *Journal of Clinical Medicine*. **12**(20):6555 (2023).
6. Chung J, Lee DY, Kim JS, and Kim YH. Effectiveness of intratympanic dexamethasone injection for tinnitus treatment: A systematic review and meta-analysis. *Clinical and Experimental Otorhinolaryngology*. **15**(1):91–99 (2022).
7. Umemoto KK, Tawk K, Mazhari N, Abouzari M, and Djalilian HR. Management of migraine-associated vestibulocochlear disorders. *Audiology Research*. **13**(4):528–545 (2023).
8. Abouzari M, and Djalilian HR. How migraine modulates tinnitus. *Bulletin (The Official Content Hub of the American Academy of Otolaryngology-Head and Neck Surgery)*. **42**(9) (September 19, 2023).
9. Brooks KA, Tawk K, Djalilian HR, and Hobson CE. Migraine management for the otolaryngologist. *Laryngoscope Investigative Otolaryngology*. **8**(4):1080–1093 (2023).
10. Barth SW, Lehner MD, Dietz GPH, and Schulze H. Pharmacologic treatments in preclinical tinnitus models with special focus on Ginkgo biloba leaf extract EGb 761. *Molecular and Cellular Neurosciences*. **116**:103669 (2021).
11. Langguth B, Kleinjung T, Schlee W, Vanneste S, and De Ridder D. Tinnitus guidelines and their evidence base. *Journal of Clinical Medicine*. **12**(9):3087 (2023).
12. Han BI, Lee HW, Ryu S, and Kim JS. Tinnitus update. *Journal of Clinical Neurology*. **17**(1):1–10 (2021).
13. Sanchez TG, Balbani AP, Bittar RS, Bento RF, and Câmara J. Lidocaine test in patients with tinnitus: Rationale of accomplishment and relation to the treatment with carbamazepine. *Auris Nasus Larynx*. **26**(4):411–417 (1999).
14. Langguth B, Elgoyhen AB, and Cederroth CR. Therapeutic approaches to the treatment of tinnitus. *Annual Review of Pharmacology and Toxicology*. **59**:291–313 (2019).

15. Chen JJ, Chen YW, Zeng BY, Hung CM, Zeng BS, Stubbs B, Carvalho AF, Thompson T, Roerecke M, Su KP, Tu YK, Wu YC, Smith L, Chen TY, Lin PY, Liang CS, Hsu CW, Hsu SP, Kuo HC, Wu MK, and Tseng PT. Efficacy of pharmacologic treatment in tinnitus patients without specific or treatable origin: A network meta-analysis of randomized controlled trials. *EClinicalMedicine*. **39**:101080 (2021).
16. Kleinjung T, and Langguth B. Pharmacotherapy of tinnitus. *Current Topics in Behavioral Neurosciences*. **51**:193–212 (2021).
17. Goshtasbi K, Abouzari M, Risbud A, Mostaghni N, Muhonen EG, Martin E, and Djalilian HR. Tinnitus and subjective hearing loss are more common in migraine: A cross-sectional NHANES analysis. *Otolaryngology-Head and Neck Surgery*. **42**(9):1329–1333 (2021).
18. Benjamin T, Gillard D, Abouzari M, Djalilian HR, and Sharon JD. Vestibular and auditory manifestations of migraine. *Current Opinion in Neurology*. **35**(1):84–89 (2022).
19. Mennink LM, Aalbers MW, van Dijk P, and van Dijk JMC. The role of inflammation in tinnitus: A systematic review and meta-analysis. *Journal of Clinical Medicine*. **11**(4):1000 (2022).
20. Singh A, Smith PF, and Zheng Y. Targeting the limbic system: Insights into its involvement in tinnitus. *International Journal of Molecular Sciences*. **24**(12):9889 (2023).
21. Wang W, Zhang LS, Zinsmaier AK, Patterson G, Leptich EJ, Shoemaker SL, Yatskievych TA, Gibboni R, Pace E, Luo H, Zhang J, Yang S, and Bao S. Neuroinflammation mediates noise-induced synaptic imbalance and tinnitus in rodent models. *PLoS Biology*. **17**(6):e3000307 (2019).
22. Jara JH, Singh BB, Floden AM, and Combs CK. Tumor necrosis factor alpha stimulates NMDA receptor activity in mouse cortical neurons resulting in ERK-dependent death. *Journal of Neurochemistry*. **100**(5):1407–1420 (2007).
23. Liu T, Jiang CY, Fujita T, Luo SW, and Kumamoto E. Enhancement by interleukin-1β of AMPA and NMDA receptor-mediated currents in adult rat spinal superficial dorsal horn neurons. *Molecular Pain*. **9**:16 (2013).
24. Singh A, and Abraham WC. Astrocytes and synaptic plasticity in health and disease. *Experimental Brain Research*. **235**(6):1645–1655 (2017).
25. Shim HJ, Song SJ, Choi AY, Hyung Lee R, and Yoon SW. Comparison of various treatment modalities for acute tinnitus. *Laryngoscope*. **121**(12):2619–2625 (2011).
26. Yener HM, Sarı E, Aslan M, Yollu U, Gözen ED, and İnci E. The efficacy of intratympanic steroid injection in tinnitus cases unresponsive to medical treatment. *Journal of International Advanced Otology*. **16**(2):197–200 (2020).
27. Olsen T, Capurro A, Švent M, Pilati N, Large C, Hartell N, and Hamann M. Sparsely distributed, pre-synaptic Kv3 K^+ channels control spontaneous

firing and cross-unit synchrony via the regulation of synaptic noise in an auditory brainstem circuit. *Frontiers in Cellular Neuroscience*. **15**:721371 (2021).

28. Lai H, Gao M, and Yang H. The potassium channels: Neurobiology and pharmacology of tinnitus. *Journal of Neuroscience Research*. **102**(1):e25281 (2024).
29. Scott LL, Lowe AS, Brecht EJ, Franco-Waite L, and Walton JP. Small molecule modulation of the large-conductance calcium-activated potassium channel suppresses salicylate-induced tinnitus in mice. *Frontiers in Neuroscience*. **16**:763855 (2022).
30. Henton A, and Tzounopoulos T. What's the buzz? The neuroscience and the treatment of tinnitus. *Physiological Reviews*. **101**(4):1609–1632 (2021).
31. Marinos L, Kouvaros S, Bizup B, Hambach B, Wipf P, and Tzounopoulos T. Transient delivery of a KCNQ2/3-specific channel activator 1 week after noise trauma mitigates noise-induced tinnitus. *Journal of the Association for Research in Otolaryngology*. **22**(2):127–139 (2021).
32. Tavares MP, and Bahmad F Jr. Analysis of gabapentin's efficacy in tinnitus treatment: A systematic review. *Annals of Otology, Rhinology & Laryngology*. **131**(3):303–311 (2022).
33. Witkin JM, Lippa A, Smith JL, Cook JM, and Cerne R. Can GABAkines quiet the noise? The GABAA receptor neurobiology and pharmacology of tinnitus. *Biochemical Pharmacology*. **201**:115067 (2022).
34. Deng D, Masri S, Yao L, Ma X, Cao X, Yang S, Bao S, and Zhou Q. Increasing endogenous activity of NMDARs on GABAergic neurons increases inhibition, alters sensory processing and prevents noise-induced tinnitus. *Scientific Reports*. **10**(1):11969 (2020).
35. Vijayakumar KA, Cho GW, Maharajan N, and Jang CH. A review on peripheral tinnitus, causes, and treatments from the perspective of autophagy. *Experimental Neurobiology*. **31**(4):232–242 (2022).
36. Kucher K, Johns D, Wagner F, Abd-Elaziz K, Derne C, Sverdlov O, Pfister CU, and Langguth B. Efficacy and safety of single- and repeated-selurampanel dosing for 2 weeks in patients with chronic subjective tinnitus: Results of a randomized, double-blind, placebo-controlled, cross-over, proof-of-concept phase IIa study. *Progress in Brain Research*. **260**:423–440 (2021).
37. Searchfield GD, Robinson JM, Skarinsky D, Wang Y, and Anderson JJ. A randomized, double-blind, placebo-controlled phase 2 study of intratympanic OTO-313 in patients with moderate to severe subjective tinnitus. *European Archives of Oto-Rhino-Laryngology*. **280**(12):5307–5318 (2023).
38. Lee A, Abouzari M, Akbarpour M, Risbud A, Lin HW, and Djalilian HR. A proposed association between subjective nonpulsatile tinnitus and migraine. *World Journal of Otorhinolaryngology-Head and Neck Surgery*. **9**(2):107–114 (2022).

https://doi.org/10.1142/9789819807345_0006

Chapter 6

Complementary and Alternative Medicine in Tinnitus

Anil Patel[*], Varun Patel[†], and Michael Seidman[‡]

[*]*Florida State University College of Medicine, Tallahassee, FL*

[†]*Department of Otolaryngology-Head and Neck Surgery, Penn State Health Milton S. Hershey Medical Center, Hershey, PA, USA*

[‡]*AdventHealth Otolaryngology-Head and Neck Surgery, Celebration, FL*

Abstract

The lack of any definitive treatment for tinnitus is often unsatisfying for patients and can precipitate emotional distress in addition to increased healthcare expenditures. Many then look to natural remedies for treatments of their tinnitus. This chapter explores the use of complementary and alternative medicines for the management of tinnitus.

Less invasive treatments for tinnitus include herbs, trace elements, and antioxidant supplements. The mechanisms of alternative treatment often rely on vaso-regulatory and antioxidant properties for which tinnitus has known causes for. Herbs such as the Ginkgo Biloba, albeit controversy abounds, is accepted as evidence-based treatment with improved tinnitus handicap inventory (THI) and visual analogue scale (VAS) scores. Some antioxidant supplements such as B-vitamins and Coenzyme Q10 have also shown results that are promising. A separate evidence-based entity is behavioral therapy and its subsets, which are not discussed in this chapter.

More invasive methods could involve electrical stimulation and acupuncture. We specifically address acupuncture as it is deeply seeded in cultural roots but has been investigated to target specific tinnitus centers such as the cochlear nucleus, resulting in increased blood flow with improved THI and VAS scores.

While evidenced-based research continues for alternative treatments, including the techniques outlined in everyday practice will take significant efforts from both the patient and the provider. Complementary and integrative medicine should be considered in addition to traditional medical and behavioral management for tinnitus.

Introduction

Tinnitus, defined as the perception of sound in the absence of any external stimuli, is a chronic condition that has reduced the quality of life in millions of patients worldwide. In addition to being a debilitating condition at times, tinnitus is also a symptom that may signify issues related to otologic, neurologic, and iatrogenic etiologies [1]. Current theories about the cause of tinnitus include spontaneous neural activity in the auditory cortex or abnormal neural activity [2]. Currently, there is no cure for tinnitus and patients are often unsatisfied with conventional medical techniques. Chronic tinnitus affects more than 50 million people in the United States alone and yields an estimated annual healthcare expenditure of $2110 per patient. It is also estimated that 80% of the population will seek medical advice about tinnitus, many of whom suffer from anxiety and sleep disturbances [3]. However, the amount of health care dollars or the number of clinic visits does not correlate proportionally with patient satisfaction with tinnitus treatment [4]. Because there is no "cure" for tinnitus, complementary medicine has evolved to address tinnitus treatment options. Current issues surrounding these complementary approaches is the general lack of clear and substantive evidence. Furthermore, it is important to remember that tinnitus experienced by patients is highly variable. Methods used for one patient may not help another, however, it is still valuable to discuss options for our patients. Tinnitus can cause a great deal of emotional stress, and health care providers should be open-minded when evaluating and treating patients with tinnitus.

Complementary and integrative medicine (CIM) consists of non-traditional methods such as herbal supplements, antioxidant supplements, and trace elements, all of which are discussed in this chapter. Other "invasive" methods including acupuncture and electrical stimulation for which acupuncture is specifically discussed.

Herbs

The Ginkgo biloba tree is a Chinese herb used for centuries as a health remedy to treat various disorders such as Alzheimer's disease, vascular dementia, and tinnitus of vascular origin. The proposed mechanisms of action include antioxidant, antiplatelet, and vaso-regulatory properties [5]. The most important chemical components of Ginkgo biloba are flavonoids and terpenes. Flavonoids have antioxidant and vasodilating actions, and terpenes have antiplatelet activity [6]. Consuming effective concentrations of flavonoids and terpenes may play a significant role in treating tinnitus. For instance, the Ginkgo biloba strain EGb761 was found to contain an appropriate concentration of flavonoids and terpenes, based on the German Commission E monograph (a comprehensive compilation of the therapeutic use of herbs) [7]. The dosage that has been established by the German Commission E Guidelines is 240 mg twice daily [8]. Adhering to this dosage recommendation for the EGb761 strain is particularly important in order to reap its health benefits [8]. Studies utilizing 50 mg or 120 mg once daily did not demonstrate any substantial effect of Ginkgo biloba compared to placebo [9, 10]. Studies have also illustrated that the EGb761 strain exhibits greater benefit compared to non-standardized extracts [10].

Ginkgo biloba in combination with antioxidants can provide marked improvement in post-treatment THI and VAS scores [11]. Additionally, Ginkgo infusion pretreatment followed by 12 weeks of oral Ginkgo has also been shown to reduce tinnitus intensity and hearing loss [12]. Numerous other randomized clinical trials have showed the effectiveness of Ginkgo in improving tinnitus perception [13]. Holstein *et al.* performed a systematic review on 19 clinical trials involving 3931 subjects using the EGb761 Ginkgo biloba strain. Eight studies had a placebo with a total of 687 patients, and 11 studies had a total of 3244 participants without use a placebo. This review demonstrated that in 11 trials there was clinical efficacy of EGb761 on both acute and chronic tinnitus [14]. However, the measurement of clinical efficacy was inconsistent among the studies.

Another comprehensive meta-analysis examined 5 randomized, placebo-controlled trials using EGb761 and found it to be superior to placebo in tinnitus treatment [15]. In disagreement with this, another randomized, double-blinded, placebo-controlled trial of 1243 patients with chronic tinnitus reported no improvement from using Ginkgo biloba [16]. One important limitation of this study, however, is that the authors used a substandard dose (150 mg per day) and also did not use the active ingredient in Ginkgo. Other limitations include a high withdrawal rate (50% of patients were not selected or withdrew) and a lack of medical or audiological exams completed on these patients.

Additionally, a previous Cochrane review showed no difference in tinnitus outcomes with Ginkgo compared to placebo [17]. Of note, three of the clinical trials included in this review used three different Gingko products. There was, however, a statistically significant reduction of subjective tinnitus in patients with concurrent vascular dementia or Alzheimer's disease. Perhaps Gingko treatment may be beneficial in those with tinnitus due to an ischemic origin [18]. Overall, the data on the Ginkgo biloba suggests that it can have a significant impact on tinnitus but is not successful for every patient [10]. Its low-risk profile may make it an attractive treatment option for those who suffer from tinnitus.

Although Ginkgo biloba has received significant attention, it is not the only herb that has been studied carefully. Other herbs that have been considered are the Korean red ginseng and the Gushen Pian. Red ginseng is known for its antioxidative and apoptotic properties. When given at a high dose, patients responded with significant improvement in their tinnitus handicap score (THI), which is the patient's perceived impact of tinnitus on daily life. The Gushen pian is a combination herbal compound consisting of Glycyrrhiza, Danshen, and Drynaria fortune [19]. Some of these compounds such as the Drynaria fortunei have already been shown to provide some benefit in ototoxicity. Gushen pian has displayed some early efficacy in the treatment of tinnitus and aural fullness [14]. These herbal remedies require further investigation before becoming a part of routine tinnitus treatment.

Vitamins, Minerals and Antioxidants

Many trace elements play a role in enzymatic and signaling cascades throughout the body. Zinc has been shown to have the highest content in the inner ear based on animal studies [20]. Naturally then, we should consider the implications of zinc on various otologic conditions such as tinnitus. Initial Cochrane reviews done on zinc supplementation did not show improvements. However, studies seem to suggest that there are greater impacts based on specific patient groups [10]. Zinc levels appear to decrease with age, and levels of zinc have shown to be inversely correlated with tinnitus severity [21]. Tinnitus loudness was found to be significantly higher in patients with low zinc levels [22].

In one study of patients with noise-induced hearing loss (NIHL), 85% of participants reported a statistically significant improvement in THI following two months of 20 mg of zinc daily, but there were no functional improvements [23]. It is important to note that zinc toxicity can occur at a dose of 200 mg or greater per day. Due to the mixed conclusions from studies, larger investigations need to be completed to evaluate zinc as a tinnitus treatment.

Reactive oxidative species (ROS) have been associated with both noise-induced and age-related hearing loss, which is partially due to loss of cochlear hair cells from the loss of mitochondrial membrane potentials [24]. Additionally, it also has been determined that total oxidant status, which is the sum of all antioxidants in the body, and the oxidative stress index, which calculates the imbalance between ROS and antioxidants, were higher in tinnitus patients [25]. This could imply that tinnitus is a result of oxidative stress and increased ROS. This conclusion is reasonable as oxygen-free radicals are responsible for many age-related changes and damage throughout the body. Using antioxidants as a complementary method is seemingly an appropriate measure to consider for some patients. It is also important to note that one out of three people on this planet have a micronutrient deficiency (such as magnesium, vitamin D, calcium, B-vitamins etc) [26]. Further, replenishing a micronutrient deficiency can take up to six months of proper supplementation [27]. Thus, it is reasonable to trial nutritional supplement strategies for at least six months before ruling out efficacy.

B vitamins have an essential role in metabolism and biochemical impacts on the nervous system. For example, neuronal demyelination can occur when the methyl donor SAM is not available, as SAM synthesis heavily depends on vitamin B12 [28]. This is one of the many functions of B-vitamins in the nervous system. Therefore, B vitamin deficiency can lead to neuronal dysfunction which could then result in hearing changes and tinnitus [29].

In a 6-week randomized double-blinded study of chronic tinnitus, patients were injected with 2500 mcg of vitamin B12 [30]. Results showed significant improvements in the visual analogue scale (VAS), which is a scale that patients use to rate their tinnitus discomfort from 0–10 [30]. One study that considered B12 deficiency to be less than 180 pg/mL found no significant difference in the vitamin B12 levels between tinnitus patients and healthy controls. However, when raising that threshold to less than 250 pg/mL, other studies have shown a higher prevalence of B12 deficiency in patients with tinnitus compared to controls. Other B complex vitamins may also play a role in tinnitus as well. Additional evidence exists based upon a survey suggesting that vitamins B2 and B3 may have an independent association or annoyance role in tinnitus patients [31].

Other antioxidants that have been studied include CoQ10, beta-carotene, vitamin C and vitamin E. CoQ10 plays a role in oxidative phosphorylation and other systemic antioxidant reactions. One study that included patients with low serum levels of CoQ10 showed significant improvement in their Tinnitus Questionnaire (TQ) and SF-36 scores following 12 weeks of 300 mg of CoQ10. Another study evaluated the supplementation of CoQ10 with multivitamins to hamper cisplatin ototoxicity. Results showed significant protection against cochlear toxicity and a lower maintained ROS status compared to controls [32]. Treatment with vitamins C and E with beta-carotene for 18 weeks suggested significant improvement in VAS for idiopathic tinnitus patients [33]. However, therapy consisting of vitamin C with alpha-lipoic acid or vitamin E with papaverine hydrochloride in patients with tinnitus with sensorineural hearing loss for 6 months showed no significant benefit. The complex and individual genetic/epigenetic variabilities related to the patient's biochemistry add additional challenges in managing people with tinnitus.

Melatonin

Produced in the pineal gland, melatonin is a natural hormone structurally similar to serotonin that plays a key role in regulating the sleep-wake cycle with additional antioxidative, anti-depressive, antidopaminergic, and vasodilatory properties [34]. It is considered a powerful neural antioxidant and its levels decline significantly over age twenty [35]. Melatonin has additionally been found to be elevated in the cochlea at night, suggesting a role in cochlear functioning.

One double-blind placebo-controlled crossover study by Rosenberg and colleagues found that the THI did not differ significantly with 3 mg of melatonin with a 7-day washout compared to controls [36]. However, they concluded that while melatonin alone did not help patients quiet their tinnitus, they found that if the patient had both sleep issues and tinnitus, it was considered beneficial. Some patients note that their tinnitus is worse if they are lacking sleep. Another study almost a decade later with a 30-day washout period found a significant improvement in self-rated tinnitus scoring and tinnitus matching [34]. This suggests that a shorter washout period affected the crossover study outcomes, but both studies have shown improvements in sleep disturbances and self-rated tinnitus scores respectively [33, 34].

Research has also compared melatonin to antidepressants for tinnitus. Abtahi and colleagues found that 3 mg of melatonin for three months compared with 50 mg of sertraline daily showed a significant decline in mean THI scores with both treatments [37]. A study by Barati and colleagues performed a similar 3-month study and the data confirmed previous results with significant improvements in THI, tinnitus loudness, and awareness scores [38]. An additional study examined melatonin compared to amitriptyline, a tricyclic antidepressant for tinnitus. VAS scores were not significantly different for either treatment, but THI scores were [39]. This confirms significant subjective improvement for each patient as described through the THI. No adverse events have been reported with the 3 mg of melatonin. Although there are positive results from many studies, there are still many questions to be answered about the use of melatonin. Continuation of double-blinded placebo trials with consistent methodology can establish firm conclusions necessary for it to be a part of routine clinical practice.

Cannabinoids and Psychedelics

Cannabinoids have gained regulatory approval for the medical treatment of many conditions such as Parkinson's disease, multiple sclerosis, amyotrophic lateral sclerosis, cancer, anxiety, and post-traumatic stress disorder (PTSD). Due to the complexity of the endocannabinoid system, the mechanisms by which cannabinoids function are still an active area of research. Two well-studied cannabinoid receptors are CB1 and CB2. CB1 is a receptor located in the central nervous system, but it also plays a role in gastrointestinal motility, metabolism, pain awareness, and energy [41]. CB2 receptors are predominately found in immunoregulatory tissues and have regulatory effects on inflammatory processes [41]. In particular, the anti-inflammatory mechanisms include effects on the JAK/STAT pathway and homeostasis of ion channels [42]. As tinnitus can be a result of neuronal inflammation, CBD may play a role in tinnitus by modulating inflammation in these cochlear pathways [41]. However, to date, no conclusive evidence is present.

A tinnitus animal model study investigating CB1 showed no effects of cannabinoids on tinnitus severity and even suggested that it may worsen it [43]. This is perhaps because CB1 may facilitate hyperactivity in the dorsal cochlear nucleus [41]. Additionally, smaller survey studies that have been completed previously show conflicting results, with objective data largely lacking due to policies on cannabinoid research [44]. The lack of scientific data on cannabinoids and tinnitus may be a reason for its cautioned use in the medical community [41, 44].

Psychedelics are another group of drugs that have been increasingly studied for the use of PTSD, depression, and anxiety. Methylenedioxymethamphetamine (MDMA) is one such compound that stimulates the release of serotonin, dopamine, and norepinephrine. A clinical trial studying MDMA showed that it was safe and effective for treating patients with PTSD with no significant side effects [45]. There are very few if any studies investigating MDMA as a tinnitus treatment. In one example, a small proof-of-concept study of five patients studied the use of methylenedioxymethamphetamine (MDMA) on tinnitus perception by utilizing fMRI. The results showed that tinnitus perception improved with MDMA, perhaps due to reduced synapses between the hippocampus, amygdala, and parahippocampal gyrus [46].

Psilocybin, the active ingredient found in "magic mushrooms", is a compound that has a high affinity for serotonin receptors. In a study of patients with advanced-terminal cancer and anxiety, psilocybin was found to help reduce anxiety and depressed mood in 80% of participants [44]. The compound has also shown promise in the management of major depression. No studies to date have evaluated the role of psilocybin in the treatment of tinnitus.

Composition of Supplements

There are important considerations concerning the herbs and supplements previously mentioned. Preparations of each supplement and herb can vary greatly. Over-the-counter minerals are often supplied in their sulfate and carbonate forms which can reduce absorption. However, linking minerals to zinc via chelation or amino acids can aid absorption [10]. Conversely, the ubiquinone type of CoQ10 is better absorbed than its ubiquinol counterpart. Other considerations include a lack of access to highly purified ingredients despite having standard compositions, as mentioned in the German Commission E guidelines for treatment.

Acupuncture

Acupuncture originated from traditional Chinese medicine where the treatment aims to alter or bring balance to the body's neurophysiology. Levels of neurotransmitters such as serotonin, dopamine, and endorphins have shown an increase in brain tissue through acupuncture. It is used in several other diseases such as trigeminal neuralgia and other pain syndromes. Its use in the treatment of tinnitus involves potential stimulation of the cochlear nucleus or increasing blood-flow to the area [47]. Data surrounding acupuncture and tinnitus is quite variable with systematic reviews remaining inconclusive [10]. A large division remains between Asian and Western studies with Asian studies showing more positive results. However, many factors such as familiarity with acupuncture sites, maneuvers, and sessions may influence results [48]. Due to the variability of acupuncture, many systematic reviews express concerns about the quality and size of trials or the presence of methodological flaws or risk bias [49, 50]. Despite the mixed results surrounding acupuncture, it may provide an avenue of treatment for tinnitus patients. A point of focus in acupuncture for tinnitus is the number of treatments and relationship to anatomic sites associated with tinnitus. Lin and Colleagues demonstrated that periauricular and distal acupuncture in combination with 17 to 24 acupuncture sessions leads to significantly improved outcomes [51]. One Brazilian study with 5 weeks of acupuncture twice a week showed statistically significant improvements in VAS and THI scores compared to a control group [52]. A study done in Denmark comparing manual and electrical acupuncture also showed some significant improvements in the subjective qualities of tinnitus [48].

Since acupuncture can boost blood flow to an area, evaluation of tinnitus centers such as the cochlea and brain have been completed. For instance, infrared thermography has been conducted before and after a single acupuncture session, and the results have suggested an increase in

Table 1. Summary of alternative treatments of tinnitus

Alternative Medicine Interventions	Study	Study Nature	Sample size	Key Findings
Ginkgo Biloba	Leutzenberg *et al.* [10] 2020	Review	—	Use of EGb761 strain of Ginkgo Biloba exhibited greater benefits at 240 mg.
	Chauhan *et al.* 2023 [11]	Clinical trial	69 patients	Ginkgo biloba in addition to antioxidants improved THI and VAS scores.
	Holsetein *et al.* 2001 [14]	Systematic review	3931 patients	Of 19 clinical trials, 11 demonstrated clinical efficacy of EGb761, but inconsistent across the trials.
	Spiegal *et al.* 2018 [15]	Meta-analysis	1972 patients	EGb761 was superior to placebo groups for tinnitus treatment in 5 randomized trials.
Zinc	Seidman *et al.* 2018 [21]	Review	—	Zinc levels decrease with age and are inversely correlated with tinnitus severity.
	Berkiten *et al.* 2015 [22]	Cross-sectional	100 patients	Zinc deficient patients displayed higher tinnitus loudness and higher hearing thresholds for air conduction.
	Yeh *et al.* 2019 [23]	Prospective trial	40 patients	Significant improvement in THI with 20 mg of zinc daily in patients with concurrent tinnitus and noise-induced hearing loss.
B-Complex Vitamins	Singh *et al.* 2016	Randomized control trial	40 patients	Results showed significant improvements in VAS after B12 treatment in patients with tinnitus and known cobalamin deficiency

	Lee & Kim. 2018	Retrospective data review	1435 patients	Vitamins B2 and B3 may have both an independent association and annoyance role in tinnitus patients respectively
Other Antioxidants (CoQ10, Beta Carotene, vitamins C and E)	Scasso *et al*. 2017 [33]	Clinical trial	26 patients	Use of CoQ10 showed significant protection against cochlear toxicity and a lower maintained ROS status compared to controls [32].
	Savastano *et al*. 2007 [33]	Prospective study	31 patients	Treatment with vitamins C and E with beta-carotene for 18 weeks suggested significant improvement in patient's VAS for idiopathic tinnitus patients [33].
Melatonin	Rosenberg *et al*. 1998 [36]	Clinical trial	30 patients	One double-blind placebo-controlled crossover study by Rosenberg and colleagues found that the THI did not differ significantly with 3 mg of melatonin with a 7-day washout compared to controls [36].
	Stefanescu *et al*. 2017 [34]	Clinical trial	20 patients	A 30-day washout period found a significant improvement in self-rated tinnitus scoring and tinnitus matching [34].
	Abtahi *et al*. 2017 [37]	Clinical trial	70 patients	Abtahi and colleagues found that 3 mg of melatonin for three months compared with 50 mg sertraline daily showed significant decline in mean THI scores with both treatments [37].
	Barati *et al*. 2022 [38]	Clinical trial	90 patients	Melatonin groups displayed significant improvements in THI, tinnitus loudness and awareness scores.

(*Continued*)

Table 1. (*Continued*)

Alternative Medicine Interventions	Study	Study Nature	Sample size	Key Findings
Cannabinoids & Psychedelics	Zheng *et al*. 2015 [43]	Animal study	50 Wistar rats	Investigating CB1 showed no effects of cannabinoids on tinnitus severity and even suggested that it may worsen it.[43]
	Narwani *et al*. 2020 [44]	Systematic review	8 studies with four animal studies, 2 survey studies, and 2 trials	Smaller survey studies that have been completed previously show conflicting results, with objective data largely lacking due to policies on cannabinoid research.[44]
	Asher 2022	Literature review	—	In a study of patients with advanced-terminal cancer and anxiety, psilocybin was found to help reduce anxiety and depressed mood in 80% of participants.[45] However, no studies have evaluated the effects on tinnitus.
	Searchfield *et al*. 2020 [46]	Clinical trial	5	proof-of-concept study of five patients studied the use of methylenedioxymethamphetamine (MDMA) on tinnitus perception by utilizing fMRI. The results showed that tinnitus perception improved with use of MDMA.[46]

Acupuncture	Wang *et al*. 2010 [48]	Randomized placebo-controlled trial	50	A study done in Denmark comparing manual and electrical acupuncture also showed some significant improvements in the subjective qualities of tinnitus.
	Doi *et al*. 2016 [52]	Randomized control trial	50	5 weeks of acupuncture twice a week showed statistically significant improvements in VAS and THI scores compared to a control group.
	Cai *et al*. 2019 [53]	Clinical trial	54	A single acupuncture session revealed significant increased cochlear blood flow and a reduction in temperature differentials.
	Yu *et al*. 2023 [54]	Clinical trial	18	Following acupuncture, the temporal lobe showed increased oxygenated hemoglobin concentration using functional near-infrared spectroscopy.

cochlear blood flow [53]. Additionally, functional near-infrared spectroscopy (fNIRS), which detects a certain concentration of oxygenated hemoglobin, has also been utilized. A study that utilized fNIRS after acupuncture showed increased oxygenated hemoglobin concentration in the temporal lobe [54]. Due to its potential positive results, acupuncture should be considered as a treatment option for tinnitus patients.

Conclusion

- Use of complementary medicines for tinnitus patients should be a holistic process.
- Physicians should not use these as a last result, but perhaps in combination with other treatment modalities.
- Treatments discussed are evolving with no singular solution but are viable options to tinnitus treatment.
- Less invasive methods include herbal supplementation, vitamins, trace elements, and antioxidants.
- Treatment with alternative methods is a trial-and-error process as each patient may not benefit in the same way.
- Complementary treatments will continue to require more stringent evaluation such as randomized double-blinded placebo-controlled trials to provide guidelines.

References

1. Han B, *et al.* Tinnitus: Characteristics, causes, mechanisms, and treatments. *Journal of Clinical Neurology*. **5**(1):11–19 (2009).
2. Ahmad N, and Seidman M. Tinnitus in the older adult: Epidemiology, pathophysiology and treatment options. *Drugs & Aging*. **21**:297–305 (2004).
3. El-Shunnar SK, Hoare DJ, Smith S, Gander PE, Kang S, Fackrell K, and Hall DA. Primary care for tinnitus: Practice and opinion among GPs in England. *Journal of Evaluation in Clinical Practice*. **17**(4):684–692 (2011).
4. Goldstein E, *et al.* Cost of care for subjective tinnitus in relation to patient satisfaction. *Otolaryngology–Head and Neck Surgery*. **152**(3):518–523 (2015).
5. Diamond BJ, *et al.* Ginkgo biloba extract: Mechanisms and clinical indications. *Archives of Physical Medicine and Rehabilitation*. **81**(5):668–678 (2000).

6. Ude C, Schubert-Zsilavecz M, and Wurglics M. Ginkgo biloba extracts: A review of the pharmacokinetics of the active ingredients. *Clinical Pharmacokinetics*. **52**(9):727–749 (2013).
7. Smith PF, Zheng Y, and Darlington CL. Ginkgo biloba extracts for tinnitus: More hype than hope? *Journal of Ethnopharmacology*. **100**:95–99 (2005).
8. Ahsan SF, and Seidman MD. Alternative medical management of tinnitus. *Brain Injury Professional*. **13** (2016).
9. Rejali D, Sivakumar A, and Balaji N. Ginkgo biloba does not benefit patients with tinnitus: A randomized placebo-controlled double-blind trial and meta-analysis of randomized trials. *Clinical Otolaryngology & Allied Sciences*. **29**(3):226–231 (2004).
10. Luetzenberg FS, Seilesh Babu, and Seidman MD. Alternative treatments of tinnitus: Alternative medicine. *Otolaryngologic Clinics of North America*. **53**(4):637–650 (2020).
11. Chauhan B, Arya S, and Chauhan K. Ginkgo biloba administered singly and combined with antioxidants in tinnitus patients. *Journal of Audiology and Otology*. **27**(1):37–44 (2023).
12. Morgenstern C, and Biermann E. The efficacy of Ginkgo special extract EGb 761 in patients with tinnitus. *International Journal of Clinical Pharmacology and Therapeutics*. **40**(5):188–197 (2002).
13. von Boetticher A. Ginkgo biloba extract in the treatment of tinnitus: A systematic review. *Neuropsychiatric Disease and Treatment*. **7**:441–447 (2011).
14. Holstein N. Ginkgo special extract EGb 761 in tinnitus therapy. An overview of results of completed clinical trials. *Fortschritte der Medizin*. **118**(4):157–164 (2001).
15. Spiegel R, Kalla R, Mantokoudis G, Maire R, Mueller H, Hoerr R, and Ihl R. Ginkgo biloba extract EGb 761® alleviates neurosensory symptoms in patients with dementia: A meta-analysis of treatment effects on tinnitus and dizziness in randomized, placebo-controlled trials. *Clinical Interventions in Aging*. **13**:1121–1127 (2018).
16. Drew S, and Davies E. Effectiveness of Ginkgo biloba in treating tinnitus: Double blind, placebo controlled trial. *BMJ*. **322**(7278):73 (2001).
17. Hilton MP, Zimmermann EF, and Hunt WT. Ginkgo biloba for tinnitus. *Cochrane Database of Systematic Reviews*. **3** (2013).
18. Posadzki P, Watson L, and Ernst E. Herb–drug interactions: An overview of systematic reviews. *British Journal of Clinical Pharmacology*. **75**(3):603–618 (2013).
19. Zhai S, Fang Y, Yang W, *et al.* Clinical investigation on the beneficial effects of the Chinese medicinal herb GushenPian on sensorineural deafness and tinnitus. *Cell Biochemistry and Biophysics*. **67**:785–793 (2013).

20. Ochi K, Kinoshita H, Kenmochi M, *et al.* Zinc deficiency and tinnitus. *Auris Nasus Larynx*. **30**:S25–S28 (2003).
21. Seidman MD. Tinnitus. *BMJ Best Practice*. (2018).
22. Berkiten G, Kumral TL, Yildirim G, *et al.* Effects of serum zinc level on tinnitus. *American Journal of Otolaryngology*. **36**:230–234 (2015).
23. Yeh CW, *et al.* Effects of oral zinc supplementation on patients with noise-induced hearing loss associated tinnitus: A clinical trial. *Biomedical Journal*. **42**(1):46–52 (2019).
24. Darrat I, Ahmad N, Seidman K, *et al.* Auditory research involving antioxidants. *Current Opinion in Otolaryngology & Head and Neck Surgery*. **15**: 358–363 (2007).
25. Ekinci A, and Kamasak K. Evaluation of serum prolidase enzyme activity and oxidative stress in patients with tinnitus. *Brazilian Journal of Otorhinolaryngology*. 405–410 (2019).
26. Tulchinsky TH. Micronutrient deficiency conditions: Global health issues. *Public Health Reviews*. **32**:243–255 (2010).
27. Dairo MD, and Ige OK. Supplementation of micronutrients in community micronutrient deficiency prevention programmes. *Annals of Ibadan Postgraduate Medicine*. **7**(1):6–9 (2009).
28. Calderón-Ospina CA, and Nava-Mesa MO. B Vitamins in the nervous system: Current knowledge of the biochemical modes of action and synergies of thiamine, pyridoxine, and cobalamin. *CNS Neuroscience & Therapeutics*. **26**(1):5–13 (2020).
29. Seidman MD, and Babu S. Alternative medications and other treatments for tinnitus: Facts from fiction. *Otolaryngologic Clinics of North America*. **36**:359–381 (2003).
30. Singh C, *et al.* Therapeutic role of Vitamin B12 in patients of chronic tinnitus: A pilot study. *Noise & Health*. **18**(81):93 (2016).
31. Lee DY, and Kim YH. Relationship between diet and tinnitus: Korea national health and nutrition examination survey. *Clinical and Experimental Otorhinolaryngology*. **11**:158–165 (2018).
32. Scasso F *et al.* Dietary supplementation of coenzyme Q10 plus multivitamins to hamper the ROS mediated cisplatin ototoxicity in humans: A pilot study. *Heliyon*. **3**(2):e00251 (2017).
33. Savastano M, Brescia G, and Marioni G. Antioxidant therapy in idiopathic tinnitus: Preliminary outcomes. *Archives of Medical Research*. **38**:456–459 (2007).
34. Stefanescu DC, *et al.* The role of melatonin in the management of patients with tinnitus. *Revista de Chimie*. **69**(5):1156–1159 (2018).
35. Zeitzer JM, Daniels JE, Duffy JF, Klerman EB, Shanahan TL, Dijk DJ, and Czeisler CA. Do plasma melatonin concentrations decline with age? *The American Journal of Medicine*. **107**(5):432–436 (1999).

36. Rosenberg SI, *et al.* Effect of melatonin on tinnitus. *The Laryngoscope.* **108**(3):305–310 (1998).
37. Abtahi SH, Hashemi SM, Mahmoodi M, *et al.* Comparison of melatonin and sertraline therapies on tinnitus: A randomized clinical trial. *International Journal of Preventive Medicine.* **8**:61 (2017).
38. Barati B, Asadi M, and Rahimipour N. Comparative evaluation of melatonin as an alternative therapy in tinnitus: A double-blind, placebo-controlled randomized trial. *Journal of Otorhinolaryngology and Facial Plastic Surgery.* **8**(1):1–6 (2022).
39. Chitsaz A, Rajol Dezfouli B, and Rezaeian A. Comparing the efficacy of amitriptyline and melatonin on tinnitus treatment. *Hospital Practices and Research.* 7(3):115–118 (2022).
40. Hurtuk A, Dome C, Holloman CH, *et al.* Melatonin: Can it stop the ringing? *Annals of Otology, Rhinology & Laryngology.* **120**:433–440 (2011).
41. Tapasak B, Edelmayer L, and Seidman MD. Endocannabinoid system and the otolaryngologist. *Otolaryngologic Clinics of North America.* **55**(5):1101–1110 (2022).
42. Perin P, *et al.* Cannabinoids, inner ear, hearing, and tinnitus: A neuroimmunological perspective. *Frontiers in Neurology.* **11**:505995 (2020).
43. Zheng Y, Reid P, and Smith PF. Cannabinoid CB1 receptor agonists do not decrease, but may increase acoustic trauma-induced tinnitus in rats. *Frontiers in Neurology.* **6**:60 (2015).
44. Narwani V, *et al.* Does cannabis alleviate tinnitus? A review of the current literature. *Laryngoscope Investigative Otolaryngology.* **5**(6):1147–1155 (2020).
45. Asher BF. Why otolaryngologists should be interested in psychedelic medicine. *Otolaryngologic Clinics of North America.* (2022).
46. Searchfield GD, Poppe TNER, Durai M, Jensen M, Kennedy MA, Maggo S, *et al.* A proof-of-principle study of the short-term effects of 3,4-methylenedioxymethamphetamine (MDMA) on tinnitus and neural connectivity. *International Journal of Neuroscience.* **130**(7):671–682 (2020).
47. Cabýoglu MT, Ergene N, and Tan U. The mechanism of acupuncture and clinical applications. *International Journal of Neuroscience.* **116**(2):115–125 (2006).
48. Wang K, Bugge J, and Bugge S. A randomised, placebo-controlled trial of manual and electrical acupuncture for the treatment of tinnitus. *Complementary Therapies in Medicine.* **18**(6):249–255 (2010).
49. Liu F, *et al.* Acupuncture in the treatment of tinnitus: A systematic review and meta-analysis. *European Archives of Oto-Rhino-Laryngology.* **273**:285–294 (2016).
50. Huang K, *et al.* Acupuncture for tinnitus: A systematic review and meta-analysis of randomized controlled trials. *Acupuncture in Medicine.* **39**(4):264–271 (2021).

51. Lin TY, Yang SW, Lee YS, *et al.* Analysis of factors influencing the efficiency of acupuncture in tinnitus patients. *Evidence-Based Complementary and Alternative Medicine.* **2019**:1318639 (2019).
52. Doi MY, *et al.* Effectiveness of acupuncture therapy as treatment for tinnitus: A randomized controlled trial. *Brazilian Journal of Otorhinolaryngology.* **82**(4):458–465 (2016).
53. Cai W, *et al.* Thermal effects of acupuncture by the infrared thermography test in patients with tinnitus. *Journal of Acupuncture and Meridian Studies.* **12**(4):131–135 (2019).
54. Yu X, *et al.* Effect of acupuncture treatment on cortical activation in patients with tinnitus: A functional near-infrared spectroscopy study. *IEEE Transactions on Neural Systems and Rehabilitation Engineering.* (2023).

https://doi.org/10.1142/9789819807345_0007

Chapter 7

Non-Invasive Neuromodulation for Tinnitus Treatment

Divya Adhia and Dirk De Ridder*

Section of Neurosurgery, Department of Surgical Sciences, University of Otago, Dunedin, New Zealand

**Dirk.deridder@otago.ac.nz*

Abstract

The pathophysiological mechanisms that underlie the generation and maintenance of tinnitus are being unraveled progressively. Based on this knowledge, a large variety of different neuromodulatory interventions have been developed, and has still been designed, adapting to the progressive mechanistic insights in the pathophysiology of tinnitus. Non-surgical or non-invasive brain stimulation involves the transcranial delivery of magnetic or electrical stimuli. Repeated transcranial magnetic stimulation (rTMS) targeting the temporal, temporoparietal and the frontal cortex has been the mainstay of non-invasive neuromodulation. Yet, the evidence is still unclear, and therefore systematic meta-analyses have been performed demonstrating that rTMS can slightly improve suffering but not loudness perception. Different forms of transcranial electrical stimulation (tDCS, tACS, tRNS), applied over the frontal and temporal cortex, have been investigated with meta-analyses in tinnitus patients as well. Most meta-analyses demonstrate that tDCS can slightly improve suffering, and some suggest that tDCS can also improve loudness, but only when applied to the left temporal area. Recently, the initial studies reporting on bimodal non-invasive stimulation have revealed promising results and it appears that targeting different sensory modalities in temporally combined manners may be more promising than single target approaches. While most neuromodulatory approaches seem promising, only rTMS and tDCS seem ready for clinical use, but further research is required to help translating the scientific outcomes of surgical and bimodal stimulation into routine clinical practice.

Introduction

Tinnitus is defined as the conscious awareness of a tonal or composite noise for which there is no identifiable corresponding external acoustic source [60]. Tinnitus becomes Tinnitus Disorder when associated with emotional distress, cognitive dysfunction, and/or autonomic arousal, leading to behavioural changes and functional disability. In other words "Tinnitus" describes the auditory or sensory component, whereas "Tinnitus Disorder" reflects the auditory component and the associated suffering [60]. Its prevalence ranges around 15% in the general population. While most individuals can habituate to the sound, approximately 20–25% of patients experience severe disruption to their quality of life [14, 264]. For these individuals, tinnitus is often associated with anxiety, depression, cognitive impairment, and sleep disturbances [19, 261, 279], and may be classified as a mental disorder. The natural evolution of tinnitus is unfortunate in that after 4 years in only about 20% of patients the tinnitus resolves spontaneously, and in those 80% who still have it, in about 10% it improves, in about 10% it worsens and in 80% it remains unaltered [51]. This suggests that improved treatments for this enigmatic symptom need to be developed.

Although a range of treatments, including pharmacological and non-pharmacological interventions, have been proposed and some are available in the market, there is a dearth of evidence for a universally effective therapy for tinnitus [78, 156]. The absence of an efficacious treatment for tinnitus is largely attributed to the heterogeneity of the condition and an incomplete understanding of the pathophysiology of its various forms [85].

Since 2011, there has been a decline in pharmaceutical interest for developing neuropharmacological products for neurological or psychiatric indications. Big pharma has reduced investment by 50% for brain-related disorders [287], due to the high cost and risk associated with it. Developing medication for brain disorders has 50% less chance of making it to the market (6.2% vs 13.3%) [115], takes 30% longer to develop (19.3 vs 14.7 months) [115], and costs 30% more than heart medication [115]. A large proportion of medications fail phase III trials, with 46% not outperforming placebo [141]. Therefore, despite repurposing existing medication [215], the prospects of a pharmaceutical solution for tinnitus and tinnitus disorder are limited. Brain stimulation, however, is a promising alternative treatment option, as the history and evolution of the understanding of the pathophysiology of tinnitus has been closely associated with the development of brain stimulation approaches.

History and Evolution of the Pathophysiology of Tinnitus

Tinnitus has been a perplexing symptom with controversial origins. In the 19th century BCE, the Ebers papyrus linked tinnitus to a bewitched ear, and until recently it was widely understood as a problem of the ear caused by hearing loss [76]. Consequently, until the 1990s, treatment focused on improving hearing function and masking sound. However, Jastreboff's 1990 neurophysiological model suggested that classical conditioning may be responsible for the phantom sound [130], leading to the development of Tinnitus Retraining Therapy (TRT). TRT involved directive counseling, hearing aids, and sound therapy to facilitate habituation of tinnitus reaction and perception. But Jastreboff did not specify which neuroanatomical structures were involved in this conditioning, so the brain components involved remained inside a black box. In 1999, Rodolfo Llinas proposed a pathophysiological model, thalamocortical dysrhythmia, that suggested the interaction between the auditory cortex and thalamus as the generator of tinnitus [173]. Clinical evidence has shown that tinnitus is mostly associated with hearing loss or abnormalities in the inner ear or the peripheral auditory pathway, leading to neuroplastic changes in auditory and non-auditory brain networks. [71]. Some forms of tinnitus, however, exist without hearing loss, and this may be linked to a different underlying pathophysiology [256, 260], requiring different neuromodulation targeting. Whether tinnitus without hearing loss is fundamentally different or a variation of the same underlying pathophysiology still remains to be proven, as tinnitus without audiometric hearing loss does not imply there is no auditory deafferentation [281]. The thalamocortical dysrhythmia model suggests that in the deafferented state, the dominant resting state alpha rhythm (8–12 Hz) slows down to theta band (4–7 Hz) frequencies [173]. Even though it is called theta, from an electrophysiological point of view it should be considered slowed alpha activity, as the neural generators of true theta activity are different from the generators of alpha and slowed alpha activity [253, 283]. In the awake state, the thalamocortical loops within the brain idle around 10 Hz (8–12 Hz), called the alpha frequency, generated by the thalamus. During light sleep, theta band activity is predominant, generated by the septal nuclei and hippocampus, and in deep sleep delta activity (1–3 Hz) predominates [32]. The theta that emerges in thalamocortical dysrhythmia is thalamicly generated and therefore more likely represents alpha activity that is slowed down to the theta frequency range. This may be due to decreased firing rates when less information needs to be processed, [24] due to the deafferentation, and that firing and oscillation rate are coupled at the thalamocortical level [49, 172]. As a consequence, $GABA_A$ mediated lateral inhibition is weakened [192], inducing gamma (>30 Hz) band activity [173] surrounding the deafferented theta area, also known as the edge effect [172]. Confirming the initial studies by Llinas [172, 173] a decrease in alpha power is associated with an increase in gamma power [176], and gamma is coupled to theta activity [282]. Other studies have confirmed the presence of both low and high frequency activity in the auditory cortex of tinnitus patients [2–4, 218], and the theta-gamma cross-frequency coupling has also been identified on electrode recordings of implanted patients [63], supporting the MEG and EEG data.

According to clinical research, negative symptoms such as hearing loss and hypoesthesia in the somatosensory system are associated with theta activity, while positive symptoms like tinnitus and pain in the somatosensory system are linked to gamma activity in diseases characterized by thalamocortical dysrhythmia [172, 174]. Decreased information processing and slowed alpha activity, which are similar to the deafferented thalamocortical columns being "asleep", are thus responsible for negative symptoms (e.g. hearing loss, hypoesthesia) [172]. It is suggested that theta can act as a long-range carrier wave [100, 101, 103] on which tinnitus information can be nested by high-frequency oscillatory activity [69, 71]. If this model is universal, all tinnitus patients could be treated with auditory cortex stimulation, either non-invasively or invasively. Initial tests of this theory were carried out using brain implants and transcranial magnetic stimulation [55, 83, 209]. Early research on TMS and implants found that not all individuals with tinnitus could be effectively treated with auditory cortex stimulation. Factors such as the duration of tinnitus [72, 145], gender [62], amount of hearing loss [145], tinnitus pitch [72], perceived laterality [72], laterality of the tinnitus generator in the brain [53, 154], and TMS dose [210] were postulated to contribute to the lack of predictable results. However, a novel pathophysiological model proposed that the tinnitus percept is not phrenologically limited to the auditory cortex, but that the auditory cortex part of a tinnitus-associated network [71, 231, 238]. Even though that network dynamically changes, adapting to external and internally generated triggers, on average it remains fairly stable, so that it could be averaged over 5 to 20 minutes recording intervals, with the current MEG and EEG technology [56, 231, 232]. Translating theoretical concepts from network science [6] to brain stimulation, the model also hypothesized that directing stimulation at more than one target could disrupt the tinnitus network more effectively than a single target. As the frontal cortex was already a somewhat successful target for transcranial direct current stimulation (tDCS) [64, 98, 242, 252, 257, 262] and TMS [61, 64, 259], the first multitarget stimulations selected the frontal cortex and temporoparietal junction (auditory cortex) [149], but later attempts also used auditory cortex and high cervical C2 area [66], using the same reasoning. Based on a network meta-analysis, it was shown that the best neuromodulation treatment for tinnitus consists of targeting the frontal cortex with direct current (tDCS) and the auditory cortex with random noise stimulation (tRNS) [40]. Network science-based models have suggested that treatment approaches should be modified to employ both random and targeted modulation on the tinnitus network [191]. Another model proposed that tinnitus could be the result of a deficient noise-canceling mechanism [159, 220], analogous to pain [90, 146]. However, according to this model the subgenual anterior cingulate cortex was critically involved in tinnitus generation, a difficult neuroanatomical target for current brain stimulation technologies. Furthermore, other groups failed to replicate the imaging data on which the model was based.

A further model was proposed that tinnitus could be the result of a Bayesian prediction error [57, 67]. According to this model tinnitus can be understood as the result of a mismatch between the predicted sound percept and the actual neural input from the ear, which is reduced because of hearing loss. It was hoped that the underlying anatomical and oscillatory correlates could be targeted with brain stimulation. It also theorized that neuroplastic changes involved in tinnitus were multiphasic, suggesting a theoretical difference between the neural generators of tinnitus in tinnitus with and without deafferentation. As such, the model attempted to combine and integrate both the deafferentation based theories and the noise canceling theories [57, 67, 71]. The neural correlates of the tinnitus prediction and prediction error still need to be elucidated before they can be targeted by neuromodulation. But the Bayesian brain model has become anatomically integrated in a model that incorporates a lateral "loudness" pathway, a medial "suffering" pathway, and a descending "inhibitory" pathway, also known as the noise canceling pathway [56].

Expanding on the Bayesian tinnitus concept, recent advancements in technology and graph theoretical analyses have led to the development of a dynamical tinnitus network (Figure 1). According to this model, the constant search for missing information by the auditory cortex results in tinnitus sound, which is distressing due to the loss of temporal flexibility in the dynamically changing network activity [189, 190]. A recent proposal suggests that tinnitus may be caused by an imbalance between bottom-up and top-down influences, which could provide clear targets for specific brain stimulation [71, 256]. This proposal aims to integrate both the deafferentation-based and noise canceling models of tinnitus. But this

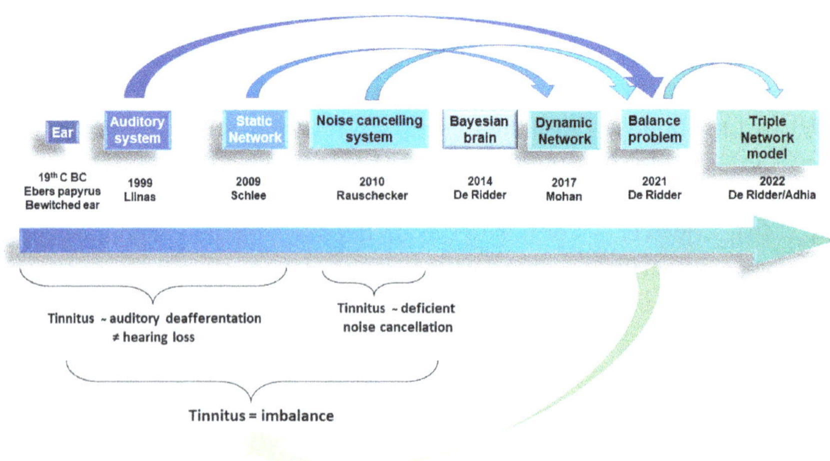

Figure 1. History of the pathophysiological models for tinnitus

leaves the problem that the unified tinnitus percept is the result of multiple overlapping and dynamically interacting networks, each generating another aspect of the unified percept {De Ridder, 2014 #6714}. With the advent of the triple network model that explains cognitive appraisal of external and internal stimuli, this triple model has been integrated in the percept of tinnitus {De Ridder, 2022 #12524}.

A Theory of Symptom Generation in the Brain

The brain is a complex adaptive system [102, 244], similar to other systems like the internet, climate, economy, and ant society [126]. To qualify as such, a system must fulfill two criteria: it must have a small world topology and embed noise. [10]. The reason for these criteria relates to the adaptiveness of the system. Network systems can be topologically structured in three ways [31]. At one extreme, the network can have a lattice or regular topology, which means that every stimulus will always result in exactly the same processing, which is both predictive and efficient but not adaptive whatsoever [37]. At the other extreme, a system can be completely random, which is inefficient and disadvantageous because every stimulus will always evoke a completely random response. An intermediate structure between regular and random topologies has a small world topology. Small world topology allows flexibility and adaptiveness to changing environments through variability, making the system able to learn [17, 134]. The brain is characterized by a small world structure and thus fulfills the first criterion [1, 17, 30, 82]. The brain is also noisy, fulfilling the second criterion, but the noise is structured, generally following a power law distribution [33], i.e., a pink ($1/f$) or brown ($1/f^2$) noise structure. This structured noise has an advantage that it has memory and can carry information, in contrast to white noise which is an unstructured completely random noise [140]. This allows the brain to learn, be flexible, and maintain stability. Implanting electrodes in an adaptive system such as the brain makes sense to modify its structure and function, but it would not make sense in a regular network or completely random system. Emergence is a fundamental characteristic of every complex adaptive system, meaning the whole is more than the sum of its parts and cannot be predicted from its constituent parts. A collection of car parts is not a car. Only when all parts are put together (i.e., connected) in a very specific way does a functional car emerge. Yet, a simple car is complex but non-adaptive. It doesn't reconfigure itself based on a changing environment.

Bees' do, ants do, our brain does and so does the internet. Emergentism in the philosophy of mind supports the belief that consciousness is an emergent property of brain function, and every thought, feeling, action is the consequence of specific connectivity patterns resulting from adaptive neuroplasticity. Every symptom or disease is the result of maladaptive plasticity [94], so using TMS or transcranial electrical stimulation, or implanting electrodes on the cortex should change or use the brain's connectivity to create a change in symptoms [59].

Neuroplasticity refers to the brain's ability to adapt to a changing environment by modifying its structure and function. However, such adaptive changes can be both beneficial and harmful, leading to both learning how to adjust to a changing environment, but maladaptive learning can also lead to symptoms. Environmental stimuli (external or internal), whether decreased or increased, can lead to neuroplasticity, which can be influenced by factors such as adrenal and gonadal hormones, neurotransmitters, growth factors, certain drugs, and aging [106]. These changes occur at various levels, including molecular, synaptic, neurogenesis, connectivity, and network levels, which can be observed through structural and functional brain imaging as changes in structure, activity and connectivity [106]. Connectivity changes can be categorized into structural, functional, and effective connectivity, which refer to anatomical changes in the brain [118], co-activation of different brain areas [105], and directional information flow changes [105], respectively. This reorganization promotes stability in constantly changing connectivity networks, leading to the emergence of altered perceptions, thoughts, emotions, actions, and symptoms.

Non-Invasive Neuromodulation for Tinnitus

Neuromodulation as targeted neuroplasticity

Neuromodulation and neurostimulation or brain stimulation are being used interchangeably. However, the term neuromodulation is now favored as it avoids the connotation of activation that comes with neurostimulation. In fact, if the electrical or magnetic stimuli reach an area of the brain with a predominance of GABA receptors, the clinical impact may be inhibitory. Neuromodulation refers to influencing brain activity and connectivity without necessarily indicating excitatory influence.

Neuromodulation refers to the induction of neuroplastic changes through targeted application of electrical, magnetic, sound (including ultrasound), optical, or pharmacological stimuli, without specifying an excitatory influence. This definition is broader than the one provided by the International Neuromodulation Society, which defines neuromodulation as technology that acts directly upon nerves by delivering electrical or pharmaceutical agents to a specific area to alter or modulate nerve activity (http://www.neuromodulation.com/about-neuromodulation).

Mechanism of action of cortex stimulation

It is crucial to gain a better understanding of the mechanisms involved in tinnitus generation in order to comprehend the beneficial effects of cortex stimulation. Symptoms are not the result of a single brain area's phrenological activity, but rather the outcome of maladaptive network activity [15, 94, 95]. Even when presented with sound, patients in a vegetative state who lack conscious awareness show functional and metabolic activation of their auditory cortex [22]. This indicates that activity in the auditory cortex alone is not adequate for conscious awareness, as demonstrated in the visual cortex [47, 187]. This has been shown for painful stimuli as well [47, 157, 158]. Only when the auditory cortex is functionally connected to consciousness-supporting networks, such as the self-representational default mode network and the attentional frontal parietal network [5], can auditory stimuli become consciously accessible. Thus, the auditory cortex must be functionally connected to these networks to allow conscious awareness of the presented stimulus.

It has been suggested that functional connections are crucial for transmitting cortically applied stimuli into wider networks associated with emergent properties such as tinnitus, which require treatment [58, 96]. The success or failure of auditory cortex stimulation via implanted electrodes depends on the functional connectivity between the auditory cortex and the parahippocampus [65], which is similar to the suggested importance of functional connectivity for anterior cingulate implants [58] and what has been shown for non-invasive brain stimulation in various neurological and psychiatric disorders [96]. In fact, studies have found that a lack of functional connectivity at certain sites rendered stimulation ineffective, and the sign of correlation was related to whether excitatory or inhibitory non-invasive stimulation was effective in treating patients. These results suggested that resting-state functional connectivity may be beneficial in optimizing treatment as well as identifying new stimulation targets [96].

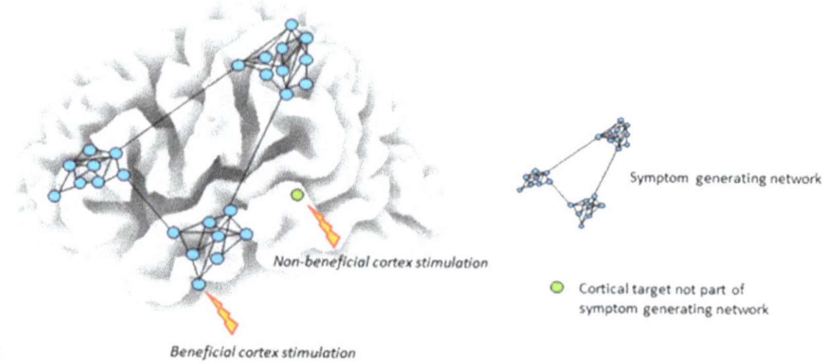

Figure 2. The hypothesized mechanism of action of cortex stimulation [59]. The electrode or non-invasive stimulation device must be positioned at a cortical target where the symptom generating network reaches the brain surface. The stimulation is thought to change the functional connectivity of the network, thereby changing its topology and its related emergent property, i.e., the symptom.

To summarize, cortical stimulation may be beneficial if the target is involved in generating the symptoms, while stimulation of a cortical target not functionally connected to the symptom-generating network may not be beneficial [59].

TMS for treating tinnitus

In 1985, a study demonstrated the possibility of depolarizing neurons in the brain using external magnetic stimulation [16]. Magnetic stimuli are advantageous over electrical stimuli because they are not attenuated by the skull's high resistance. The skull has a resistivity of 8–15 times that of soft tissues [16], resulting in 75% of applied electrical current being blocked by the skull [271]. Magnetic stimulation of the cortex is particularly effective because of the ability of the magnetic field to pass through high-resistance structures. TMS produces a magnetic field of the same size as that of a MRI scanner, i.e., 1 to 3 Tesla, but that only lasts for about a millisecond [274]. According to Faraday's law, the rapidly changing magnetic field (i.e., from zero to a very high value, then back to zero again in 1 ms) induces electrical currents in the brain area beneath the coil. Essentially, the magnetic field transfers the electrical stimulus through the scalp and skull into the brain [222]. The induced current pulse, which is similar in amplitude to that generated by a conventional stimulator directly applied to the brain surface, lasts about 200 μs and activates the axons of neurons in the cortex and subcortical white matter [222]. It is thought to activate the axons of neurons in the cortex and subcortical white matter, rather than the cell bodies of cortical neurons (which have a much higher threshold) [222] requiring a longer pulse width (>1000 μs) [218]. More recent proposals suggest that rTMS exerts its clinical effect via two mechanisms. On the one hand the induced electrical field modulates the release of neurotransmitters and changes synaptic plasticity via LTP and LTD-like influences, and on the other hand, the intrinsic magnetic field exerts an effect on the genetic expression of neurons and glial cells, and has neurotrophic and neuroprotective effects [43].

There are two primary limitations of rTMS. Firstly, the magnetic field attenuates quickly with distance from the coil surface, restricting direct stimulation to the superficial parts of the cerebral cortex immediately under the skull. Secondly, the site of stimulation is not as focused as direct electrical stimulation with implanted electrodes [222]. However, when using neuronavigation or robotic neuronavigated positioning, rTMS can achieve a precision of 2 mm [233], or even 1.3 mm [112], respectively, which matches the intrinsic resolution of structural and functional imaging. High-field MRI machines can resolve brain details as small as 1 millimetre, and at 7 Tesla, as small as 0.5 millimetre [203].

Magnetic coils come in various shapes, including round, figure of eight, double cone, and H-coils. Round coils are powerful but less focal than figure-eight-shaped coils, which have a maximal current at the intersection of their two round components, making them more focal [224]. Double cone and H-coils are capable of penetrating deeper and reaching areas such as the anterior and posterior cingulate cortex [36, 120]. Although the magnetic field declines rapidly with increasing distance from the coil, stimulation effects can propagate to functionally connected remote areas. Furthermore, depending on the frequency of stimulation rTMS can strengthen or weaken functional connectivity: Low frequency rTMS can increase functional connectivity, while high frequency rTMS can decrease functional connectivity [96].

Electromyographic recordings from TMS studies of the motor cortex have revealed a double effect of TMS. After a single TMS pulse, there is an initial burst of activity that lasts 5 to 10 milliseconds [52], followed by a suppression period lasting 100 to 200 milliseconds [222]. The effect of the TMS pulse is brain state dependent, as well as dependent on the position and orientation of the stimulation coil and the exact site of stimulation [222]. For example, if a TMS stimulus is given during sleep, anesthesia or coma, the stimulus will only exert a local effect and will not spread through the brain, in contrast to an awake state. Furthermore, TMS efficacy seems to be dependent on the stimulated person's genetic polymorphism. Certain genes, such as BDNF and 5-HT (1A) influence the sensitivity to non-invasive stimulation, both TMS and tDCS [39, 178]. In view of the inter-individual anatomical variability of the brain, neuronavigated TMS based on individual brain structure obtained from imaging has been suggested to improve efficacy [91]. The interindividual variability in the location of the sylvian sulcus and superior temporal sulcus that border the auditory cortex is 1.5–2 cm [245], suggesting that indeed a navigated approach makes sense. Concomitant intake of medication such as benzodiazepines [73, 126], psychostimulants [126] or neuroleptics [73] can influence the effect of rTMS, but other medication, such as anti-migraine medication does not seem to influence the treatment effect by TMS [8]. For example, whereas concomitant psychostimulants increases the efficacy [126], antiepileptics and lithium do not alter rTMS efficacy [121]. For concomitant benzodiazepines the data are conflicting. Some studies suggest that benzodiazepines decrease the efficacy of TMS as a treatment for depression [73, 126], whereas others do not [90] {Tran, 2024 #14421}.

The impact and potential adverse effects of TMS are influenced by device-related factors. The type of coil employed can also affect the consistency of TMS outcomes [91]. Additionally, stimulation parameters, including frequency and amplitude, are critical determinants of TMS effects [242].

In summary, many patient-dependent and device-related factors determine the outcome of transcranial magnetic stimulation (Table 1).

High frequency stimulation (HFS) and low frequency stimulation (LFS) are believed to have opposing effects (Table 2). Functional imaging studies have shown that HFS increases metabolism, while LFS decreases metabolism [143, 242]. Additionally, their effects on excitability and plasticity are also opposite: HFS appears to increase excitability [204] and have an LTP-like effect [275, 276], while LFS generates a decrease in

Table 1. Factors influencing TMS and tES effects.

	tES influencing factors	TMS influencing factors
Person/patient dependent	brain structure	brain structure
	brain state/function	brain state/function
	history of activity in the stimulated area	history of activity in the stimulated area
	brain area	brain area
	genetic polymorphism	genetic polymorphism
	medication	medication
	soft tissue and bone structure	Skull-cortex distance
Device/protocol dependent	electrodes size	coil type
	electrodes positions	coil orientation
	frequency in tACS/tRNS	frequency
	intensity	intensity
	electrode polarity	stimulation pattern (burst/tonic)
	duration	duration
		inter-train interval (intermittent/ continuous)
	number of sessions	number of sessions
	interval between sessions	interval between sessions
		pulse form

Table 2. Different effects of high (≥5 Hz) and low (1 Hz) frequency transcranial magnetic stimulation.

	High frequency TMS	Low frequency TMS
Frequency	>5 Hz	1 Hz
Excitability	Increased	Decreased
PET metabolism	Increased metabolism	Decreased metabolism
EEG	Upper alpha & beta synchronisation	Lower alpha & beta synchronisation
Molecular	GABA and glutamate unchanged	GABA and glutamate increased
Plasticity	LTP-like	LTD-like

excitability [42] and an LTD-like effect [275, 276]. Furthermore, the effect on oscillatory activity is different [26, 106, 206], as is the effect on neurotransmitter release [134, 135, 288] (Table 2).

TMS has a widespread effect beyond the area of stimulation, which has been demonstrated by research [142]. The mechanism behind this phenomenon involves the transmission of the stimulus via structural connectivity [195], leading to changes in functional and effective connectivity [237]. As mentioned, low and high frequency stimulation have different effects on functional connectivity. Moreover, TMS can also alter directional functional connectivity, which is also known as effective connectivity, and this property can be changed by TMS to exert its clinical effect [113]. In summary, TMS can modify the functional and effective connectivity of the stimulated network, resulting in changes in its emergent properties, leading to clinical benefits.

The regimen, parameter, and efficacy of TMS

rTMS has been used in tinnitus research in two different ways: single and repeated sessions.

Single sessions

Single sessions of rTMS have been utilized for three purposes. First, to evaluate the pathophysiology and anatomy of tinnitus by targeting brain areas associated with tinnitus and inducing transient reductions in tinnitus perception. This helps determine if these areas are causally involved in tinnitus generation, as demonstrated by studies targeting the temporal cortex [71, 82, 208], frontal cortex [60], parietal cortex [267], and anterior cingulate cortex [258]. Second, to determine if TMS could be a prognostic test for selecting proper candidates for more invasive permanent solutions like cortical brain implants [54, 55, 58, 68]. Third, to delineate optimal stimulation parameters that can be used for multiple rTMS sessions as a potential treatment [52, 150, 233].

Repeated sessions

It has been suggested that repeated sessions of low frequency rTMS targeting the temporal (auditory) cortex could be a promising approach for treating tinnitus, based on the assumption that tinnitus is associated with increased neuronal activity in this area [84]. However, studies examining the efficacy of low frequency TMS for tinnitus have produced inconsistent results [233], possibly due to high variability and gender differences (i.e., women respond better than men) in response to rTMS [165]. While some meta-analyses have demonstrated improvements in tinnitus-related suffering, [41, 165, 169, 241], others have not [79]. Overall, analysis of the literature indicates that low frequency rTMS may be effective in reducing tinnitus suffering, but the clinical effect is usually partial and temporary [162], with a THI improvement of about 7–8 points [169], which is the minimum for clinical efficacy [295], lasting up to 6 months [169]. Evidence-based guidelines have previously recommended low frequency (1 Hz) rTMS of the left temporoparietal cortex as a possible therapy for tinnitus (a level C recommendation, possible efficacy) [163], and a recent follow-up study has confirmed these findings [175]. However, none of the national tinnitus treatment guidelines include rTMS as a possible treatment, likely due to a delay in adopting novel evidence in guidelines that are only updated from time to time [223].

In addition to variations in study design and gender response differences, different stimulation parameters have also been employed. A recent systematic analysis investigated the relationship between stimulation parameters and treatment outcome, and found a higher success rate for lower stimulation intensities [233], which supports the evidence-based guidelines [163].

Recently, a personalized approach has been explored to reduce individual variability by considering the heterogeneity of tinnitus generators in the brain. A pilot study tested this concept by delivering rTMS at various frequencies over the left and right dorsolateral prefrontal (DLPFC) or temporoparietal (TPC) cortex in a preliminary test session. The type of protocol was then selected based on the results of the test session and applied for several days [150]. The personalized protocol resulted in greater benefit than the standard protocol (20 Hz-rTMS over left DLPFC followed by 1 Hz-rTMS over the left TPC), suggesting that a tailored rTMS protocol may be more clinically beneficial.

In summary, rTMS likely provides some benefit in the treatment of tinnitus. However, if rTMS is to become routine clinical practice, it is essential to look for mechanisms that may boost or potentiate the therapeutic effect. One such way is to enhance the efficacy of rTMS pharmacologically, e.g., by adding psychostimulants, as mentioned above, or ketamine. Early findings suggest that ketamine-enhanced rTMS can have a sustained beneficial effect on depression for up to two years [18], making it a potential treatment option for tinnitus patients who also suffer from depression or anxiety.

Moreover, the side effects of rTMS are rare, as reported in a meta-analysis of studies, and mostly consist of mild discomforts such as headaches, muscle twitching, neck contractions, and worsening of tinnitus [79]. There were no reports of seizures, possibly because rTMS is delivered within established safety guidelines, which show that it leads to fewer than one seizure per 60,000 sessions in individuals without risk factors [167].

tES for treating tinnitus

There are three distinct types of tES, which vary in how the electrical current is administered to the brain (see Figure 3): transcranial direct current stimulation (tDCS), transcranial alternating current stimulation (tACS), and transcranial random noise stimulation (tRNS), which is a variant of tACS [206, 263].

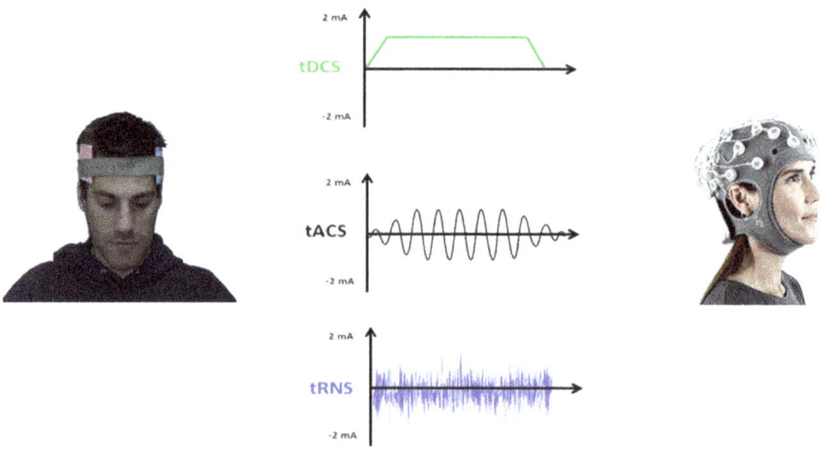

Figure 3. Different forms of tES: tDCS delivers direct current, tACS and tRNS deliver alternating current. tES can be applied with 2 electrodes or more. When more electrodes are used it is called high-definition transcranial electrical stimulation (HD-tES).

Transcranial direct current stimulation (tDCS)

Conventional tDCS

The use of direct electric current for medical purposes dates back centuries. As early as 43–48 A.D., Scribonius Largus, physician to Roman Emperor Claudius, reported using a live torpedo fish to deliver a strong direct current to the scalp as a treatment for pain [93]. In the 11th century, Ibn-Sidah recommended placing a live electrical catfish on the frontal bone to treat epilepsy patients [139]. In the 18th century, Galvani's introduction of the electrical battery [109, 110] led to the recognition that varying durations of electrical stimuli can produce different physiological effects [294]. Direct current stimulation is often referred to as Galvanic stimulation in honor of Galvani. In the 19th century, researchers such as Aldini (Galvani's nephew) [7] and Arndt [13] used transcranial electrical stimulation to treat melancholia and depression. This method had a brief resurgence in the 1960s and 1970s [171, 222], and a more sustained revival in the early 2000s [116].

Conventional tDCS involves placing one anode and one cathode electrode on the scalp to stimulate a specific brain region by creating a controlled electrical current that flows from the anode to the cathode. However, due to the high electrical resistance of the skull [18], only 25% to 50% of the direct current applied transcranially actually reaches the brain, while the rest is diverted through extracranial soft tissues. This has been demonstrated through calculations on realistic head models and validated by both animal and human experiments [229, 272] and human cadaver studies [81, 272].

tDCS modulates the cellular membrane potential, which can facilitate or inhibit spontaneous neuronal activity [198, 201]. When using anodal stimulation, inward current flow is produced, leading to depolarization of pyramidal cortical neurons and hyperpolarization of apical dendrites. Conversely, cathodal stimulation typically produces outward current flow, resulting in somatic hyperpolarization of pyramidal cortical neurons and depolarization of apical dendrites [271, 292]. The depolarization under the anode increases firing and excitability, while the firing rate and excitability are reduced under the cathode [21, 203].

Despite its immediate effect on neuronal activity, tDCS often leads to a delayed clinical response [108, 247]. To explain this, researchers have suggested that two additional mechanisms might be involved: glial and stem cell modulation. Astrocytes, a type of glial cell, are believed to be affected by tDCS [197, 228]. These cells regulate the formation, function, and elimination of synapses through various signals and can thereby affect neural circuit development and function [46]. Additionally, microglia, another type of glial cell that prunes synapses, might also be involved [188]. Studies have shown that tDCS activates microglia under both the anode and cathode [226]. Therefore, it is possible that tDCS modulates glial cells, leading to synapse formation and/or elimination, which may take time to manifest and explain the delayed effects of tDCS.

Moreover, tDCS-induced activation of microglia and astrocytes has been suggested to have a neuroimmunomodulatory impact [112] by modifying the expression of genes involved in immune regulation [216].

In addition to modulating neurons, including pyramidal and interneurons, and glial cells such as astrocytes and microglia, tDCS may also induce delayed effects through the activation of stem cells. Studies suggest that tDCS can recruit proliferating neural stem cells under the cathode [226], which could potentially lead to regenerative capabilities and further delayed clinical effects of tDCS. Animal models have demonstrated the neurogenesis and improved rehabilitation effects of tDCS. [296].

The effects of tDCS are influenced by a variety of factors, including patient-related and device-related factors (Table 1). Certain factors, such as the resistance of cephalic structures (e.g., skin, skull, blood vessels, brain tissue), cannot be controlled [29, 184, 198, 274]. Device-related factors, on the other hand, include electrode polarity, size, and placement, as well as stimulation intensity (measured in mA) and duration (typically ranging from 20-40 min in most studies) [29, 184, 198, 274]. By varying these parameters, tDCS protocols can be tailored to achieve specific effects on cortical activity and excitability in terms of direction, strength, focality, and duration [29, 202].

Typically, tDCS application is not associated with any serious adverse events, as reported in the contemporary tDCS literature, which includes investigations on more than 10,000 subjects between 1998 and 2014 [28, 104]. The safety profile of tDCS is influenced by various factors, such as the strength of current, size of the electrodes, contact media, and duration of the stimulation [128, 213]. This has been confirmed, andasno serious adverse effects or irreversible injuries have been reported in over 33,200 sessions involving 1000 subjects with repeated sessions protocols in human trials [20]. The maximum current used in these trials was 4 milliamperes, and the total charge delivered was no more than 7.2 Coulombs, for a maximum duration of 40 minutes. In a safety study conducted on rats, it was found that the current density needed to induce brain damage was at least 100 times higher than the current density used in tDCS trials [170]. The most common side effects reported include tingling and itching sensations predominantly under the anode, headache, moderate fatigue, and burning sensation [28, 88, 217].

Studies have demonstrated that tDCS can modify not only the regions underlying the anodal and cathodal stimulation [12, 29, 75, 183, 211], but also functional and effective connectivity [9, 44, 138, 185, 186, 208, 212, 245, 257, 280]. This may alter the emergent property of the network responsible for generating symptoms [177]. Moreover, such changes in functional and effective connectivity allow for the modulation of regions beyond the effects of tDCS under the stimulation electrodes [153].

The impact of tDCS on the physiology of various cortical regions has been examined, including the motor cortex [29], the visual cortex [12], the somatosensory cortex [75, 183], and the auditory cortex [290].

Studies have utilized single sessions of tDCS, tACS, and tRNS to shed light on the role of particular brain networks in the pathophysiology of tinnitus, while repeated sessions have been explored as a promising therapeutic strategy for tinnitus patients. In contrast to the extensive research on rTMS, tES has been the subject of fewer investigations, primarily examining the impact of single tDCS sessions. The auditory cortex, temporoparietal cortex, and dorsolateral prefrontal cortex have been the main targets for stimulation.

Initial studies found that anodal tDCS, but not cathodal tDCS, over the temporoparietal cortex resulted in temporary tinnitus reduction, but this effect could not be consistently replicated by subsequent studies [288]. As a result, left anodal tDCS and bilateral tRNS over the auditory cortex have emerged as the most promising single-session approaches. However, controlled studies testing the potential of repeated sessions for these treatments yielded disappointing results, as tDCS did not outperform sham stimulation [164] and bilateral tRNS did not surpass a control treatment [151].

In addition to studies that have focused on the temporal and temporoparietal regions, several investigations have targeted the DLPFC, primarily using a bifrontal electrode setup. Findings showed that a single session of bifrontal anode right/cathode left tDCS led to a reduction in tinnitus intensity or distress for roughly one-third of patients, while anode left/cathode right tDCS did not produce any effects [266]. In another study, the same group explored whether the efficacy of bifrontal tDCS could be enhanced by utilizing gamma connectivity from EEG measurements to inform the electrode polarity [70], but this was not the case. The therapeutic potential of the bifrontal tDCS protocol with anode right and cathode left was also evaluated in uncontrolled pilot studies using repeated sessions, revealing preliminary promising effects [99, 237].

According to earlier meta-analyses [278], tDCS has the potential to alleviate tinnitus distress but not loudness perception in tinnitus patients. These findings are consistent with an earlier meta-analytic study [242]. However, a different meta-analysis found no significant benefits [165]. In contrast, more recent meta-analyses with larger number of included studies do show a benefit on suffering and loudness perception [181], but loudness is only modulated by tDCS of the left temporal area [181]. Furthermore, a recent network meta-analysis discovered that among all non-invasive neuromodulatory approaches, combining cathodal tDCS over the left DLPFC with tRNS over the bilateral auditory cortex resulted in the greatest improvement in tinnitus severity and quality of life compared to controls [40].

Researchers have combined tDCS with various types of auditory stimulation [160, 235, 249]. However, according to these investigations, tDCS did not improve the therapeutic effects of hearing aids on tinnitus complaints [235] or the effects of notched music training, which is a customized form of auditory stimulation [249].

In conclusion, tDCS has meta-analytic evidence for improving suffering on tinnitus, like what has been observed in rTMS studies. The improvement in loudness is less evident, even though meta-analytic data suggests that one particular area, i.e., the left auditory cortex, may give some benefit. Given the multitude of factors that can influence tDCS outcomes, these findings are not unexpected, and further research is necessary to refine the more promising tDCS approaches.

High definition tDCS

Recently, High-Definition tDCS (HD-tDCS) has been introduced to enhance spatial accuracy by utilizing arrays of smaller electrodes instead of the two large pad electrodes used in conventional tDCS [50, 77, 116, 124, 236, 271] (Figure 2, right image). For High-Definition tDCS, studies using 4x1 ring configuration have demonstrated its tolerability in both healthy [23, 152, 200] and patient populations [80, 271], with intensities up to 2 mA for up to 20 minutes. No significant differences have been found between 2 and 3 mA, suggesting that the safety limits of HD-tDCS can be extended to 3 mA [221]. HD-tDCS has higher spatial resolution and permits more focal targeting [23]. Additionally, HD-tDCS allows for simultaneous multifocal stimulation, which can facilitate the development of network stimulation [227].

Studies on tinnitus have utilized HD-tDCS, and it has been observed that longer stimulation duration and higher intensities are linked with increased benefit [236]. However, a clinical distinction between the advantages of conventional tDCS and HD-tDCS has not been found [129].

Transcranial alternating current stimulation (tACS)

tACS is thought to affect brain physiology mainly through frequency-specific entrainment, where externally applied oscillating electric currents phase-align endogenous brain oscillations [251, 285, 291], and spike-timing dependent plasticity modulation [211, 273]. Alpha-tACS, for instance, has been shown to increase individual alpha frequency, [291], but the functional outcome varies with background activity [133]. Similarly, tACS at 140 Hz enhances motor excitability [198], while at 15 Hz it reduces it [293]. tACS has been demonstrated to affect perception [133], memory [180], motor function [26] and higher-order cognition [230]. However, studies suggest that tACS of the auditory cortex does not improve tinnitus [263, 269], neither in single, nor in multiple sessions [45].

Transcranial random noise stimulation (tRNS)

tRNS, a variant of tACS, utilizes alternating current with a white noise structure and Gaussian amplitude to stimulate the brain. The range of frequencies used is between 0.1–640 Hz, with low frequency tRNS defined as 0.1–100 Hz and high frequency tRNS limited to 100–640 Hz. tRNS has a higher perception threshold (1200 μA) than tDCS (400 μA) [11]. Interestingly, both tACS and high frequency tRNS have opposite effects at low and high amplitudes, with both being inhibitory at 0.4 mA and switching to excitatory modulation at 1 mA [195].

While the mechanisms of action for high frequency tRNS are still unknown, it is thought to increase excitability through stochastic resonance effect [250], preventing homeostasis of the system [89], or through desynchronization of pathological rhythms [206], but none of the above-mentioned mechanisms of action have been proven. tRNS modulates perception [224], memory [199], learning [25, 255] and other cognitive functions [35] possibly by NMDA-receptor independent but sodium-channel blocker and benzodiazepines sensitive plasticity [38].

In the clinical treatment of tinnitus, low and high frequency tRNS have been found to have beneficial effects on reducing loudness perception and distress [132, 263]. In fact, tRNS was the only effective transcranial electrical stimulation method for tinnitus suppression in a head-to-head comparison with tDCS and tACS [263]. Interestingly, both low and high frequency tRNS showed efficacy, but the combined low + high frequency tRNS was not effective in suppressing tinnitus [132]. A pilot study compared tACS and tRNS applied bilaterally over the temporal cortices and found that tRNS resulted in transient suppression of tinnitus, while tACS did not show any benefit. The study further showed that repeated sessions of tRNS were more effective than single sessions [45]. Yet, when performing daily sessions of HF-tRNS, in contrast to the more traditional 2-3 weekly sessions, no benefit was obtained [151], suggesting that too much stimulation may be counterproductive. Combining bifrontal DLPFC tDCS with auditory cortex tRNS was found to be more effective in suppressing tinnitus than using only auditory cortex tRNS [252]. This multisite tRNS approach has been supported by confirmatory studies [192, 193].

Transcutaneous electrical nerve stimulation (TENS)

Transcutaneous electrical nerve stimulation (TENS) is a technique that involves the use of a small, battery-operated device which delivers low-voltage electrical impulses to the body through electrodes placed on the skin. The most common sites chosen for TENS treatment include the C2 dermatome [168, 267] or the vagus nerve around the ear (also known as transcunateous vagal nerve stimulation, discussed in detailed below) [161, 254]. Applying TENS in the vicinity of the ear region is thought to heighten the activation of the dorsal cochlear nucleus along the somatosensory pathway, which may enhance the inhibitory function of the nucleus within the central nervous system, potentially alleviating tinnitus. Multiple systematic reviews demonstrate that TENS has a clinically meaningful effect on THI and VAS scores in the short term [25, 34, 123].

Transcutaneous vagal nerve stimulation (tVNS)

A non-invasive approach called transcutaneous vagus nerve stimulation (tVNS) has been recently developed. This approach takes advantage of the afferent branch of the vagus nerve that is located medial to the tragus at the entry of the acoustic meatus and innervates the external ear canal. Specialized devices have been developed to provide electrical stimulation for reliable stimulation of the auricular branch of the vagal nerve. The tVNS is typically performed on the left side to minimize potential effects on cardiac rhythm, similar to invasive VNS [147]. A pilot study investigated the feasibility and safety of tVNS in patients with tinnitus for six months. While the stimulation was well-tolerated, it did not lead to a significant improvement in tinnitus complaints [148]. However, treated patients did show a trend towards a reduction in depressive symptoms [148]. Further research found that tVNS can attenuate the sympathetic activation of tinnitus patients by investigating the effect of tVNS on heart rate activity in 173 tinnitus patients [286]. This supports the notion that tVNS induces stress attenuation.

Two small pilot studies have investigated the combination of tVNS with auditory stimulation, which showed promising effects on tinnitus [116, 239]. These findings are consistent with the animal data [87], which highlighted the importance of pairing vagus nerve stimulation with tones that exclude the tinnitus frequency for tinnitus reduction. In contrast, vagus nerve stimulation alone did not show any effect on tinnitus-related behaviour. However, tVNS has also been performed without auditory pairing and demonstrated some benefits [248], again raising questions about the relevance of paired sound presentation.

Neurofeedback

Neurofeedback, also known as brain computer interface training, biofeedback or neurotherapy, is a type of therapy that uses real-time monitoring of brainwave activity or functional connectivity to teach self-regulation of brain function [240]. This neuromodulation method draws upon principles and practices rooted in the psychology of learning, specifically operant conditioning. It revolves around acquiring the ability to regulate brain activity by modifying the neuronal processes linked to the circuits responsible for processing and interpreting symptoms. This approach utilizes brainwave detection to exert control over brain function through a feedback loop.

Different types of neurofeedback methods include: electrophysiologically based techniques encompassing electroencephalography (EEG), magnetoencephalography (MEG), and invasive electrocorticography (ECoG); and haemodynamic imaging methods including functional MRI (fMRI) and functional near-infrared spectroscopy (fNIRS) [240]. The literature on neurofeedback treatment for tinnitus is sparse, with 7 studies evaluating the EEG-based neurofeedback method and only one trial investigating the fMRI-based neurofeedback training.

EEG-based neurofeedback: Both sensor-based and source-based neurofeedback training protocols has been investigated. Most training protocols have targeted increasing alpha wave (8 Hz–12 Hz) activity [48, 117, 120] while inhibiting either delta wave (2 Hz–4 Hz) activity [48, 117, 142], beta activity [256] and gamma wave (48 Hz–54 Hz) activity [265]. Sensor-based protocols have used following electrode locations for targeting frontal lobe (left: FC1 and F3, and right: FC2 and F4) and temporal lobe (left: T3); whereas source-based training protocols haves used standardized low-resolution brain electromagnetic tomography (sLORETA) or ConSole to target the posterior cingulate cortex and the Lingual circumvolution. A recent systematic review [270] highlighted that irrespective of the training protocol evaluated, all studies demonstrated that neurofeedback training for tinnitus decreases symptom perception and related consequences. To determine the specific effects of neurofeedback training protocol, a recent trial [131] compared the superiority of the alpha/delta ratio training and the beta/theta ratio training to the non-neurofeedback minimal treatment intervention group. Their findings indicated that while alpha/delta ratio neurofeedback proved effective, it was not the sole factor contributing to positive outcomes on both the Tinnitus Handicap Inventory and Tinnitus Magnitude Index, in comparison to beta/theta ratio neurofeedback. However, the clinical outcomes data suggested that there was a discernible inclination toward the specificity of alpha/delta ratio neurofeedback. Consequently, the authors concluded that it might be premature to dismiss the potential efficacy of alpha/delta ratio neurofeedback in tinnitus treatment.

fmRI-based neurofeedback: Two initial pilot studies [86, 119] have been undertaken to better understand and refine the foundational mechanisms and configuration of fMRI neurofeedback for tinnitus treatment. A recent prospective randomized controlled trial [111] evaluated the effectiveness of real-time fMRI neurofeedback for reducing chronic tinnitus distress. The authors compared 15 weekly sessions of real-time fMRI neurofeedback training (downregulation of bilateral auditory cortex) to 10 weekly sessions of group cognitive behavioural therapy. Their findings demonstrated that the fMRI neurofeedback therapy led to a greater reduction in Tinnitus Handicap Inventory scores compared with cognitive behavioural therapy 6 months after the intervention; with improvements in depression and general functioning only evident in neurofeedback group.

References

1. Achard S, Salvador R, Whitcher B, Suckling J, and Bullmore E. A resilient, low-frequency, small-world human brain functional network with highly connected association cortical hubs. *Journal of Neuroscience.* **26**(1):63–72 (2006).
2. Adamchic I, Hauptmann C, and Tass PA. Changes of oscillatory activity in pitch processing network and related tinnitus relief induced by acoustic CR neuromodulation. *Frontiers in Systems Neuroscience.* **6**:18 (2012).
3. Adamchic I, Toth T, Hauptmann C, and Tass PA. Reversing pathologically increased EEG power by acoustic coordinated reset neuromodulation. *Human Brain Mapping.* **35**(5):2099–2118 (2014).
4. Adjamian P, Sereda M, Zobay O, Hall DA, and Palmer AR. Neuromagnetic indicators of tinnitus and tinnitus masking in patients with and without hearing loss. *Journal of the Association for Research in Otolaryngology.* **13**(5):715–731 (2012).
5. Akeju O, Loggia ML, Catana C, Pavone KJ, Vazquez, R, Rhee J, and Purdon PL. Disruption of thalamic functional connectivity is a neural correlate of dexmedetomidine-induced unconsciousness. *Elife.* **3**:e04499 (2014).
6. Albert R, Jeong H, and Barabasi AL. Error and attack tolerance of complex networks. *Nature.* **406**(6794):378–382 (2000).
7. Aldini G. *Essai theoretique et experimental sur le galvanisme.* Paris. (1804).
8. Almaraz AC, Dilli E, and Dodick DW. The effect of prophylactic medications on TMS for migraine aura. *Headache.* **50**(10):1630–1633 (2010).
9. Alon G, Roys SR, Gullapalli RP, and Greenspan JD. Non-invasive electrical stimulation of the brain (ESB) modifies the resting-state network connectivity of the primary motor cortex: A proof of concept fMRI study. *Brain Research.* **1403**:37–44 (2011).
10. Amaral LA, Diaz-Guilera A, Moreira AA, Goldberger AL, and Lipsitz LA. Emergence of complex dynamics in a simple model of signaling networks.

Proceedings of the National Academy of Sciences of the United States of America. **101**(44):15551–15555 (2004).

11. Ambrus GG, Paulus W, and Antal A. Cutaneous perception thresholds of electrical stimulation methods: Comparison of tDCS and tRNS. *Clinical Neurophysiology*. **121**(11):1908–1914 (2010).
12. Antal A, Kincses TZ, Nitsche MA, Bartfai O, and Paulus W. Excitability changes induced in the human primary visual cortex by transcranial direct current stimulation: Direct electrophysiological evidence. *Investigative Ophthalmology & Visual Science*. **45**(2):702–707 (2004).
13. Arndt, R. Die electricitat in der psychiatrie. *Arch Psychiat Nervenkrankh*. **2**:259–261 (1869).
14. Axelsson A, and Ringdahl A. Tinnitus — A study of its prevalence and characteristics. *British Journal of Audiology*. **23**(1):53–62 (1989).
15. Barabasi AL, Gulbahce N, and Loscalzo J. Network medicine: A network-based approach to human disease. *Nature Reviews Genetics*. **12**(1):56–68 (2011).
16. Barker AT, Jalinous R, and Freeston IL. Non-invasive magnetic stimulation of human motor cortex. *Lancet*. **1**(8437):1106–1107 (1985).
17. Bassett DS, Meyer-Lindenberg A, Achard S, Duke T, and Bullmore E. Adaptive reconfiguration of fractal small-world human brain functional networks. *Proceedings of the National Academy of Sciences of the United States of America*. **103**(51):19518–19523 (2006).
18. Best SRD, Pavel DG, and Haustrup N. Combination therapy with transcranial magnetic stimulation and ketamine for treatment-resistant depression: A long-term retrospective review of clinical use. *Heliyon*. **5**(8):e02187 (2019).
19. Bhatt JM, Bhattacharyya N, and Lin HW. Relationships between tinnitus and the prevalence of anxiety and depression. *Laryngoscope*. **127**(2):466–469 (2017).
20. Bikson M, Grossman P, Thomas C, Zannou AL, Jiang J, Adnan T, Mourdoukoutas AP, Kronberg G, Truong D, Boggio P, Brunoni AR, and Woods AJ. Safety of transcranial direct current stimulation: Evidence based update 2016. *Brain Stimulation*. **9**(5):641–661 (2016).
21. Bindman LJ, Lippold OC, and Redfearn JW. The action of brief polarizing currents on the cerebral cortex of the rat (1) during current flow and (2) in the production of long-lasting after-effects. *Journal of Physiology*. **172**:369–382 (1964).
22. Boly M, Faymonville ME, Peigneux P, Lambermont B, Damas F, Luxen A, Lamy M, Moonen G, Maquet P, and Laureys S. Cerebral processing of auditory and noxious stimuli in severely brain injured patients: Differences between VS and MCS. *Neuropsychological Rehabilitation*. **15**(3-4):283–289 (2005).
23. Borckardt JJ, Bikson M, Frohman H, Reeves ST, Datta A, Bansal V, Madan A, Barth K, and George MS. A pilot study of the tolerability and effects of

high-definition transcranial direct current stimulation (HD-tDCS) on pain perception. *Journal of Pain.* **13**(2):112–120 (2012).

24. Borst A, and Theunissen FE. Information theory and neural coding. *Nature Neuroscience.* **2**(11):947–957 (1999).
25. Bousema EJ, Koops EA, van Dijk P, and Dijkstra PU. Effects of physical interventions on subjective tinnitus, a systematic review and meta-analysis. *Brain Science.* **13**(2) (2023).
26. Brignani D, Manganotti P, Rossini PM, and Miniussi C. Modulation of cortical oscillatory activity during transcranial magnetic stimulation. *Human Brain Mapping.* **29**(5):603–612 (2008).
27. Brignani D, Ruzzoli M, Mauri P, and Miniussi C. Is transcranial alternating current stimulation effective in modulating brain oscillations? *PLoS One.* **8**(2):e56589 (2013).
28. Brunoni AR, AmaderaJ, Berbel B, Volz MS, Rizzerio BG, and Fregni F. A systematic review on reporting and assessment of adverse effects associated with transcranial direct current stimulation. *International Journal of Neuropsychopharmacology.* **14**(8):1133–1145 (2011).
29. Brunoni AR, Nitsche MA, Bolognini N, Bikson M, Wagner T, Merabet L, Edwards DJ, Valero-Cabre A, Rotenberg A, Pascual-Leone A, and Ferrucci R. Clinical research with transcranial direct current stimulation (tDCS): Challenges and future directions. *Brain Stimulation.* **5**:175–195.
30. Bullmore E, and Sporns O. Complex brain networks: Graph theoretical analysis of structural and functional systems. *Nature Reviews Neuroscience.* **10**(3):186–198 (2009).
31. Bullmore E, and Sporns O. The economy of brain network organization. *Nature Reviews Neuroscience.* **13**(5):336–349 (2012).
32. Buzsaki G. *Rhythms of the Brain.* Oxford: Oxford University Press. (2006).
33. Buzsaki G, and Mizuseki K. The log-dynamic brain: How skewed distributions affect network operations. *Nature Reviews Neuroscience.* **15**(4):264–278 (2014).
34. Byun YJ, Lee JA, Nguyen SA, Rizk HG, Meyer TA, and Lambert PR. Transcutaneous electrical nerve stimulation for treatment of tinnitus: A systematic review and meta-analysis. *Otology & Neurotology.* **41**(7):e767–e775 (2020).
35. Cappelletti M, Gessaroli E, Hithersay R, Mitolo M, Didino D, Kanai R, .., Walsh, V. Transfer of cognitive training across magnitude dimensions achieved with concurrent brain stimulation of the parietal lobe. *Journal of Neuroscience.* **33**(37):14899–14907 (2013).
36. Carmi L, Tendler A, Bystritsky A, Hollander E, Blumberger DM, Daskalakis J, Ward H, Lapidus K, Goodman W, Casuto L, and Feifel D. Efficacy and safety of deep transcranial magnetic stimulation for obsessive-compulsive

disorder: A prospective multicenter randomized double-blind placebo-controlled trial. *American Journal of Psychiatry*. **176**(11):931–938 (2019).
37. Catania KC. Tentacled snakes turn C-starts to their advantage and predict future prey behavior. *Proceedings of the National Academy of Sciences of the United States of America (PNAS)*. **106**(27):11183–11187 (2009).
38. Chaieb L, Antal A, and Paulus W. Transcranial random noise stimulation-induced plasticity is NMDA-receptor independent but sodium-channel blocker and benzodiazepines sensitive. *Frontiers Neuroscience*. **9**:125 (2015).
39. Cheeran B, Talelli P, Mori F, Koch G, Suppa A, Edwards M, Houlden H, Bhatia K, Greenwood R, and Rothwell JC. A common polymorphism in the brain-derived neurotrophic factor gene (BDNF) modulates human cortical plasticity and the response to rTMS. *The Journal of Physiology*. **586**(Pt 23):5717–5725 (2008).
40. Chen JJ, Zeng BS, Wu CN, Stubbs B, Carvalho AF, Brunoni AR, Su KP, Tu YK, Wu YC, Chen TY, and Lin PY. Association of central noninvasive brain stimulation interventions with efficacy and safety in tinnitus management: A Meta-analysis. *JAMA Otolaryngology–Head & Neck Surgery*. **146**(9):801–809 (2020).
41. Chen R, Classen J, Gerloff C, Celnik P, Wassermann EM, Hallett M, and Cohen LG. Depression of motor cortex excitability by low-frequency transcranial magnetic stimulation. *Neurology*. **48**(5):1398–1403 (1997).
42. Chervyakov A V, Chernyavsky AY, Sinitsyn DO, and Piradov MA. Possible mechanisms underlying the therapeutic effects of transcranial magnetic stimulation. *Frontiers in Human Neuroscience*. **9**:303 (2015).
43. Chib VS, Yun K, Takahashi H, and Shimojo S. Noninvasive remote activation of the ventral midbrain by transcranial direct current stimulation of prefrontal cortex. *Transl Psychiatry*. **3**:e268 (2013).
44. Claes L, Stamberger H, Van de Heyning P, De Ridder D, and Vanneste S. Auditory cortex tACS and tRNS for tinnitus: Single versus multiple sessions. *Neural Plasticity*. **2014**:436713 (2014).
45. Clarke LE, and Barres BA. Emerging roles of astrocytes in neural circuit development. *Nat Rev Neurosci*. *14*(5):311–321 (2013).
46. Crick F, and Koch C. Are we aware of neural activity in primary visual cortex? *Nature*. **375**(6527):121–123 (1995).
47. Crocetti A, Forti S, and Del Bo L. Neurofeedback for subjective tinnitus patients. *Auris Nasus Larynx*. **38**(6):735–738 (2011).
48. Crunelli V, Lőrincz ML, Connelly WM, David F, Hughes SW, Lambert RC, Leresche N, and Errington AC. Dual function of thalamic low-vigilance state oscillations: Rhythm-regulation and plasticity. *Nature Reviews Neuroscience*. **19**(2):107–118 (2018).

49. Datta A, Bansal V, Diaz J, Patel J, Reato D, and Bikson M. Gyri-precise head model of transcranial direct current stimulation: Improved spatial focality using a ring electrode versus conventional rectangular pad. *Brain Stimulation.* **2**(4):201–207 e201 (2009).
50. Dawes P, Newall J, Stockdale D, and Baguley DM. Natural history of tinnitus in adults: A cross-sectional and longitudinal analysis. *BMJ Open.* **10**(12):e041290 (2020).
51. Day BL, Rothwell JC, Thompson PD, Dick JP, Cowan JM, Berardelli A, and Marsden CD. Motor cortex stimulation in intact man. 2. Multiple descending volleys. *Brain.* **110** *(Pt 5)*:1191–1209 (1987).
52. De Ridder D. Should rTMS for tinnitus be performed left-sided, ipsilaterally or contralaterally, and is it a treatment or merely investigational? *European Journal of Neurology.* **17**(7):891–892 (2010).
53. De Ridder D, De Mulder G, Verstraeten E, Van der Kelen K, Sunaert S, Smits M, ..., Moller AR. Primary and secondary auditory cortex stimulation for intractable tinnitus. *ORL: Journal for Oto-Rhino-Laryngology and its Related Specialties.* **68**(1):48–54 (2006).
54. De Ridder D, De Mulder G, Walsh V, Muggleton N, Sunaert S, and Moller A. Magnetic and electrical stimulation of the auditory cortex for intractable tinnitus. Case report. *Journal of Neurosurgery.* **100**(3):560–564 (2004).
55. De Ridder D, Elgoyhen AB, Romo R, and Langguth B. Phantom percepts: tinnitus and pain as persisting aversive memory networks. *Proceedings of the National Academy of Sciences of the United States of America.* **108**(20):8075–8080 (2011).
56. De Ridder D, Joos K, and Vanneste S. The enigma of the tinnitus-free dream state in a Bayesian world. *Neural Plasticity.* **2014**:612147 (2014).
57. De Ridder D, Joos K, and Vanneste S. Anterior cingulate implants for tinnitus: Report of 2 cases. *Journal of Neurosurgery.* **124**(4):893–901 (2016).
58. De Ridder D, Perera S, and Vanneste S. State of the Art: Novel applications for cortical stimulation. *Neuromodulation.* **20**(3):206–214 (2017).
59. De Ridder D, Schlee W, Vanneste S, Londero A, Weisz N, Kleinjung T, ..., Langguth B. Tinnitus and tinnitus disorder: Theoretical and operational definitions (an international multidisciplinary proposal). *Progress in Brain Research.* **260**:1–25 (2021).
60. De Ridder D, Song, JJ, and Vanneste, S. (2012). Frontal cortex TMS for tinnitus. *Brain Stimul.* **6**(3), 355–362.
61. De Ridder D, van der Loo E, Van der Kelen K, Menovsky T, van de Heyning P, and Moller A. Theta, alpha and beta burst transcranial magnetic stimulation: Brain modulation in tinnitus. *International Journal of Medical Sciences.* **4**(5):237–241 (2007).

62. De Ridder D, van der Loo E, Vanneste S, Gais S, Plazier M, Kovacs S, ..., van de Heyning P. Theta-gamma dysrhythmia and auditory phantom perception. *Journal of Neurosurgery*. **114**(4):912–921 (2011).
63. De Ridder D, and Vanneste, S. EEG Driven tDCS Versus Bifrontal tDCS for Tinnitus. *Frontiers in Psychiatry*. **3**:84 (2012).
64. De Ridder D, and Vanneste S. Targeting the parahippocampal area by auditory cortex stimulation in tinnitus. *Brain Stimulation*. 7(5):709–717 (2014).
65. De Ridder D, and Vanneste S. Multitarget surgical neuromodulation: Combined C2 and auditory cortex implantation for tinnitus. *Neuroscience Letter*. **591**:202–206 (2015).
66. De Ridder D, Vanneste S, and Freeman W. The Bayesian brain: Phantom percepts resolve sensory uncertainty. *Neuroscience & Biobehavioral Reviews*. **44C**:4–15.
67. De Ridder D, Vanneste S, Kovacs S, Sunaert S, Menovsky T, van de Heyning P, and Moller A. Transcranial magnetic stimulation and extradural electrodes implanted on secondary auditory cortex for tinnitus suppression. *Journal of Neurosurgery*. **114**(4):903–911 (2011).
68. De Ridder D, Vanneste S, Langguth B, and Llinas R. Thalamocortical dysrhythmia: A theoretical update in tinnitus. *Frontiers in Neurology*. **6**:124 (2015).
69. De Ridder D, Vanneste S, Plazier M, Menovsky T, van de Heyning P, Kovacs S, and Sunaert S. Dorsolateral prefrontal cortex transcranial magnetic stimulation and electrode implant for intractable tinnitus. *World Neurosurgery*. **77**(5-6):778–784 (2012).
70. De Ridder D, Vanneste S, Weisz N, Londero A, Schlee W, Elgoyhen AB, and Langguth B. An integrative model of auditory phantom perception: Tinnitus as a unified percept of interacting separable subnetworks. *Neuroscience & Biobehavioral Reviews*. **44**:16–32 (2014).
71. De Ridder D, Verstraeten E, Van der Kelen K, De Mulder G, Sunaert S, Verlooy J, ..., Moller A. Transcranial magnetic stimulation for tinnitus: Influence of tinnitus duration on stimulation parameter choice and maximal tinnitus suppression. *Otolaryngology–Head and Neck Surgery*. **26**(4):616–619 (2005).
72. Demertzi A, Soddu A, and Laureys S. Consciousness supporting networks. *Current Opinion in Neurobiology*. **23**(2):239–244 (2013).
73. Deppe M, Abdelnaim M, Hebel T, Kreuzer P, Poeppl T, Langguth B, and Schecklmann M. Concomitant lorazepam use and antidepressive efficacy of repetitive transcranial magnetic stimulation in a naturalistic setting. *European Archives of Psychiatry and Clinical Neuroscience*. **271**:61–67 (2020).
74. Dieckhofer A, Waberski TD, Nitsche M, Paulus W, Buchner H, and Gobbele R. Transcranial direct current stimulation applied over the somatosensory

cortex — Differential effect on low and high frequency SEPs. *Clinical Neurophysiology*. **117**(10):2221–2227 (2006).

75. Dietrich S. Earliest historic reference of 'tinnitus' is controversial. *Journal of Laryngology & Otology*. **118**(7):487–488 (2004).
76. Dmochowski JP, Datta A, Bikson, M, Su Y, and Parra LC. (2011). Optimized multi-electrode stimulation increases focality and intensity at target. *Journal of Neural Engineering*. **8**(4):046011.
77. Dobie RA. A review of randomized clinical trials in tinnitus. *Laryngoscope*. **109**(8):1202–1211 (1999).
78. Dong C, Chen C, Wang T, Gao C, Wang Y, Guan X, and Dong X. Low-frequency repetitive transcranial magnetic stimulation for the treatment of chronic tinnitus: A systematic review and meta-analysis of randomized controlled trials. *Biomedical Research International*. **2020**:3141278.
79. Donnell A, TDN, Lawrence M, Gupta V, Zieba T, Truong DQ, ..., DaSilva AF. High-definition and non-invasive brain modulation of pain and motor dysfunction in chronic TMD. *Brain Simulation*. (2015).
80. Dymond AM, Coger RW, and Serafetinides EA. Intracerebral current levels in man during electrosleep therapy. *Biological Psychiatry*. **10**(1):101–104 (1975).
81. Eguiluz VM, Chialvo DR, Cecchi GA, Baliki M, and Apkarian AV. Scale-free brain functional networks. *Physical Review Letters*. **94**(1):018102 (2005).
82. Eichhammer P, Langguth B, Marienhagen J, Kleinjung T, and Hajak G. Neuronavigated repetitive transcranial magnetic stimulation in patients with tinnitus: A short case series. *Biological Psychiatry*. **54**(8):862–865 (2003).
83. Elgoyhen AB, Langguth B, De Ridder D, and Vanneste S. Tinnitus: Perspectives from human neuroimaging. *Nature Review Neuroscience*. **16**(10):632–642 (2015).
84. Emmert K, Kopel R, Koush Y, Maire R, Senn P, Van De Ville D, and Haller S. Continuous vs. intermittent neurofeedback to regulate auditory cortex activity of tinnitus patients using real-time fMRI — A pilot study. *NeuroImage Clinical*. **14**:97–104 (2017).
85. Engineer ND, Riley JR, Seale JD, Vrana WA, Shetake JA, Sudanagunta SP, ..., Kilgard MP. Reversing pathological neural activity using targeted plasticity. *Nature*. **470**(7332):101–104 (2011).
86. Fertonani A, Ferrari C, and Miniussi C. What do you feel if I apply transcranial electric stimulation? Safety, sensations and secondary induced effects. *Clinical Neurophysiology*. (2015).
87. Fertonani A, Pirulli C, and Miniussi C. Random noise stimulation improves neuroplasticity in perceptual learning. *Journal of Neuroscience*. **31**(43): 15416–15423 (2011).

88. Fields H. State-dependent opioid control of pain. *Nature Reviews Neuroscience*. **5**(7):565–575 (2004).
89. Fitzgerald PB, Daskalakis ZJ, and Hoy KE. Benzodiazepine use and response to repetitive transcranial magnetic stimulation in major depressive disorder. *Brain Simulation*. **13**(3):694–695 (2020).
90. Fleming MK, Sorinola IO, Newham DJ, Roberts-Lewis SF, and Bergmann JH. The effect of coil type and navigation on the reliability of transcranial magnetic stimulation. *IEEE Transactions on Neural Systems and Rehabilitation Engineering (TNSRE)*. **20**(5):617–625 (2012).
91. Fodstad H, and Hariz M. Electricity in the treatment of nervous system disease. *Acta Neurochirurgica Supplement*. **97**(Pt 1):11–19 (2007).
92. Fornito A, and Bullmore ET. Connectomics: A new paradigm for understanding brain disease. *European Neuropsychopharmacology*. **25**(5):733–748 (2014).
93. Fornito A, Zalesky A, and Breakspear M. The connectomics of brain disorders. *Nature Review Neuroscience*. **16**(3):159–172 (2015).
94. Fox MD, Buckner RL, Liu H, Chakravarty MM, Lozano AM, and Pascual-Leone A. Resting-state networks link invasive and noninvasive brain stimulation across diverse psychiatric and neurological diseases. *Proceedings of the National Academy of Sciences of the United States of America*. 111(41):E4367–E4375 (2014).
95. Fox MD, Halko MA, Eldaief MC, and Pascual-Leone A. Measuring and manipulating brain connectivity with resting state functional connectivity magnetic resonance imaging (fcMRI) and transcranial magnetic stimulation (TMS). *Neuroimage*. **62**(4):2232–2243 (2012).
96. Frank E, Schecklmann M, Landgrebe M, Burger J, Kreuzer P, Poeppl TB, ..., Langguth B. Treatment of chronic tinnitus with repeated sessions of prefrontal transcranial direct current stimulation: Outcomes from an open-label pilot study. *Journal of Neurology*. **259**(2):327–333 (2012).
97. Freeman WJ. The wave packet: An action potential for the 21st century. *Journal of Integrative Neuroscience*. **2**(1):3–30 (2003).
98. Freeman WJ. A field-theoretic approach to understanding scale-free neocortical dynamics. *Biol Cybern*. **92**(6):350–359 (2005).
99. Freeman WJ, Kozma R, and Werbos PJ. Biocomplexity: Adaptive behavior in complex stochastic dynamical systems. *Biosystems*. **59**(2):109–123 (2001).
100. Freeman WJ, and Rogers LJ. Fine temporal resolution of analytic phase reveals episodic synchronization by state transitions in gamma EEGs. *Journal of Neurophysiology*. **87**(2):937–945 (2002).
101. Fregni F, Nitsche MA, Loo CK, Brunoni AR, Marangolo P, Leite J, ..., Bikson M. regulatory considerations for the clinical and research use of

transcranial direct current stimulation (tDCS): Review and recommendations from an expert panel. *Clinical Research and Regulatory Affairs.* **32**(1):22–35 (2015).

102. Friston KJ. Functional and effective connectivity: A review. *Brain Connectivity.* **1**(1):13–36 (2011).
103. Fuchs E, and Flugge G. Adult neuroplasticity: More than 40 years of research. *Neural Plasticity.* **2014**:541870 (2014).
104. Fuggetta G, Pavone EF, Fiaschi A, and Manganotti P. Acute modulation of cortical oscillatory activities during short trains of high-frequency repetitive transcranial magnetic stimulation of the human motor cortex: A combined EEG and TMS study. *Human Brain Mapping.* **29**(1):1–13 (2008).
105. Fujiyama H, Hyde J, Hinder MR, Kim SJ, McCormack GH, Vickers JC, and Summers JJ. Delayed plastic responses to anodal tDCS in older adults. *Frontiers in Aging Neuroscience.* **6**:115 (2014).
106. Galvani L. *De Viribus Electricitatis in Motu Muscolari Animalium.* Bologna. (1791).
107. Galvani L, Spallanzani L, and Aldini G. *Memorie Sulla Elettricità Animale.* 1797.
108. Gninenko N, Trznadel S, Daskalou D, Gramatica L, Vanoy J, Voruz F, ..., Haller S. Functional MRI neurofeedback outperforms cognitive behavioral therapy for reducing tinnitus distress: A prospective randomized clinical trial. *Radiology.* **310**(2):e231143 (2024).
109. Goerigk S, Cretaz E, Sampaio-Junior B, Vieira ELM, Gattaz W, Klein I, ..., Brunoni AR. Effects of tDCS on neuroplasticity and inflammatory biomarkers in bipolar depression: Results from a sham-controlled study. *Progress in Neuropsychopharmacology and Biological Psychiatry.* **105**:110119 (2021).
110. Goetz SM, Kozyrkov IC, Luber B, Lisanby SH, Murphy DLK, Grill WM, and Peterchev AV. Accuracy of robotic coil positioning during transcranial magnetic stimulation. *Journal of Neural Engineering.* **16**(5):054003 (2019).
111. Grefkes C, Nowak DA, Wang LE, Dafotakis M, Eickhoff SB, and Fink GR. Modulating cortical connectivity in stroke patients by rTMS assessed with fMRI and dynamic causal modeling. *Neuroimage.* **50**(1):233–242 (2010).
112. Gribkoff VK, and Kaczmarek LK. The need for new approaches in CNS drug discovery: Why drugs have failed, and what can be done to improve outcomes. *Neuropharmacology.* **120**:11–19 (2017).
113. Guleyupoglu B, Schestatsky P, Edwards D, Fregni F, and Bikson M. Classification of methods in transcranial electrical stimulation (tES) and evolving strategy from historical approaches to contemporary innovations. *Journal of Neuroscience Methods.* **219**(2):297–311.

114. Güntensperger D, Thüring C, Kleinjung T, Neff P, and Meyer M. Investigating the efficacy of an individualized Alpha/Delta neurofeedback protocol in the treatment of chronic tinnitus. *Neural Plasticity*. **2019**:3540898 (2019).
115. Hagmann P, Cammoun L, Gigandet X, Meuli R, Honey CJ, Wedeen VJ, and Sporns O. Mapping the structural core of human cerebral cortex. *PLoS Biology*. **6**(7):e159 (2008).
116. Haller S, Birbaumer N, and Veit R. Real-time fMRI feedback training may improve chronic tinnitus. *European Radiology*. **20**(3):696–703 (2010).
117. Hartmann T, Lorenz I, Müller N, Langguth B, and Weisz N. The effects of neurofeedback on oscillatory processes related to tinnitus. *Brain Topography*. **27**(1):149–157 (2014).
118. Hayward G, Mehta MA, Harmer C, Spinks TJ, Grasby PM, and Goodwin GM. Exploring the physiological effects of double-cone coil TMS over the medial frontal cortex on the anterior cingulate cortex: An H2(15)O PET study. *European Journal of Neuroscience*. **25**(7):2224–2233 (2007).
119. Hebel T, Abdelnaim MA, Deppe M, Kreuzer PM, Mohonko A, Poeppl TB, ..., Schecklmann M. Antidepressant effect of repetitive transcranial magnetic stimulation is not impaired by intake of lithium or antiepileptic drugs. *European Archives of Psychiatry and Clinical Neuroscience*. **271**(7):1245–1253 (2021).
120. Heiland LD, Owen JM, III, Nguyen SA, Labadie RF, Lambert PR, and Meyer TA. Neuromodulation for treatment of tinnitus: A systematic review and meta-analysis. *Otolaryngology Head and Neck Surgery*. **170**(5):1234–1245 (2024).
121. Heimrath K, Breitling C, Krauel K, Heinze HJ, and Zaehle T. Modulation of pre-attentive spectro-temporal feature processing in the human auditory system by HD-tDCS. *European Journal of Neuroscience*. **41**(12):1580–1586 (2015).
122. Herpich F, Melnick M, Huxlin K, Tadin D, Agosta S, and Battelli L. Transcranial random noise stimulation enhances visual learning in healthy adults. *Journal of Vision*. **15**(12):40 (2015).
123. Holland J. *Complexity*. Oxford: Oxford University Press, (2014).
124. Hunter AM, Minzenberg MJ, Cook IA, Krantz DE, Levitt JG, Rotstein NM, ..., Leuchter AF. Concomitant medication use and clinical outcome of repetitive transcranial magnetic stimulation (rTMS) treatment of major depressive disorder. *Brain Behavior*. **9**(5):e01275 (2019).
125. Iyer MB, Mattu U, Grafman J, Lomarev M, Sato S, and Wassermann EM. Safety and cognitive effect of frontal DC brain polarization in healthy individuals. *Neurology*. **64**(5):872–875 (2005).
126. Jacquemin L, Shekhawat GS, Van de Heyning P, Mertens G, Fransen E, Van Rompaey V, ..., Gilles A. Effects of electrical stimulation in tinnitus

patients: Conventional versus high-definition tDCS. *Neurorehabil Neural Repair*. **32**(8), 714–723 (2018).

127. Jastreboff PJ. Phantom auditory perception (tinnitus): Mechanisms of generation and perception. *Neuroscience Research*. **8**(4):221–254.
128. Jensen M, Alanis JCG, Hüttenrauch E, Winther-Jensen M, Chavanon ML, Andersson G, and Weise C. Does it matter what is trained? A randomized controlled trial evaluating the specificity of alpha/delta ratio neurofeedback in reducing tinnitus symptoms. *Brain Communication*. **5**(4), fcad185 (2023).
129. Joos K, De Ridder D, and Vanneste S. The differential effect of low-versus high-frequency random noise stimulation in the treatment of tinnitus. *Exp Brain Res*. **233**(5):1433–1440 (2015).
130. Kanai R, Chaieb L, Antal A, Walsh V, and Paulus W. Frequency-dependent electrical stimulation of the visual cortex. *Current Biology*. **18**(23):1839–1843 (2008).
131. Karuza EA, Thompson-Schill SL, and Bassett DS. Local patterns to global architectures: Influences of network topology on human learning. *Trends in Cognitive Sciences*. **20**(8):629–640 (2016).
132. Keck M E, Welt T, Muller MB, Erhardt A, Ohl F, Toschi N, ..., Sillaber I. Repetitive transcranial magnetic stimulation increases the release of dopamine in the mesolimbic and mesostriatal system. *Neuropharmacology*. **43**(1):101–109 (2002).
133. Keck ME, Welt T, Post A, Muller MB, Toschi N, Wigger A, ..., Engelmann M. Neuroendocrine and behavioral effects of repetitive transcranial magnetic stimulation in a psychopathological animal model are suggestive of antidepressant-like effects. *Neuropsychopharmacology*. **24**(4):337–349 (2001).
134. Keeser D, Meindl T, Bor J, Palm U, Pogarell O, Mulert C, ..., Padberg F. Prefrontal transcranial direct current stimulation changes connectivity of resting-state networks during fMRI. *J Neurosci*. **31**(43):15284–15293 (2011).
135. Keeser D, Padberg F, Reisinger E, Pogarell O, Kirsch V, Palm U, ..., Mulert C. Prefrontal direct current stimulation modulates resting EEG and event-related potentials in healthy subjects: A standardized low resolution tomography (sLORETA) study. *Neuroimage*. **55**(2):644–657 (2011).
136. Kelloway P. The part played by the electrical fish in the early history of bioelectricity and electrotherapy. *Bulletin of the History of Medicine*. **20**:112–137 (1946).
137. Keshner MS. 1/f Noise. *Proceedings of the IEEE*. **70**(3):212–218.
138. Kesselheim AS, Hwang TJ, and Franklin JM. Two decades of new drug development for central nervous system disorders. *Nature Reviews Drug Discovery*. **14**(12):815–816 (2015).
139. Khoramzadeh S, Saki N, Davoodi I, Nosratabadi M, and Yadollahpour A. Investigating the therapeutic efficacy of neurofeedback treatment on the

severity of symptoms and quality of life in patients with tinnitus. *International Journal of Mental Health and Addiction.* **14**(6):982–992 (2016).

140. Kimbrell TA, Dunn RT, George MS, Danielson AL, Willis MW, Repella JD, ..., Wassermann EM. Left prefrontal-repetitive transcranial magnetic stimulation (rTMS) and regional cerebral glucose metabolism in normal volunteers. *Psychiatry Research.* **115**(3):101–113 (2002).
141. Kimbrell TA, Little JT, Dunn RT, Frye MA, Greenberg BD, Wassermann EM, ..., Post RM. Frequency dependence of antidepressant response to left prefrontal repetitive transcranial magnetic stimulation (rTMS) as a function of baseline cerebral glucose metabolism. *Biological Psychiatry.* **46**(12):1603–1613 (1999).
142. Kleinjung T, Steffens T, Sand P, Murthum T, Hajak G, Strutz J, ..., Eichhammer P. Which tinnitus patients benefit from transcranial magnetic stimulation? *Otolaryngology–Head and Neck Surgery.* **137**(4):589–595 (2007).
143. Kong J, Loggia ML, Zyloney C, Tu P, Laviolette P, and Gollub RL. Exploring the brain in pain: Activations, deactivations and their relation. *Pain.* **148**(2):257–267 (2010).
144. Kreuzer PM, Landgrebe M, Husser O, Resch M, Schecklmann M, Geisreiter F, ..., Langguth B. Transcutaneous vagus nerve stimulation: Retrospective assessment of cardiac safety in a pilot study. *Front Psychiatry.* **3**:70 (2012).
145. Kreuzer PM, Landgrebe M, Resch M, Husser O, Schecklmann M, Geisreiter F, ..., Langguth B. Feasibility, safety and efficacy of transcutaneous vagus nerve stimulation in chronic tinnitus: An open pilot study. *Brain Simulation.* **7**(5):740–747 (2014).
146. Kreuzer PM, Landgrebe M, Schecklmann M, Poeppl TB, Vielsmeier V, Hajak G, ..., Langguth B. Can temporal repetitive transcranial magnetic stimulation be enhanced by targeting affective components of tinnitus with frontal rTMS? A randomized controlled pilot trial. *Frontiers in Systems Neuroscience.* **5**:88 (2011).
147. Kreuzer PM, Poeppl TB, Rupprecht R, Vielsmeier V, Lehner A, Langguth B, and Schecklmann M. Individualized repetitive transcranial magnetic stimulation treatment in chronic tinnitus? *Frontiers in Neurology.* **8**:126 (2017).
148. Kreuzer PM, Poeppl TB, Rupprecht R, Vielsmeier V, Lehner A, Langguth B, and Schecklmann M. Daily high-frequency transcranial random noise stimulation of bilateral temporal cortex in chronic tinnitus — A pilot study. *Scientific Reports.* **9**(1):12274 (2019).
149. Kuo HI, Bikson M, Datta A, Minhas P, Paulus W, Kuo MF, and Nitsche MA. Comparing cortical plasticity induced by conventional and high-definition 4 x 1 ring tDCS: A neurophysiological study, *Brain Simulation.* **6**(4):644–648 (2013).

150. Lang N, Siebner HR, Ward NS, Lee L, Nitsche MA, Paulus W, ..., Frackowiak RS. How does transcranial DC stimulation of the primary motor cortex alter regional neuronal activity in the human brain? *European Journal of Neuroscience*. **22**(2):495–504 (2005).
151. Langguth B, Eichhammer P, Kreutzer A, Maenner P, Marienhagen J, Kleinjung T, ..., Hajak G. The impact of auditory cortex activity on characterizing and treating patients with chronic tinnitus–first results from a PET study. *Acta Otolaryngol*. Suppl(556):84–88 (2006).
152. Langguth B, Kleinjung T, Schlee W, Vanneste S, and De Ridder D. Tinnitus guidelines and their evidence base. *Journal of Clinical Medicine*. **12**(9) (2023).
153. Langguth B, Kreuzer PM, Kleinjung T, and De Ridder D. Tinnitus: Causes and clinical management. *The Lancet Neurology*. **12**(9):920–930 (2013).
154. Laureys S, Faymonville ME, Degueldre C, Fiore GD, Damas P, Lambermont B, ..., Maquet P. Auditory processing in the vegetative state. *Brain*. **123**(Pt 8):1589–1601 (2000).
155. Laureys S, Faymonville ME, Peigneux P, Damas P, Lambermont B, Del Fiore G, ..., Maquet P. Cortical processing of noxious somatosensory stimuli in the persistent vegetative state. *Neuroimage*. **17**(2):732–741 (2002).
156. Leaver AM, Renier L, Chevillet MA, Morgan S, Kim HJ, and Rauschecker JP. Dysregulation of limbic and auditory networks in tinnitus. *Neuron*. **69**(1): 33–43 (2011).
157. Lee HY, Choi MS, Chang DS, and Cho CS. Combined bifrontal transcranial direct current stimulation and tailor-made notched music training in chronic tinnitus. *Journal of Audiology & Otology*. **21**(1):22–27 (2017).
158. Lee SK, Chung H, Chung JH, Yeo SG, Park MS, Byun JY. Effectiveness of transcutaneous electrical stimulation for chronic tinnitus. *Acta Otolaryngologica*. **134**(2):159–167 (2014).
159. Lefaucheur JP, Aleman A, Baeken C, Benninger DH, Brunelin J, Di Lazzaro V, ..., Ziemann U. Evidence-based guidelines on the therapeutic use of repetitive transcranial magnetic stimulation (rTMS): An update (2014–2018). *Clinical Neurophysiology*. **131**(2):474–528 (2020).
160. Lefaucheur JP, Andre-Obadia N, Antal A, Ayache SS, Baeken C, Benninger DH, ..., Garcia-Larrea L. Evidence-based guidelines on the therapeutic use of repetitive transcranial magnetic stimulation (rTMS). *Clinical Neurophysiology*. **125**(11):2150–2206 (2014).
161. Lefaucheur JP, Antal A, Ayache SS, Benninger DH, Brunelin J, Cogiamanian F, ..., Paulus W. Evidence-based guidelines on the therapeutic use of transcranial direct current stimulation (tDCS). *Clinical Neurophysiology*. **128**(1):56–92 (2017).

162. Lefebvre-Demers M, Doyon N, and Fecteau S. Non-invasive neuromodulation for tinnitus: A meta-analysis and modeling studies. *Brain Simulation.* **14**(1):113–128 (2020).
163. Lehtimaki J, Hyvarinen P, Ylikoski M, Bergholm M, Makela JP, Aarnisalo A, ..., Ylikoski J. Transcutaneous vagus nerve stimulation in tinnitus: A pilot study. *Acta Otolaryngologica.* **133**(4):378–382 (2012).
164. Lerner AJ, Wassermann EM, and Tamir DI. Seizures from transcranial magnetic stimulation 2012–2016: Results of a survey of active laboratories and clinics. *Clinical Neurophysiology.* **130**(8):1409–1416 (2019).
165. Li L, Shi H, and Wang M. A pilot randomized controlled trial of transcutaneous electrical nerve stimulation for patients with acute tinnitus. *Medicine.* **98**(1):e13793 (2019).
166. Liang Z, Yang H, Cheng G, Huang L, Zhang T, and Jia H. Repetitive transcranial magnetic stimulation on chronic tinnitus: A systematic review and meta-analysis. *BMC Psychiatry.* **20**(1):547 (2020).
167. Liebetanz D, Koch R, Mayenfels S, Konig F, Paulus W, and Nitsche MA. Safety limits of cathodal transcranial direct current stimulation in rats. *Clinical Neurophysiology.* **120**(6):1161–1167 (2009).
168. Lippold OC, and Redfearn JW. Mental changes resulting from the passage of small direct currents through the human brain. *The British Journal of Psychiatry.* **110**:768–772.
169. Llinas R, Urbano FJ, Leznik E, Ramirez RR, and van Marle HJ. Rhythmic and dysrhythmic thalamocortical dynamics: GABA systems and the edge effect. *Trends Neuroscience.* **28**(6):325–333 (2005).
170. Llinas RR, Ribary U, Jeanmonod D, Kronberg E, and Mitra PP. Thalamocortical dysrhythmia: A neurological and neuropsychiatric syndrome characterized by magnetoencephalography. *Proceedings of the National Academy of Sciences of the United States of America.* **96**:15222–15227 (1999).
171. Llinas RR, and Steriade M. Bursting of thalamic neurons and states of vigilance. *Journal of Neurophysiology.* **95**:3297–3308 (2006).
172. Londero A, Bonfils P, and Lefaucheur JP. Transcranial magnetic stimulation and subjective tinnitus. A review of the literature, 2014–2016. *European Annals of Otorhinolaryngology, Head and Neck Diseases.* **135**:51–58 (2018).
173. Lorenz I, Muller N, Schlee W, Hartmann T, and Weisz N. Loss of alpha power is related to increased gamma synchronization — A marker of reduced inhibition in tinnitus? *Neuroscience Letters.* **453**:225–228 (2009).
174. Luft CD, Pereda E, Banissy MJ, and Bhattacharya J. Best of both worlds: promise of combining brain stimulation and brain connectome. *Frontiers in Systems Neuroscience.* **8**:132 (2014).
175. Ma X, Wang F, Zhang C, Shen W, and Yang S. Self-directed neurofeedback treatment for subjective tinnitus patients evaluated by multimodal functional imaging. *Neural Plasticity.* **2022**:5114721 (2022).

176. Malaguti A, Rossini D, Lucca A, Magri L, Lorenzi C, Pirovano A, ..., Zanardi R. Role of COMT, 5-HT(1A), and SERT genetic polymorphisms on antidepressant response to transcranial magnetic stimulation. *Depression and Anxiety*. **28**:568–573 (2011).
177. Marshall L, Helgadottir H, Molle M, and Born J. Boosting slow oscillations during sleep potentiates memory. *Nature*. **444**:610–613 (2006).
178. Martins ML, Souza DDS, Cavalcante M, Barboza HN, de Medeiros JF, Dos Santos Andrade SMM, ..., da Rosa MRD. Effect of transcranial direct current stimulation for tinnitus treatment: A systematic review and meta-analysis. *Neurophysiologie Clinique*. **52**:1–16 (2022).
179. Massimini M, Ferrarelli F, Huber R, Esser SK, Singh H, and Tononi G. Breakdown of cortical effective connectivity during sleep. *Science*. **309**:2228–2232 (2005).
180. Matsunaga K, Nitsche MA, Tsuji S, and Rothwell JC. Effect of transcranial DC sensorimotor cortex stimulation on somatosensory evoked potentials in humans. *Clinical Neurophysiology*. **115**:456–460 (2004).
181. Medeiros LF, de Souza IC, Vidor LP, de Souza A, Deitos A, Volz MS., ..., Torres IL. Neurobiological effects of transcranial direct current stimulation: A review. *Frontiers in Psychiatry*. **3**:110 (2012).
182. Meinzer M, Antonenko D, Lindenberg R, Hetzer S, Ulm L, Avirame K., .., Floel A. Electrical brain stimulation improves cognitive performance by modulating functional connectivity and task-specific activation. *Journal of Neuroscience*. **32**:1859–1866 (2012).
183. Meinzer M, Lindenberg R, Antonenko D, Flaisch T, and Floel A. Anodal transcranial direct current stimulation temporarily reverses age-associated cognitive decline and functional brain activity changes. *Journal of Neuroscience*. **33**:12470–12478 (2013).
184. Melloni L, Molina C, Pena M, Torres D, Singer W, and Rodriguez E. Synchronization of neural activity across cortical areas correlates with conscious perception. *Journal of Neuroscience*. **27**:2858–2865 (2007).
185. Mishima T, Nagai T, Yahagi K, Akther S, Oe Y, Monai H, ..., Hirase H. Transcranial direct current stimulation (tDCS) induces adrenergic receptor-dependent microglial morphological changes in mice. *eNeuro*. **6**:1–5 (2019).
186. Mohan A, Alexandra SJ, Johnson CV, De Ridder D, and Vanneste, S. Effect of distress on transient network dynamics and topological equilibrium in phantom sound perception. *Progress in Neuro-Psychopharmacology and Biological Psychiatry*. **84**:79–92 (2018).
187. Mohan A, De Ridder D, Idiculla R, CDS, and Vanneste S. Distress-dependent temporal variability of regions encoding domain-specific and domain-general behavioral manifestations of phantom percepts. *European Journal of Neuroscience*. **48**:1743–1764 (2018).

188. Mohan A, De Ridder D, and Vanneste S. Robustness and dynamicity of functional networks in phantom sound. *NeuroImage*. **146**:171–187 (2017).
189. Mohsen S, Mahmoudian S, Talebian S, and Pourbakht, A. Prefrontal and auditory tRNS in sequence for treating chronic tinnitus: A modified multi-site protocol. *Brain Stimulation*. **11**:1177–1179 (2018).
190. Mohsen S, Mahmoudian S, Talebian S, and Pourbakht A. Multisite transcranial random noise stimulation (tRNS) modulates the distress network activity and oscillatory powers in subjects with chronic tinnitus. *Journal of Clinical Neuroscience*. **67**:178–184 (2019).
191. Moliadze V, Antal A, and Paulus, W. Boosting brain excitability by transcranial high frequency stimulation in the ripple range. *Journal of Physiology*. **588**:4891–4904 (2010).
192. Moliadze V, Atalay D, Antal A, and Paulus W. Close to threshold transcranial electrical stimulation preferentially activates inhibitory networks before switching to excitation with higher intensities. *Brain Stimulation*. **5**:505–511 (2012).
193. Momi D, Ozdemir RA, Tadayon E, Boucher P, Shafi MM, Pascual-Leone A, and Santarnecchi E. Network-level macroscale structural connectivity predicts propagation of transcranial magnetic stimulation. *NeuroImage*. **229**:117698 (2020).
194. Monai H, and Hirase H. Astrocytes as a target of transcranial direct current stimulation (tDCS) to treat depression. *Neuroscience Research*. **126**:15–21 (2018).
195. Moreno-Duarte I, Gebodh N, Schestatsky P, Guleyupoglu B, Reato D, Bikson M, and Fregni F. Transcranial electrical stimulation: Transcranial direct current stimulation (tDCS), transcranial alternating current stimulation (tACS), transcranial pulsed current stimulation (tPCS), and transcranial random noise stimulation (tRNS). In Kadosh RC, (Ed.) *The Stimulated Brain: Cognitive Enhancement Using Non-Invasive Brain Stimulation*. London: Academic Press (2014).
196. Mulquiney PG, Hoy KE, Daskalakis ZJ, and Fitzgerald PB. Improving working memory: Exploring the effect of transcranial random noise stimulation and transcranial direct current stimulation on the dorsolateral prefrontal cortex. *Clinical Neurophysiology*. **122**:2384–2389 (2011).
197. Nikolin S, Loo CK, Bai S, Dokos S, and Martin DM. Focalised stimulation using high definition transcranial direct current stimulation (HD-tDCS) to investigate declarative verbal learning and memory functioning. *NeuroImage*. **117**:11–19 (2015).
198. Nitsche MA, Cohen LG, Wassermann EM, Priori A, Lang N, Antal A., .., Pascual-Leone A. Transcranial direct current stimulation: State of the art 2008. *Brain Stimulation*. **1**:206–223 (2008).

199. Nitsche MA, Kuo M, Paulus W, and Antal A. Transcranial direct current stimulation: Protocols and physiological mechanisms of action. In Knotkova H, Rasche D, (Ed.) *Textbook of Neuromodulation: Principles, Methods and Clinical Application*. New York: Springer (2015).
200. Nitsche MA, and Paulus W. Sustained excitability elevations induced by transcranial DC motor cortex stimulation in humans. *Neurology*. **57**:1899–1901 (2001).
201. Nowogrodzki A. The world's strongest MRI machines are pushing human imaging to new limits. *Nature*. **563**:24–26 (2018).
202. Pascual-Leone A, Grafman J, and Hallett M. Modulation of cortical motor output maps during development of implicit and explicit knowledge. *Science*. **263**:1287–1289 (1994).
203. Paulus W. Transcranial electrical stimulation (tES — tDCS; tRNS, tACS) methods. *Neuropsychological Rehabilitation*. **21**:602–617 (2011).
204. Paus T, Sipila PK, and Strafella AP. Synchronization of neuronal activity in the human primary motor cortex by transcranial magnetic stimulation: An EEG study. *Journal of Neurophysiology*. **86**:1983–1990 (2001).
205. Pena-Gomez C, Sala-Lonch R, Junque C, Clemente IC, Vidal D, Bargallo N, *et al.* Modulation of large-scale brain networks by transcranial direct current stimulation evidenced by resting-state functional MRI. *Brain Stimulation*. **5**:252–263 (2012).
206. Plewnia C, Bartels M, and Gerloff C. Transient suppression of tinnitus by transcranial magnetic stimulation. *Annals of Neurology*. **53**:263–266 (2003).
207. Plewnia C, Reimold M, Najib A, Brehm B, Reischl G, Plontke SK, and Gerloff C. Dose-dependent attenuation of auditory phantom perception (tinnitus) by PET-guided repetitive transcranial magnetic stimulation. *Human Brain Mapping*. **28**:238–246 (2007).
208. Polania R, Nitsche MA, Korman C, Batsikadze G, and Paulus W. The importance of timing in segregated theta phase-coupling for cognitive performance. *Current Biology*. **22**:1314–1318 (2012).
209. Polania R, Nitsche MA, and Paulus W. Modulating functional connectivity patterns and topological functional organization of the human brain with transcranial direct current stimulation. *Human Brain Mapping*. **32**:1236–1249 (2011).
210. Polania R, Paulus W, and Nitsche MA. Modulating cortico-striatal and thalamo-cortical functional connectivity with transcranial direct current stimulation. *Human Brain Mapping*. **33**:2499–2508 (2012).
211. Poreisz C, Boros K, Antal A, and Paulus W. Safety aspects of transcranial direct current stimulation concerning healthy subjects and patients. *Brain Research Bulletin*. **72**:208–214 (2007).

212. Pushpakom S, Iorio F, Eyers PA, Escott KJ, Hopper S, Wells A, *et al.* Drug repurposing: Progress, challenges and recommendations. *Nature Reviews Drug Discovery.* **18**:41–58 (2019).
213. Rabenstein M, Unverricht-Yeboah M, Keuters MH, Pikhovych A, Hucklenbroich J, Vay SU, *et al.* Transcranial current stimulation alters the expression of immune-mediating genes. *Frontiers in Cellular Neuroscience.* **13**:461 (2019).
214. Radman T, Ramos RL, Brumberg JC, and Bikson M. Role of cortical cell type and morphology in subthreshold and suprathreshold uniform electric field stimulation *in vitro. Brain Stimulation.* **2**:215–228 (2009).
215. Ramirez RR, Kopell BH, Butson CR, Gaggl W, Friedland DR, and Baillet S. Neuromagnetic source imaging of abnormal spontaneous activity in tinnitus patient modulated by electrical cortical stimulation. *Conference Proceedings IEEE Engineering in Medicine and Biology Society.* **1**:1940–1944 (2009).
216. Ranck JB Jr. Which elements are excited in electrical stimulation of mammalian central nervous system: A review. *Brain Research.* **98**:417–440 (1975).
217. Rauschecker JP, Leaver AM, and Muhlau M. Tuning out the noise: Limbic-auditory interactions in tinnitus. *Neuron.* **66**:819–826 (2010).
218. Reckow J, Rahman-Filipiak A, Garcia S, Schlaefflin S, Calhoun O, DaSilva AF, *et al.* Tolerability and blinding of 4×1 high-definition transcranial direct current stimulation (HD-tDCS) at two and three milliamps. *Brain Stimulation.* **11**:991–997 (2018).
219. Redfearn JW, Lippold OC, and Costain R. A preliminary account of the clinical effects of polarizing the brain in certain psychiatric disorders. *British Journal of Psychiatry*, **110**:773–785 (1964).
220. Ridding MC, and Rothwell JC. Is there a future for therapeutic use of transcranial magnetic stimulation? *Nature Reviews Neuroscience.* **8**:559–567 (2007).
221. Romanska A, Rezlescu C, Susilo T, Duchaine B, and Banissy MJ. High-frequency transcranial random noise stimulation enhances perception of facial identity. *Cerebral Cortex.* **25**:4334–4340 (2015).
222. Rossini PM, Burke D, Chen R, Cohen LG, Daskalakis Z, Di Iorio R, *et al.* Non-invasive electrical and magnetic stimulation of the brain, spinal cord, roots and peripheral nerves: Basic principles and procedures for routine clinical and research application. *Clinical Neurophysiology.* **126**:1071–1107 (2015).
223. Rueger MA, Keuters MH, Walberer M, Braun R, Klein R, Sparing R, *et al.* Multi-session transcranial direct current stimulation (tDCS) elicits inflammatory and regenerative processes in the rat brain. *PLoS One.* **7**:e43776 (2012).

224. Ruffini G, Fox MD, Ripolles O, Miranda PC, and Pascual-Leone A. Optimization of multifocal transcranial current stimulation for weighted cortical pattern targeting from realistic modeling of electric fields. *NeuroImage*. **89**:216–225 (2014).
225. Ruohonen J, and Karhu J. tDCS possibly stimulates glial cells. *Clinical Neurophysiology*. **123**:2006–2009 (2012).
226. Rush S, and Driscoll DA. Current distribution in the brain from surface electrodes. *Anesthesia & Analgesia*. **47**:717–723 (1968).
227. Santarnecchi E, Polizzotto NR, Godone M, Giovannelli F, Feurra M, Matzen L, *et al.* Frequency-dependent enhancement of fluid intelligence induced by transcranial oscillatory potentials. *Current Biology*. **23**:1449–1453 (2013).
228. Schlee W, Hartmann T, Langguth B, and Weisz N. Abnormal resting-state cortical coupling in chronic tinnitus. *BMC Neuroscience*. **10**:11 (2009).
229. Schlee W, Mueller N, Hartmann T, Keil J, Lorenz I, and Weisz N. Mapping cortical hubs in tinnitus. *BMC Biology*. **7**:80 (2009).
230. Schoisswohl S, Agrawal K, Simoes J, Neff P, Schlee W, Langguth B, and Schecklmann M. RTMS parameters in tinnitus trials: A systematic review. *Scientific Reports*. **9**:12190 (2019).
231. Schonfeldt-Lecuona C, Thielscher A, Freudenmann RW, Kron M, Spitzer M, and Herwig U. Accuracy of stereotaxic positioning of transcranial magnetic stimulation. *Brain Topography*. **17**:253–259 (2005).
232. Shekhawat GS, Searchfield GD, and Stinear CM. Randomized trial of transcranial direct current stimulation and hearing aids for tinnitus management. *Neurorehabilitation and Neural Repair*. **28**:410–419 (2014).
233. Shekhawat GS, Sundram F, Bikson M, Truong D, De Ridder D, Stinear CM, *et al.* Intensity, duration, and location of high-definition transcranial direct current stimulation for tinnitus relief. *Neurorehabilitation and Neural Repair*. **29**:591–603 (2015).
234. Shekhawat GS, and Vanneste S. Optimization of transcranial direct current stimulation of dorsolateral prefrontal cortex for tinnitus: A non-linear dose-response effect. *Scientific Reports*. **8**:8311 (2018).
235. Shen K, Hutchison RM, Bezgin G, Everling S, and McIntosh AR. Network structure shapes spontaneous functional connectivity dynamics. *Journal of Neuroscience*. **35**:5579–5588 (2015).
236. Shim HJ, Kwak MY, An YH, Kim DH, Kim YJ, and Kim HJ. Feasibility and safety of transcutaneous vagus nerve stimulation paired with notched music therapy for the treatment of chronic tinnitus. *Journal of Audiology & Otology*. **19**(3):159–167 (2015).
237. Sitaram R, Ros T, Stoeckel L, Haller S, Scharnowski F, Lewis-Peacock J, ..., Sulzer J. Closed-loop brain training: The science of neurofeedback. *Nature Reviews Neuroscience*. **18**(2):86–100 (2017).

238. Soleimani R, Jalali MM, Hasandokht, T. Therapeutic impact of repetitive transcranial magnetic stimulation (rTMS) on tinnitus: A systematic review and meta-analysis. *European Archives of Oto-Rhino-Laryngology*. **273**(7): 1663–1675 (2016).
239. Song JJ, Vanneste S, Van de Heyning P, De Ridder D. Transcranial direct current stimulation in tinnitus patients: A systemic review and meta-analysis. *Scientific World Journal*. **2012**:427941 (2012).
240. Speer AM, Kimbrell TA, Wassermann EM, JDR, Willis MW, Herscovitch P, Post RM. Opposite effects of high and low frequency rTMS on regional brain activity in depressed patients. *Biological Psychiatry*. **48**(12):1133–1141 (2000).
241. Sporns O, Chialvo DR, Kaiser M, Hilgetag CC. Organization, development and function of complex brain networks. *Trends in Cognitive Sciences*. **8**(9):418–425 (2004).
242. Stagg CJ, Lin RL, Mezue M, Segerdahl A, Kong Y, Xie J, Tracey I. Widespread modulation of cerebral perfusion induced during and after transcranial direct current stimulation applied to the left dorsolateral prefrontal cortex. *Journal of Neuroscience*. **33**(28):11425–11431 (2013).
243. Steinmetz H, Furst G, Freund HJ. Variation of perisylvian and calcarine anatomic landmarks within stereotaxic proportional coordinates. *American Journal of Neuroradiology*. **11**(6):1123–1130 (1990).
244. Stramaccia DF, Penolazzi B, Sartori G, Braga M, Mondini S, and Galfano G. Assessing the effects of tDCS over a delayed response inhibition task by targeting the right inferior frontal gyrus and right dorsolateral prefrontal cortex. *Experimental Brain Research*. **233**(8):2283–2290 (2015).
245. Suk WC, Kim SJ, Chang DS, and Lee HY. Characteristics of stimulus intensity in transcutaneous vagus nerve stimulation for chronic tinnitus. *Journal of International Advanced Otology*. **14**(2):267–272 (2018).
246. Teismann H, Wollbrink A, Okamoto H, Schlaug G, Rudack C, and Pantev C. Combining transcranial direct current stimulation and tailor-made notched music training to decrease tinnitus-related distress — A pilot study. *PLoS One*. **9**(2):e89904 (2014).
247. Terney D, Chaieb L, Moliadze V, Antal A, and Paulus W. Increasing human brain excitability by transcranial high-frequency random noise stimulation. *Journal of Neuroscience*. **28**(52):14147–14155 (2008).
248. Thut G, Schyns PG, and Gross J. Entrainment of perceptually relevant brain oscillations by non-invasive rhythmic stimulation of the human brain. *Frontiers in Psychology*. **2**:170 (2011).
249. To WT, Ost J, Hart J, Jr, De Ridder D, and Vanneste S. The added value of auditory cortex transcranial random noise stimulation (tRNS) after bifrontal transcranial direct current stimulation (tDCS) for tinnitus. *Journal of Neural Transmission*. **124**(1):79–88 (2017).
250. Tsanov M. Septo-hippocampal signal processing: Breaking the code. *Prog Brain Res*. **219**:103–120 (2015).

251. Tutar B, Atar S, Berkiten G, Üstün O, Kumral TL, and Uyar Y. The effect of transcutaneous electrical nerve stimulation (TENS) on chronic subjective tinnitus. *American Journal of Otolaryngology*. **41**(1):102326 (2020).
252. Tyler S, Conto F, and Battelli L. Rapid effect of high-frequency tRNS over the parietal lobe during a temporal perceptual learning task. *Journal of Vision*. **15**(12):393 (2015).
253. Vanneste S, Alsalman O, and De Ridder D. Top-down and bottom-up regulated auditory phantom perception. *Journal of Neuroscience*. **39**(2): 364–378 (2019).
254. Vanneste S, and De Ridder D. Bifrontal transcranial direct current stimulation modulates tinnitus intensity and tinnitus-distress-related brain activity. *European Journal of Neuroscience*. **34**(4):605–614 (2011).
255. Vanneste S, and De Ridder D. Differences between a single session and repeated sessions of 1 Hz TMS by double-cone coil prefrontal stimulation for the improvement of tinnitus. *Brain Stimul*. **1642**:170–179 (2012a).
256. Vanneste S, and De Ridder D. The involvement of the left ventrolateral prefrontal cortex in tinnitus: A TMS study. *Experimental Brain Research*. **221**(3):345–350 (2012b).
257. Vanneste S, and De Ridder D. Deafferentation-based pathophysiological differences in phantom sound: Tinnitus with and without hearing loss. *Neuroimage*. **129**:80–94 (2015).
258. Vanneste S, Faber M, Langguth B, and De Ridder D. The neural correlates of cognitive dysfunction in phantom sounds. *Brain Research*. **6**(2):155–159 (2016).
259. Vanneste S, Focquaert F, Van de Heyning P, and De Ridder D. Different resting state brain activity and functional connectivity in patients who respond and not respond to bifrontal tDCS for tinnitus suppression. *Experimental Brain Research*. **210**(2):217–227 (2011).
260. Vanneste S, Fregni F, and De Ridder D. Head-to-Head comparison of transcranial random noise stimulation, transcranial AC stimulation, and transcranial DC stimulation for tinnitus. *Frontiers in Psychiatry*. **4**:158 (2013).
261. Vanneste S, Joos K, Langguth B, To WT, and De Ridder D. Neuronal correlates of maladaptive coping: An EEG-Study in tinnitus patients. *PLoS One*. **9**(2):e88253 (2014).
262. Vanneste S, Joos K, Ost J, and De Ridder D. Influencing connectivity and cross-frequency coupling by real-time source localized neurofeedback of the posterior cingulate cortex reduces tinnitus related distress. *Neurobiology Stress*. **8**:211–224 (2018).
263. Vanneste S, Plazier M, Ost J, van der Loo E, Van de Heyning P, and De Ridder D. Bilateral dorsolateral prefrontal cortex modulation for tinnitus by transcranial direct current stimulation: A preliminary clinical study. *Experimental Brain Research*. **202**(4):779–785 (2010).

264. Vanneste S, Plazier M, Van de Heyning P, and De Ridder D. Transcutaneous electrical nerve stimulation (TENS) of upper cervical nerve (C2) for the treatment of somatic tinnitus. *Experimental Brain Research.* **204**(2):283–287 (2010).

265. Vanneste S, van der Loo E, Plazier M, and De Ridder D. Parietal double-cone coil stimulation in tinnitus. *Experimental Brain Research.* **221**(3): 337–343 (2012).

266. Vanneste S, Walsh V, Van De Heyning P, and De Ridder D. Comparing immediate transient tinnitus suppression using tACS and tDCS: A placebo-controlled study. *Experimental Brain Research.* **226**(1):25–31 (2013).

267. Varela Barrenechea F. Efficacy of neurofeedback as a treatment for people with subjective tinnitus in reducing the symptom and related consequences: A systematic review from 2010 to 2020. *Acta Otorrinolaringológica Española (English Edition)*. **74**(4):253–262 (2023).

268. Villamar MF, Volz MS, Bikson M, Datta A, Dasilva AF, and Fregni F. Technique and considerations in the use of 4×1 ring high-definition transcranial direct current stimulation (HD-tDCS). *Journal of Visualized Experiments*. (77): e50309 (2013).

269. Voroslakos M, Takeuchi Y, Brinyiczki K, Zombori T, Oliva A, Fernandez-Ruiz A., ..., Berenyi A. Direct effects of transcranial electric stimulation on brain circuits in rats and humans. *Nature Communications.* **9**(1):483 (2018).

270. Vossen A, Gross J, and Thut G. Alpha power increase after transcranial alternating current stimulation at alpha frequency (alpha-tACS) reflects plastic changes rather than entrainment. *Brain Simulation.* **8**(3):499–508 (2015).

271. Wagner T, Fregni F, Fecteau S, Grodzinsky A, Zahn M, and Pascual-Leone A. Transcranial direct current stimulation: A computer-based human model study. *Neuroimage.* **35**(3):1113–1124 (2007).

272. Walsh V, and Rushworth M. A primer of magnetic stimulation as a tool for neuropsychology. *Neuropsychologia.* **37**(2):125–135 (1999).

273. Wang H, Wang X, and Scheich H. LTD and LTP induced by transcranial magnetic stimulation in auditory cortex. *Neuro Report.* 7(2):521–525 (1996).

274. Wang H, Wang X, Wetzel W, and Scheich H. Rapid-rate transcranial magnetic stimulation in auditory cortex induces LTP and LTD and impairs discrimination learning of frequency-modulated tones. *Electro-encephalography and Clinical Neurophysiology Supplements.* **51**:361–367 (1999).

275. Wang TC, Tyler RS, Chang TY, Chen JC, Lin CD, Chung HK, and Tsou YA. Effect of transcranial direct current stimulation in patients with tinnitus: A meta-analysis and systematic review. *Annals of Otology, Rhinology & Laryngology.* **127**(2):79–88 (2018).

276. Wang Y, Zhang JN, Hu W, Li JJ, Zhou JX, Zhang JP, ..., Li M. The characteristics of cognitive impairment in subjective chronic tinnitus. *Brain and Behavior*. **8**(3):e00918 (2018).
277. Weber MJ, Messing SB, Rao H, Detre JA, and Thompson-Schill SL. Prefrontal transcranial direct current stimulation alters activation and connectivity in cortical and subcortical reward systems: A tDCS-fMRI study. *Human Brain Mapping*. **35**(8):3673–3686 (2014).
278. Weisz N, Hartmann T, Dohrmann K, Schlee W, and Norena A. High-frequency tinnitus without hearing loss does not mean absence of deafferentation. *Hearing Research*. **222**(1-2):108–114 (2006).
279. Weisz N, Muller S, Schlee W, Dohrmann K, Hartmann T, and Elbert T. The neural code of auditory phantom perception. *Journal of Neuroscience*. **27**(6):1479–1484 (2007).
280. Whittington MA, Traub RD, and Adams NE. A future for neuronal oscillation research. *Brain Neuroscience Advances*. **2**:2398212818794827 (2018).
281. Witkowski M, Garcia-Cossio E, Chander BS, Braun C, Birbaumer N, Robinson SE, and Soekadar SR. Mapping entrained brain oscillations during transcranial alternating current stimulation (tACS). *Neuroimage*. **140**:89–98 (2015).
282. Yang T, Zhang J, Wang B, Zhang W, Xu M, Yang S, and Liu H. Electrical stimulation to treat tinnitus: A meta-analysis and systemic review of randomized controlled trials. *Therapeutic Advances in Chronic Disease*. **12**: 20406223211041069 (2021).
283. Ylikoski J, Lehtimaki J, Pirvola U, Makitie A, Aarnisalo A, Hyvarinen P, and Ylikoski M. Non-invasive vagus nerve stimulation reduces sympathetic preponderance in patients with tinnitus. *Acta Otolaryngologica*. **137**(4): 426–431 (2017).
284. Yokley BH, Hartman M, and Slusher BS. Role of academic drug discovery in the quest for new CNS therapeutics. *ACS Chemical Neuroscience*. **8**(3):429–431 (2017).
285. Yuan T, Yadollahpour A, Salgado-Ramirez J, Robles-Camarillo D, and Ortega-Palacios R. Transcranial direct current stimulation for the treatment of tinnitus: A review of clinical trials and mechanisms of action. *BMC Neuroscience*. **19**(1):66 (2018).
286. Yue L, Xiao-lin H, and Tao S. The effects of chronic repetitive transcranial magnetic stimulation on glutamate and gamma-aminobutyric acid in rat brain. *Brain Research*. **1260**:94–99 (2009).
287. Zaehle T, Beretta M, Jancke L, Herrmann CS, and Sandmann P. Excitability changes induced in the human auditory cortex by transcranial direct current stimulation: Direct electrophysiological evidence. *Experimental Brain Research*. **215**(2):135–140 (2011).

288. Zaehle T, Rach S, and Herrmann CS. Transcranial alternating current stimulation enhances individual alpha activity in human EEG. *PLoS One*. **5**(11): e13766 (2010).
289. Zaghi S, Acar M, Hultgren B, Boggio PS, and Fregni F. Noninvasive brain stimulation with low-intensity electrical currents: Putative mechanisms of action for direct and alternating current stimulation. *The Neuroscientist*. **16**(3):285–307 (2010).
290. Zaghi S, de Freitas Rezende L, de Oliveira LM, El-Nazer R, Menning S, Tadini L, and Fregni F. Inhibition of motor cortex excitability with 15Hz transcranial alternating current stimulation (tACS). *Neuroscience Letter*. **479**(3):211–214 (2010).
291. Zago S, Ferrucci R, Fregni F, and Priori A. Bartholow, sciamanna, alberti: Pioneers in the electrical stimulation of the exposed human cerebral cortex. *The Neuroscientist*. **14**(5):521–528 (2008).
292. Zeman F, Koller M, Figueiredo R, Aazevedo A, Rates M, Coelho C., ..., Landgrebe M. Tinnitus handicap inventory for evaluating treatment effects: Which changes are clinically relevant? *Otolaryngology–Head and Neck Surgery*. **145**(2):282–287 (2011).
293. Zhang K, Guo L, Zhang J, Rui G, An G, Zhou Y, ..., Ding G. tDCS Accelerates the rehabilitation of MCAO-Induced motor function deficits via neurogenesis modulated by the notch1 signaling pathway. *Neurorehabilitation and Neural Repair*. **34**(7):640–651 (2020).

https://doi.org/10.1142/9789819807345_0008

Chapter 8

Invasive Electric Stimulation for Tinnitus Treatment

Cynthia Tsang[*], Fan-Gang Zeng[*,†], and Mehdi Abouzari[*,‡]

[*]*Department of Otolaryngology-Head and Neck Surgery, University of California, Irvine, CA, USA*

[†]*Hearing and Speech (HESP) Laboratory, Center for Hearing Research, University of California, Irvine, CA, USA*

[‡]*mabouzar@hs.uci.edu*

Abstract

Given the chronicity, prevalence, and morbidity of tinnitus, major efforts have been applied to intervene with symptoms. Studies indicating central nervous system involvement with tinnitus have led to more invasive therapies over time, including electrical modalities such as vagus nerve, transtympanic, auditory cortex, and deep brain stimulation. Dozens of studies have explored the optimal parameters, target regions, and conditions from which a greater understanding of efficacy can be discussed here. For each intervention, we describe the route of administration, history of use, and significant studies that articulate and inform modern understandings and clinical applications. Literature on auditory cortex stimulation included various reports and case studies, the most thorough of which found no significant benefits of the treatment. Understanding of transtympanic and tinnitus distress network stimulation remains ongoing, though preliminary studies show a variety of responses that require further exploration. In a cohort of 30, vagus nerve stimulation has shown positive results on symptoms. As of today, deep brain stimulation for tinnitus in a cohort of 122 affected and matched controls showed a strong reduction in symptoms, leading to phase 1 studies and beyond. The wide berth of available invasive electrical techniques for tinnitus serves as an insight into the exciting frontier of available therapies, bound to alleviate the daily burden of this disease in millions.

Introduction

Tinnitus is a phantom perception of sound without a corresponding identifiable external auditory source. It affects 10–15% of the population and can significantly interfere with the quality of life by increasing annoyance, anxiety, depression, sleep disturbance, and even the risk for suicide [1]. Despite intense research, a consensus on the pathophysiology of tinnitus remains elusive. Popular modern models propose tinnitus as the brain's response to auditory deafferentation, with increased central gain or central noise or both [2].

In the past, researchers believed the etiology of tinnitus could be found in the inner ear or the cochlear nerve. This school of thought was challenged when Jackson showed that sectioning of the eighth cranial nerve did not terminate tinnitus [3]. A new wave of research began to implicate the central nervous system as a key modulator of tinnitus generation. In modern understanding of tinnitus, the involvement of the central nervous system is undeniable, as evidenced by the fact that in many metanalyses, most guidelines recommend cognitive behavioral therapy as an important treatment for chronic disabling tinnitus. One theory that persists today finds the source of tinnitus in peripheral lesions of cochlear hair cells resulting in a loss of central auditory input. The best evidence for this theory is in studies that demonstrate the efficacy of cochlear implantation in patients with single-sided hearing loss with tinnitus [4, 5]. Early thoughts about the central nervous system's involvement in tinnitus heavily suspected that erroneous auditory cortex organization or reorganization was at the nucleus of symptom presentation. However, animal models and functional mapping of the tinnitus brain since then have led to the understanding that while the auditory cortex does play a significant role in tinnitus symptoms, non-auditory brain regions (particularly the brain's limbic regions, responsible for experiencing sensation) are involved in symptoms present in chronic tinnitus patients [6–8]. The involvement of a non-auditory network in patients suffering from chronic tinnitus was first introduced in 1990 by Jastreboff, who proposed that tinnitus was caused by peripheral noise-induced hearing loss and succeeded by central auditory pathway changes [9]. Extensive research has aimed to map out the associated networks that operate parallel to the auditory cortex. These areas involve multiple brain regions and have been coined as the perception, salience, distress, and memory networks [10].

Integral to the modern understanding of tinnitus pathophysiology is the concept of thalamocortical dysrhythmia (TCD), which considers how alterations of neuro-oscillatory patterns can manifest as a wide host of neurological conditions [11]. Both sensory deafferentation (bottom-up) and noise-canceling deficit (top-down) theories have been proposed to alter thalamocortical auditory signal transmission resulting in TCD (Figure 1). In support of the theory of sensory deafferentation, magnetoencephalography and electroencephalography studies have shown a slowing down of resting-state alpha activity associated with an increase

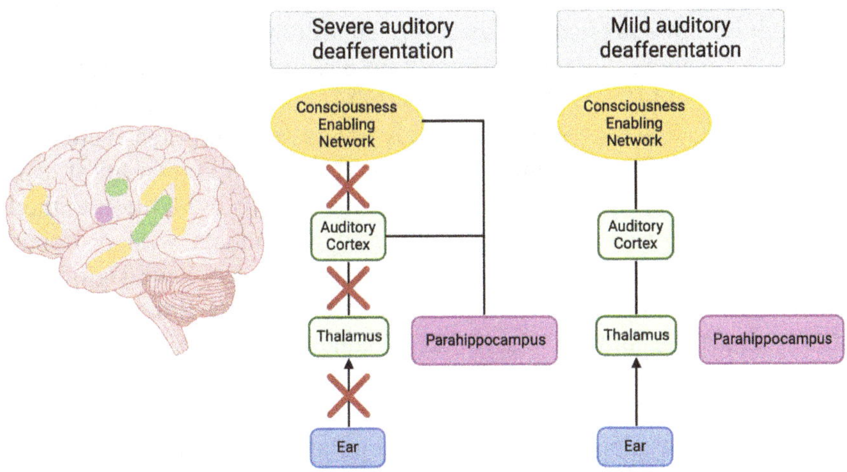

Figure 1. Auditory deafferentation pathway and mechanism.

in gamma band activity in the auditory cortex [12–15]. Gamma activity was shown to be active during instances of auditory stimuli, and *y*-band activity in the sensory cortex correlated with tinnitus perception and intensity [12–15]. Gamma-y band coordinated activity was disrupted when electrical stimulation of the auditory cortex suppresses tinnitus in patients [16]. Increased gamma signaling in tinnitus was shown to correlate with theta activity, resulting in cross-frequency coupling. Theta firing increases neural network recruitment and synchrony that could present as a carrier wave to integrate tinnitus-related gamma activity with regions associated with consciousness [17]. The global workspace model suggests that focal brain activity does not reach consciousness, except if the activity becomes coupled to a conscious-enabling network. In the sensory deafferentation model, the proposed mechanism connecting tinnitus activity with conscious perception is via theta burst firing. The other prominent theory is called the noise cancellation deficit theory of tinnitus which proposes a dysfunction in the brain's ability to cancel out or suppress internal neural noise [18]. This system primarily operates through alpha frequencies, so malfunction of this mechanism would suggest theta and gamma wave predomination [17]. Research on noise cancellation deficit remains scarce, thus it is usually proposed as an auxiliary process operating in concert with other tinnitus theories [19]. Regardless, these two effects are at the forefront of our tinnitus understanding and reflect the culmination of decades of research into this persistently disabling condition.

Unfortunately, there are currently no devices or drugs approved by the US Food and Drug Administration (FDA) for tinnitus treatment. The lack of efficacious interventions likely stems from tinnitus' heterogeneous presentation and complex etiology. It has been known for at least 40 years that cochlear implants can suppress tinnitus in addition to their success in restoring functional hearing to individuals with profound deafness [20]. The issue is that most tinnitus sufferers have significant or even normal hearing and thus are not suitable for cochlear implantation because the implant would likely damage the hearing. Therefore, other types of non-invasive and invasive electric stimulation have been developed for tinnitus treatment [21, 22]. Recent developments have attempted transtympanic stimulation, stimulation of the auditory nerve, auditory cortex, tinnitus distress network, vagus nerve, and deep brain. Table 1 summarizes the main characteristics of these different types of stimulation.

Table 1. Summary of invasive electrical stimulation.

Invasive Electrical Modality	Route of Delivery	Efficacy
Transtympanic stimulation	Application of electrode to middle ear (no implantation)	Studies show various conclusions on efficacy, with some patients reporting complete tinnitus suppression and others describing a range of no to moderate improvement or return to baseline after a period of time. Variable results can be contributed to the diversity of study design and the plethora of available stimulation sites, devices, and amplitude–frequency conditions.
Round window or promontory stimulation	Application of electrode to round window or promontory, sometimes involving implantation	Studies have shown improvement in at least one aspect of tinnitus measures with some stimulation patterns totally abolishing tinnitus in most subjects. Particularly beneficial for high-frequency tinnitus.
Auditory cortex stimulation	Implantation of electrode along the auditory circuit	In a randomized, double-blinded, crossover study, investigators found no significant improvement in chronic, severe tinnitus with cortical stimulation. A handful of case reports and studies report significant or no effect with this treatment modality.
Tinnitus distress network stimulation	Implantation of bilateral electrodes over dorsal anterior cingulate cortex	Few studies have been done to extrapolate true conclusions on efficacy. Of the two participants, one experienced benefits and the other did not from this intervention.
Vagus nerve stimulation (VNS)	Implantation of device capable of electrical stimulation in chest wall	Of the 30 participants, the experimental cohort receiving VNS showed significant clinical improvement and better THI scores ($p = 0.0012$).
Deep brain stimulation	Implantation of electrode in brain	A large study of 122 tinnitus patients and matched controls found a significant reduction in THI scores, leading to a phase I trial indicating promising results.

Vagus Nerve Stimulation

Vagus nerve stimulation (VNS) is an FDA-approved procedure to treat epilepsy and depression. It involves surgically implanting a pacemaker-like device in the chest wall with wires that extend upwards into the patient's neck to stimulate the vagal nerve within the carotid sheath. The implanted device delivers electrical signals to the vagus nerve at encoded intervals, which can be adjusted by the participants or researchers. Electrical stimulation has been shown to modulate neural activity; thus, VNS can aim to alter individuals' vagal nerve and parasympathetic brain activity. After a neuromodulation technique on rats showed a reversal of tinnitus-associated behaviors, researchers have started testing the therapeutic effects of pairing VNS with acoustic stimulation in human subjects [23].

An early case report on this subject considered a 59-year-old man with bilateral tinnitus for 14 years [24]. He had tried pharmacotherapy, transcranial magnetic stimulation, direct current stimulation, and bilateral auditory cortex stimulation, all of which did not improve his symptoms. A vagal nerve stimulator was implanted, and he received daily VNS paired with tones. After four weeks, the patient described a significant reduction in symptoms up to two months after cessation of therapy [24]. Specifically, his Tinnitus Handicap Inventory (THI) was reduced by 48% and his Tinnitus Reaction Questionnaire (TRQ) was reduced by 68% [24]. Three months after ending therapy, placebo stimulation was performed, where the only tone was presented without vagal nerve electrical stimulation. The patient experienced a gradual relapse to his baseline state, without any improvement, undermining the potential benefits of VNS as a reliable therapy for tinnitus [24]. De Ridder *et al.* followed 10 adult patients who were treated with simultaneous VNS and acoustic stimulation. This study was different in that it specifically excluded the application of tinnitus-matched tones [25]. VNS was delivered at 0.8 mA (it could be reduced to 0.05 mA if the patient desired), 100 μsec pulse width at 30 Hz, every 30 seconds. From Monday to Friday for four weeks, patients underwent 2.5 hours of paired VNS and tone stimulation in a clinic setting. Five of the 10 patients were on medications unrelated to tinnitus (muscarinic antagonists, norepinephrine agonists, and γ-amino butyric acid agonists) that investigators hypothesized interfered with the neurochemicals known to be released in response to VNS [25]. These five patients saw little to no improvement in their tinnitus symptoms as compared to their medication-free counterparts [25]. On the THI, patients not taking drugs showed an average decrease of 28.17% in their scores, while patients taking drugs showed an increase of 0.97% immediately after treatment. At follow-up, patients not taking drugs had an overall decrease of 26.06%, whereas the group taking drugs only demonstrated a decrease of 1.94% in survey scores compared with baseline. Overall, tinnitus symptoms improved in 40% of participants and remained stable for at least two months. This study shows that VNS paired with tone stimulation excluding tinnitus-matched frequency is safe [25]. The results also indicate that this therapy may benefit drug-free patients, though larger studies are necessary to confirm this conclusion [25]. Tyler *et al.* pilot trial expanded on the De Ridder *et al.* study by tripling the sample size. Thirty patients with chronic tinnitus were implanted with a vagal nerve stimulator. For the first six weeks of the study, participants were randomized such that 16 received paired VNS and auditory stimulation (paired VNS group) and 14 received auditory stimulation only (control group) [26]. This study had stimulators and leads placed internally, so stimulation could be delivered at home. Participants continued therapy for over a year, and investigators checked in quarterly. After this randomized initial six-week period, control participants crossed over, so all participants received paired VNS. At the end of the first six-week, randomized portion of the study, the paired VNS group improved on the THI ($p = 0.0012$) while the control group did not [26]. At one year, 50% of all participants had a clinically beneficial response to the therapy. Investigators noted that paired VNS had greater benefits in participants with tonal and non-blast-induced tinnitus [26]. Thus, the study suggests that vagal nerve stimulation has greater benefits than purely auditory stimulation interventions. Wichova *et al.* focused on 20 tinnitus-positive patients who received VNS implantation devices for medically refractory seizures. They found significant results in tinnitus loudness assessed through the visual analog scale (VAS) [27]. While all participants continued to have tinnitus before and after treatment, 16 out of 20 patients noted a reduction in tinnitus loudness by at least one point ($p < 0.001$) [27].

Recent human studies for VNS pair the therapy with acoustic stimulation stripped of individual participants' specific tinnitus frequency. Future studies with curated therapies may become increasingly popular as the diversity of tinnitus symptoms and affected frequencies is well established. Results indicate that VNS-acoustic-coupled stimulation treatment may improve tinnitus-related symptoms, though this improvement is especially realized in certain subgroups of tinnitus patients (drug-free participants). The path forward requires further research with larger cohorts to draw conclusions about the efficacy of VNS and to bring the promise of this therapy to wider circles of patients seeking respite from their chronic tinnitus.

Transtympanic Stimulation

Transtympanic stimulation is considered a minimally invasive intervention for tinnitus and is relatively uncommon as a treatment option. The protocol for electrical stimulation transtympanically involves applying an electrode to the middle ear, particularly the promontory and the round window. The round window is found on the medial wall of the tympanic cavity. It is the exit point for vibratory acoustic energy from the cochlea and is anatomically significant for the tympanic nerve that runs over it (Figure 2).

The first mention of using electrical stimulation to suppress tinnitus was in 1801 by Grappenglesser. He found that direct current caused tinnitus in normal-hearing patients, but in patients with tinnitus, electrical stimulation suppressed their symptoms [28]. The first known invasive transtympanic stimulation study was done in 1977 by Graham and Hazel who inserted a needle through the tympanic membrane in 13 patients and applied alternating current to explore its effects on tinnitus. Of the 13, four had current-induced vertigo [29]. The other nine had somatic sensations,

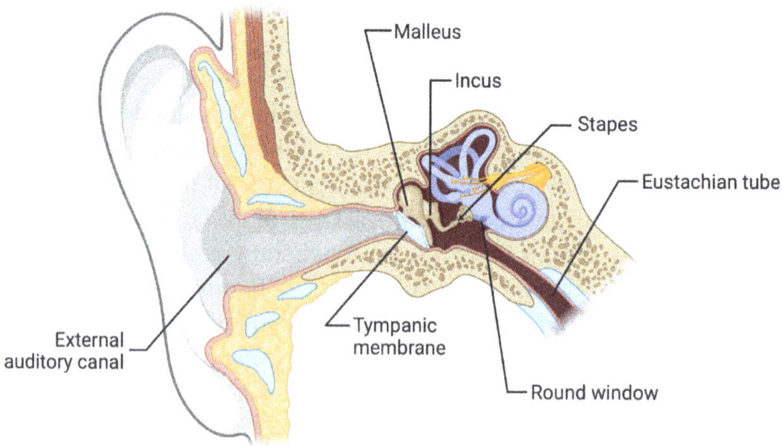

Figure 2. Anatomy of the outer and middle ear connected by the tympanic membrane.

and of the nine, two (22%) described tinnitus suppression [29]. A year later, Cazals *et al.* studied 16 patients with profound deafness, six of whom had accompanying tinnitus. Stimulation of the round window was slightly more invasive, involving skin flap removal and drilling of the external acoustic meatus to access the round window [30]. They found that in five of the six patients with tinnitus, positive electrical stimulation of the promontory and round window abolished phantom sounds [30]. In one patient, this effect could only be produced during the stimulation of the round window [30]. In a seventh, the subject only realized he had tinnitus when a resultant "silence" was induced by electrical stimulation [30]. These results offered promising possibilities for tinnitus therapies. The benefits of electrical stimulation were reproducible in Portmann *et al.*'s study on the treatment of hearing loss with electrical stimulation. They found that positive electrical stimulation directly resulted in tinnitus suppression in 12 of the 15 participants [31]. More recent studies by Konopka *et al.* and Rubinstein *et al.* conducted in 2001 and 2003, respectively, found that electrical promontory stimulation resulted in a tinnitus reduction rate in about 50% of participants. Specifically, Konopka found subjective tinnitus improvement in 62 out of 108 participants (57%), with two participants reporting complete tinnitus abolishment [32]. Rubinstein found a substantial or complete tinnitus suppression in five out of 11 (45%) transtympanic subjects [33].

The earliest modern studies involving invasive tympanic electrical stimulation have been done alongside cochlear implants (CI). There have been a plethora of reports describing tinnitus suppression through CI in bilateral sensorineural hearing loss. This supports the theory that auditory deprivation is one component of tinnitus symptom manifestation, thus the restoration of auditory function through CI is a therapeutic avenue commonly taken for patients suffering from both hearing loss and chronic tinnitus [34–36].

To highlight a few, Chang *et al.* sought 13 tinnitus subjects and found that nine (70%) responded to electrical stimulation well, reporting a minimum of 30% reduction in tinnitus loudness. The other four participants (30%) did not respond to electrical stimulation well, reporting a less than 30% tinnitus loudness reduction in response to any stimulated condition tested [37]. Overall, the participants who responded to at least one stimulus parameter (9/13) were more likely to respond to other stimulus conditions [37]. Remo *et al.* performed intracochlear electrical stimulation on patients with CI in 11 subjects in 2015. Using electrical stimulation generated from the CI, investigators aimed to assess how varied current levels and anatomical stimulation sites inside the cochlea affected tinnitus in the short term. They found that pitch-matched electrical stimulation did not have a major effect on tinnitus suppression [38]. There was a positive relationship between the current level and the percentage of tinnitus suppression [38]. While optimal efficacious conditions for each patient were highly subject-specific, tinnitus loudness was reduced by at least 30% for participants and nine of 11 subjects (82%) achieved a successful condition that suppressed their symptoms of tinnitus [38]. The most recent study on electrical stimulation in CI patients was conducted in 2021 by Kloostra *et al.* and involved 19 adult participants who had unilateral CI due to severe bilateral hearing loss. All patients suffered from tinnitus. The subjects were stimulated with a single CI electrode for 300 seconds [39]. During this time, the position of cochlear stimulation, stimulation rate, and stimulation amplitude were varied to create a total of 12 conditions [39]. Participants rated their tinnitus loudness on a VAS promptly after each stimulus condition was administered. Researchers found that the application of moderate-intensity electrical stimulation created the most tinnitus reduction (30%) compared to stimuli at near-threshold levels (18%) with a p-value < 0.01 [39]. Interestingly, different stimulation sites and pulse rates had no effect on tinnitus levels. The effect on tinnitus loudness was small, with a change in VAS less than three points in all cases [39]. Overall, the effect of unilateral electrical stimulation in CI patients had an insignificant effect on tinnitus suppression and symptoms.

In 2009, Di Nardo *et al.* delivered electrical promontory stimulation to patients with chronic, disabling tinnitus. A total of 11 participants, who were affected with single-sided or bilateral hearing loss, were enrolled [40]. The electrostimulation was provided by a direct, continuous positive current through a needle electrode connected to a promontory stimulator device [40]. Immediately after stimulation, five patients (45.4%) reported complete tinnitus suppression and four (36.4%) reported tinnitus attenuation [40]. At one-month follow-up, a reduction in tinnitus handicap inventory (THI) occurred in five patients (45.4%) and remained unchanged in the other six patients (54.6%) [40].

Punte *et al.* aimed to explore how electrical stimulation near the round window could reduce severe tinnitus in patients with single-sided deafness, building on the foundation that CI has significantly reduced tinnitus in this patient population. To do this, researchers compared the levels of tinnitus suppression reached with stimulation of the four basal-most electrodes (closest to the round window) of intracochlear CI and compared this with stimulation of all 12 CI electrodes [41]. The study focused on seven patients who had severe tinnitus due to single-sided deafness and were implanted with a Med-El CI and a FlexSoft or Flex24 electrode [41]. The protocol involved activating one pair of basal electrodes each week until all four were activated [41]. Subsequently, all CI electrodes were activated. Tinnitus symptoms were assessed before CI placement and before each electrode pair was activated [41]. Following the complete activation of the CI electrodes, participants were followed for 1, 3, and 6 months using the VAS for loudness and the Tinnitus Questionnaire (TQ) for the effects of tinnitus on quality of life [41]. They found no tinnitus suppression from basal stimulation (first four pairs of electrodes nearest the round window) but did find significantly decreased tinnitus symptoms with complete CI activation [41]. Pre-implantation, the average tinnitus loudness score was 8.2/10, but at six months post-implantation, the loudness score was reduced to 4.1/10 [41]. The TQ scores went from 60/84 to 39/82 during this same period [41]. This contrasts with previous studies on conventional CI placement which did not identify an effect with varying sites of stimulation in a total of 43 participants [37–39]. These studies did identify that unique stimulation patterns could totally abolish tinnitus perception in many patients, proposing that electrical stimulation through CI is an intervention worth pursuing and the main challenge ahead to expanding its efficacy is identifying and optimizing the stimulus conditions.

Another important contribution to advancing transtympanic stimulation was a study conducted by Wenzel *et al.* in 2015 which used a non-penetrating stimulation electrode connected to a Neurelec CI in three patients with single-sided deafness, but normal hearing contralaterally. A Tinnelec implant encompassed the stimulation electrode and was positioned into the round window after the round window membrane had been exposed [42]. The electrode did not penetrate the cochlea like a conventional implantation technique would. Participants could modulate stimulation conditions based on setting: one was created for normal daily circumstances and another for silent conditions [42]. Participants were instructed to use the stimulation device for at least four hours daily, with stimulation settings adjusted to individual needs (amplitude range: 260–950 uA, frequency range: 70.7–5924 Hz, transmission power: 28–35%, stimulus width: 57 us–3.0 ms) [42]. Patients were followed up for years and were asked to report scores on the VAS (on loudness, annoyance, mood) and through the THI [42]. All patients benefited from electrical stimulation (THI-score improvement of 20–70), though at different degrees [42]. All patients improved in at least one aspect of the outcome measured, though eventually all three had the Tinnelec implant removed and replaced with a conventional CI [42].

Finally, a study in 2015 conducted by Perez *et al.* considered 10 patients with severe unilateral tinnitus and offered electrical stimulation of the promontory during three consecutive 30-minute sessions every other day. The charged-balanced electrostimulation pulses were delivered by a transtympanic needle. This study was unique in the therapeutic timing, with alternating days of treatment, likely attributed to the invasiveness of the procedure. It also included seven patients with normal or mild hearing loss (70%), which deviates from previous studies that focus on patients with single-sided or bilateral hearing deficits [43]. Overall, they found no long-term adverse safety outcomes [43]. Through VAS scores, subjects showed significant tinnitus reduction during or following treatment sessions by more than two points in five patients (50%), who returned to baseline four weeks following treatment ($p = 0.048$) [43]. A statistically significant decrease in tinnitus handicap inventory was also observed four weeks after treatment [43].

Auditory Cortex Stimulation

Auditory cortex stimulation involves stimulating a surgically placed electrode in the auditory pathway. Specifically, this can include the auditory nerve, cochlear nuclei, inferior colliculus, or even the auditory cortex directly. The rationale behind this therapy is that increased activity in the auditory cortex is involved in generating tinnitus, so the stimulation of the cortex will alleviate symptoms.

Many studies have been conducted on the stimulation of the auditory pathway in non-invasive forms (transcranially, transcutaneously, extradurally, etc.) and have indicated diverse efficacy. More invasive forms of auditory cortex electrical stimulation began in 2004 with a case of intractable tinnitus [44]. A 32-year-old patient had undergone three surgeries for hemifacial spasms. After the third operation, her hemifacial spasm symptoms resolved, but she noticed total hearing loss in one year. Three weeks later, she began to develop tinnitus that disturbed her in her sleep every two to three hours. Pre-operative physical exam revealed high-pitched and low-pitched tinnitus. One year after the tinnitus symptom onset, the patient had an extradural octopolar electrode (Lamitrode 44) implanted for auditory cortex electrical stimulation [44]. Five hours post-operatively, the electrode was activated. It completely abolished her high-pitched tinnitus and reduced her low-pitched tinnitus by 60% [44]. The stimulation settings were modified before discharge to better suppress her tinnitus while in the prone position *vs*. sitting upright [44]. Unfortunately, three weeks post-implantation, the high-pitched tinnitus reappeared. Maximal stimulation could only abolish the low-pitched tinnitus. The return of the high-pitched frequency was hypothesized to be a result of cortical plasticity in response to constant stimulation to suppress the high-pitched phantom sounds [44]. Researchers introduced an alternation stimulation program between different poles of the implanted electrode to bypass the cortical plasticity observed. Ten months after this adjustment, the patient remained tinnitus-free [44].

De Ridder *et al.* recruited 12 patients who underwent primary and secondary auditory cortex electrode implantation after chronic tinnitus was successfully suppressed by >50% in transcranial magnetic stimulation (TMS) application. Before implantation, TMS was found to have an average effect of 77% pure tone tinnitus reduction and 67% white noise reduction [45]. Electrical stimulation through the implanted primary and secondary auditory cortex electrodes resulted in a mean 97% reduction of pure tone tinnitus and 24% white noise suppression [45]. Post-operatively, VAS scores were decreased from 9.5 to 1.5 for pure tone and from 8.8 to 6.8 for white noise perception [45]. This study used a non-invasive tinnitus treatment to predict the efficacy of an invasive, surgical treatment and found that the electrical stimulation of primary and secondary auditory cortex regions primarily ameliorates pure tone tinnitus sounds [45].

Another study considered a specific secondary auditory cortex region: the posterior superior temporal gyrus. Friedland *et al.* recruited eight adults with constant tinnitus (primarily unilateral with a frequency <8,000 Hz) to explore the safety and effects of implanting an epidural cortical stimulator on intractable tinnitus. Participants underwent electrode implantation, and after one week of recovery, were subject to a randomized two-week stimulation period alternated with a two-week sham period [46]. During this period, they rated their tinnitus loudness and device effectiveness using scales ranging from 0 to 100 [46]. These four weeks were then followed by continuous stimulation at parameters adjusted to individual patient requests to maximize tinnitus suppression and benefits for patients [46]. Of the eight subjects, four noted a change during the switch between the two treatment arms. Two reported a change during the stimulation period, and the other two reported a response during the sham period [46]. Overall, there were no effects of stimulation during the four-week blinded period. With continuous stimulation, though, two patients had persistent pure-tone reduction of tinnitus and six had a period of complete tinnitus suppression [46]. After 12 weeks of treatment, 7/8 patients noted improved scores on distributed surveys. The Beck Depression Inventory also showed improvement in all but one participant. Overall, the study showed the safety of electrical stimulation of the secondary auditory cortex [46].

Seidman *et al.* presented and subsequently wrote about two unique patients with debilitating tinnitus refractory to conventional therapy trials over three months. Measures were first conducted to determine the frequency and pitch of each patient's tinnitus [47]. The first patient (RP) had complained of bilateral tinnitus for two years [47]. The presentation of RP's tones during magnetoencephalography and functional magnetic resonance imaging (fMRI) determined the tonotopic map in Heschl's gyrus. RP was then implanted with quadripolar electrodes at these tonotopic sites. In the second patient (MV), who had experienced unilateral tinnitus for seven years, fMRI was performed to determine their tonotopic map, and a region was selected for bipolar electrode implantation [47]. In the first patient, all post-operative surveys showed improvement. For the first several months, RP described significant improvement, though on some days the tinnitus increased to 2/10 on the VAS [47]. If the tinnitus was quiet, RP would turn off the pulse generator and note residual inhibition lasting hours to several days [47]. Three months after surgery, RP reported that his tinnitus was essentially resolved when the electrode was active [47]. This benefited his tinnitus bilaterally. Two years post-operatively, RP rated his tinnitus 0 to 1 on the VAS and did not present with any change in hearing after surgery [47]. In the first four months after MV's surgery, a variety of stimulation regimens failed to suppress her tinnitus [47]. After four months, she underwent additional surgery to replace the electrode with a subdural two-contact electrode, one placed directly on Heschl's gyrus [47]. Post-operatively, attempts to alter the stimulation settings resulted in a maximal 30–35% decrease in tinnitus symptoms [47]. At her two-year follow-up, her tinnitus returned to its original level, indicating a lack of sustained tinnitus inhibition with electrical stimulation [47].

A key paper on this topic comes from De Ridder and Vanneste in 2014, where researchers use a multitarget method of neuromodulation, implanting a combined C2 and auditory cortex electrode to observe effects in tinnitus. This was a case study done on a 54-year-old man with traumatically induced chronic tinnitus on his right side for the last 1.5 years [48]. The patient experienced a pure tone and noise-like component tinnitus [48]. His pure tone tinnitus was completely abolished with cortical stimulation but had no effect on the noise-like component [48]. After transcutaneous electrical nerve stimulation successfully treated his noise tinnitus, a wire electrode was inserted subcutaneously in the C2 dermatome [48]. Stimulation of this electrode improved his noise-like tinnitus by 50%. This suppressive effect was maintained up to the last follow-up, five years post-operatively [48]. Through this patient, De Ridder and Vanneste suggested that multitarget stimulation may be useful in intractable tinnitus.

One of the most expansive, long-term studies on electrical auditory cortex stimulation as a treatment for tinnitus had results that dampened expectations. In 2014, Engelhardt *et al.* published a prospective, randomized, double-blinded trial on nine chronic, severe tinnitus patients. Chronic was defined as greater than two years of symptoms, and severe was defined as a Strukturierte Tinnitus Interview (STI) score of greater than 19 [49]. After implantation of a single, four contact-epidural electrode and a stimulator placed in the pectoral region, stimulation parameters were adjusted for four months per patient request [49]. These four months were followed by two weeks of no stimulation. Subjects were then randomized into two groups: the first group began with two weeks of significant stimulation followed by two weeks of washout and two weeks of non-significant stimulation [49]. Group B received the inverse protocol. Before each cross-over phase, patients were administered all relevant tinnitus surveys, including the STI [49]. A decrease of at least 35% in scores on the STI was considered clinically significant [49]. One patient was explanted before trial completion due to severe post-operative psychiatric decompensation [49]. Three patients were explanted at the end of the study due to lack of effect and developing breast cancer in one patient under the stimulator [49]. During the six-week double-blinded phase, none of the participants experienced improvement in symptoms (>35% improvement in STI score). Of the five that remained stimulated after the six-week period, three felt a slight to great subjective effectiveness and the remaining two reported a clinically significant decrease in STI score (>35% reduction) [49]. Of the two patients that reported significant benefits, one reported an STI improvement of 52% and a tinnitus attenuation of 60–70% and stated that they "could not live now without the stimulation". The other reported a 55% improvement in STI score, though described no specific improvement in tinnitus [49]. In conclusion, the researchers did not find an objective improvement in chronic, severe tinnitus with cortical stimulation. The investigators noted that this result was discordant with other open-evaluation studies, and they ascribed these differences to a placebo effect of surgery, a double-blinded randomized phase that was too short in their own study, a poorly defined target for cortical implantation, or outcome measures that were insufficient at measuring improvements [49].

The combination of this literature reveals a few conclusions: There are different types of tinnitus, different brain regions involved, and a variety of different stimulation settings and preferences that best ameliorate individual patients' symptoms. These studies also show that invasive auditory cortex stimulation can be effective and life-changing in some patients or have little impact in others. Why these dichotomous results occur remains at large. Regardless, this intervention remains at an early stage and greater investigation is required. As far as invasive neurosurgical procedures go, with their various risks, costs, and recovery times, cortical stimulation as a potential therapeutic option will likely always be limited to a small subset of patients with extreme, unbearable, intractable tinnitus.

Tinnitus Distress Network Stimulation

Tinnitus often causes distress in patients; this specific distress can be linked to anatomical brain regions or abnormal functional connectivity between brain regions. Specifically, the anterior cingulate cortex, amygdala, insula, and parahippocampal area have been associated with increased signaling with tinnitus distress.

De Ridder *et al.* implanted bilateral electrodes over the dorsal anterior cingulate cortex (dACC) in two participants. One participant responded to dACC stimulation dramatically, with a significant reduction in tinnitus loudness and distress [50]. These effects were sustained for more than two years after implantation. The other patient did not benefit from electrode insertion, even after trialing various stimulation parameters. Further investigation showed that the participants who responded positively had increased functional connectivity from the dACC to the tinnitus distress network [50]. The non-responder had decreased functional connectivity between these same areas, suggesting that functional connectivity is required to effectively deliver brain stimulation through implanted electrodes [50].

Deep Brain Stimulation

Deep brain stimulation (DBS) involves the implantation of an electrode within the brain and a high-frequency stimulation (>100 Hz) intended to disrupt pathological neuronal signaling (Figure 3). It is an established treatment for patients with epilepsy, movement disorders, and psychiatric conditions. It is not FDA-approved for tinnitus, but patients with approved disorders and comorbid tinnitus can decide to treat with DBS.

The first study that looked at patients with movement disorders and coexisting tinnitus was by Shi *et al.,* where an electrode was implanted in the ventralis intermedius nucleus of the thalamus of seven patients [51]. When DBS was turned on, three of these patients reported reduced tinnitus loudness [51]. Cheung and Larson studied six Parkinson's disease patients with concomitant tinnitus who all underwent electrode implantation in the subthalamic or ventral intermediate nucleus. The lead traversed through or was adjacent to the locus of caudate neurons (LC), associated with the striatal network [52]. In the five patients where the DBS lead traversed LC, tinnitus loudness in both ears was suppressed [52]. In one patient, where the DBS lead was adjacent, thus outside, area LC, tinnitus was unchanged [52].

Two case studies worth discussing describe the effect of DBS on intractable tinnitus. The first considered six adult subjects, three with and three without chronic tinnitus, all undergoing surgery for movement

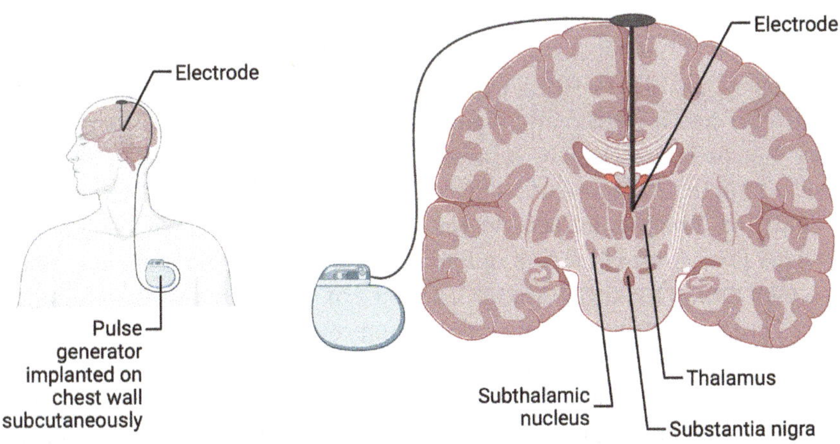

Figure 3. In deep brain stimulation, high frequency stimulation acts to resynchronize subthalamic nuclei activity to increase dopamine release in the substantia nigra.

disorders [53]. DBS short-term stimulation of LC triggered new phantom tones in five subjects [53]. From this study, investigators speculated that LC neuromodulation can trigger temporary phantom percept gate dysfunction and perceive new tinnitus tones [53]. They propose long-term DBS stimulation of LC to control this gating problem. In another study, Dijkstra *et al.* describe a 41-year-old woman with intractable tinnitus since age 35. The constant hissing sound was so burdensome that the patient developed a severe depression and performed a suicide attempt during admission at a psychiatric facility [54]. The patient was approved for DBS and underwent surgical implantation of electrodes in the ventral anterior limb of the internal capsule, a region previously demonstrated to control mood and anxiety [54]. Her settings were stabilized at 5 V. One year after chronic DBS, her symptoms had improved significantly [54]. On the TFI, her score improved by 63.2% compared to the pre-surgical one [54]. On the THI, her scores improved by 73.7%. The patient reported that while the tinnitus persisted, she was now able to ignore the sound, greatly improving her mood and energy [54]. When the DBS was unintentionally discontinued due to battery depletion, the patient noticed an immediate recurrence of tinnitus sound and an inability to ignore it. Her tinnitus symptoms and mood worsened considerably until the original effects of DBS were reinstated with the replacement of the neurostimulator [54].

Smit *et al.* increased the scope of the investigation, considering the clinical impact of DBS on tinnitus in a retrospective multicenter study that included 122 tinnitus patients and matched controls. The study found that stimulation of the subthalamic nucleus was most beneficial in tinnitus suppression, though investigators speculated that other nuclei linked to this circuitry would also be worth exploring [55]. Based on THI scores, patients with DBS had a statistically significant reduction in scores compared to their matched controls, who represented the natural course of tinnitus over time without DBS intervention [55]. These exciting results spurred a phase I study on treating intractable tinnitus with DBS of bilateral caudate nucleus [56]. Smit *et al.* first optimized five patients' stimulation settings and then monitored them through 24 weeks of continuous caudate stimulation. Three of the five patients reported improved tinnitus distress and four of the five patients reported statistically significantly improved scores on Tinnitus Functional Index (greater than a 13-point decrease) and Tinnitus Handicap Inventory (greater than a 20-point decrease) [56]. No significant safety concerns were discovered or described.

Mechanisms: Masking vs. Suppression

The underlying effect of electrical stimulation on tinnitus is still not fully understood, but two main mechanisms have been proposed: the masking effect and the suppression effect [57]. In the masking effect, acoustic and electrical stimulations induce sound that can reduce the disparity between the tinnitus signal and silence, leading to a decrease in tinnitus perception [58]. Interestingly, multiple studies, including some covered later in this book, show that inaudible stimulation can also suppress tinnitus symptoms [38, 59, 60]. Thus, another mechanism must be involved. These findings are explained through the suppression effect, which is believed to modulate auditory cortex activity and suspend the generation of tinnitus phantom sounds. Some studies suggest that electrical stimulation is only effective if the source of tinnitus is found in specific locations [61]. Others link mechanisms of tinnitus to varying electrical patterns along the auditory pathway, thus high pulse rate stimulation may restore abnormal activity and suppress tinnitus [33]. In conflict with this speculation is that low or moderate pulse rate studies have been shown to induce greater central activity, which is expected to interfere with abnormal tinnitus-induced activity [62]. With each new study on the effects of electrical stimulation on tinnitus suppression, the questions grow, and our relative knowledge of tinnitus feels stagnant. It is important to collate a fundamental understanding of these concepts in order to generate new conceptual avenues to pursue. Particularly, fields such as neuroimaging or electrophysiology may help facilitate better understanding and optimization of tinnitus suppression through electrical stimulation. In this chapter, we examine invasive electric stimulation treatment options available for tinnitus and consider existing evidence on their efficacies and shortcomings.

Conclusion

- Tinnitus pathophysiology, presentation, and resulting therapies are complex with a diverse array of stimulation parameters that could be used, targets that could be identified, and interventions that could be applied.
- Studies with larger sample sizes are necessary to advance our knowledge of tinnitus, but there are many patients who stand to benefit from innovative treatments.

- Of invasive electrical stimulation modalities, vagus nerve stimulation and deep brain stimulation serve as today's most exciting and efficacious forms of therapy for tinnitus relief.
- Few studies have been carried out on the tinnitus distress network. Transtympanic stimulation has shown a wide range of benefits in patients, while the most thorough study on auditory cortex stimulation has indicated no significant improvement for tinnitus patients.
- More sophisticated regimens and parameters of electrical stimulation should be an area of focus with the existing technology so that dynamic regions of the brain at a variety of frequencies and intensities can be applied and physiologically tailored to unique manifestations of tinnitus in patients so that the efficacy of therapies can be maximized.

References

1. Joos K, *et al.* Polarity specific suppression effects of transcranial direct current stimulation for tinnitus. *Neural Plasticity*. **2014**:930860 (2014).
2. Zeng FG. Tinnitus and hyperacusis: Central noise, gain and variance. *Current Opinion Physiology*. **18**:123–129 (2020).
3. Nicole P and Kleinjung T. Neuromodulation for tinnitus treatment: An overview of invasive and non-invasive techniques. *Journal of Zhejiang University*. **20**:116–130 (2019).
4. Van de Heyning P, *et al.* Incapacitating unilateral tinnitus in single-sided deafness treated by cochlear implantation. *Annals of Otology, Rhinology & Laryngology*. **117**:645–652 (2008).
5. Punte AK, *et al.* Cochlear implantation as a durable tinnitus treatment in single-sided deafness. *Cochlear Implants International*. **12**:S26–S29 (2011).
6. Schlee W, *et al.* Mapping cortical hubs in tinnitus. *BMC Biology*. **7**:80 (2009).
7. Schlee W, *et al.* Using auditory steady state responses to outline the functional connectivity in the tinnitus brain. *PLoS One*. **3**:e3720 (2008).
8. Vanneste S, *et al.* The neural correlates of tinnitus-related distress. *NeuroImage*. **52**:470–480 (2010).
9. Jastreboff PJ. Phantom auditory perception (tinnitus): Mechanisms of generation and perception. *Neuroscience Research*. **8**:221–254 (1990).
10. De Ridder D, *et al.* Phantom percepts: Tinnitus and pain as persisting aversive memory networks. *Proceedings of the National Academy of Sciences*. **108**: 8075–8080 (2011).
11. Vanneste S, Song J-J, and De Ridder D. Thalamocortical dysrhythmia detected by machine learning. *Nature Communications*. **9**:1103 (2018).
12. Crone NE, *et al.* Induced electrocorticographic gamma activity during auditory perception. Brazier Award-winning article, 2001. *Clinical Neurophysiology*. **112**:565–582 (2001).

13. Joliot M, Ribary U, and Llinás R. Human oscillatory brain activity near 40 Hz coexists with cognitive temporal binding. *Proceedings of the National Academy of Sciences of the United States of America*. **91**:11748–11751 (1994).
14. Llinás RR, *et al.* Thalamocortical dysrhythmia: A neurological and neuropsychiatric syndrome characterized by magnetoencephalography. *Proceedings of the National Academy of Sciences of the United States of America*. **96**:15222–15227 (1999).
15. Weisz N, *et al.* The neural code of auditory phantom perception. *The Journal of Neuroscience*. **27**:1479–1484 (2007).
16. De Ridder D, *et al.* Theta-gamma dysrhythmia and auditory phantom perception. *Journal of Neurosurgery*. **114**:912–921 (2011).
17. De Ridder D, *et al.* Thalamocortical dysrhythmia: A theoretical update in tinnitus. *Frontiers Neurological*. **6**:124 (2015).
18. Rauschecker JP, Leaver AM and Mühlau M. Tuning out the noise: Limbic-auditory interactions in tinnitus. *Neuron*. **66**:819–826 (2010).
19. De Ridder D, *et al.* An integrative model of auditory phantom perception: Tinnitus as a unified percept of interacting separable subnetworks. *Neuroscience & Biobehavioral Reviews*. **44**:16–32 (2014).
20. House JW, and Brackmann DE. Tinnitus: Surgical treatment. *Ciba Foundaton Symposius*. **85**:204–216 (1981).
21. Zeng FG, *et al.* Tinnitus treatment using noninvasive and minimally invasive electric stimulation: Experimental design and feasibility. *Trends in Hearing*. **23**:2331216518821449 (2019).
22. Zeng FG, Djalilian H and Lin H. Tinnitus treatment with precise and optimal electric stimulation: Opportunities and challenges. *Current Opinion Otolaryngology & Head and Neck Surgery*. **23**:382–387 (2015).
23. Engineer ND, *et al.* Reversing pathological neural activity using targeted plasticity. *Nature*. **470**:101–104 (2011).
24. De Ridder D, *et al.* Placebo-controlled vagus nerve stimulation paired with tones in a patient with refractory tinnitus: A case report. *Otology & Neurotology*. **36**:575–580 (2015).
25. De Ridder D, *et al.* Safety and efficacy of vagus nerve stimulation paired with tones for the treatment of tinnitus: A case series. *Neuromodulation*. **17**:170–179 (2014).
26. Tyler R, *et al.* Vagus nerve stimulation paired with tones for the treatment of tinnitus: A prospective randomized double-blind controlled pilot study in humans. *Scientific Reports*. **7**:11960 (2017).
27. Wichova H, *et al.* Tinnitus perception in patients after vagal nerve stimulator implantation for epilepsy. *American Journal of Otolaryngology*. **39**:599–602 (2018).
28. Balkany T, *et al.* Workshop: Direct electrical stimulation of the inner ear for the relief of tinnitus. *American Journal of Otolaryngology*. **8**:207–212 (1987).

29. Graham JM, and Hazell JW. Electrical stimulation of the human cochlea using a transtympanic electrode. *British Journal of Audiology*. **11**:59–62 (1977).
30. Cazals Y, Negrevergne M, and Aran J-M. Electrical stimulation of the cochlea in man: Hearing induction and tinnitus suppression. *Ear and Hearing*. **3**:209–213 (1978).
31. Portmann M, *et al.* Temporary tinnitus suppression in man through electrical stimulation of the cochlea. *Acta Oto-Laryngologica*. **87**:294–299 (1979).
32. Konopka W, *et al.* Tinnitus suppression by electrical promontory stimulation (EPS) in patients with sensorineural hearing loss. *Auris Nasus Larynx*. **28**:35–40 (2001).
33. Rubinstein JT, *et al.* Electrical suppression of tinnitus with high-rate pulse trains. *Otology & Neurotology*. **24**:478–485 (2003).
34. Baguley DM and Atlas MD. Cochlear implants and tinnitus. *Progress in Brain Research*. **166**:347–355 (2007).
35. Olze H, *et al.* Cochlear implantation has a positive influence on quality of life, tinnitus, and psychological comorbidity. *The Laryngoscope*. **121**:2220–2227 (2011).
36. Di Nardo W, *et al.* Tinnitus modifications after cochlear implantation. *European Archives of Oto-Rhino-Laryngology*. **264**:1145–1149 (2007).
37. Chang JE and Zeng FG. Tinnitus suppression by electric stimulation of the auditory nerve. *Frontiers in Systems Neuroscience*. **6**:19 (2012).
38. Arts RA, *et al.* Optimizing intracochlear electrical stimulation to suppress tinnitus. *Ear & Hearing*. **36**:125–135 (2015).
39. Kloostra FJJ, *et al.* Changes in tinnitus by cochlear implantation: A parametric study of the effect of single-electrode stimulation. *Audiology & Neuro-otology*. **26**:140–148 (2021).
40. Di Nardo W, *et al.* Transtympanic electrical stimulation for immediate and long-term tinnitus suppression. *International Tinnitus Journal*. **15**:100–106 (2009).
41. Punte AK, De Ridder D and Van de Heyning P. On the necessity of full length electrical cochlear stimulation to suppress severe tinnitus in single-sided deafness. *Hearing Research*. **295**:24–29 (2013).
42. Wenzel GI, *et al.* Non-penetrating round window electrode stimulation for tinnitus therapy followed by cochlear implantation. *European Archives of Oto-Rhino-Laryngology*. **272**:3283–3293 (2015).
43. Perez R, *et al.* Multiple electrostimulation treatments to the promontory for tinnitus. *Otology & Neurotology*. **36**:366–372 (2015).
44. De Ridder D, *et al.* Magnetic and electrical stimulation of the auditory cortex for intractable tinnitus. Case report. *Journal of Neurosurgery*. **100**:560–564 (2004).

45. De Ridder D, *et al.* Primary and secondary auditory cortex stimulation for intractable tinnitus. *Journal for Oto-Rhino-Laryngology and its Related Specialties*. **68**:48–55 (2006).
46. Friedland DR, *et al.* Feasibility of auditory cortical stimulation for the treatment of tinnitus. *Otology & Neurotology*. **28**:1005–1012 (2007).
47. Seidman MD, *et al.* Direct electrical stimulation of heschl's gyrus for tinnitus treatment. *The Laryngoscope*. **118**:491–500 (2008).
48. De Ridder D, and Vanneste S. Multitarget surgical neuromodulation: Combined C2 and auditory cortex implantation for tinnitus. *Neuroscience Letters*. **591**:202–206 (2015).
49. Engelhardt J, *et al.* Effect of chronic cortical stimulation on chronic severe tinnitus: A prospective randomized double-blind cross-over trial and long-term follow up. *Brain Stimulation*. **7**:694–700 (2014).
50. De Ridder D, Joos K, and Vanneste S. Anterior cingulate implants for tinnitus: Report of 2 cases. *Journal of Neurosurgery*. **124**:893–901 (2016).
51. Shi Y, *et al.* Deep brain stimulation effects in patients with tinnitus. *Otolaryngology Head Neck Surgery*. **141**:285–287 (2009).
52. Cheung SW, and Larson PS. Tinnitus modulation by deep brain stimulation in locus of caudate neurons (area LC). *Neuroscience*. **169**:1768–1778 (2010).
53. Larson PS, and Cheung SW. Deep brain stimulation in area LC controllably triggers auditory phantom percepts. *Neurosurgery*. **70**:398–406 (2012).
54. Dijkstra E, *et al.* Effective deep brain stimulation of intractable tinnitus: A case study. *Brain Stimulation*. **11**:1205–1207 (2018).
55. Smit JV, *et al.* The impact of deep brain stimulation on tinnitus. *Surgical Neurology International*. **7**:848–854 (2016).
56. Cheung SW, *et al.* Phase I trial of caudate deep brain stimulation for treatment-resistant tinnitus. *Journal of Neurosurgery*. 1–10 (2019).
57. Assouly KKS, *et al.* Systematic review on intra- and extracochlear electrical stimulation for tinnitus. *Brain Science*. **11**:1394 (2021).
58. Knipper M, *et al.* The neural bases of tinnitus: Lessons from deafness and cochlear implants. *Journal of Neuroscience*. **40**:7190–7202 (2020).
59. Okusa M, *et al.* Tinnitus suppression by electrical promontory stimulation in sensorineural deaf patients. *Acta Oto-Laryngologica. Supplementum*. **501**:54–58 (1993).
60. Rothera M, *et al.* Evaluation of patients for cochlear implant by promontory stimulation. Psychophysical responses and electrically evoked brainstem potentials. *British Journal of Audiology*. **20**:25–28 (1986).
61. Aran JM, and Cazals Y. Electrical suppression of tinnitus. *Ciba Foundation Symposium*. **85**:217–231 (1981).
62. Zeng FG, *et al.* Tinnitus suppression by low-rate electric stimulation and its electrophysiological mechanisms. *Hearing Research*. **277**:61–66 (2011).

https://doi.org/10.1142/9789819807345_0009

Chapter 9

Bimodal Neurostimulation in Tinnitus Treatment

Ryka Vahidi[*], Yalda Yazdani[*], and Hamid R. Djalilian[*,†,‡,§]

[*]*Department of Otolaryngology-Head and Neck Surgery, University of California, Irvine, CA, USA*

[†]*Department of Biomedical Engineering, University of California, Irvine, CA, USA*

[‡]*Department of Neurosurgery, University of California, Irvine, CA, USA*

[§]*hdjalili@hs.uci.edu*

Abstract

Bimodal neurostimulation, an innovative neuromodulatory approach combining auditory and somatosensory stimuli, has emerged as a promising intervention for tinnitus management. Devices such as Lenire™, Neosensory Duo™, and the device developed at the University of Michigan (a.k.a., "Dr. Susan Shore Device") utilize this approach by delivering synchronized auditory and electrical stimulation to the tongue, wrist, or neck, respectively. These techniques attempt to modulate imbalances in neuronal activity thought to underly the etiology of tinnitus symptoms. Clinical trials have demonstrated significant reductions in tinnitus severity, with sustained benefits observed in many patients. This chapter explores the neurophysiological mechanisms, clinical efficacy, and therapeutic applications of bimodal neurostimulation in tinnitus management. It delves into the underlying neuroplastic changes, including the modulation of neural plasticity and the reorganization of auditory pathways. Additionally, it discusses the design and outcomes of key clinical trials, highlighting the safety profile and tolerability of bimodal neurostimulation. The potential for integrating this therapy into standard clinical practice and its implications for future research are also considered. By providing a comprehensive overview of bimodal neurostimulation, this chapter aims to elucidate its role in advancing tinnitus treatment and improving patient outcomes.

Introduction

Tinnitus, often described as a phantom auditory perception in the absence of an external acoustic source, is a common otologic condition with a significant impact on quality of life. Affecting approximately 10–15% of the adult population, tinnitus can lead to various psychophysiological disturbances, including sleep disorders, anxiety, and depression [3–5, 8, 9]. Traditional therapeutic approaches, such as cognitive-behavioral therapy (CBT), acoustic therapy, and pharmacological interventions, offer varying degrees of symptomatic relief but are not universally effective. This condition is often associated with hearing loss but can also occur in individuals with normal audiograms, suggesting that it arises from pathological neural activity in the central auditory pathway [10–13]. This brings about the concept of bimodal neurostimulation. The advent of bimodal neurostimulation represents a novel neuromodulatory avenue, leveraging neuroplasticity to mitigate tinnitus symptoms [13]. Bimodal neurostimulation involves the simultaneous application of two distinct sensory modalities, typically combining auditory and somatosensory inputs. This approach aims to modulate neural activity in the auditory system and associated cortical regions, potentially reversing the maladaptive neuroplastic changes thought to underlie tinnitus perception. The growing body of research in this field has led to the development of innovative neuromodulatory devices and treatment protocols, offering new hope for individuals struggling with chronic tinnitus.

Neurophysiological Mechanisms of Bimodal Neurostimulation

To understand the potential mechanisms of bimodal stimulation, we have to first understand some of the relevant neuroscience. The underlying mechanism of bimodal neurostimulation is rooted in the concept of neuroplasticity — the brain's ability to reorganize itself by forming new neural connections [9]. In the context of tinnitus, it is believed that dysfunctional neuroplastic changes in the auditory cortex and associated brain regions contribute to the persistent perception of phantom sounds. Bimodal neurostimulation aims to harness the brain's neuroplasticity to reverse these maladaptive changes.

Dorsal Cochlear Nucleus Plasticity

The dorsal cochlear nucleus (DCN) is the first nucleus in the auditory pathway and integrates auditory nerve input from the cochlea with somatosensory input from the dorsal column and trigeminal systems [14–17]. Experimental changes in the DCN can result in tinnitus, leading to the conclusion that the DCN is a key structure in the pathologic basis of tinnitus [18]. Fusiform cells, the principal output neurons of the DCN, receive auditory nerve input on their basal dendrites and indirectly receive somatosensory input via granule cell axons on their apical dendrites [19–20]. The apical dendrite synapses are plastic, allowing their ability to stimulate fusiform cells and overall strength to sum over time. [19–20]. Auditory and somatosensory integration by fusiform cells involves spike-timing-dependent plasticity (STDP), such that the order and frequency between the auditory and somatosensory stimuli determines changes in fusiform-cell excitability [21–23]. When the temporal relationship between auditory and somatosensory stimulation increases fusiform-cell excitability, the process is referred to as long-term potentiation (LTP), while decreases in fusiform-cell excitability are termed long-term depression (LTD) [21–23].

Bimodal Stimulation Via the Vagal Nerve

Animal studies

The vagus nerve plays a crucial role in neuromodulation, with about 80% of its afferent fibers terminating in the nucleus tractus solitarius (NTS)

[24]. The NTS projects to key deep brain nuclei, including the locus coeruleus, nucleus basalis, and raphe nucleus, which regulate the release of important neuromodulators. Research involving rats found that stimulating abdominal vagal afferents increases catecholamine neuron activity in the NTS, triggering the release of noradrenaline, acetylcholine, and serotonin from the cerebral cortex [25–29]. These neuromodulators are essential for regulating plasticity changes [25, 26]. While deep brain stimulation of the nucleus basalis can induce significant functional changes in the auditory cortex, vagus nerve stimulation (VNS) can activate similar pathways non-invasively [24, 30, 31]. VNS, when paired with certain frequencies, induces excitability in the primary auditory system, as well as cortex synchrony [24]. Notably, VNS engages not only the nucleus basalis-acetylcholine pathway but also the locus coeruleus-noradrenaline and raphe nucleus-serotonin pathways, potentially leveraging plasticity more than nucleus basalis stimulation alone [25, 26]. In animal models of tinnitus, VNS matched with auditory frequencies has shown promise in reversing tonotopic map changes. However, recent findings suggest that these map reorganizations may be more related to hearing loss than tinnitus itself [32].

Based on the knowledge that cortical plasticity may be affected by Vagus nerve stimulation [33], a study was conducted in which animals underwent a 20-day protocol involving repeated Vagus nerve stimulation paired with the presentation of a mid-frequency sound. This resulted in an increased number of neurons in the area which responds to mid-frequency tones. In a follow-up study, the same protocol was conducted with a high-frequency tone, which yielded the same results for neurons in the high frequency area, but not the mid-frequency region. This suggests that the pairing of certain sounds with Vagus nerve stimulation can induce frequency-specific changes in the primary auditory cortex [33]. Based on this, a study was conducted by Engineer *et al.* which attempted to treat tinnitus by using specific tones to increase the number of neurons at frequencies other than the tinnitus frequency in order to diminish the tinnitus frequency [33]. This study aimed to reverse the plasticity changes which cause tinnitus [34, 35]. The study treated rats with a tinnitus percept by assigning them to receive 4 weeks of vagus nerve and sound stimulation or a sound-only therapy and found that the tinnitus percept of rats in the treatment group was fully eliminated within 2 weeks of treatment. At 3 months follow up, these study rats still had no tinnitus percept, suggesting that the tonotopic map had been permanently changed through this treatment.

Human studies

Following these preclinical results, a study involving 10 tinnitus patients was conducted in which all participants received daily vagus nerve stimulation combined with non-tinnitus frequencies for 2.5 hours during a span of 20 days [36]. Four patients presented with clinically significant improvements in their tinnitus, measured through the Tinnitus Handicap Inventory and minimum masking level. More than 2 months later, these participants' improvements were still stable. Of the 6 patients who did not experience improvements in their tinnitus, 5 of them were on medications that were found to interfere with norepinephrine and acetylcholine release by the vagus nerve, which is crucial to induce neuroplasticity. The objective of another prospective, randomized, double-blind, controlled pilot study was to investigate the impact of paired vagus nerve stimulation in individuals with persistent tinnitus [37]. While not statistically significant, 50% of individuals in the paired vagus nerve stimulation group exhibited clinically significant improvements, compared to 28% in the control group, with half of the subjects still experiencing a clinically significant improvement after one year. A recent systematic review of the relevant literature could not draw definitive conclusions regarding Vagus nerve stimulation for the treatment of tinnitus and larger scale studies were recommended to further establish treatment efficacy [38].

VNS and bimodal auditory-somatosensory stimulation have demonstrated similar effects on spontaneous firing rates and synchrony in auditory structures, suggesting a shared mechanism of action. Various studies have shown that pairing sounds with electrical stimulation of the neck, back, ear, limbs, or tongue can drive extensive neural plasticity across the inferior colliculus, cochlear nucleus, and auditory cortex [12, 13, 39–43]. However, further research is needed to understand the unique and shared mechanisms of these approaches. Since all these parameters are interwoven in a complex manner, optimizing the stimuli provided to patients in bimodal treatment continues to be a significant challenge, especially considering the heterogeneity of conditions like tinnitus. The efficacy of these treatments may depend on tailoring the approach to individual patient characteristics.

Auditory and somatosensory integration

A second method of bimodal stimulation seeks to influence the cochlear nucleus, a deep brain structure. The cochlear nucleus is a region that combines somatosensory information from the dorsal column and the trigeminal systems with auditory information from the cochlea [16, 17]. Because of the spiketiming-dependent plasticity involved in this integration of auditory and somatosensory data, neuron excitability grows or decreases depending on the sequence and temporal interval between auditory and somatosensory inputs [21]. A study by Marks *et al.* found that recordings from the dorsal cochlear nucleus of guinea pigs exhibiting tinnitus-like behaviors show that tinnitus is associated with long-term potentiation, enhanced synchronization, and spontaneous activity [13]. In the study, six different conditions were examined at random using transcutaneous electrical stimulation of the neck and tone presentation (somatosensory stimulation first, then auditory stimulation, or vice versa, at intervals of 5, 10, and 20 ms each). For 25 days, this protocol was conducted on 25 guinea pigs for 20 minutes each day. Their results demonstrated that the greatest long-term depression (LTD) in the dorsal cochlear nucleus cells resulted from auditory stimulation that was administered five milliseconds ahead of somatosensory stimulation. The authors also found a marked drop in tinnitus, which was accompanied by a decline in synchronization and spontaneous firing rate. These results led to the hypothesis that inducing long-term depression should cause a reduction in synchronization and spontaneous firing, which would accompany the suppression of tinnitus. This same study by Marks *et al.* sought to examine a similar treatment protocol in 20 adult human subjects. This was a double-blind, sham-controlled study with either unimodal auditory treatment or bimodal auditory-somatosensory treatment [13]. After four weeks of treatment, the bimodal treatment group experienced a statistically significant −8 dB SPL reduction in tinnitus, whereas the unimodal treatment group did not have a significant change. These results confirm that bimodal but not unimodal stimulation induces LTD to reduce fusiform cell synchrony and spontaneous activity. Another study by Spencer *et al.* also found a statistically significant reduction in TFI scores using a similar protocol in a broader group of tinnitus patients [44]. In this study, all 29 subjects received 30 minutes of bimodal stimulation treatment for 6 weeks. The treatment consisted of electrical stimulation via electrodes on the jaw or neck, based on the subject's ability to modulate their symptoms via movement of these regions. Auditory stimulation was delivered via headphones at the subject's individual tinnitus frequency, which was achieved using a pitch-matching technique. The study found a statistically significant decrease in average TFI scores (6.9 points) at the follow up (9–12 weeks) [44].

Tongue Stimulation

Another method of bimodal stimulation combines tone therapy and electrical tongue stimulation, which is based on earlier studies on the role of the somatosensory system in tinnitus. Forty-five patients with chronic tinnitus participated in an open-label pilot trial [45] where they were given electrical stimulation of the tongue twice a day for 30 minutes along with a broad-band spectrum noise with temporal fluctuations over a period of 10 weeks. Average reductions of 8.1 dB for low masking levels and 5.4 dB for tinnitus loudness match were observed with this intervention, along with a significant average reduction of 8.6 points on the tinnitus handicap questionnaire. Overall, 64% of the participants had a minimal important difference (≥5.3 points) for minimal masking levels. Another large, randomized, double-blind clinical study involving 326 patients with tinnitus was conducted following this pilot study. Three arms, each with varied parameter settings of tongue stimulation, were randomly assigned to participants [46]. In order to examine the long-term therapeutic effects following the conclusion of treatment, participants were assessed three times throughout the 12-week treatment period and once again up to a year later. During the course of the 12-week treatment, 84% of participants used the device following the minimum compliance level of 36 hours, and 75–88% of these individuals had improved tinnitus handicap inventory or tinnitus functional index scores by the end of the program. In total, 63% of all enrolled individuals had improvements in their tinnitus handicap inventory (THI) or tinnitus functional index scores, with an average difference of almost 14 points in both surveys between the beginning and conclusion of therapy. The improvement achieved by the 12-week treatment persisted for another 12 months after it was completed for most participants. This is slightly better compared to a 10.91 reduction in the THI for cognitive behavioral therapy based on a recent Cochrane review [47].

Modulation of Neural Circuits

The synchronized delivery of auditory and somatosensory stimuli is thought to modulate neural circuits within the auditory cortex and brainstem. This modulation may involve:

(1) *Spike-timing-dependent plasticity (STDP)*: The precise temporal correlation of inputs from different sensory modalities can strengthen or weaken synaptic connections, potentially altering the neural representation of tinnitus [22, 23].
(2) *Lateral inhibition*: Bimodal stimulation may enhance lateral inhibition within the auditory system, helping to suppress the aberrant neural activity associated with tinnitus [11].
(3) *Attentional modulation*: By providing novel sensory inputs, bimodal neurostimulation may help shift attention away from the tinnitus percept, reducing its salience and intrusiveness [9].
(4) *Neuromodulator release*: The combined stimulation may trigger the release of neuromodulators such as dopamine and acetylcholine, which play crucial roles in neural plasticity and attentional processes [24].

Devices and Technology

Clinical evidence

Lenire

Lenire, invented by Hubert Lim, Ph.D. at the University of Minnesota, developed by Neuromod Devices, is the first FDA-approved bimodal neurostimulation device for tinnitus treatment. The system consists of three main components:

(1) Bluetooth headphones: Deliver customized audio stimulation.
(2) Tonguetip®: A non-invasive intra-oral device that provides mild electrical stimulation to the tongue.
(3) Controller: Coordinates the delivery of auditory and somatosensory stimuli.

The Lenire system is designed for home use, with a typical treatment protocol involving two daily 30 minute sessions over a period of 12 weeks.

The device settings can be adjusted by healthcare professionals to optimize treatment for individual patients.

Several trials have investigated the efficacy of the Lenire device in tinnitus treatment. The most notable among these are the TENT-A1, TENT-A2, and TENT-A3 trials, which have provided some modest evidence supporting the effectiveness of this neuromodulatory approach.

TENT-A1 trial

The TENT-A1 trial demonstrated the potential of bimodal neurostimulation for tinnitus treatment [32]. This double-blind, randomized, parallel-arm study included 326 adults with chronic tinnitus. The subjects were given a treatment protocol which included Bluetooth headphones for auditory stimulation, Tonguetip® for electrical stimulation of the tongue, and a handheld controller for managing stimuli delivery. Subjects underwent 60 minutes of daily treatment for 12 weeks, as well as a 12-month follow-up. All three groups were treated with sound and tongue stimulation. Each arm differed both in the frequency of sound stimulation and the timing between electrical and auditory stimuli. The treatment group utilized the Lenire device (PS1 stimulation setting), and the outcome of the treatment was measured utilizing the Tinnitus Handicap Inventory (THI) and Tinnitus Functional Index (TFI). The key findings were that 86.2% of treatment-compliant participants experienced a significant reduction in tinnitus severity after the 12-week treatment period. The improvement in TFI and THI scores ranged between 74.7 and 88% across the three treatment arms. Additionally, 80.1% of treatment-compliant participants (83.7%) maintained improvements for at least 12 months post-treatment. The treatment was well-tolerated, with no serious adverse events reported.

TENT-A2 trial

The TENT-A2 trial built on the success of TENT-A1, focusing on optimizing stimulation parameters and investigating the effects of changing treatment stimuli [47]. This study included 191 adults with chronic tinnitus, in which subjects underwent the same treatment protocol as in TENT-A1, but the stimulation settings were changed after 6 weeks of treatment. The different stimulation settings used were classified as PS1, PS2, and PS3. PS1 consisted of simultaneous bimodal stimulation, PS2 had short interstimulus delays (30–50 ms), and PS3 used lower frequency tones with longer interstimulus delays (550–950 ms). The study achieved an 83.8% treatment compliance, as 36 subjects did not appear for follow up clinic visits. The notable results were that 95% of treatment-compliant participants reported a reduction in tinnitus severity. Additionally, 91% of treatment-compliant participants maintained improvements for at least 12 months after the end of treatment. The TENT-A2 trial demonstrated that changing stimulation settings midway through treatment resulted in greater improvements compared to TENT-A1. The study found that participants who received PS1 stimulation for the first 6 weeks followed by PS2 stimulation for the next 6 weeks showed the greatest improvements in tinnitus symptoms.

TENT-A3 trial

The TENT-A3 trial compared bimodal neurostimulation with sound therapy alone, and included 112 adults, in which all participants underwent sound-only treatment for 6 weeks, followed by auditory stimulation paired with electrical stimulation of the tongue for another 6 weeks [32]. Of the 112 initial participants, 83.4% remained in the study and were treatment-compliant by the end of 12 weeks. Of these subjects, 79.4% of them experienced clinically significant improvements (a decrease of 7 or more points in TFI), wherein bimodal neurostimulation showed superior efficacy compared to sound therapy alone for 70.5% of patients with moderate and above tinnitus. These results support earlier findings in the University of Michigan Study [13], in which it was found that bimodal stimulation resulted in significantly greater improvements in TFI scores, and tinnitus loudness compared to unimodal stimulation. The above data builds on an earlier study by Vanneste *et al.*, 2010, which utilized transcutaneous electrical stimulation combined with sound therapy, resulting in a statistically significant reduction in TFI scores.

A recent case series on the Lenire Device in a group of 20 subjects who used the device 60 minutes daily for 10 weeks showed a mean decrease in THI score of 10.4 points [48].

Limitations

The Lenire device's clinical trials revealed several significant limitations that impact the interpretation of its efficacy [32, 47]. Notably, the trials lacked a proper placebo control group, which is essential for differentiating between actual physiological effects and psychological effects due to patient expectations. Instead, the control group used an audio-only stimulus (PS9), which does not provide a valid comparison to standard sound therapy. The small size of the control group (only 13 participants out of 191 in TENT-A2) further undermines the statistical power of the study. Additionally, the trials did not compare Lenire to other established tinnitus treatments, such as cognitive behavioral therapy (CBT) or tinnitus retraining therapy (TRT). The protocol's requirement for participants to engage in 30 minutes of relaxation twice daily while listening to calming audio files introduces a confounding variable, as stress reduction alone could account for some of the observed benefits. The reduction in the TFI and THI in the Lenire studies is not substantially different from sound therapy or CBT studies.

The trials were described as randomized and double-blinded, but the randomization and blinding were not clinically meaningful, as participants were randomized into groups with slightly different device settings rather than into a true treatment and control group. Furthermore, the absence of an intention-to-treat (ITT) analysis and the focus on questionnaire-based outcomes without adequate subjective measures of patient satisfaction limit the robustness of the findings. Lastly, a significant proportion of participants reported worsening tinnitus symptoms during the treatment period, raising concerns about the device's safety and tolerability for all patients.

The Lenire studies have shown a reduction in the THI of between 10.4 to 18.5 which is not significantly better than studies using sound therapy alone. There may be a subset of patients who achieve significant benefit from the Lenire device, but that population of patients is currently undefined.

Neosensory Duo

Another device utilizing bimodal neurostimulation is Neosensory Duo, which delivers vibratory stimulation to the wrist instead of the tongue. While much less extensively studied than Lenire, preliminary research has shown promising results for this approach. Key features of Duo include:

- Wrist-worn stimulator for ease of use.
- Customizable sound therapy delivered through headphones.
- Adjustable stimulation intensity.

Only one clinical study has investigated the efficacy of bimodal stimulation through the Neosensory Duo device, which combines auditory and somatosensory (vibratory) stimuli, for the treatment of tinnitus [49]. The study utilized a wristband with vibratory motors paired with sound stimulation delivered through a smartphone application [49]. This approach aimed to explore an alternative to the traditional bimodal stimulation methods that typically involve electrical stimulation of the tongue.

Methodology

The study included 45 participants with Tinnitus Functional Index (TFI) scores of 25 and above that were randomly assigned to either the treatment group receiving bimodal stimulation (sound paired with wrist vibration), or the control group, receiving sound-only stimulation. Both groups received the same type of auditory stimulation, in which tones were played to match the subject's main tinnitus frequency. The treatment protocol consisted of 10-minute daily sessions over an 8-week period. The auditory stimuli were centered on each participant's tinnitus frequency, and the wristband vibrated in spatially and temporally correlated patterns with the tones [49].

Results

The study's primary outcome measure was the change in TFI scores. Results indicated a significantly greater reduction in TFI scores for the experimental group compared to the control group [49]. Specifically, participants in the experimental group with baseline TFI scores of 50 or above showed an average reduction of 21.8 points, while the control group showed an average reduction of 4.6 points. Overall, 91% of participants in the experimental group exhibited improvements, with 64% achieving clinically significant reductions (a decrease of 13 or more points in TFI scores [49].

Limitations

The study had several limitations. The sample size was relatively small, which may limit the generalizability of the findings. Additionally, the study did not include a long-term follow-up to assess the durability of the treatment effects beyond the 8-week period. Future studies with larger sample sizes, longer follow-up periods, and the inclusion of sham controls would help to further validate the efficacy of this bimodal stimulation approach. Also, patients with a higher TFI are expected to have a larger change due to the "regression to the mean" effect. In this effect that is well-known in clinical trials, patients with more extreme initial scores tend to have a larger change than those who are closer to the mean.

Auricle Inc. Device

The Auricle Inc. device utilizes electrical stimulation through skin electrodes placed on the neck or jaw paired with auditory stimulation delivered 5 milliseconds after each electrical pulse. The device is not yet FDA approved or available on the market like the Lenire and Neosensory devices. Figure 1 demonstrates a schematic representation of the three mentioned devices, utilized in bimodal neurostimulation.

Preclinical studies

Preclinical studies at the University of Michigan under Susan Shore, Ph.D. using animal models demonstrated that auditory neurons exhibit increased spontaneous firing rates and synchronization along the auditory pathway, correlating with behavioral measures of tinnitus [22, 50, 51]. The dorsal cochlear nucleus (DCN) fusiform cells, which integrate auditory and somatosensory information, showed significant reductions in spontaneous firing rates, bursting, and synchrony with bisensory stimulation. These findings informed the development of the treatment parameters for human clinical trials.

Clinical evidence

The efficacy of the Auricle device was investigated through a double-blind, crossover, single-center randomized clinical trial conducted from March 2019 to July 2022 [52]. The study included 99 participants with symptomatic tinnitus, normal to moderate hearing loss, and no other tinnitus treatments in the prior six months. Participants were randomized into two groups: one receiving bisensory treatment first, followed by auditory-only treatment, and the other receiving the treatments in reverse order. Each treatment phase lasted six weeks, followed by a six-week washout phase. Participants used the device for 30 minutes daily during treatment phases.

Results

The bisensory treatment resulted in clinically significant decreases in Tinnitus Functional Index (TFI) scores at week 6 of phase 1 (intent to treat (ITT): –12.0 points; per protocol (PP): –13.2 points) and further reductions at week 12 of phase 2 (ITT: –14.1 points; PP: –17.0 points) [52]. Tinnitus loudness levels decreased by more than 6 dB sensation level (SL) at week 6 and up to 11 dB SL at week 12. These reductions extended into the washout phase, indicating a prolonged treatment effect. In contrast, the auditory-only treatment showed minimal changes in TFI scores and tinnitus loudness levels.

Limitations

The study had a few minor limitations, including a possible carryover effect due to the prolonged treatment effect, a relatively modest sample size, and the possibility of a placebo effect. Due to the potential long-term effect of this device, a cross-over design is not ideal. Additionally, a significant number of participants withdrew from the study due to the burden of weekly in-clinic visits, and there was no long-term follow-up to assess the durability of the treatment effects beyond the 12-week period. Table 1 provides a comparison of the mentioned devices in different characteristics such as cost, efficacy, risks, side effects, FDA approval, treatment duration, stimulation type, and ease of use.

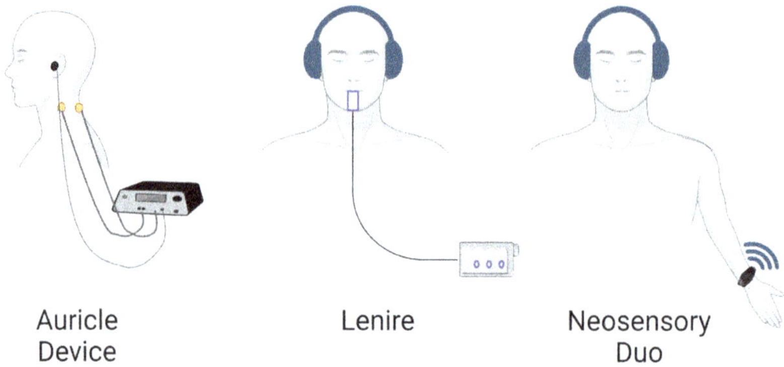

Figure 1. Bimodal Stimulation Devices. Created with BioRender.com.

Comparison of Devices

Table 1. Lenire vs Neosensory Duo

	Lenire	Neosensory Duo	Auricle Inc. Device
Cost	\$4,000–\$6,000	\$249 per month for two months, with the option to purchase after rental	Unknown. Not available to purchase
Efficacy	80–91% of participants reported significant reduction in tinnitus severity	87% of users reported symptom relief within three weeks	Tinnitus loudness levels decreased by more than 6 dB sensation level (SL) at week 6 and up to 11 dB SL at week 12
Risks/Side Effects	Mild and transient (tongue tingling, temporary taste changes, mild headache)	Mild and transient (wrist discomfort, mild skin irritation)	Mild and transient (neck/jaw discomfort, mild skin irritation)
FDA Approval	FDA-approved (March 2023)	Not FDA-approved, but endorsed by leading scientists	Not FDA-approved
Treatment Duration	12 weeks (30–60 min daily sessions)	8 weeks (10 min daily sessions)	6 weeks (30 min daily sessions)
Stimulation Type	Auditory (headphones) + Tongue stimulation	Auditory (headphones) + Wrist stimulation	Auditory (headphones) + Neck/jaw stimulation
Ease of Use	User-friendly, at-home use with professional guidance	Wrist-worn stimulator, user-friendly, app-based	Requires professional guidance to be used by patients at home
Duration of Effect	Sustained for at least 12 months post-treatment	Effects reported to last beyond the 8-week program	No follow-up to test effectiveness beyond 12-week period
Availability	Available in the US	Available in the US	Not available

Research and Development

Research and development in bimodal neurostimulation for tinnitus treatment continues to advance in several key areas. Researchers are focusing on optimizing stimulation parameters by investigating the effects of different patterns, intensities, and durations to maximize treatment efficacy. This includes exploring various combinations of auditory stimuli and somatosensory stimulation, as demonstrated in studies using pure tones, wideband noise, and electrical stimulation of different body regions such as the tongue, face, neck, and ear [38]. Personalization of treatment is another crucial area, with efforts to develop algorithms that can tailor stimulation protocols based on individual tinnitus characteristics and treatment responses. Combination therapies are being explored to investigate potential synergistic effects of bimodal neurostimulation with other tinnitus treatments, such as cognitive behavioral therapy or pharmacological interventions. Long-term efficacy studies are underway to assess the durability of treatment effects and determine optimal maintenance protocols, with some studies showing sustained benefits for up to 12 months post-treatment [20]. Advanced neuroimaging techniques are being utilized to better understand the neural mechanisms underlying the effects of bimodal neurostimulation, providing insights into how this approach modulates brain activity within auditory regions associated with tinnitus.

Innovative approaches are also being developed to enhance the effectiveness and accessibility of bimodal neurostimulation. These include miniaturization efforts to create more compact and discreet devices, improving patient comfort and compliance. Integration of advanced wireless technologies is being pursued to enable seamless data tracking and remote monitoring, while some researchers are exploring the potential of combining bimodal stimulation with immersive virtual reality environments to enhance patient engagement and potentially improve outcomes. The development adaptive stimulation protocols based on physiological feedback is another promising avenue. Additionally, investigations into alternative locations for somatosensory stimulation are ongoing, aiming to optimize treatment effects and patient comfort.

Conclusion

- Bimodal neurostimulation combines auditory and somatosensory stimuli to modulate neural activity in the auditory system and associated brain regions, targeting the underlying pathophysiology of tinnitus.
- This approach leverages neuroplasticity to potentially reverse maladaptive changes associated with tinnitus, aiming to reduce hyperactivity and hypersynchrony in neural circuits.
- Large-scale clinical trials have demonstrated the safety and efficacy of bimodal neurostimulation, with a majority of participants reporting significant reductions in tinnitus severity.

- Bimodal stimulation has shown superior efficacy compared to sound therapy alone, offering hope for patients who do not respond well to conventional tinnitus treatments.
- Ongoing research focuses on optimizing stimulation parameters, personalizing treatment protocols, and exploring potential synergies with other tinnitus therapies.
- The Lenire, Neosensory Duo, and Susan Shore devices utilize distinct bimodal neurostimulation paradigms for tinnitus intervention.
- All three devices exhibit favorable safety profiles with only mild adverse effects reported and have shown promising results in attenuating tinnitus severity through neuromodulatory mechanisms.

References

1. Axelsson A, and Ringdahl A. Tinnitus — A study of its prevalence and characteristics. *British Journal of Audiology*. **23**(1):53–62 (1989).
2. Shargorodsky J, Curhan GC, and Farwell WR. Prevalence and characteristics of tinnitus among US adults. *American Journal of Medicine*. **123**(8):711–718 (2010).
3. Langguth B, Landgrebe M, Kleinjung T, Sand GP, and Hajak G. Tinnitus and depression. *World Journal of Biological Psychiatry*. **12**(7):489–500 (2011 Oct).
4. Coco L, Hooker ER, Gilbert TA, Harker GR, Clark KD, Reavis KM, Henry JA, Zaugg TL, and Carlson KF. The impact of tinnitus severity on work functioning among U.S. military veterans with tinnitus. *Seminars in Hearing*. **45**(1):40–54 (2023 Jul 24).
5. Demoen S, Cardon E, Jacquemin L, Timmermans A, Van Rompaey V, Gilles A, and Michiels S. Health-related quality of life in subjective, chronic tinnitus patients: A scoping review. *Journal of the Association for Research in Otolaryngology*. **25**(2):103–129 (2024 Apr).
6. Eggermont JJ, and Roberts LE. The neuroscience of tinnitus. *Trends in Neurosciences*. **27**(11):676–682 (2004 Nov).
7. Elgoyhen AB, Langguth B, De Ridder D, and Vanneste S. Tinnitus: Perspectives from human neuroimaging. *Nature Reviews Neuroscience*. **16**(10):632–642 (2015 Oct).
8. Baguley D, McFerran D, and Hall D. Tinnitus. *The Lancet*. **382**(9904):1600–1607 (2013).
9. Shore SE, Roberts LE, and Langguth B. Maladaptive plasticity in tinnitus — Triggers, mechanisms and treatment. *Nature Reviews Neurology*. **12**(3):150–160 (2016).

10. Schaette R, and McAlpine D. Tinnitus with a normal audiogram: Physiological evidence for hidden hearing loss and computational model. *Journal of Neuroscience*. **31**(38):13452–13457 (2011).
11. Roberts LE, Eggermont JJ, Caspary DM, Shore SE, Melcher JR, and Kaltenbach JA. Ringing ears: The neuroscience of tinnitus. *Journal of Neuroscience*. **30**(45):14972–14979 (2010).
12. Gu JW, Herrmann BS, Levine RA, and Melcher JR. Brainstem auditory evoked potentials suggest a role for the ventral cochlear nucleus in tinnitus. *Journal of the Association for Research in Otolaryngology*. **13**(5):819–833 (2012).
13. Marks KL, Martel DT, Wu C, Basura GJ, Roberts LE, Schvartz-Leyzac KC, and Shore SE. Auditory-somatosensory bimodal stimulation desynchronizes brain circuitry to reduce tinnitus in guinea pigs and humans. *Science Translational Medicine*. **10**(422):eaal3175 (2018).
14. Itoh K, Kamiya H, Mitani A, Yasui Y, Takada M, and Mizuno N. Direct projections from the dorsal column nuclei and the spinal trigeminal nuclei to the cochlear nuclei in the cat. *Brain Research*. **400**(1):145–150 (1987).
15. Kanold PO, and Young ED. Proprioceptive information from the pinna provides somatosensory input to cat dorsal cochlear nucleus. *Journal of Neuroscience*. **21**(19):7848–7858 (2001).
16. Shore SE, El Kashlan H, and Lu J. Effects of trigeminal ganglion stimulation on unit activity of ventral cochlear nucleus neurons. *Neuroscience*. **119**(4):1085–1101 (2003).
17. Zhou J, and Shore SE. Convergence of spinal trigeminal and cochlear nucleus projections in the inferior colliculus of the guinea pig. *Journal of Comparative Neurology*. **495**(1):100–112 (2006).
18. Baizer JS, Manohar S, Paolone NA, Weinstock N, and Salvi RJ. Understanding tinnitus: The dorsal cochlear nucleus, organization and plasticity. *Brain Research*. **1485**:40–53 (2012 Nov 16).
19. Osen KK, Storm-Mathisen J, Ottersen OP, and Dihle B. Glutamate is concentrated in and released from parallel fiber terminals in the dorsal cochlear nucleus: A quantitative immunocytochemical analysis in guinea pig. *Journal of Comparative Neurology*. **357**(4):482–500 (1995).
20. Fujino K, and Oertel D. Bidirectional synaptic plasticity in the cerebellum-like mammalian dorsal cochlear nucleus. *Proceedings of the National Academy of Sciences*. **100**(1):265–270 (2003).
21. Tzounopoulos T, Kim Y, Oertel D, and Trussell LO. Cell-specific, spike timing-dependent plasticities in the dorsal cochlear nucleus. *Nature Neuroscience*. **7**(7):719–725 (2004).
22. Koehler SD, Shore SE. Stimulus-timing dependent multisensory plasticity in the guinea pig dorsal cochlear nucleus. *PLoS One*. **8**(3):e59828 (2013).

23. Koehler SD, and Shore SE. Stimulus timing-dependent plasticity in dorsal cochlear nucleus is altered in tinnitus. *Journal of Neuroscience.* **33**(50):19647–19656 (2013).
24. Engineer ND, Moller AR, and Kilgard MP. Directing neural plasticity to understand and treat tinnitus. *Hearing Research.* **295**:58–66 (2013).
25. Hulsey DR, Hays SA, Khodaparast N, Ruiz A, Das P, Rennaker RL 2nd, and Kilgard MP. Reorganization of motor cortex by vagus nerve stimulation requires cholinergic innervation. *Brain Stimulation.* **9**(2):174–181 (2016).
26. Hulsey DR, Shedd CM, Sarker SF, Kilgard MP, and Hays SA. Norepinephrine and serotonin are required for vagus nerve stimulation directed cortical plasticity. *Experimental Neurology.* **320**:112975 (2019).
27. Manta S, Dong J, Debonnel G, and Blier P. Enhancement of the function of rat serotonin and norepinephrine neurons by sustained vagus nerve stimulation. *Journal of Psychiatry & Neuroscience.* **34**(4):272–280 (2009).
28. Sumal KK, Blessing WW, Joh TH, Reis DJ, and Pickel VM. Synaptic interaction of vagal afferents and catecholaminergic neurons in the rat nucleus tractus solitarius. *Brain Research.* **277**(1):31–40 (1983).
29. Van Bockstaele EJ, Peoples J, and Telegan P. Efferent projections of the nucleus of the solitary tract to peri-locus coeruleus dendrites in rat brain: Evidence for a monosynaptic pathway. *Journal of Comparative Neurology.* **412**(3):410–428 (1999).
30. Kilgard MP, and Merzenich MM. Cortical map reorganization enabled by nucleus basalis activity. *Science.* **279**(5357):1714–1718 (1998).
31. Kilgard MP, and Merzenich MM. Plasticity of temporal information processing in the primary auditory cortex. *Nature Neuroscience.* **1**(8):727–731 (1998).
32. Langers DR, de Kleine E, and van Dijk P. Tinnitus does not require macroscopic tonotopic map reorganization. *Frontiers in Systems Neuroscience.* **6**:2 (2012).
33. Engineer ND, Riley JR, Seale JD, Vrana WA, Shetake JA, Sudanagunta SP, and Kilgard MP. Reversing pathological neural activity using targeted plasticity. *Nature.* **470**(7332):101–104 (2011).
34. Komiya H, and Eggermont JJ. Spontaneous firing activity of cortical neurons in adult cats with reorganized tonotopic map following pure-tone trauma. *Acta Otolaryngologica.* **120**(6):750–756 (2000).
35. Muhlnickel W, Elbert T, Taub E, and Flor H. Reorganization of auditory cortex in tinnitus. *Proceedings of the National Academy of Sciences of the United States of America.* **95**(17):10340–10343 (1998).
36. De Ridder D, Kilgard MP, Engineer ND, and Vanneste S. Placebo-controlled vagus nerve stimulation paired with tones in a patient with refractory tinnitus: A case report. *Otolaryngology-Head and Neck Surgery.* **36**(4):575–580 (2015).

37. Tyler RS, Cacace AT, Stocking C, Tarver B, Engineer ND, Martin J, *et al.* Vagus nerve stimulation paired with tones for the treatment of tinnitus: A prospective randomized double-blind controlled pilot study in humans. *Scientific Reports*. **7**(1):11960 (2017).
38. Stegeman I, Velde HM, Robe PAJT, Stokroos RJ, and Smit AL. Tinnitus treatment by vagus nerve stimulation: A systematic review. *PLoS One*. **16**(3): e0247221 (2021 Mar 11).
39. Basura GJ, Koehler SD, and Shore SE. Multi-sensory integration in brainstem and auditory cortex. *Brain Research*. **1485**:95–107 (2012).
40. Gloeckner CD, Smith BT, Markovitz CD, and Lim HH. A new concept for noninvasive tinnitus treatment utilizing multimodal pathways. *Conference Proceedings: IEEE Engineering in Medicine and Biology Society*. **2013**:3122–3125 (2013).
41. Markovitz CD, Smith BT, Gloeckner CD, and Lim HH. Investigating a new neuromodulation treatment for brain disorders using synchronized activation of multimodal pathways. *Scientific Reports*. **5**:9462 (2015).
42. Vanneste S, Plazier M, Van de Heyning P, and De Ridder D. Transcutaneous electrical nerve stimulation (TENS) of upper cervical nerve (C2) for the treatment of somatic tinnitus. *Experimental Brain Research*. **204**(2):283–287 (2010).
43. Conlon B, Hamilton C, Hughes S, Meade E, Hall DA, Vanneste S, and Langguth B. Different bimodal neuromodulation settings reduce tinnitus symptoms in a large randomized trial. *Scientific Reports*. **12**(1):1–15 (2022).
44. Spencer S, Mielczarek M, Olszewski J, Sereda M, Joossen I, Vermeersch H, Gilles A, and Michiels S. Effectiveness of bimodal auditory and electrical stimulation in patients with tinnitus: A feasibility study. *Frontiers in Neuroscience*. **16**:971633 (2022 Aug 24).
45. Hamilton C, D'Arcy S, Pearlmutter BA, Crispino G, Lalor EC, and Conlon BJ. An investigation of feasibility and safety of bi-modal stimulation for the treatment of tinnitus: An open-label pilot study. *Neuromodulation*. **19**(8):832–837 (2016).
46. Conlon B, Langguth B, Hamilton C, Hughes S, Meade E, Connor CO, and Lim HH. Bimodal neuromodulation combining sound and tongue stimulation reduces tinnitus symptoms in a large randomized clinical study. *Science Translational Medicine*. **12**(564): (2020).
47. Fuller T, Cima R, Langguth B, Mazurek B, Vlaeyen JW, and Hoare DJ. Cognitive behavioural therapy for tinnitus. *Cochrane Database of Systematic Reviews*. **1**:CD012614 (2020).
48. Buechner A, Lesinski-Schiedat A, Becker P, and Lenarz T. Real-world clinical experience with bimodal neuromodulation for the treatment of tinnitus — A case series. *Brain Stimulation*. **15**(2):383–384 (2022 Mar–Apr).

49. Perrotta MV, Kohler I, and Eagleman DM. Bimodal stimulation for the reduction of tinnitus using vibration on the skin. *International Tinnitus Journal*. **27**(1):1–5 (2023).
50. Riffle TL, Martel DT, Jones GR, and Shore SE. Bimodal auditory-electrical stimulation for the treatment of tinnitus: Pre-clinical and clinical studies. *Frontiers in Neurology*. **12**:8058117 (2021).
51. Wu C, Martel DT, and Shore SE. Increased synchrony and bursting of dorsal cochlear nucleus fusiform cells correlate with tinnitus. *Journal of Neuroscience*. **36**(6):2068–2073 (2016).
52. Jones GR, Martel DT, Riffle TL, Errickson J, Souter JR, Basura GJ, Stucken E, Schvartz-Leyzac KC, and Shore SE. Reversing synchronized brain circuits using targeted auditory-somatosensory stimulation to treat phantom percepts: A randomized clinical trial. *JAMA Network Open*. **6**(6):e2315914 (2023 Jun 1).

https://doi.org/10.1142/9789819807345_0010

Chapter 10

Neural Therapy (Botulinum Toxin) for Tinnitus

Martha Lucía Gutiérrez Pérez[*,†], Ella J. Lee[*], and Mehdi Abouzari[*,‡]

[*]*Department of Otolaryngology-Head and Neck Surgery, University of California, Irvine, CA, USA*

[†]*Universidad El Bosque, Colombian School of Medicine, Otorhinolaryngology Interest Group UEB (ORLIG-UEB), Bogotá, Colombia*

[‡]*mabouzar@hs.uci.edu*

Abstract

Tinnitus, the perception of sound originating internally, poses a significant challenge for millions worldwide. Despite its complexity, emerging therapies offer hope for relief. One such approach, neural therapy with botulinum toxin (BTX), targets the neuroplastic changes within the central auditory system involved in tinnitus perception. This chapter provides a comprehensive overview of the rationale, efficacy, and clinical evidence of BTX therapy in tinnitus management. By disrupting maladaptive neuroplastic processes, BTX injections hold promise in restoring the balance of excitatory and inhibitory neurotransmissions within auditory circuits. Clinical trials, along with multiple case reports, demonstrate encouraging outcomes, particularly in cases of middle ear myoclonic tinnitus (MEMT) and palatal myoclonus (PM), where BTX therapy provides sustained symptom relief and improved quality of life with minimal adverse effects. Furthermore, existing evidence supports the use of BTX in managing the symptoms of migraine, a common comorbidity and often a contributing factor to tinnitus. Intramuscular injections of Onabotulinumtoxin A (OnaA) into craniocervical muscles have also shown promise in alleviating tinnitus associated with chronic migraine (CM), suggesting broader therapeutic applications for neurological comorbidities. Overall, neural therapy with BTX offers a novel, individualized approach to tinnitus management, addressing specific subtypes and underlying pathophysiological mechanisms. With ongoing research and clinical trials, BTX emerges as a safe and effective treatment option for tinnitus, including refractory cases, improving patient outcomes and quality of life.

Introduction

Understanding tinnitus

Tinnitus, characterized by the perception of sound originating within the body rather than from an external source, is a prevalent condition experienced by nearly everyone at some point. According to the American Tinnitus Association, approximately 10 million individuals have experienced or currently face tinnitus, with cases also reported in non-industrialized countries [1]. Tinnitus can arise from various origins (Table 1) and can be categorized as either subjective or objective. The majority of tinnitus cases are subjective, meaning they cannot be detected by an examiner or measured using conventional tools. A widely agreed notion is that subjective tinnitus originates from cochlear dysfunction, including hearing loss of any cause, that triggers an aberrant central neuroplastic response [2, 3]. Atypical migraines associated with inner ear dysfunction, including vestibular migraine and cochlear migraine, are more recently explored etiologies of subjective tinnitus. These forms of migraine are thought to induce or exacerbate tinnitus through heightened sensory sensitivity to auditory stimuli during attacks [4]. On the other hand, objective tinnitus can be heard by an examiner, and can be caused

Table 1. Tinnitus etiology [1, 4, 6].

Tinnitus Causes	
Noise trauma	• Most common cause • Particularly prevalent in occupations exposed to loud environments
Certain diseases and syndromes	• Aneurysm • Diabetes • Heart disease • Hypertension
Drugs	• Aspirin at high doses (resolves upon discontinuation)
Atypical migraine	• Vestibular migraine; cochlear migraine, basilar-type migraine • Enhanced sensory sensitivity during migraine attacks • Trigeminal nerve stimulation disrupting cochlear function
Ear diseases	• Ménière's disease • Lesions affecting the eighth cranial nerve • Hearing loss • Eustachian tube dysfunction
Chiari malformation	• Infrequent cause • Due to the downward displacement of the cerebellar tonsils, causing tension on the auditory nerve

by vascular or neurological disorders, eustachian tube dysfunction, or, less commonly, spasms of the middle ear or soft palate muscles [5]. In this chapter, we focus on tinnitus etiologies related to muscular dysfunctions in the head and neck, as well as neurological disorders.

The pathophysiology of tinnitus is yet to be fully understood, but it is known to involve various physiological and psychological factors. Although stress itself may not directly cause tinnitus, the distressing nature of the condition can activate the body's fight-or-flight response, akin to stress, thereby exacerbating the overall experience. Magnetic Resonance Imaging (MRI) studies have identified multiple brain regions implicated in tinnitus processing, including cognitive, emotional, and auditory centers, with initial sound perception routed through the amygdala [1].

Neural Therapy with Botulinum Toxin

Rationale for botulinum toxin therapy in tinnitus

Botulinum toxin (BTX) is a potent neurotoxin widely utilized in various medical contexts. It acts by targeting the cholinergic nerve endings and inhibiting the secretion of acetylcholine from synaptic vesicles through the disruption of the soluble *N*-ethylmaleimide-sensitive factor attachment protein receptor (SNARE) complex at the neuromuscular junction. This results in neuromuscular blockade and consequent muscle relaxation [7].

One proposed mechanism for using BTX in tinnitus management is the modulation of neuroplasticity within the central auditory system. Neuroplasticity, the brain's ability to reorganize and adapt in response to sensory information and experiences, plays a critical role in tinnitus pathophysiology. Maladaptive neuroplastic changes, such as aberrant synaptic connectivity and altered neurotransmitter dynamics, contribute to the generation and persistence of tinnitus perception [8]. By modulating synaptic transmission and altering neuronal firing patterns, BTX injections could potentially induce long-lasting changes in the neural circuitry underlying tinnitus, leading to symptom relief and improved auditory perception [9].

Moreover, BTX inhibits the release of pro-inflammatory and excitatory neurotransmitters and neuropeptides, such as substance P, calcitonin gene-related peptide, and glutamate, from primary sensory fibers [10]. These substances are responsible for transmitting pain signals and contribute to the development of peripheral and central sensitization — a process closely linked to the onset of migraine, which is frequently associated with tinnitus [11, 12]. In other words, BTX may exert a direct antinociceptive effect through its influence on the autonomic nervous system, in addition to its neuromuscular effect [13].

This dual action — targeting both the neuromuscular system and the autonomic nervous system — suggests a broader therapeutic potential for BTX in managing tinnitus symptoms.

Efficacy, Considerations, and Clinical Evidence

Currently, there is a lack of extensive randomized controlled trials evaluating the efficacy of BTX as a treatment for tinnitus. However, existing studies and numerous case series suggest its prospective benefits. Evidence supports BTX's promise in managing both objective tinnitus related to myoclonus of the middle ear and palate, and subjective tinnitus associated with neurological and vestibulocochlear disorders.

In an animal study and a nonrandomized controlled clinical trial in humans, Park *et al.* demonstrated that BTX injection serves as a novel therapeutic modality for treating MEMT [14]. In their animal study, male Sprague-Dawley rats received an intratympanic injection of 1 unit (U)/ 100 μL of BTX-A, while control ears were injected with normal saline. Analysis of hearing thresholds, cochlear morphometry, electron microscopy examination, and immunofluorescence of middle ear muscles (MEMs) revealed that intratympanic BTX-A effectively induces chemodenervating effects on the MEMs, resulting in reduced tinnitus symptoms [14]. Their clinical trial involved 10 patients with refractory MEMT. The affected ear was injected intratympanically with 0.4 cc of 1 U/100 μL BTX-A (Figure 1). Three patients received 0.1 U of BTX, three received 0.5 U, and the remaining four patients received 1 U. Pure-tone thresholds remained stable in all patients over the three-month period, confirming the procedure's safety (Wilcoxon signed-rank test, $p > 0.05$). Additionally, 75% of patients with initially perturbed acoustic reflex decay tympanometry showed improvement with a stabilized waveform three months after injection [14].

Statistically significant improvements were observed in the average tinnitus visual analog scale (VAS) scores (Wilcoxon signed-rank test,

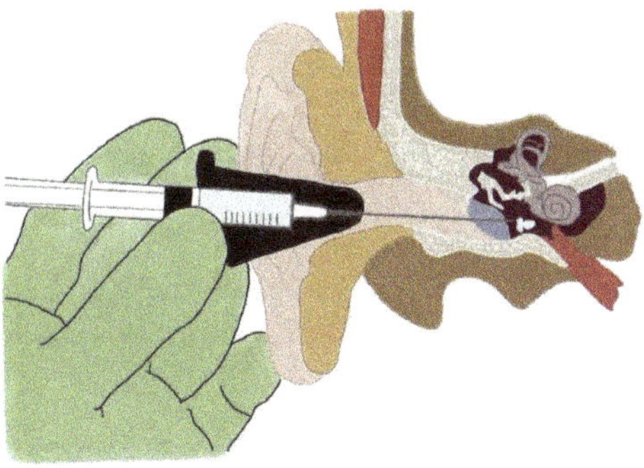

(a)

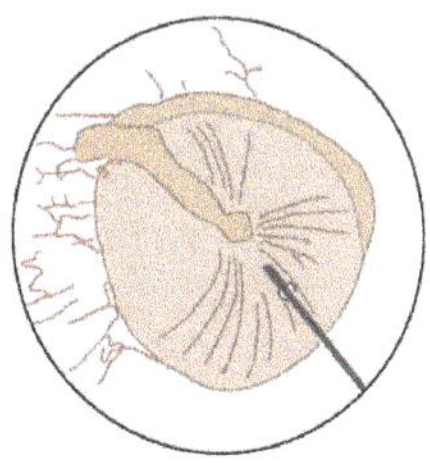

(b)

Figure 1. The right ear is depicted: (a) coronal schematic view showing otomicroscopy followed by BTX injection in the posterior inferior quadrant; (b) otomicroscopic view of the tympanic membrane during the procedure.

$p < 0.05$). The impact of BTX became apparent within two weeks, peaked at one month, and gradually diminished but persisted for up to three months. The most noticeable changes were in the awareness score. While treatment outcomes assessed by the Tinnitus Handicap Inventory (THI) mirrored changes in the four tinnitus distress VAS scores, the magnitude of change did not reach statistical significance, likely due to initially low THI scores. Patients who received the highest concentration (1 U) of the injection exhibited the most favorable outcomes on both VAS and THI. There were no immediate or delayed complications observed during the three-month follow-up, except in one case (a patient who received 0.1 U BTX), who experienced throat foreign body sensation immediately after the procedure, which subsided after two days [14].

Palatal myoclonus (PM) is a rare cause of objective tinnitus characterized by involuntary rhythmical contraction of the soft palate [15]. Several case reports have documented the effectiveness of BTX treatment in alleviating objective tinnitus secondary to PM. In general, complete symptom resolution follows one or two BTX injections, with reported doses ranging from 2.5 U to 30 U. These injections typically target the tensor veli palatini (TVP) muscle and/or the levator veli palatini (LVP) muscle (Figure 2).

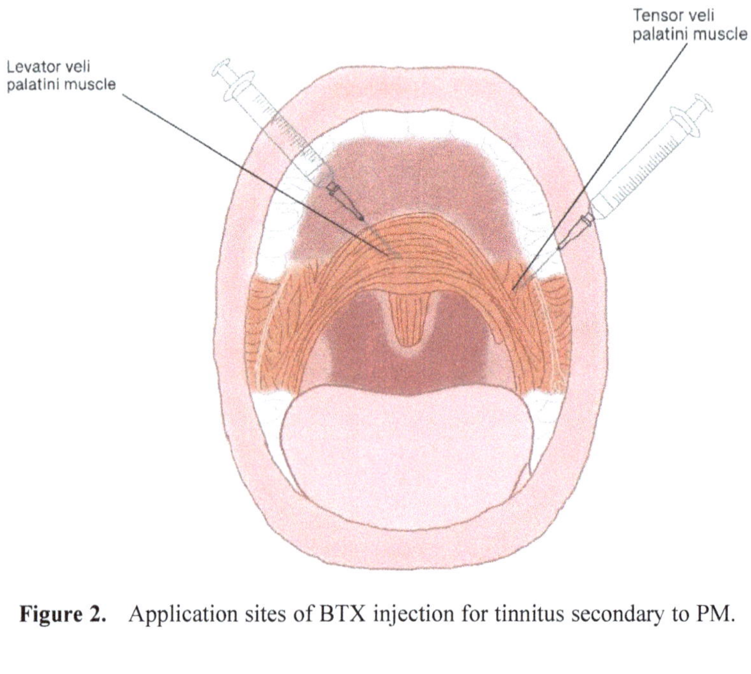

Figure 2. Application sites of BTX injection for tinnitus secondary to PM.

Treatment leads to the cessation of soft palate involuntary movement and subsequent relief from tinnitus [16, 17]. One case report described a young female with persistent tinnitus lasting four months, which was attributed to PM. Administration of a minimum effective dose of 2.5 U BTX-A into these areas under electromyographic (EMG) guidance resulted in complete relief of tinnitus after two weeks [16].

In another case report involving a patient with PM causing objective tinnitus, a total of 10 U of BTX-A were administered, with 5 U injected into both the TVP and LVP muscles of the soft palate bilaterally. Tinnitus ceased within seven days after the injection, and by day 15, bilateral paralysis of the TVP and LVP muscles was observed with only slight contraction of the left soft palate and no associated clinical symptoms. A transient side effect observed during the first week of treatment was open rhinolalia, causing nasal fluid reflux on two occasions. Follow-up examinations revealed restored laryngeal function. At the five-month mark, although the tinnitus recurred, its intensity was lower than initially perceived, prompting a second dose of BTX-A (30 U). This subsequent dose did not notably extend the duration of the treatment effect [18]. A comparable initial dose (10–10.5 U) of BTX-A injection into the same areas has shown benefits to PM patients who were unresponsive to previous anxiolytic and antidepressant pharmacotherapy [15, 17]. Similar positive outcomes have also been observed in a pediatric patient who received a total of 20 U BTX injections at four different sites near the insertion of TVP [19].

BTX-A is indicated for conditions associated with muscular hyperactivity and autonomic dysfunction. Its peak effects usually become apparent within 7 to 14 days, with muscle functionality returning within three to six months. Multiple administrations do not result in muscle wasting or permanent degeneration. The sole limitation lies in the temporary nature of the toxin's efficacy, necessitating repeat injections every five to six months [18].

Tinnitus is frequently associated with various types of migraine, including atypical migraine [20]. Multiple studies have reported this comorbidity, and treating the underlying migraine has led to improvements in tinnitus outcomes in several cases. Pharmacologic treatments commonly used to manage migraine symptoms, including anticonvulsants, tricyclic antidepressants, and vasodilators, have been shown effective in treating tinnitus [21]. Similarly, BTX is a promising option for tinnitus arising from migraine-related conditions due to its demonstrated efficacy in managing both chronic and atypical forms of migraine. Systematic reviews indicate that BTX treatments can reduce the frequency of migraine attacks per month and diminish their severity while maintaining a favorable safety profile [22, 23]. It has also been shown to be beneficial for those with vestibular migraines who are resistant to conventional pharmacological therapies [24, 25]. Subjective tinnitus often intensifies with migraine episodes, and a significant correlation exists between the laterality of tinnitus and headaches [12, 26, 27]. Given these connections, employing BTX to treat migraines is expected to provide symptomatic relief by addressing the related cause.

The use of BTX rarely results in adverse effects. Reported side effects include flu-like symptoms, muscle weakness, and pain. In some cases, the toxin may spread and cause weakness in muscles that were not directly targeted. Autonomic side effects are more often associated with type B toxin [7].

Contraindications to BTX treatment include known hypersensitivity to the toxin, Eaton-Lambert syndrome, myasthenia gravis, pregnancy (category C), or simultaneous administration of aminoglycoside antibiotics [7].

Clinical Outcomes

A prospective, double-blinded clinical study evaluated the effect of BTX-A on tinnitus in 30 patients [28]. Patients either received the initial BTX-A at doses of either 20 or 50 U, or a saline injection, with the alternate treatment provided four months later. BTX-A or placebo was injected subcutaneously into three sites around the ear: 1 cm above the superior aspect of the auricle, 1 cm behind the auricle at the 2 o'clock position, and 1 cm behind the auricle at the 5 o'clock position. Of the 26 patients who completed both injections, seven showed improvement, three worsened, and 16 remained unchanged with BTX treatment. In comparison, following the placebo injection, two patients showed improvement, seven worsened, and 17 remained unchanged. Tinnitus and hearing were assessed using THI, audiograms, Pure Tone Average-2 (PTA-2), and Speech Discrimination Scores (SDS) before and after each treatment. Statistically significant differences between treatment and placebo groups were noted ($p < 0.05$), and there was a significant reduction in THI scores from pre-treatment to four months post-BTX-A injection ($p = 0.0422$) [28].

A study conducted by Ranoux and Levine focused on patients with CM revealed a serendipitous finding related to tinnitus [29]. In this study, 57 patients with CM who were refractory to pharmacologic therapy received OnaA injections at doses ranging from 10 U to 40 U into cranio-cervical muscles. The injection sites were tailored based on myofascial trigger points and headache distribution, following the "follow-the-pain" technique. Remarkably, while the primary aim was to alleviate migraine symptoms, some patients also reported substantial improvements in their tinnitus. Among the five patients with pre-existing tinnitus, OnaA completely abolished tinnitus in two cases. One patient experienced a permanent resolution after a single treatment, with no recurrence even when her headaches returned. The remaining patients reported a reduction in tinnitus loudness by over 70%, with improvement persisting for as long as the migraine relief, typically lasting three to four months. Subsequent examination of injection sites in patients with transient improvement suggested that targeting the temporalis muscle likely contributed to the reduction in tinnitus severity [29]. These findings support the dorsal cochlear nucleus hypothesis for craniocervical somatic tinnitus, and highlight the established association between migraine and tinnitus.

The evidence as a whole indicates that BTX therapy, whether given intratympanically or intramuscularly, may provide a safe and reliable treatment choice for managing tinnitus, particularly in cases where conventional treatments have been ineffective.

Conclusions

- Neural therapy with BTX injection offers a novel approach to tinnitus management by modulating neuroplastic changes within the central auditory system.
- Tinnitus is influenced by intricate physiological and psychological interactions, with stress exacerbating symptoms.
- The effectiveness of BTX therapy depends on the injection technique, dosage, and frequency, with higher doses generally yielding better outcomes.
- Clinical evidence supports the efficacy of BTX in reducing tinnitus severity and improving related symptoms in conditions associated with muscular hyperactivity and autonomic dysfunction, including MEMT and PM, with peak impact typically observed within 7 to 14 days.

- BTX, a well-established treatment for migraines, can improve tinnitus associated with chronic and atypical forms of migraines, suggesting a potential treatment approach for tinnitus linked to neurological conditions.
- Adverse effects of BTX therapy are rare, and typically mild and transient, with serious complications being uncommon.
- Case studies and clinical trials offer valuable insights into the real-world application and viability of BTX in treating tinnitus.
- Contraindications to BTX treatment include hypersensitivity to the toxin, certain neuromuscular disorders, pregnancy, or concurrent use of certain antibiotics.
- These findings collectively suggest that BTX therapy, whether administered intratympanically or intramuscularly, is a promising and safe treatment option for managing tinnitus, particularly in cases refractory to conventional therapies.

References

1. Grossan M, and Peterson DC. Tinnitus. In *StatPearls*. Treasure Island (FL): StatPearls Publishing; March 20, 2023.
2. Park KW, Kullar P, Malhotra C, and Stankovic KM. Current and emerging therapies for chronic subjective tinnitus. *Journal of Clinical Medicine.* **12**(20):6555 (2023).
3. Shapiro SB, Noij KS, Naples JG, and Samy RN. Hearing loss and tinnitus. *Medical Clinics of North America.* **105**(5):799–811 (2021).
4. Lee A, Abouzari M, Akbarpour M, Risbud A, Lin HW, and Djalilian HR. A proposed association between subjective nonpulsatile tinnitus and migraine. *World Journal of Otorhinolaryngology — Head and Neck Surgery.* **9**(2):107–114 (2022).
5. Chari DA and Limb CJ. Tinnitus. *Medical Clinics of North America.* **102**(6):1081–1093 (2018).
6. Baguley D, McFerran D, and Hall D. Tinnitus. *The Lancet.* **382**(9904):1600–1607 (2013).
7. Wheeler A, and Smith HS. Botulinum toxins: Mechanisms of action, antinociception and clinical applications. *Toxicology.* **306**:124–146 (2013).
8. Landgrebe M, Langguth B, Rosengarth K, *et al.* Structural brain changes in tinnitus: Grey matter decrease in auditory and non-auditory brain areas. *NeuroImage.* **46**(1):213–218 (2009).
9. Folmer RL, Theodoroff SM, Martin WH, and Shi Y. Experimental, controversial, and futuristic treatments for chronic tinnitus. *Journal of the American Academy of Audiology.* **25**(01):106–125 (2014).

10. Burstein R, Blumenfeld AM, Silberstein SD, Manack Adams A, and Brin MF. Mechanism of action of Onabotulinumtoxin A in chronic migraine: A narrative review. *Headache.* **60**(7):1259–1272 (2020).
11. Iyengar S, Ossipov MH, and Johnson KW. The role of calcitonin gene–related peptide in peripheral and central pain mechanisms including migraine. *Pain.* **158**(4):543–559 (2017).
12. Lugo A, Edvall NK, Lazar A, *et al.* Relationship between headaches and tinnitus in a Swedish study. *Scientific Reports.* **10**:8494 (2020).
13. Láinez MJ, and Piera A. Botulinum toxin for the treatment of somatic tinnitus. Prog Brain Res. 2007;**166**:335–338.
14. Park JM, Kim MJ, Han JS, Park SY, and Park SN. Intratympanic botulinum toxin injection as a new therapeutic modality for middle ear myoclonic tinnitus. *Otolaryngology — Head and Neck Surgery.* **169**(2):348–357 (2023).
15. Saeed SR, and Brookes GB. The use of *Clostridium botulinum* toxin in palatal myoclonus. A preliminary report. *The Journal of Laryngology and Otology.* **107**(3):208–210 (1993).
16. Tatar EÇ, Günay MM, Öcal B, Özdek A, and Korkmaz MH. An electromyography guided botulinum toxin injection is effective treatment for objective tinnitus as an office procedure: A case report. *Egyptian Journal of Ear, Nose, Throat and Allied Sciences.* **17**(3):159–161 (2016).
17. Anis MM, and Pollak N. Treatment of palatal myoclonus with botulinum toxin injection. *Case Reports in Otolaryngology.* **2013**:231505 (2013).
18. Conill Tobías N, de Paula Vernetta C, García Callejo FJ, and Marco Algarra J. Objective tinnitus from palatal myoclonus. Use of botulinum toxin: A case report. *Acta Otorrinolaringologica Espanola.* **63**(5):391–392 (2012).
19. Jero J. Palatal myoclonus and clicking tinnitus in a 12-year-old girl-case report. *Acta Oto-Laryngologica.* **120**(543):61–62 (2000).
20. Benjamin T, Gillard D, Abouzari M, Djalilian HR, and Sharon JD. Vestibular and auditory manifestations of migraine. *Current Opinion in Neurology.* **35**(1):84–89 (2022).
21. Kim SH, Kim D, Lee JM, Lee SK, Kang HJ, and Yeo SG. Review of pharmacotherapy for tinnitus. *Healthcare (Basel).* **9**(6):779 (2021).
22. Herd CP, Tomlinson CL, Rick C, *et al.* Cochrane systematic review and meta-analysis of botulinum toxin for the prevention of migraine. *BMJ Open.* **9**(7):e027953 (2019).
23. Shaterian N, Shaterian N, Ghanaatpisheh A, *et al.* Botox (OnabotulinumtoxinA) for treatment of migraine symptoms: A systematic review. *Pain Research and Management.* **2022**:3284446 (2022).
24. Oh SY, Kang JJ, Kim S, Lee JM, Kim JS, and Dieterich M. A preliminary trial of botulinum toxin type A in patients with vestibular migraine: A longitudinal fMRI study. *Frontiers in Neurology.* **13**:955158 (2022).
25. Liu YF, Macias D, Donaldson L, Dornhoffer JR, and Rizk HG. Pharmacotherapy failure and progression to botulinum toxin injection in

vestibular migraine. *The Journal of Laryngology and Otology*. **134**(7):586–591 (2020).
26. Goshtasbi K, Abouzari M, Risbud A, *et al*. Tinnitus and subjective hearing loss are more common in migraine: A cross-sectional NHANES analysis. *Otology and Neurotology*. **42**(9):1329–1333 (2021).
27. Langguth B, Hund V, Busch V, *et al*. Tinnitus and headache. *BioMed Research International*. **2015**:797416 (2015).
28. Stidham KR, Solomon PH, and Roberson JB. Evaluation of botulinum toxin A in treatment of tinnitus. *Otolaryngology-Head and Neck Surgery*. **132**(6):883–889 (2005).
29. Ranoux D, and Levine RA. Botulinum toxin can abolish and/or quiet tinnitus associated with chronic migraine: Serendipidous observations. *The International Tinnitus Journal*. **25**(2):133–136 (2021).

https://doi.org/10.1142/9789819807345_0011

Chapter 11

Hearing Aid Fitting for Tinnitus Management

Troy Cascia

Department of Audiology, UCSF,
San Francisco, CA, USA

troy.cascia@ucsf.edu

Abstract

Tinnitus, often linked with hearing loss, arises from decreased auditory stimulation to the brain, leading to hyperactive neural firing in the auditory cortex. This chapter explores how hearing aids can mitigate tinnitus by reintroducing sound stimulation, thereby reducing the perception of tinnitus and the stress associated with it. Hearing aids amplify ambient sounds, providing the auditory stimulation needed to counteract neural hyperactivity, and reduce listening effort that can exacerbate tinnitus symptoms. Many users report significant tinnitus reduction with hearing aid use, though evidence of permanent resolution remains insufficient.

Understanding tinnitus involves recognizing its origins in cochlear damage and subsequent neural plasticity in the auditory cortex and other areas of the brain. The distress linked to tinnitus is primarily mediated by the limbic system, particularly the amygdala, which amplifies the perceived threat of tinnitus, creating a cycle of stress and heightened perception. Effective tinnitus management aims to interrupt this cycle, fostering neural pathways that allow the brain to habituate to tinnitus.

Hearing aids support this process by amplifying ambient sounds, which reduces the brain's focus on tinnitus, and by utilizing built-in noise generators that can provide continuous sound therapy. Modern hearing aids offer a range of sound options, including broad- and narrow-band noises, proprietary sounds like fractal tones, and notched acoustic stimuli. Hearing aids with Bluetooth capability offer additional sound therapy options, further enhancing their utility.

The role of listening effort is also instrumental in tinnitus management. Hearing aids reduce the cognitive effort required for central processing of acoustic stimuli, preserving cognitive resources and reducing fatigue that can further exacerbate tinnitus symptoms.

By integrating amplification with sophisticated sound therapy features, hearing aids provide a multi-faceted approach to tinnitus management. Properly programmed and supported by counseling, hearing aids can help tinnitus sufferers achieve reduced perception and distress, ultimately improving their quality of life.

Tinnitus has long been associated with hearing loss [1], whether such deficits to the auditory system are measurable, as with conventional behavioral or objective test measures, or not [2]. Changes to the auditory system, particularly those resulting in decreased auditory stimulation to the brain, are believed to trigger hyperactive and synchronized neural firing in the auditory cortex [3]. When this neural firing is brought to the level of consciousness, tinnitus becomes audible, and when the audible percept triggers a negative response from the body's limbic and sympathetic nervous systems, the tinnitus becomes problematic and distressing [4]. It stands to reason that reintroducing the sound stimulation that the auditory cortex is expecting but is no longer receiving can quell the hyperactive neural firing, resulting in reduced tinnitus perception [5]. Hearing aids have been found to be an effective way to deliver this sound to the auditory system when programmed to an intensity level sufficient to compensate for the auditory deprivation associated with hearing loss [6].

Most hearing aid users with tinnitus report some degree of reduction in their tinnitus while wearing their hearing aids, occasionally to such a degree that some individuals find that their tinnitus resolves completely and becomes temporarily inaudible while wearing their hearing aids [7] Use of hearing aids provides access to the ambient environmental sounds that would normally stimulate the auditory system, but which by themselves are not loud enough to be perceived due to the individual's hearing loss. Unfortunately, despite their ability to temporarily decrease the perception of tinnitus, there is insufficient evidence to suggest that the use of hearing aids could be expected to cause tinnitus to permanently resolve [8].

To understand how hearing aids can be effective tinnitus management tools, it helps to understand the nature of tinnitus. While there are many factors that can cause or contribute to tinnitus, including pathologies or injuries to the ear, various infectious disease processes, or side effects from potentially ototoxic medications, hearing loss is known to be the most common cause of tinnitus [9]. The most popular pathophysiological model for tinnitus explains that tinnitus is the product of enhanced neural activity in the auditory pathway, particularly in the auditory cortex, in response to cochlear damage [11]. The sensory deprivation that this peripheral damage causes to the central auditory system has been shown to drive neural plasticity, triggering heightened sensitivity and spontaneous neural firing in the auditory cortex [12].

While the hyperactive and spontaneous firing of neurons in the auditory cortex is believed to be the origin of the tinnitus percept, it is not the source of tinnitus distress. In fact, most individuals with tinnitus are not debilitated, or even significantly annoyed by their tinnitus [13]. The primary neural structures responsible for how the brain and body react to tinnitus are the limbic system and, more specifically, the amygdala [14].

As the brain receives vast amounts of environmental stimuli of different types, (e.g., visual, auditory, tactile, etc.), the amygdala, a structure in the brain's limbic system, is responsible for determining which stimuli are important to attend to and which do not need to be brought to the level of consciousness [15]. Any benign stimuli in the environment, such as the sensation of the shoes on one's feet, are deemed by the amygdala to be unimportant and the brain quickly habituates, not being consciously aware of them. Conversely, any stimuli that are judged to be unpleasant or otherwise threatening will trigger a stress response from the amygdala whereby the autonomic nervous system is activated, causing physiological changes in the body preparing the individual for fight or flight. The tinnitus itself is often exacerbated by this stress response, with both tinnitus loudness and annoyance levels being enhanced. The increased tinnitus percept reinforces the limbic system response which creates a vicious circle of stress and anxiety [16]. There is emerging evidence that links the central sensitivity of migraine to this increased attention to tinnitus [17].

Prolonged limbic system engagement may result in further neuroplastic changes that encourage the limbic system to remain hypervigilant and to resist habituation [18]. The primary goal of tinnitus management is to interrupt the maladaptive neuroplastic changes that are preventing habituation from occurring and to create healthier neural pathways that allow the tinnitus to fade into the background. Once the limbic system is no longer interested in monitoring the tinnitus or reacting negatively to it, the tinnitus is no longer perceived as problematic and tinnitus distress decreases [1].

While the limbic system is being trained to habituate to the tinnitus, many individuals also find that tinnitus perception can be temporarily decreased by the presence of ambient sound. In the absence of sound stimulation, the brain increases its central gain and enhances any sound that it can detect, which in this case is the tinnitus. Conversely, the brain tends to decrease its awareness of tinnitus when other environmental sounds are audible [19]. An apt analogy involves a lighted candle in a dark room. Without any other light sources in the darkness, the light of the candle is the only thing that can be seen and one's attention is immediately drawn to it. However, the same candle in a well-lit room is less noticeable. In this scenario, there are more visual stimuli to attend to, and although the candle is still there, its light is not as prominent in contrast to the other things that catch the eye. Similarly, tinnitus is more noticeable in silence, and becomes less noticeable when contrasted with a sound-rich environment.

There are several ways in which the use of hearing aids can help with tinnitus management. The first and most direct way is by amplifying the ambient sound that is already available in the environment, but which is not audible to the individual with hearing loss [6]. By providing access to environmental sounds, hearing aids act to restore the sound stimulation that the brain is expecting to receive but that it has been deprived of due to hearing loss. As described above, this sound enrichment reduces central gain and minimizes the contrast between the tinnitus perception and ambient sound, helping the brain to focus less on tinnitus, and allowing it to fade into the background through the process of habituation.

In addition to providing amplification, most modern hearing aids also contain a built-in noise generator for tinnitus management. This noise generator provides constant sound stimulation to the auditory system, which is particularly necessary when there are no ambient sounds in the environment for the hearing aids to amplify. A quiet and comfortable sound is produced by the hearing aids to fill in any periods of environmental silence, so the auditory system receives constant sound stimulation throughout the day. The quality and types of sounds that are preferred and most effective vary among individuals. Common and popular sounds that are used by various hearing aid manufacturers including white noise, pink noise, red noise, shaped noise, amplitude modulated noise, notched noise, and fractal tones. The salient characteristics of any sounds that are to be used for sound therapy are that they must be quiet but audible so as to be effective, non-intrusive so as not to attract attention, and comfortable or emotionally neutral so as not to trigger a negative response from the limbic system [20, 21].

While some users choose to set their hearing aids up with a separate, manually accessible tinnitus management program that is only activated as needed, it may be preferable for the hearing care professional to activate the noise generator in the default program of the hearing aid so that the therapy sound is automatically activated. This is to minimize the amount of time spent consciously thinking about sound therapy and, by extension, the tinnitus. The therapy sounds are set by the hearing care professional in the clinic at a level that is just barely audible to the user. The user may notice that the perception of the tinnitus changes in the presence of the therapy sound, but the tinnitus should still be perceptible, as the limbic system is unable to habituate to a sound that it cannot detect [1]. Again, the purpose of the therapy sound is not to cover up or mask the tinnitus, but rather to provide contrast so the tinnitus is not the only sound the person can hear. If the sound generators are constantly active during hearing aid use, the user relies on access to ambient, amplified sound when such environmental sounds are present, but the generated sounds being fed through the hearing aids are automatically present to fill the gaps when the user is in quiet environments. In this way, the individual is never "left alone in silence" with their tinnitus. Additionally, the user does not have to constantly monitor the sound environment to ensure that sound enrichment is constantly present, as it is being fed into the ears during all waking hours. Over-monitoring of the sound environment and by extension, the tinnitus, reinforces the importance of the tinnitus to the limbic system and serves to discourage habituation [22]. The "set and forget" approach, whereby the user sets the level of the sound generator at the beginning of the day in a quiet location and maintains this setting throughout the day, minimizes the amount of time the individual spends attending to their tinnitus.

Virtually all of the major hearing aid brands have developed and adopted a variety of tinnitus support sounds that have been built into their products. Some of these sounds are general and universally available, while others are proprietary. For all manufacturers, the built-in sound generators can be selected and activated through the programming software of each manufacturer. Additionally, the user may be given access to various controls to activate or deactivate the tinnitus support sounds they wish to use, or to increase or decrease the loudness of these sounds. Such access is available either through on-board buttons built into the hearing aid itself, through remote control accessories that are available for purchase separately, or through free smartphone applications that are designed by the respective hearing aid manufacturer and are readily available for download.

The most basic type of tinnitus support sound, and one that is available in most current hearing aid models, is broadband noise. Broadband noise has been well established as an effective masking or partial masking tool for tinnitus management [23, 21]. The most basic and most commonly used broadband noise is white noise. White noise is an acoustic signal that contains all of the frequencies in the audible sound spectrum with equal amounts of energy at each frequency, analogous to white light containing all frequencies of light waves. Psychoacoustically, however, the high frequencies of white noise are often perceived as being particularly prominent, which can be off-putting to some individuals, particularly those who are experiencing recruitment, hyperacusis, or other sound sensitivities [24].

In contrast, pink noise is a broadband noise that contains a high frequency roll-off of -3 dB/octave, which reduces the high frequency energy in the sound stimulus and gives the perception of a more acoustically balanced signal. As such, many people find pink noise to be more comfortable and emotionally neutral than white noise, which is an important factor in the selection of therapeutically successful tinnitus support sounds [21].

Red noise contains a still greater high frequency roll-off of –6 dB/octave, twice that of pink noise, and has a decidedly lower frequency emphasis and percept. Continuing with the analogy of white noise and white light both containing all frequencies of sound and light in their respective spectra, red noise is analogous to red light, in which both reside at the lower ends of their frequency ranges. Red noise is also acoustically similar to many sounds found in nature with a prominent low frequency component, such as waves and wind. Such sounds often produce positive emotional responses in listeners, and as such are particularly useful in tinnitus management [25].

Acoustically shaped broadband noise is noise that is modified to match the individual's hearing thresholds as entered in the programming software. Because audiometric threshold data are required to provide appropriate and adequate amplification with hearing aids, which is their primary purpose, these data have typically already been entered into the programming software prior to programming. One advantage of this type of sound stimulus is that it ensures that the auditory system is receiving the sound at all, or at least most, frequencies. This is particularly important for individuals with a significant enough degree of hearing impairment that they may not be able to hear the full acoustic signal [26]. Additionally, acoustically shaped noise is usually able to be modified further and customized by the health care provider by increasing or decreasing the intensity of the broadband noise at various frequencies. With subjective feedback from the user, this ensures that the sound stimulus is comfortable and effective.

Broadband noise presented with amplitude modulation, mimicking the undulation of ocean waves, has been found to be an effective tool for providing sound stimulation while simultaneously promoting relaxation and stress relief [27], and has been shown in some cases to suppress tinnitus more effectively than steady broadband noise [28]. Several hearing aid manufacturers allow for the creation of amplitude modulated noise with adjustable frequency modulation, i.e., the rate at which the "ocean waves" are presented. The faster the modulation, usually the more audible and noticeable it is to the user. The frequency of wave presentation is selected on a case-by-case basis with subjective feedback from the user.

One hearing aid manufacturer on the market, while still offering a traditional broadband white noise, also has developed a proprietary sound stimulus for tinnitus support using fractal mathematics that has been found to be effective for many people [29]. This stimulus consists of a series of tones that have a quality often described as being similar to wind chimes. These fractal tones are generated using a randomizing algorithm that ensures that the same series of tones is never repeated exactly, avoiding monotony and minimizing user fatigue [30]. The frequency range of the tones can be adjusted, as well as the speed and briskness with which the tones are presented and the overall volume of the signal. There are several pre-programmed settings to choose from, but they can also be manually adjusted with the user's subjective input regarding their sound preferences. It is also possible to overlay these sounds over a background of independently adjusted white noise, pair them with specifically selected musical passages, or pair them with nature sounds like ocean waves.

Another treatment for tinnitus that has piqued the interest of hearing healthcare providers in recent years is notched sound therapy. A form of neuromodulation, notched sound therapy seeks to drive neurological changes, particularly in the auditory cortex and other auditory pathways, to actively reduce the perception, perceived loudness, and emotional distress levels of tonal tinnitus [31]. Through repeated stimulation of targeted neurons in the brain, neuromodulation takes advantage of the brain's ability to build new neural pathways. This intentional driving of neuroplasticity encourages the creation of new neural pathways that function in a desired manner rather than allowing the previous organically created, maladaptive pathways to persist and strengthen further [32].

Notched sound therapy exploits the neurological phenomenon of lateral inhibition to suppress hyperactive and synchronously firing neurons in the auditory cortex, which are understood to be the primary origin of tinnitus. Lateral inhibition refers to the process whereby certain sensory neurons, when stimulated, will release neurotransmitters to suppress the firing of neighboring neurons [33]. This behavior has been observed in neural activity in both the visual and auditory cortices and is believed to be an adaptation of our sensory systems to aid with boundary discrimination [34]. Notched sound therapy involves first identifying the perceived pitch of the individual's tinnitus, and then targeting and stimulating the tonotopically arranged neurons adjacent to the tinnitus pitch. By stimulating these neighboring neurons, the therapy aims to trigger lateral inhibition to suppress the firing of the neurons responsible for the tinnitus perception. Through persistent targeted stimulation, the neuroplastic changes in the auditory cortex can result in lasting changes to the tinnitus perception [35]. There have been several commercially available tinnitus therapies on the market that have used this technique, and there is one hearing aid manufacturer that also offers this as a proprietary treatment option through their programming software by providing notched sound amplification. If notched sound amplification is selected as an option, the hearing aid provides amplification at all frequencies where the user has hearing loss, with the exception of the frequency correlated with the pitch of the tinnitus, which is notched out. The user typically cannot perceive that the amplified signal has a notch filter at the target frequency, however the stimulation at adjacent frequencies provides the sound contrast necessary to provide short term tinnitus relief while the notch filter drives longer term changes through lateral inhibition. While notched sound amplification is a proprietary feature, it is possible with some manufacturers that allow manual adjustment of on-board noise generators to create an acoustically shaped broadband noise and reduce the volume for the frequency band corresponding to the tinnitus pitch. In this way, the benefits of notched sound therapy can be obtained using other tinnitus support sounds.

One drawback of notched sound therapy is that lateral inhibition has only been shown to be effective with tinnitus that is perceived as a single tone [36]. Most individuals with tinnitus do have this type, typically perceived as a ringing sound [13]. However, tinnitus can be perceived in a very broad range of ways; as buzzing, chirping, narrow or broad band "static" noise, or complex tinnitus involving the perception of multiple layered sounds. Notched sound therapy is not effective for these varieties of tinnitus. Additionally, it can be challenging and time consuming to match the pitch even of tonal tinnitus. And while instruments do exist to attempt tinnitus pitch matching in 1 Hz increments, most hearing healthcare providers rely on a clinical audiometer for pitch matching. This is despite the fact that most audiometers are able to match pitch only to the closest half-octave, resulting in an imprecise measurement that could theoretically result in poorer treatment outcomes.

Another way in which hearing aids help with tinnitus management, albeit indirectly, is by reducing the cognitive effort being expended by the hearing-impaired individual in the seemingly simple act of listening.

All listeners, even those with normal hearing sensitivity, occasionally miss phonemes and words in conversational speech; however, the brain is able to predict the missing content based on semantic, lexical, and phonemic context as well as visual indicators such as lipreading and various aspects of non-verbal communication. For the hearing-impaired listener, however, the gaps in the speech signal are often larger as they are unable to hear a larger number of phonemes [37]. Additionally, hearing impaired listeners experience more difficulty with listening tasks involving higher level auditory processing, such as separating speech from background noise, filtering out competing speakers, and understanding rapid speech [37]. This additional listening effort that must be expended by hearing impaired individuals increases fatigue, which often results in increased or more poorly managed tinnitus [38]. The use of hearing aids that have been properly selected and programmed helps to provide as much auditory information as possible to the user, making listening and following conversational speech less effortful, and thereby reducing cognitive load and preserving the cognitive resources necessary to manage tinnitus more effectively [39].

In addition to built-in sound generators that produce various sounds within the hearing aids themselves that can be used to provide tinnitus sound therapy, many modern hearing aids are also equipped with Bluetooth technology for wireless streaming of audio output from compatible smartphones. Bluetooth therefore widens the range of available therapy sounds to virtually any sound that can be downloaded to the user's phone or found on the internet. Music, nature sounds, broadband and narrowband noises, and more can be streamed from a smartphone through the hearing aids, directly into the user's ears. This technology can be particularly helpful for individuals who dislike or have difficulty habituating to the limited selection of sounds built into the hearing aids themselves but have been able to find sounds they like elsewhere. Additionally, there are a large number of applications available for download to most smartphones that provide a wide variety of sounds that can be used for tinnitus sound therapy. Some of them allow the user to combine multiple sounds that can be presented simultaneously to create custom therapy sounds for maximum comfort. Each of the major hearing aid manufacturers has developed their own proprietary apps for use with tinnitus. Most are available for free download, although some have additional content for a nominal fee, and none require the use of hearing aids, as they can be just as easily used with Bluetooth or hardwire headphones. In addition to therapy sounds, several of the apps also contain relaxation tips and other tinnitus management suggestions to be used in conjunction with sound therapy.

While hearing aids are intended for use by hearing-impaired individuals to ameliorate the effects of their hearing loss, including those with tinnitus, there is some research to suggest that hearing aids may also be beneficial when used in an off-label manner by individuals with tinnitus who have normal hearing thresholds. In one study involving participants with an audiometric pure tone average (average threshold at 500, 1000, and 2000 Hz) of less than 30 dBHL, tinnitus distress was found to be significantly decreased when using hearing aids with only minimal amplification, and with maximum output limited to prevent noise induced hearing loss [40].

In addition to providing enhancement of ambient sounds for sound enrichment as a tinnitus management strategy, hearing aid users with normal hearing thresholds may also take advantage of the other features discussed previously that can be useful as tinnitus management tools, namely built-in noise generators and Bluetooth streaming technology. Even if the individual does not use the hearing aids for amplification and the volume is programmed down to the lowest level possible, the devices can be used as simple ear level noise generators. True ear level noise generators — ear worn devices that look like hearing aids, but which only produce white or pink noise and do not amplify — already exist in the marketplace. However, the benefit of using hearing aids instead of noise generators is that in addition to producing white and pink noise (or other types of built-in tinnitus support sounds), many hearing aids are also equipped with Bluetooth technology which allows streaming of virtually any tinnitus support sound from a compatible smartphone. This capability expands the repertoire of available sounds far beyond that of a simple noise generator. The hearing aids can always be reprogrammed later to provide amplification if needed or desired. Additionally, because an individual who does not present with hearing loss has an essentially normally functioning cochlea, it is often not necessary for the user to spend more money on hearing aids with more advanced technologies and features. Basic, entry level hearing aids with simpler technology usually provide all the functionality the person needs, and at a similar price point as the ear level noise generators that provide less flexibility.

Modern hearing aids are multi-faceted, flexible tools that can be an important part of a patient's tinnitus management toolkit. Besides their primary function of amplifying sound for better audibility, which alone helps many people to decrease the awareness of their tinnitus, hearing aids also utilize a variety of features to minimize tinnitus awareness and help with the habituation process. Examples include built-in noise generators that provide a variety of comfortable and soothing sounds, Bluetooth audio streaming technology, free downloadable apps, and notched amplification filters. Amplification also helps to reduce listening effort and preserve cognitive resources to better manage tinnitus.

When properly programmed and accompanied by appropriate counseling (see Table 1), hearing aids can help tinnitus sufferers to improve their quality of life by decreasing the perceived loudness of their tinnitus, regain a sense of control, and take positive steps toward the ultimate goal of habituation.

Table 1. Considerations and steps in programming hearing aids for effective tinnitus management.

Assess Hearing Loss and Tinnitus	Conduct comprehensive audiometric testing to determine hearing thresholds.
	For select sound therapy options (e.g., notched therapy) perform tinnitus pitch matching to identify the perceived frequency of tinnitus.
Program Amplification Settings	Set hearing aids to provide sufficient amplification to compensate for the individual's hearing loss.
	Ensure that ambient environmental sounds are audible to reduce central gain and minimize tinnitus contrast.
Activate Built-in Noise Generators	Select and activate appropriate tinnitus support sounds (white noise, pink noise, red noise, fractal tones, etc.) through the hearing aid programming software.
	Customize the noise generators based on user feedback to ensure comfort and effectiveness.
Activate Notched Sound Therapy (if applicable)	Using the previously identified tinnitus pitch, apply a notched filter in the amplification settings to stimulate lateral inhibition and promote neuroplastic changes.
Set Up Sound Therapy Programs	Decide whether to create a separate tinnitus management program or integrate it into the default program for ease of use.
	Ensure the therapy sounds are set at a barely audible level to provide contrast without masking the tinnitus completely.

Table 1. (*Continued*)

Utilize Bluetooth Streaming	Enable Bluetooth functionality to allow the streaming of additional therapeutic sounds from compatible smartphones.
	Recommend specific tinnitus management apps that offer a variety of sound options and additional management tips.
Provide Counseling and Support	Offer guidance on the "set and forget" approach to minimize the user's focus on tinnitus.
	Educate users on the importance of continuous sound stimulation, even in quiet environments, to prevent over-monitoring and reinforce habituation.
Adjust Based on User Feedback	Regularly follow up with the user to assess the effectiveness of the tinnitus management settings.
	Make necessary adjustments to amplification levels, noise generator settings, and other sound therapy features to optimize comfort and efficacy.
Promote Cognitive and Emotional Strategies	Advise on relaxation techniques and strategies to reduce tinnitus-related stress.
	Encourage the use of hearing aids to reduce listening effort and preserve cognitive resources for better tinnitus management.

Conclusion

- *Tinnitus and Hearing Loss*: Tinnitus is commonly associated with hearing loss, resulting from measurable or non-measurable deficits in the auditory system. Decreased auditory stimulation triggers hyperactive neural firing in the auditory cortex, leading to the conscious perception of tinnitus and potential distress through the limbic and sympathetic nervous systems.
- *Role of Hearing Aids*: Hearing aids reintroduce sound stimulation, compensating for auditory deprivation and reducing hyperactive neural firing, thus lowering tinnitus perception. Many hearing aid users report significant tinnitus relief while using their devices, although there is insufficient evidence for permanent tinnitus resolution.
- *Sound Therapy*: In addition to amplification of ambient sound, hearing aids can also provide various sound therapies through built-in noise generators, including white, pink, and red noise, and proprietary sounds like fractal tones and notched sound therapy. Modern hearing aids with Bluetooth technology also offer extensive options for streaming therapeutic sounds, enhancing flexibility in tinnitus management.

- *Cognitive Load Reduction*: Properly programmed hearing aids reduce listening effort for individuals with hearing loss, preserving cognitive resources and reducing listening fatigue that can further exacerbate tinnitus symptoms.
- *Habituation and Counseling*: Effective tinnitus management through hearing aids involves interrupting maladaptive neural plasticity and fostering habituation, supported by appropriate counseling. When properly used, hearing aids help tinnitus sufferers reduce tinnitus perception and distress, improving overall quality of life.

References

1. Hazell JW, and Jastreboff PJ. Tinnitus. I: Auditory mechanisms: a model for tinnitus and hearing impairment. *Journal of Otolaryngology*. **19**(1):1–5 (1990).
2. Kehrle HM, Costa F, Schneider R, De Carvalho R, and Conti C. Characteristics of tinnitus in the elderly. *International Tinnitus Journal*. **26**(1):57–62 (2022).
3. Eggermont JJ, and Tass PA. Maladaptive neural synchrony in tinnitus: Origin and restoration. *Frontiers in Neurology*. **17**(6):29 (2015).
4. Noreña AJ. Revisiting the cochlear and central mechanisms of tinnitus and therapeutic approaches. *Audiology and Neurotology*. **20**(Suppl 1):53–59 (2015).
5. Henry JA. Sound therapy to reduce auditory gain for hyperacusis and tinnitus. *American Journal of Audiology*. **31**(4):1067–1077 (2022).
6. Trotter MI, Møller AR, and Andersson G. Hearing aids and tinnitus therapy: A 25-year experience. *Journal of Laryngology & Otology*. **122**(10):1052–1056 (2008).
7. Lee HJ, Lee H, Jang H, *et al.* Hearing aid effects and satisfaction in patients with tinnitus. *Journal of Clinical Medicine*. **11**(4):1096 (2022).
8. Shapiro SB, Semenov Y, Verhoef R, *et al.* Hearing loss and tinnitus. *Medical Clinics of North America*. **105**(5):799–811 (2021).
9. Leske MC. Prevalence estimates of communication disorders in the U.S. Language, hearing and vestibular disorders. *American Speech-Language-Hearing Association*. **23**:229–237 (1981).
10. Knipper M, Schrott-Fischer A, Rösch S, *et al.* The neural bases of tinnitus: Lessons from deafness and cochlear implants. *Journal of Neuroscience*. **40**(38):7190–7202 (2020).
11. Esmaili AA, and Renton J. A review of tinnitus. *Australian Journal of General Practice*. **47**(4):205–208 (2018).
12. Eggermont JJ. The auditory cortex and tinnitus — A review of animal and human studies. *European Journal of Neuroscience*. **41**(5):665–676 (2015).

13. Jarach CM, Silva M, Ribeiro M, *et al.* Global prevalence and incidence of tinnitus: A systematic review and meta-analysis. *JAMA Neurology.* **79**(9):888–900 (2022).
14. Møller AR. Pathophysiology of tinnitus. *Otolaryngologic Clinics of North America.* **36**(2):249–266 (2003).
15. Fonseca R, Soares A, Costa M, *et al.* Resilience to fear: The role of individual factors in amygdala response to stressors. *Molecular Cell Neuroscience.* **110**:103582 (2021).
16. Zenner HP, Mazurek B, Haupt H, *et al.* Tinnitus sensitization: Sensory and psychophysiological aspects of a new pathway of acquired centralization of chronic tinnitus. *Otolaryngology–Head and Neck Surgery.* **27**(8):1054–1063 (2006).
17. Lee A, Abouzari M, Akbarpour M, Risbud A, Lin HW, and Djalilian HR. A proposed association between subjective nonpulsatile tinnitus and migraine. *World Journal of Otorhinolaryngology–Head and Neck Surgery.* **9**(2):107–114 (2022).
18. Shore SE, Elgoyhen AB, and Salvi RJ. Maladaptive plasticity in tinnitus — triggers, mechanisms and treatment. *Nature Reviews Neurology.* **12**(3):150–160 (2016).
19. Auerbach BD, Rodrigues PV, and Salvi RJ. Central gain control in tinnitus and hyperacusis. *Frontiers in Neurology.* **5**:206 (2014).
20. Jastreboff PJ, and Jastreboff MM. Neurophysiological approach to tinnitus patients. *American Journal of Otology.* **17**(2):236–240 (1996).
21. Jastreboff PJ. Tinnitus retraining therapy. *Progress in Brain Research.* **166**:415–423 (2007).
22. Henry JA. Directed attention and habituation: Two concepts critical to tinnitus management. *American Journal of Audiology.* **31**(1):1–8 (2023).
23. Vernon J, and Griest S. Attributes of tinnitus and the acceptance of masking. *American Journal of Otology.* **11**(1):44–50 (1990).
24. Barozzi S, Trevisani G, and Bertelli S. Effects of tinnitus retraining therapy with different colours of sound. *International Tinnitus Journal.* **21**(2):139–143 (2017).
25. Alvarsson JJ, Nilsson ME, Rosenqvist U, *et al.* Stress recovery during exposure to nature sound and environmental noise. *International Journal of Environmental Research and Public Health.* **7**:1036–1046 (2010).
26. Schaette R, McAlpine D. Acoustic stimulation treatments against tinnitus could be most effective when tinnitus pitch is within the stimulated frequency range. *Hearing Research.* **269**(1–2):95–101 (2010).
27. Searchfield GD, Waddington C, and McFerran DJ. Sound therapies and instrumentation for tinnitus therapy. *New Zealand Medical Journal.* **123**(1311): 112–125 (2010).

28. Reavis KM, Shupak N, Han JH, *et al.* Temporary suppression of tinnitus by modulated sounds. *Journal of the Association for Research in Otolaryngology.* **13**:561–571 (2012).
29. Herzfeld M, Sutherland S, and Schwindt K. Clinical trial on the effectiveness of Widex Zen Therapy. *Hearing Review.* **21**(11):1–8 (2014).
30. Sweetow RW, Sabes JH, Albert A, *et al.* Effects of acoustical stimuli delivered through hearing aids on tinnitus. *Journal of the American Academy of Audiology.* **21**(7):461–473 (2010).
31. Huang B, Xu J, Han Z, *et al.* Notched sound alleviates tinnitus by reorganization emotional center. *Frontiers in Human Neuroscience.* **15**:762492 (2022).
32. Nahum M, Nelken I, Bar-Haim Y, *et al.* Principles of neuroplasticity-based rehabilitation. *Progress in Brain Research.* **207**:141–171 (2013).
33. Gerken GM. Central tinnitus and lateral inhibition: An auditory brainstem model. *Hearing Research.* **97**(1–2):75–83 (1996).
34. Van Tonder GJ, and Kruger JJ. Shape encoding: A biologically inspired method of transforming boundary images into ensembles of shape-related features. *IEEE Transactions on Systems, Man, and Cybernetics B: Cybernetics.* **27**(5):749–759 (1997).
35. Therdphaothai J, Pradithisarn N, and Pongpirul WA. A randomized controlled trial of notched music therapy for tinnitus patients. *Journal of International Advanced Otology.* **17**(3):221–227 (2021).
36. Stein A, Glickman G, Williams W, *et al.* Clinical trial on tonal tinnitus with tailor-made notched music training. *BMC Neurology.* **16**:38 (2016).
37. Zekveld AA, Kramer SE, and Festen JM. Cognitive load during speech perception in noise: The influence of age, hearing loss, and cognition on the pupil response. *Ear and Hearing.* **32**(4):498–510 (2011).
38. Miguel GS, Navarro M, Garcia MA, *et al.* The effect of insomnia on tinnitus. *Annals of Otology, Rhinology, and Laryngology.* **123**(10):696–700 (2014).
39. Sanders ME, Kant E, Smit AL, and Stegeman I. The effect of hearing aids on cognitive function: A systematic review. *PLoS One* **16**(12):e0261207 (2021).
40. Suzuki N, Kondo A, Kurozumi T, *et al.* Effectiveness of hearing aids in treating patients with chronic tinnitus with average hearing levels of <30 dBHL and no inconvenience due to hearing loss. *Acta Otolaryngologica.* **141**(8): 773–779 (2021).

https://doi.org/10.1142/9789819807345_0012

Chapter 12

Hyperacusis and Misophonia Treatment

Madelyn Frank and Harrison W. Lin*

Department of Otolaryngology-Head and Neck Surgery, University of California, Irvine, CA, USA

**harriswl@hs.uci.edu*

Abstract

Both hyperacusis and misophonia are common disorders of sound tolerance. Hyperacusis is a reduced tolerance to sound level. Misophonia is a reduced tolerance to specific sounds or stimuli associated with specific sounds. People with hyperacusis typically respond to intolerable loudness levels with pain or discomfort. The response in people with misophonia to trigger sounds is often emotional such as rage or disgust. While there are no current established diagnostic criteria, multiple questionnaires are used to diagnose and monitor the severity of hyperacusis and misophonia. Hyperacusis may be experienced by 0.2–17.2% of the general population and misophonia has been found in 19.9–49% of students. Hyperacusis is often associated with tinnitus, Williams syndrome, and autism spectrum disorder. Misophonia is frequently comorbid with psychiatric disorders. Cognitive behavioral therapy (CBT) is an appropriate first-line and evidence-supported method for treatment. Programs of CBT specific to misophonia have been developed. Auditory therapy consists of passive sound attenuators, sound maskers or generators, and tinnitus retraining therapy (TRT). For patients with tinnitus, TRT may successfully alleviate disease burden of hyperacusis. There are no routine medical or surgical procedures provided for misophonia or hyperacusis at this time. Surgical intervention for intractable hyperacusis includes round and oval window reinforcement. Medical management for misophonia has been included in case reports with drugs commonly used for ADHD, OCD, psychosis, and panic disorder. Multi-modal migraine prophylaxis has successfully alleviated symptoms of hyperacusis. This consists of both lifestyle modifications, supplementation, and escalating treatment with nortriptyline, verapamil, and/or topiramate. As more people suffering from misophonia and hyperacusis seek diagnoses and treatments, there are opportunities for social support and future research.

Introduction

Definition

Hyperacusis and misophonia can be classified under the umbrella term of decreased sound tolerance [2]. (Table 1) The term "hyperacusis" means literally excessive sound and the term "misophonia" means hatred of sound. The first appearance of hyperacusis in the literature was in 1938 in reference to paralysis of the stapedius muscle [3]. A Delphi-based consensus definition of hyperacusis is as follows: "A reduced tolerance to sound(s) that are perceived as normal to the majority of the population or were perceived as normal to the person before their onset of hyperacusis, where 'normal' refers to sounds that are generally well tolerated [4]". Patients may report physical discomfort or pain in response to sound

Table 1. Comparison of misophonia and hyperacusis.

	Misophonia	**Hyperacusis**
Decreased tolerance to…	Specific auditory or visual stimuli	Degree of loudness
Primary response is …	Emotional	Pain or discomfort
Strong association with …	Psychiatric conditions	Tinnitus
Commonalities	• Normal audiometry and subjective hearing • Common among those with autism • CBT is first-line evidence-based treatment	

louder than their threshold of tolerance. People with hyperacusis have lower level of sound tolerance, or loudness discomfort level (LDL), and may tolerate only up to 60–70 dB which is equivalent to the volume of typical speech. Some reports in the literature include subjects with uncomfortable loudness levels (ULLs) as low as 30 dB [5]. For comparison, in the general population the threshold for discomfort is 110 dB, which is approximately the loudness of a rock concert or nightclub, and the threshold for pain is 130 dB, which is roughly the level of a jet engine.

Misophonia is a more recently recognized phenomenon with the word coined in 2000 by Margaret Jastreboff and Pawel Jastreboff [2]. Misophonia has been previously considered to be a subset of hyperacusis although more recently has been studied as a distinct entity [6]. Within the Delphi consensus of hyperacusis definition, 66.7% of people agreed that misophonia was a feature of hyperacusis [4]. The two disease processes are distinct as hyperacusis is characterized by intolerance to loudness, while misophonia is intolerance to a pattern of sound. Clinician consensus by the Delphi process defined misophonia as: "a disorder of decreased tolerance to specific sounds or stimuli associated with such sounds [7]". The specific stimuli are referred to as "triggers" and evoke disproportionate negative emotional, physiological, and behavioral responses that are not seen in the general population. The triggers are typically repetitive and often produced by another individual. Nasal and oral functions are well-recognized triggers including chewing, breathing, throat clearing, sniffing, typing, and foot tapping. People can have triggers that are not auditory although these are usually derived from auditory triggers. Visual triggers can include jaw movement, leg jiggling, hands touching face, putting food into mouth, and yawning. Olfactory and kinesthetic triggers (e.g., fabric textures) are rarely described. The person experiencing misophonia may produce their own triggers although this often does not induce a negative response. Mimicking sounds may even function as a coping mechanism [8]. Notably, the sources of the triggers are more often isolated to friends and family closer to the individual rather than strangers or acquaintances. Responses to triggers are varied and may be physical and emotional. Physical responses include muscle constriction, perspiration, and increased heart rate. An increase in body temperature, physical pain, and shortness of breath have been reported as well [9]. Anger is a common emotional response along with disgust, irritation, panic, and aggression. In distinction from PTSD, fear is usually not associated with triggers in misophonia [8]. Most individuals with misophonia are aware that their reactions to triggers are abnormal and are conscious of maladaptive behavior.

Assessment and diagnosis

At this time there are no established diagnostic criteria for either hyperacusis or misophonia. Questionnaires have been developed that aid in providing a diagnosis and measuring severity. Investigations into prevalence and treatment outcomes utilize these questionnaires as well. The two most adopted questionnaires for hyperacusis evaluation are the Hyperacusis Questionnaire (HQ) sometimes referred to as a Khalfa score [10] and the Multiple-Activity Scale for Hyperacusis (MASH) [11]. The HQ is formatted with three "dimensions" including attentional, social, and emotional for a total of 14 questions with a 4-point rating scale. The maximum score is 42 with a score above 28 suggesting hyperacusis compared to a mean score of 15 found in control groups. The MASH questionnaire contains 14 activities with which the participant ranks their annoyance level on a 0–10 scale. In conjunction with symptom-based scales, audiological assessment can be useful for the diagnosis and assessment of post-treatment outcomes of hyperacusis. It has been shown that of those with a ULL value equal to or below 77 dB, 95% of patients also scored above 22 on the HQ [12]. LDL and ULL cut-offs indicative of hyperacusis are variable but values between 70–100 dB are often used in practice.

There are multiple self-assessment tools for misophonia. The most widely used is the three-part Misophonia Questionnaire (MQ) designed by Wu *et al* [13]. The sections include symptoms, emotions and behaviors, and severity scale. The Amsterdam Misophonia Scale (A-MISO) is a six-item scale consisting of questions on time spent on misophonia or avoidance of triggers, social function, anger level, impulse resistance, and control over thoughts or anger [14]. It is an adapted version of the Yale-Brown Obsessive-Compulsive Scale which has also been successfully adapted for pathological gambling and body dysmorphic disorder. MisoQuest, developed in Poland, is a reliable questionnaire with excellent internal validation consisting of 14 questions with a focus on anger and source of triggers [15]. The Duke Misophonia Questionnaire is notable for including 86 items within 8 subscales (trigger frequency, affective responses, physiological responses, cognitive responses, coping before, coping during, coping after, impairment, and belief) and 2 derived composite scales for symptom severity and coping [16]. The Berlin Misophonia Questionnaire contains 20 symptom-oriented scales including 73 items which can be utilized to study specific symptoms of interest [17]. In part due to the accessibility of these questionnaires, misophonia is more commonly a self-diagnosis than a physician-led one. There are no objective audiometric findings that aid in the diagnosis of misophonia at this time.

Epidemiology

A scoping review by Ren *et al* reports hyperacusis is experienced by 0.2–17.2% of the general population, 3.8–67% of musicians, music students, and teachers, and 4.7–95% of those with associated conditions [18]. A population-based study out of Sweden reported those with hyperacusis were older and predominantly female, and that those that self-reported their hyperacusis had a higher level of education when compared to those without hyperacusis [19].

Our understanding of the prevalence of misophonia is currently and largely based on online survey population studies of students. These find misophonia reported by 19.9% of US undergraduates [13], 20% of Chinese undergraduates [20], and 49% of UK medical students [21]. A household sample from Turkey found a prevalence of 12.8% where those with misophonia diagnosis were more likely to be younger, have a family member with misophonia, and have previous contact with mental health services [22].

Comorbidities

As conditions with both auditory and emotional-behavioral regulatory aspects, hyperacusis and misophonia often occur alongside other otologic or cognitive conditions. There is a strong association between tinnitus and hyperacusis. Hyperacusis may be present in as high as 80% of persons with tinnitus [11]. Tinnitus patients with hyperacusis have been reported to be younger, experience higher levels of mental and general distress, and have associated pain or vertigo [23]. Proposed etiologies for both conditions are closely linked. Hyperacusis may lead to abnormal central gain in response to sound-evoked auditory pathway. A suggested mechanism of tinnitus is spontaneous central neural activity that acts similarly to the sound-evoked auditory pathway [24]. The connection between tinnitus and hyperacusis can explain that treatment focused primarily on tinnitus can subsequently lessen the symptom burden of hyperacusis.

Another condition with high rates of hyperacusis is Williams syndrome. Williams syndrome is caused by a microdeletion in chromosome 7q11.23 and is characterized by intellectual differences, unique and friendly personalities, and elevated musicality. The prevalence of hyperacusis in Williams syndrome has been reported to range from 4.7% to 95% [18]. Those with hyperacusis in William's syndrome often have high-frequency sensorineural hearing loss [25]. These symptoms may be a consequence of congenital auditory nerve dysfunction that may lead to a deficiency in the acoustic reflex.

Autism spectrum disorder is a multifaceted developmental disorder defined by persistent differences in social communication and interaction presenting with restricted and repetitive behaviors. Those with autism process an increased amount of information compared to the general population and can experience hypersensitivity to visual and auditory stimuli. The prevalence of hyperacusis among autistic people ranges from 18% to 69% [26]. Autistic individuals can have lower LDLs with one study reporting 37% of those with autism to have an LDL greater than three standard deviations below those without autism [27].

Misophonia similarly has an association with autism spectrum disorder. Of those with misophonia, 3% were found to be diagnosed with autism in a clinical sample [28] and in an online sample of people with self-reported misophonia, 3.6% had an autism diagnosis [29]. Even when not formally diagnosed with misophonia, many autistic individuals report strong emotional and behavioral reactions to common misophonic trigger sounds. Similarly, those with misophonia have more autism-related traits (by the Autism Spectrum Quotient) than peers [30]. Misophonia among those with tic disorders has been described where an increase in tics and abrupt emotional dysregulation was associated with sound triggers. Those with tic disorders can have a higher sensory awareness which could explain exacerbations of misophonia responses [31].

Psychiatric disorders are commonly comorbid with misophonia, with 28% of those with misophonia reporting concomitant post-traumatic stress disorder (PTSD), obsessive-compulsive disorder (OCD), major depressive disorder (MDD), anorexia, or others [28]. Having any psychiatric disorder is associated with misophonia presenting with more severe symptoms [32]. PTSD and misophonia share avoidant behaviors, hypervigilance, and hyperarousal in reaction to trigger sounds; some patients with both PTSD and misophonia can have shared triggers. However, a PTSD reaction is primarily of fear while a misophonia response is characterized by anger or irritation. OCD and misophonia both are characterized by impulsive behavior that the person recognizes as irrational. For people with misophonia, a compulsion can manifest as avoidance of triggers. Additionally, people with misophonia may perform their triggers as a coping mechanism. This mimicking of the repetitive behaviors causing one stress may appear as acting a compulsion or possibly a tic. Those with eating disorders can be particularly attuned to auditory stimuli of eating or chewing. A three-case series described a young woman with anorexia nervosa and misophonia reported a high daily lifestyle impact due to situation avoidance [33]. For patients with commonly comorbid conditions associated with hyperacusis or misophonia, assessment and treatment for these conditions may be prudent as improvement of symptoms for either diagnosis may improve overall well-being and functional status.

Treatment

Cognitive behavioral therapy

As definitive or curative treatment methods for hyperacusis and misophonia are currently absent, symptomatic management and alleviation of related distress is the primary modality for treatment. There is a strong focus on the depression and anxiety often related to these conditions as well as improvement of functional status. Cognitive behavioral therapy (CBT) is a tool to help patients modify their emotions, thoughts, and behavior and has been applied with success in the treatment of both hyperacusis and misophonia.

The CBT method was developed in the 1960s by Aaron Beck with discrete protocols that have grown into an empirically validated psychological treatment. In part due to the structured and didactic nature of this therapy method, CBT has been applied and studied in a wide range of issues such as depression, pain, OCD, and PTSD, among other diagnoses. The goal of CBT is to discover and modify patterns of distress to indirectly reduce negative emotions. Patients aim to recognize negative patterns and learn to replace them with less distressful thoughts and behaviors. CBT requires the patient to have strong intrinsic motivation, time commitment, social support, and a threshold level of education. An advantage of CBT is the ability to simultaneously target comorbid depression and anxiety to reduce overall symptom burden. However, CBT has been criticized for its focus on present emotion and not root causes arguably to the point of equating symptoms to cognitive cause [34]. CBT, while goal-oriented and empirically sound, may primarily act as a temporizing symptom management strategy rather than an ultimate cure.

The CBT model maps well onto the symptomatology of hyperacusis and misophonia. For example, the distress associated with misophonia or hyperacusis can initiate an emotional response such as anger or disgust or physical sensation following a trigger. Subsequent negative thoughts and actions like the need to leave the setting can lead to additional negative emotions including frustration and shame, which provide an opportunity for evaluative thoughts in relation to the sound [35].

A randomized controlled trial investigating the use of CBT for hyperacusis conducted by Juris *et al.* [36] consisted of two treatment groups of 30 subjects each. Patients either participated in 6 sessions of psychologist-led CBT or were placed on a 6-month waiting list. CBT treatment included psychoeducation, exposure therapy, applied relaxation, and behavioral activation. Patients were educated on the physiology of the auditory system, sound levels, and the CBT model. Pre-treatment patients ranked sounds by difficulty to endure from 0–100. Exposure therapy started at level 30 while patients were instructed to complete a task. During challenging exposure therapy sessions, patients were prompted through relaxation strategies. Additionally, patients were encouraged to gradually resume activities they avoided due to hyperacusis. This protocol showed success with a statistically and clinically significant improvement in the LDLs as well as HQ score, HADS depression scale, and Quality of Life Inventory. While this study supports the positive effect of CBT therapy for hyperacusis, it is limited by comparison to a non-treatment group rather than a group receiving non-CBT therapy.

Long-term CBT for hyperacusis and tinnitus was implemented and studied by Nolan *et al.* in a 10-year treatment program [37]. Therapy protocol began with psychoeducation and continued with individual weekly therapy with a psychiatrist. Additional modalities included cognitive restructuring, physical therapy of the shoulder-neck-jaw region, ear acupuncture, and relaxation techniques with the use of creative therapy. For those with tinnitus, treatment of hyperacusis with CBT improved overall wellness and lifestyle. The role of CBT for misophonia has been addressed by a randomized controlled trial produced by Jager *et al.* [28]. Patients were randomized to a 3-month waiting list or to receiving CBT by clinical psychologists in combination with psychotherapy and Psychomotor Therapy. Clinical improvement as determined by the Clinical Global Impression-Improvement (CGI-I) was found in 37% that received any CBT and in 56% of those that completed all 3 months of weekly group sessions. Additionally, there was a statistically significant decrease in the A-MISO among those in the CBT treatment group. Improvement of misophonia-related symptoms was sustained after 12 months after the conclusion of treatment. In a previous open-label clinical trial by Schroder *et al.* [38] measurements of CGI-I and A-MISO improved following CBT. One of the positive response predictors was a high baseline A-MISO score indicative of more severe misophonia. Also, patients who primarily experience disgust rather than anger to triggers had higher rates of positive response to the CBT treatment. In case studies, CBT was found to reduce the feeling of anger in female students with misophonia [39] and lead to a reduction in symptoms in teenage female patients [40].

There are four elements of CBT therapy for misophonia that were introduced and evaluated by Schroger *et al.* [38]. The first is a task consisting of concentration exercises. The goal of these exercises is to improve attentional shifting and to remove the attention bias that is experienced by those with misophonia in response to a trigger. Patients will learn how to focus on sensory input separate from their triggers. For example, one can learn to shift their attention from a commuter chewing gum to a phone-based game. This is best done first in a neutral setting such as with the psychiatrist and then practiced in a setting with high levels of triggers such as a dinner with family members. The next CBT technique that can be used for misophonia is counterconditioning. This is a method of systematic desensitization through stimulus substitution that is often used for those with PTSD or phobias. This can be done either *in vitro* in which the patient imagines the trigger or *in vivo* in which the patient is exposed to the trigger itself. By including a pleasant stimulus alongside the trigger, the patient trains to associate the trigger with the emotion or neutrality normally experienced by the pleasant stimulus. Stimulus manipulation is another technique in which patients can simulate control over a trigger. Misophonia triggers tend to be repetitive but usually not continuous. The unpredictability and therefore lack of control the patient feels may contribute to their unique emotional response. One cannot reasonably ask for someone to stop breathing and may not be able to remove themselves from the situation. With stimulus manipulation, patients practice by modulating sounds with home computer-based tools.

This imitates a feeling of control over the trigger sounds that can carry into real-life scenarios. Finally, a core component of CBT for misophonia is relaxation exercises. The physical arousal that results from irritability or panic stemming from a trigger can heighten these negative emotions. Decreasing this physical arousal can be practiced with deep breathing, progressive muscle relaxation, autogenic training, and guided imagery.

Group therapy for CBT is often preferred over individual therapy due to both convenience and opportunity for peer support. Jager *et al.* recently developed a group CBT treatment protocol [41] based on the four primary components addressed above. It is designed for a maximum of 9 patients and consists of 7 weekly 180-minute meetings. The benefits of group therapy include the ability to simulate social situations, mutual support, learning techniques from each other, and cost efficacy. Internet-based CBT has been under development and much psychiatric treatment has shifted to telehealth secondary to the COVID-19 pandemic. This move to a virtual setting may additionally reduce barriers to access and provide wider reach for CBT treatment.

Other psychological treatment

Dialectical behavior therapy (DBT) is another therapeutic treatment that focuses on the acceptance of anger rather than the reduction of negative emotions through exposure. DBT has been suggested for use in misophonia that does not respond to CBT or has anger as a primary trigger response. One case report of a 16-year-old-girl reported a reduction of misophonia symptoms and global psychological symptoms following DBT [42].

Among evidence-based psychological treatments, eye movement desensitization and reprocessing therapy (EMDR) is also a promising treatment for misophonia. The purpose of EMDR is to reprocess traumatic or dysfunctional stored memories and forge more adaptive associations. The reprocessing sessions of EMDR include repetitive bilateral stimulation most commonly performed by visually tracking an object while focusing on a negative stimulus. Jager *et al.* found EMDR to be feasible with a reduction of A-MISO and CGI-I scores in 2 of 8 patients trialed on EMDR therapy. A hurdle for improved misophonia-specific EMDR protocol is identifying negative misophonia-related target memories [43].

Auditory therapy

Methods to reduce sound exposure have been used as treatment strategies given hyperacusis is essentially a decreased tolerance to sound levels. Similarly, those with misophonia may wear devices to block exposure to trigger sounds. Wearing passive attenuators including earmuffs or headphones is often the first course of action for people with hyperacusis and misophonia when entering loud or trigger-filled settings. Electronic loudness suppression devices have been created and have been found to have a perceived benefit in hyperacusis patients [44]. Conversely to sound suppression, sound maskers or generators are also useful for the management of hyperacusis. For those with hyperacusis or misophonia and comorbid hearing loss, hearing aid amplification can help to mask loud or triggering sounds. Low-level acoustic stimulation also known as sound therapy has been applied to patients with tinnitus with some success and has been translated to a hyperacusis therapy. Sound therapy adds neutral auditory signals that may interfere with elevated noise signals to decrease enhanced gain in the auditory pathways. Both broad-band and narrow-band noise have been used with indeterminate efficacy. Formby *et al.* found the use of broad-band noise generators by hearing-impaired individuals at 50 dB for 2 weeks increased LDLs by 6–8 dB [45]. Acoustic training sessions in patients with Williams syndrome has be performed with long-lasting remission of symptoms with the use of narrow-band noise at 60 dB increased to 95 dB in 5 dB increments [46]. Another study reported that passive exposure at a just audible level for 1–3 hours with an enriched acoustic environment decreased both auditory sensitivity and HQ scores [47]. The enriched acoustic environment was individually generated from participants' audiograms and consisted of pure tones of 100 ms at an average interval of 100 ms. Another device that may have a role in hyperacusis treatment is low-power laser therapy. In a randomized controlled trial, the continuous wave laser TinniTool improved LDLs compared to placebo in patients with tinnitus and hyperacusis [48]. For 20-minute intervals, participants donned a headset fitted with a laser probe to provide a total of 6 J onto the tympanic membrane.

Tinnitus retraining therapy (TRT) consists of sound therapy with the addition of concurrent counseling that was initially targeted to reduce tinnitus symptoms in those without hearing loss. This method has been used for those with hyperacusis with positive results measured by decreased HQ scores and increased LDLs [2, 49]. Long-term results are possible with one study finding LDLs to be at the normal range above 100 dB in 60% of patients at 2 years post-treatment [50]. TRT is based on habituation in which the participant becomes accustomed to a sound and has a diminished response. The goal is to achieve habituation of both the perceived sound and the subsequent negative response to the sound. An important initial step of TRT is for the patient to accept the neutrality of hyperacusis which can be achieved through patient education. A positive relationship with the audiologist is essential before beginning the sound exposure component.

Surgical treatments

Hyperacusis relieving surgery was recently introduced in patients with superior semicircular canal syndrome. In these patients, conductive hyperacusis was often a result of a "third window" diminishing the impedance of the inner ear fluids. By reinforcing the round window, the impedance can be increased, and as a result, symptomatic hyperacusis can be reduced. A transcanal approach has been reported with post-operative symptom improvement and no complications in two patients [51]. Of note, in patients with superior semicircular canal dehiscence, blockage of the round window can increase sound pressure towards the dehiscence which may worsen pre-existing Tullio phenomenon (sound-induced vertigo). Patients may furthermore benefit from this procedure from the potential creation of conductive hearing loss but should be avoided in patients experiencing the Tullio phenomenon.

In patients with hyperacusis without superior canal dehiscence syndrome, round and oval window reinforcement has been described using a minimally invasive technique. The authors noted that stapes hypermobility was frequently observed intraoperatively, and accordingly they provided soft tissue reinforcement around the oval window and the stapes. Short-term overall improvement has been reported in 60–80% of patients and as well as statistically significant increased LDLs post-surgery [52]. Younger age was found to be a positive prognostic factor for surgery while a history of tinnitus may negatively influence outcomes [52]. For patients with unilateral sensorineural hearing loss, severe tinnitus, and hyperacusis, cochlear implants can reduce symptoms [53]. One study assessing the hyperacusis severity 12 months after cochlear implant surgery found a reduction of HQ in 6 of 7 patients [54].

Medical management

Case reports have highlighted treatment modalities that can be successful in the management of misophonia. (Table 2) For a 14-year-old male with misophonia and inattention, methylphenidate provided aid in relieving symptoms [55]. This patient had misophonia manifesting as negative emotional responses to sounds specifically associated with his mother's voice that worsened distractibility resulting in school difficulties. The patient was trialed on methylphenidate as well as misophonia-specific CBT treatment with improved distractibility and tolerance to misophonia triggers. The authors suggested that the frontostriatal network contributing to distractibility for those with ADHD could be acting with the frontocortical region associated with misophonia.

In a 14-year-old female, fluoxetine was associated with a 40% decreased A-MISO score and a 70% decreased C-GAS score on the four-month follow-up [56]. The patient had multiple misophonia triggers, particularly whispering, that lead to anger outbursts that were heightened during the period of pandemic isolation. After the initiation of fluoxetine, she would have no negative reaction to her siblings whispering. Fluoxetine has been approved for use in OCD and this case may suggest a treatment pathway for OCD can be translated for misophonia. Risperidone improved misophonia symptoms in a 32-year-old male patient with concomitant autism spectrum disorder, generalized anxiety disorder, and schizoaffective disorder previously treated with olanzapine [57]. The anti-psychotic was initiated during admission to the psychiatric unit following an episode of acute paranoia. After beginning risperidone, he maintained a substantially increased tolerance to his numerous triggers. The mechanism of D2 dopaminergic antagonism may be responsible, however risperidone binds to multiple additional receptors as well.

A 16-year-old male was treated successfully with propranolol [58]. The patient first experienced misophonia at the age of 9 with his sister's open mouth eating as the only trigger. This trigger generated flight-or-fight physical responses ultimately limiting his ability to participate in family dinners. A hearing device was used to mask sounds during meals; however, he still experienced visual triggers. Propranolol was taken 1 hour before meals. The patient reported 100% efficacy when used in addition to the masking hearing aid and he was able to eat meals with family free of symptoms. Beta-blockers are used for PTSD and panic disorders prophylactically when patients must experience known trigger scenarios.

Table 2. Medications reported by case report with successful treatment of misophonia or hyperacusis.

Misophonia	Hyperacusis
Methylphenidate	Carbamazepine
Fluoxetine	Pyroxidine
Risperidone	Multi-model migraine prophylaxis
Propranolol	• Nortriptyline
	• Verapamil
	• Topiramate

For patients that experience misophonia in specific environments, medications that depress sympathetic activity can alleviate symptoms.

As with misophonia, case series have proposed successful medication management for hyperacusis (Table 2). For two patients with Lyme disease-induced hyperacusis, the anticonvulsant medication carbamazepine improved sound tolerance and symptoms worsened when the drug was discontinued [59]. Carbamazepine is commonly used to treat trigeminal neuralgia which is highly associated with Lyme disease. In addition to hyperacusis, these patients experienced vertigo, headaches, photophobia, nystagmus, and neuropsychiatric symptoms after the onset of Lyme disease, and a pathologic auditory-limbic link may be the culprit. Pyroxidine is another treatment for neuropathy that has been found to improve sound hypersensitivity in patients with developmental disorders [60]. For a 22-year-old woman with hyperacusis in the absence of other auditory or developmental disorders, lithium and counseling improved depressive symptoms and increased her LDL from 30 to 100 dB [61].

Multi-modal migraine prophylaxis can be a useful treatment option for hyperacusis. A prospective study by Abouzari *et al.* of 25 patients with hyperacusis treated with multi-modal migraine prophylaxis regimen found audiometric and self-reported improvement in symptoms [62]. Patients were included with an LDL less than 85 dB and HQ score above 28. Migraine prophylaxis therapy included lifestyle modifications consisting of avoidance of trigger foods and supplementation with magnesium and riboflavin as well as regular meal and sleep schedules. Pharmacologic management was done in a step-wise manner. Nortriptyline was used as the initial medication with verapamil added if dose escalation was ineffective. For those on an SSRI or with hypertension, verapamil was initiated first. Topiramate was added and dose escalated in the absence of symptom improvement with first-line medications. Following treatment, 88% of patients self-reported symptom improvement, increased QOL scores, and the average LDL showed statistically significant increases. Migraines are often associated with sensitization to smell, light, and sound; hyperacusis symptoms are often referred to as phonophobia in this context. Given the successful reduction of hyperacusis symptoms with evidence-based multimodal migraine prophylaxis, the pathophysiology of hyperacusis may be similar to migraine or have overlapping comorbidities.

Social support, patient perception, and future directions

Misophonia and hyperacusis can be isolating disorders. There are multiple websites, Facebook groups, Reddit pages, and social networks that provide patient education and peer support for those with hyperacusis and misophonia. Annual conventions are hosted by the non-profit Misophonia Association founded in 2013. Another nonprofit is soQuiet which provides free information cards, peer-support groups, and graduate student research grants. Adeel Ahmad's Misophonia Podcast interviews guests with misophonia and incorporates sound editing to mask sound triggers for their listeners. As these resources are easily available online, many patients may self-diagnose prior to or in absence of meeting with a physician. As many as 80% of patients that self-report having misophonia have not been diagnosed by a professional [63]. Therefore, up-to-date, and accessible patient information on the internet is important.

Given limited evidence-based treatment options, patient perception of treatment efficacy can be significant and drive future investigations. A large sample of adults and parents of children with misophonia have provided information about treatment experiences [63]. Lifestyle modifications and audiologic interventions including noise canceling methods were deemed most appropriate by patients. Overall, most patients were not satisfied with the treatment options they trialed. Particularly, patients perceived exposure therapy as the least appropriate. CBT, which includes a modern formulation of exposure therapy as a core component, is often provided as first-line treatment for misophonia. As CBT requires high levels of patient involvement and motivation, it is unlikely to yield success if patients do not feel this treatment appropriate. Overall, a multidisciplinary approach with both psychologists and audiologists is best to provide the most individually appropriate audiological, medical, and psychological therapy for both hyperacusis and misophonia.

While there still exist knowledge gaps in our understanding of hyperacusis and misophonia, new innovations and advancements are being made at a fast pace. For hyperacusis, further research priorities can focus on the efficacy of TRT, sound therapy intervention, additional surgical treatment, and pharmaceutical prophylaxis. Misophonia treatment targets also include tailored CBT and management of comorbid psychiatric disorders.

Conclusion

- Hyperacusis is a reduced tolerance to sound level. Misophonia is a reduced tolerance to specific sounds or stimuli associated with specific sounds.
- Multiple questionnaires are used to diagnose and monitor the severity of hyperacusis and misophonia. Hyperacusis may be experienced by 0.2–17.2% of the general population. Misophonia has been found in 19.9–49% of students.
- Hyperacusis is often associated with tinnitus, Williams syndrome, and ASD. Misophonia is frequently comorbid with psychiatric disorders.
- Cognitive behavioral therapy is a common and evidence-supported method for treatment.
- Auditory therapy consists of passive sound attenuators, sound maskers or generators, and tinnitus retraining therapy.
- Surgical intervention for intractable hyperacusis includes round and oval window reinforcement.
- Medical management for misophonia has been included in case reports with drugs commonly used for ADHD, OCD, psychosis, and panic disorder. Multi-modal migraine prophylaxis has successfully alleviated symptoms of hyperacusis.
- There are opportunities for social support for those with misophonia and hyperacusis.

References

1. Ben-Sasson A, Hen L, Fluss R, Cermak SA, Engel-Yeger B, and Gal E. A meta-analysis of sensory modulation symptoms in individuals with autism spectrum disorders. *Journal of Autism and Developmental Disorders*. **39**(1):1–11 (2009).

2. Jastreboff PJ, and Jastreboff MM. Tinnitus retraining therapy (TRT) as a method for treatment of tinnitus and hyperacusis patients. *Journal of the American Academy of Audiology*. **11**(3):162–177 (2000).
3. Perlman HB. LXXIX Hyperacusis. *Annals of Otology, Rhinology & Laryngology*. **47**(4):947–953 (1938).
4. Adams B, Sereda M, Casey A, Byrom P, Stockdale D, and Hoare DJ. A Delphi survey to determine a definition and description of hyperacusis by clinician consensus. *International Journal of Audiology*. **60**(8):607–613 (2021).
5. Aazh H, and Moore BCJ. Effectiveness of audiologist-delivered cognitive behavioral therapy for tinnitus and hyperacusis rehabilitation: Outcomes for patients treated in routine practice. *American Journal of Audiology*. **27**(4):547–558 (2018).
6. Tyler RS, Pienkowski M, Roncancio ER, *et al*. A review of hyperacusis and future directions: Part I. Definitions and manifestations. *American Journal of Audiology*. **23**(4):402–419 (2014).
7. Swedo SE, Baguley DM, Denys D, *et al*. Consensus definition of misophonia: A delphi study. *Frontiers in Neuroscience*. **16**:841816 (2022).
8. Edelstein M, Brang D, Rouw R, and Ramachandran VS. Misophonia: Physiological investigations and case descriptions. *Frontiers in Human Neuroscience*. **7**:296 (2013).
9. Ferrer-Torres A, and Giménez-Llort L. Misophonia: A systematic review of current and future trends in this emerging clinical field. *International Journal of Environmental Research and Public Health*. **19**(11) (2022).
10. Khalfa S, Dubal S, Veuillet E, Perez-Diaz F, Jouvent R, and Collet L. Psychometric normalization of a hyperacusis questionnaire. *Otolaryngology–Head and Neck Surgery*. **64**(6):436–442 (2002).
11. Dauman R, and Bouscau-Faure F. Assessment and amelioration of hyperacusis in tinnitus patients. *Acta Oto-Laryngologica*. **125**(5):503–509 (2005).
12. Aazh H, Moore BC, Lammaing K, and Cropley M. Tinnitus and hyperacusis therapy in a UK national health service audiology department: Patients' evaluations of the effectiveness of treatments. *International Journal of Audiology*. **55**(9):514–522 (2016).
13. Wu MS, Lewin AB, Murphy TK, and Storch EA. Misophonia: Incidence, phenomenology, and clinical correlates in an undergraduate student sample. *Journal of Clinical Psychology*. **70**(10):994–1007 (2014).
14. Schröder A, Vulink N, and Denys D. Misophonia: Diagnostic criteria for a new psychiatric disorder. *PLoS One*. **8**(1):e54706 (2013).
15. Siepsiak M, Śliwerski A, and Łukasz Dragan W. Development and psychometric properties of MisoQuest — A new self-report questionnaire for misophonia. *International Journal of Environmental Research and Public Health*. **17**(5):1797 (2020).

16. Rosenthal MZ, Anand D, Cassiello-Robbins C, *et al.* Development and initial validation of the duke misophonia questionnaire. *Frontiers in Psychology.* **12**:709928 (2021).
17. Remmert N, Schmidt KMB, Mussel P, Hagel ML, and Eid M. The Berlin misophonia questionnaire revised (BMQ-R): Development and validation of a symptom-oriented diagnostical instrument for the measurement of misophonia. *PLoS One.* **17**(6): e0269428 (2022).
18. Ren J, Xu T, Xiang T, *et al.* Prevalence of hyperacusis in the general and special populations: A scoping review. *Frontiers in Neurology.* **12**:706555 (2021).
19. Paulin J, Andersson L, and Nordin S. Characteristics of hyperacusis in the general population. *Noise & Health.* **18**(83):178–184 (2016).
20. Zhou X, Wu MS, and Storch EA. Misophonia symptoms among Chinese university students: Incidence, associated impairment, and clinical correlates. *Journal of Obsessive-Compulsive and Related Disorders.* **14**:7–12 (2017).
21. Naylor J, Caimino C, Scutt P, Hoare DJ, and Baguley DM. The prevalence and severity of misophonia in a UK undergraduate medical student population and validation of the Amsterdam misophonia scale. *Psychiatric Quarterly.* **92**(2):609–619 (2021).
22. Kılıç C, Öz G, Avanoğlu KB, and Aksoy S. The prevalence and characteristics of misophonia in ankara, Turkey: Population-based study. *BJPsych Open.* **7**(5):e144 (2021).
23. Schecklmann M, Landgrebe M, Langguth B, and Group TRIDS. Phenotypic characteristics of hyperacusis in tinnitus. *PLoS One.* **9**(1):e86944 (2014).
24. Diehl PU, and Schaette R. Abnormal auditory gain in hyperacusis: Investigation with a computational model. *Frontiers in Neurology.* **6**:157 (2015).
25. Gothelf D, Farber N, Raveh E, Apter A, and Attias J. Hyperacusis in Williams syndrome. Characteristics and associated neuroaudiologic abnormalities. *Neurology.* **66**(3):390–395 (2006).
26. Danesh AA, Howery S, Aazh H, Kaf W, and Eshraghi AA. Hyperacusis in autism spectrum disorders. *Audiology Research.* **11**(4):547–556 (2021).
27. Demopoulos C, and Lewine JD. Audiometric profiles in autism spectrum disorders: Does subclinical hearing loss impact communication? *Autism Research.* **9**(1):107–120 (2016).
28. Jager IJ, Vulink NCC, Bergfeld IO, van Loon A, and Denys D. Cognitive behavioral therapy for misophonia: A randomized clinical trial. *Depression and Anxiety.* **38**(7):708–718 (2020).
29. Claiborn JM, Dozier TH, Hart SL, and Lee J. Self-Identified misophonia phenomenology, impact, and clinical correlates. *Psychological Thought.* **13**:349–375 (2020).
30. Rinaldi LJ, Simner J, Koursarou S, and Ward J. Autistic traits, emotion regulation, and sensory sensitivities in children and adults with Misophonia. *Journal of Autism and Developmental Disorders.* **53**(3):1162–1174 (2023).

31. Robinson S, Hedderly T, Conte G, Malik O, and Cardona F. Misophonia in children with tic disorders: A case series. *Journal of Developmental & Behavioral Pediatrics*. **39**(6):516–522 (2018).
32. Erfanian M, Kartsonaki C, and Keshavarz A. Misophonia and comorbid psychiatric symptoms: A preliminary study of clinical findings. *Nordic Journal of Psychiatry*. **73**(4–5):219–228 (2019).
33. Kluckow H, Telfer J, and Abraham S. Should we screen for misophonia in patients with eating disorders? A report of three cases. *International Journal of Eating Disorders*. **47**(5):558–561 (2014).
34. Gaudiano BA. Cognitive-behavioural therapies: Achievements and challenges. *Evidence Based Mental Health*. **11**(1):5–7 (2008).
35. Aazh H, Landgrebe M, Danesh AA, and Moore BC. Cognitive behavioral therapy for alleviating the distress caused by tinnitus, hyperacusis, and misophonia: Current perspectives. *Psychology Research and Behavior Management*. **12**:991–1002 (2019).
36. Jüris L, Andersson G, Larsen HC, and Ekselius L. Cognitive behaviour therapy for hyperacusis: A randomized controlled trial. *Behaviour Research and Therapy*. **54**:30–37 (2014).
37. Nolan DR, Gupta R, Huber CG, and Schneeberger AR. An effective treatment for tinnitus and hyperacusis based on cognitive behavioral therapy in an inpatient setting: A 10-year retrospective outcome analysis. *Frontiers in Psychiatry*. **11**:25 (2020).
38. Schröder AE, Vulink NC, van Loon AJ, and Denys DA. Cognitive behavioral therapy is effective in misophonia: An open trial. *Journal of Affective Disorders*. **217**:289–294 (2017).
39. Roushani K, and Mehrabizadeh Honarmand M. The effectiveness of cognitive-behavioral therapy on anger in female students with misophonia: A single-case study. *Iranian Journal of Medical Sciences*. **46**(1):61–67 (2021).
40. McGuire JF, Wu MS, and Storch EA. Cognitive-behavioral therapy for 2 youths with misophonia. *Journal of Clinical Psychiatry*. **76**(5):573–574 (2015).
41. Jager I, Vulink N, van Loon A, *et al*. Synopsis and qualitative evaluation of a treatment protocol to guide systemic group-cognitive behavioral therapy for misophonia. *Frontiers in Psychiatry*. **13**:794343 (2022).
42. Kamody RC, and Del Conte GS. Using dialectical behavior therapy to treat misophonia in adolescence. *Primary Care Companion for CNS Disorders*. **19**(5):17102105 (2017).
43. Jager I, Vulink N, de Roos C, and Denys D. EMDR therapy for misophonia: A pilot study of case series. *European Journal of Psychotraumatology*. **12**(1):1968613 (2021).
44. Sammeth CA, Preves DA, and Brandy WT. Hyperacusis: Case studies and evaluation of electronic loudness suppression devices as a treatment approach. *Scandinavian Audiology*. **29**(1):28–36 (2000).

45. Formby C, Sherlock LP, Hawley ML, and Gold SL. A sound therapy-based intervention to expand the auditory dynamic range for loudness among persons with sensorineural hearing losses: Case evidence showcasing treatment efficacy. *Seminars in Hearing*. **38**(1):130–150 (2017).
46. Miani C, Passon P, Bracale AM, Barotti A, and Panzolli N. Treatment of hyperacusis in Williams syndrome with bilateral conductive hearing loss. *European Archives of Oto-Rhino-Laryngology*. **258**(7):341–34 (2001).
47. Noreña AJ, and Chery-Croze S. Enriched acoustic environment rescales auditory sensitivity. *Neuroreport*. **18**(12):1251–1255 (2007).
48. Teggi R, Bellini C, Piccioni LO, Palonta F, and Bussi M. Transmeatal low-level laser therapy for chronic tinnitus with cochlear dysfunction. *Audiology and Neurotology*. **14**(2):115–120 (2009).
49. Park JM, Kim WJ, Ha JB, Han JJ, Park SY, and Park SN. Effect of sound generator on tinnitus and hyperacusis. *Acta Oto-Laryngologica*. **138**(2):135–139 (2018).
50. Fackrell K, Potgieter I, Shekhawat GS, Baguley DM, Sereda M, and Hoare DJ. Clinical interventions for hyperacusis in adults: A scoping review to assess the current position and determine priorities for research. *Biomedical Research International*. **2017**:2723715 (2017).
51. Nikkar-Esfahani A, Whelan D, and Banerjee A. Occlusion of the round window: a novel way to treat hyperacusis symptoms in superior semicircular canal dehiscence syndrome. *Journal of Laryngology and Otology*. **127**(7):705–707 (2013).
52. Silverstein H, Kellermeyer B, and Martinez U. Minimally invasive surgery for the treatment of hyperacusis: New technique and long term results. *American Journal of Otolaryngology*. **41**(1):102319 (2020).
53. Ramos Macías A, Falcón González JC, Manrique M, *et al*. Cochlear implants as a treatment option for unilateral hearing loss, severe tinnitus and hyperacusis. *Audiology and Neurotology*. **20** Suppl 1:60–66 (2015).
54. Kompis M, Pelizzone M, Dillier N, Allum J, DeMin N, and Senn P. Tinnitus before and 6 months after cochlear implantation. *Audiology and Neurotology*. **17**(3):161–168 (2012).
55. Osuagwu FC, Osuagwu VC, and Machoka AM. Methylphenidate ameliorates worsening distractibility symptoms of misophonia in an adolescent male. *Primary Care Companion for CNS Disorders*. **22**(5):19102553 (2020).
56. Sarigedik E, and Yurteri N. Misophonia successfully treated with fluoxetine: A case report. *Clinical Neuropharmacology*. **44**(5):191–192 (2021).
57. Pan EJ, Weleff J, Anand A, and Barnett BS. Treatment of misophonia with risperidone in a patient with autism spectrum disorder. *Case Reports in Psychiatry*. **2022**:3169834 (2022).
58. Webb J. β-Blockers for the treatment of misophonia and misokinesia. *Clinical Neuropharmacology*. **45**(1):13–14 (2022).

59. Nields JA, Fallon BA, and Jastreboff PJ. Carbamazepine in the treatment of Lyme disease-induced hyperacusis. *Journal of Neuropsychiatry and Clinical Neurosciences*. **11**(1):97–99 (1999).
60. Kamiyama M, Kuriyama S, and Watanabe M. A clinical study of pyridoxine treatment for pervasive developmental disorders with hypersensitivity to sound. *No to Hattatsu*. **38**(4):277–282 (2006).
61. Attri D, and Nagarkar AN. Resolution of hyperacusis associated with depression, following lithium administration and directive counselling. *Journal of Laryngology and Otology*. **124**(8):919–921 (2010).
62. Abouzari M, Tan D, Sarna B, *et al*. Efficacy of multi-modal migraine prophylaxis therapy on hyperacusis patients. *Annals of Otology, Rhinology & Laryngology*. **129**(5):421–427 (2020).
63. Smith EEA, Guzick AG, Draper IA, *et al*. Perceptions of various treatment approaches for adults and children with misophonia. *Journal of Affective Disorders*. **316**:76–82 (2022).

https://doi.org/10.1142/9789819807345_0013

Chapter 13

Treatment of Pulsatile Tinnitus

Erika Lipford and David J. Eisenman*

Department of Otorhinolaryngology-Head and Neck Surgery,
University of Maryland School of Medicine,
Baltimore, MD, USA

**deisenman@som.umaryland.edu*

Abstract

Pulsatile tinnitus (PT) is defined as the self-perception of one's blood flow, accounting for about 4% of all patients with a chief complaint of tinnitus. PT etiologies can be classified into different categories, including abnormal sound production such as sigmoid sinus wall and transverse sinus anomalies, abnormal sound perception like third mobile window syndromes and Eustachian tube dysfunction, a combination of both, or causes of unknown mechanisms such as migraine and Meniere's disease. It is crucial to explain to patients that PT is a symptom of an underlying disease. Based on the findings of a comprehensive review of the patient's medical history, focusing on PT and its associated otologic and non-otologic symptoms, physical examination, and hearing evaluation, appropriate imaging can be ordered to aid in the diagnosis. Parasympathetic paragangliomas, specifically jugulotympanic paraganglia, are associated with PT due to their location within the middle ear and/or jugular foramen. In addition, PT is one of the most common presenting symptoms in patients with idiopathic intracranial hypertension (IIH). Furthermore, sigmoid sinus wall abnormalities and transverse sinus stenosis, frequently associated with IIH, can also cause PT. Dural vascular lesions, specifically type I lesions of the transverse sigmoid sinus, are an important cause of arterial PT. A dehiscence in the superior semicircular canal is another cause of PT by increasing inner ear compliance and sensitivity to bone-conducted sounds as well as potentially having the third window transmitting dural pulsations to the cochlea. Finally, in patients with a normal imaging workup and normal neuro-ophthalmologic exam, otologic migraine should be considered as a potential etiology for PT. For each of these conditions, several treatment options are available. However, when a specific cause cannot be found, non-specific treatments, such as tinnitus habituation therapies, masking, and cognitive therapies, like cognitive behavioral therapy, can be employed.

Introduction

Pulsatile tinnitus (PT) is defined as the self-perception of one's blood flow, a vascular somatosound. It should more properly be called pulse-synchronous tinnitus, but the term PT has become so ensconced in medical parlance and the literature, and even has its own ICD-10 code, that it is the default, commonly accepted terminology. Unlike the far more common non-rhythmic tinnitus, which is a centrally generated electrochemical phenomenon most commonly arising from auditory pathway pathology, true PT is a mechanical phenomenon usually arising from a source outside the auditory pathway, and often more amenable to direct surgical or endoluminal intervention. In that respect, the relationship between PT and central, non-rhythmic tinnitus, is loosely analogous to that between conductive hearing loss and sensorineural hearing loss, where the underlying problem can often be directly remedied for the former, rather than employing treatments that are largely workarounds, as for the latter.

Not all rhythmic tinnitus is vascular or even always mechanical. Middle ear myokymias, tensor tympani or stapedius muscle spasms, and palatal myoclonus, produce a rhythmic tinnitus as well, though not pulse synchronous. Additionally, and very importantly, central tinnitus such as is commonly associated with sensorineural hearing loss or acoustic trauma can also sometimes present as a rhythmic sound synchronous with the pulse, for unclear reasons, and it is critical to try to distinguish this from a true PT that is due to the perception of one's blood flow, before undertaking an expensive and fruitless diagnostic evaluation for a source of sound production that does not exist.

This chapter will first review the epidemiology of PT and then outline an approach to diagnosis and treatment. It will start with general principles of diagnosis that apply to anybody with PT, and then discuss specific etiologies and their treatment. It is important to understand and explain to patients from the outset that PT is a symptom of an underlying problem and not the primary problem itself. The goal of the diagnostic evaluation is to uncover the cause of the symptom. If and when that is accomplished, then attention can be given to the treatment of that etiology. When a specific cause cannot be found, non-specific treatments, such as tinnitus habituation therapies, masking and cognitive therapies, like cognitive behavioral therapy, can be employed just as they can be for central tinnitus.

Epidemiology

PT likely accounts for about 4% of all patients with a chief complaint of tinnitus. This estimate is based on data from Stouffer and Tyler's study of 528 patients with tinnitus asking them to characterize the sound they heard [1]. Their study cohort had slightly more male than female participants (293 vs. 235), and 52% were aged 50 years or older. 3.8% described their tinnitus as a "pulsing" (2% of males and 6% of females), and 0.6% described it as a "whooshing" (0.3% of males and 0.9% of females), for a total of 4.4% of the entire cohort. Both descriptors had a female preponderance, consistent with what is known about PT now.

The epidemiology of PT varies depending on the specific etiology. Patients' demographics differ for symptomatic temporal bone paragangliomas, superior semicircular canal dehiscence, and sigmoid sinus wall anomalies, for example. Nevertheless, when PT is viewed as a presenting symptom, the Epic Cosmos database has some revealing demographic data. A search of the nearly 184,000,000 patients in that database showed that over a 6-year period ending in March 2023, there were 19,362 with a new ICD-10 diagnosis of PT, either unilateral or bilateral. The majority of these, in a roughly 3:1 ratio, are female, largely ranging from 30 to 75 years of age (Figures 1a and 1b; unpublished data from Epic Cosmos database accessed on March 23, 2023).

The same database shows a high preponderance of elevated BMI in that population, but one that is limited to females; BMI was relatively evenly distributed in the male population (Figures 2a and 2b).

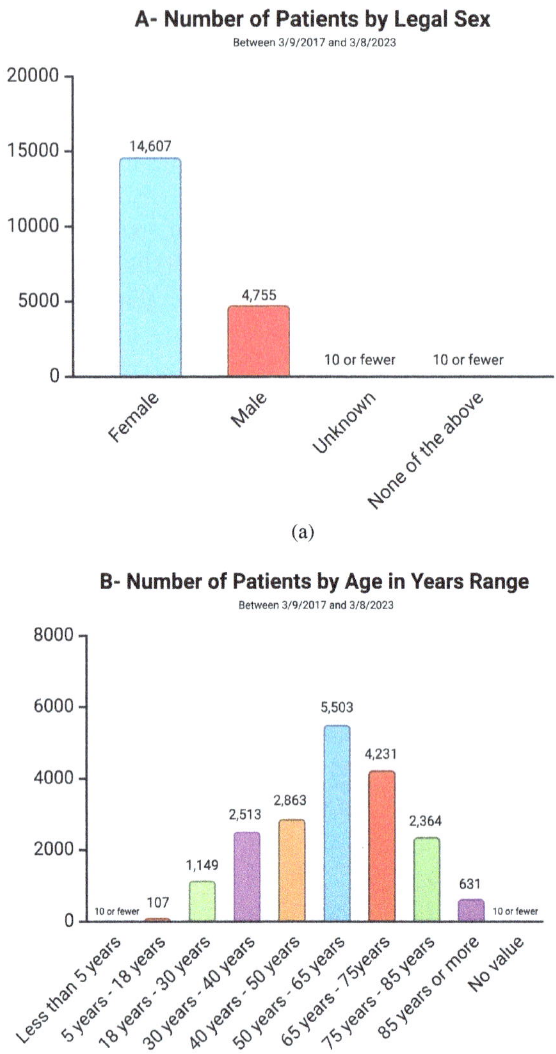

(b)

Figure 1. (a) Column chart illustrating the sex distribution of patients with pulsatile tinnitus in the Epic Cosmos database. (b) Column chart displaying the number of patients with pulsatile tinnitus, categorized by age in years range.

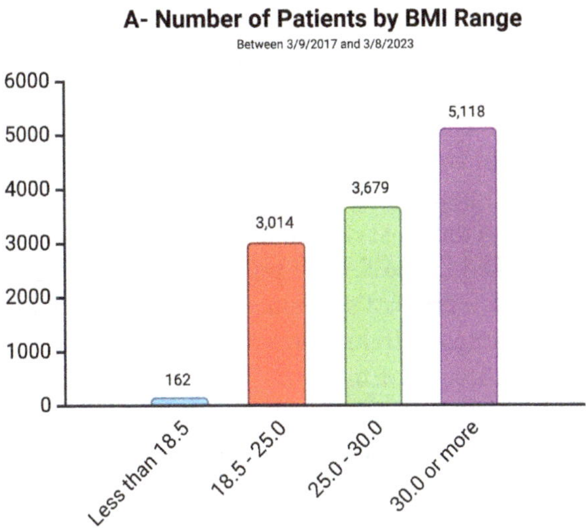

(a)

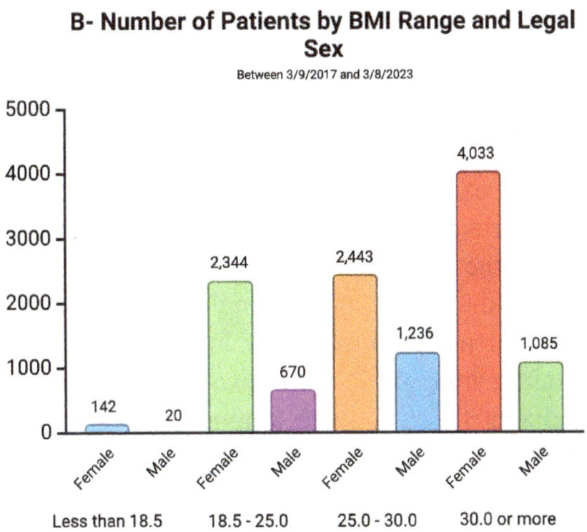

(b)

Figure 2. (a) Column chart illustrating patients with pulsatile tinnitus in the Epic Cosmos database across different BMI categories. (b) Column chart comparing sex distribution across different BMI categories.

These data are significant since those dates coincide with the known demographics of United States patients with sigmoid sinus wall anomalies (SSWA), transverse sinus stenosis (TSS), and idiopathic intracranial hypertension (IIH), 3 entities known to overlap and cause PT either independently or in combination [2]. This suggests that these patients are driving the overall demographics of the population with the symptom of PT, and they are therefore the most common group affected by PT, though this trend is mostly limited to the female population. However, Cosmos data need to be taken with a grain of salt, since there could be widely divergent coding practices amongst different providers. Some providers may not be familiar with the existence of the ICD-10 codes for PT, some might use them erroneously, and some might not use them at all when a more specific diagnosis and diagnostic code, e.g., paraganglioma or carotid artery stenosis, have been identified. This is a limitation of any diagnostic code-based study of PT or any disease process. Nevertheless, in this circumstance, the data are consistent with known and previously established findings that further support their validity.

Differential Diagnosis

Results from studies of the most common causes of PT vary because of regional patient differences, study methodology, and practice patterns. For instance, regions of the country with lower BMI may have a lower incidence of SSWAs. Studies based on imaging retrieval, rather than a clinical diagnostic code, will underestimate the incidence of small middle ear paragangliomas since imaging is not indicated for those. Clinical centers that initiate the diagnosis of PT with magnetic resonance-based imaging may underestimate the incidence of SSWAs. If they diagnose a TSS, they may stop there and not order a computed tomography (CT) scan, thinking that TSS alone suffices to explain the symptom. These and other practice pattern variations could lead to divergent results and conclusions. A comparison of the results of different studies underscores this challenge.

A retrospective study by Lynch *et al.* [3] of 251 patients with an ICD-10 diagnosis of PT over a 5 ½ year period found that roughly half of the patients had an identifiable cause. These were equally divided between vascular and non-vascular etiologies. The most common causes of PT in their population were neoplasms (primarily paragangliomas and acoustic neuromas), arteriopathies (arteriovenous malformations, aneurysms and dissections, atherosclerotic plaque), and venopathies (IIH, SSWAs, major venous thrombosis). This study has numerous limitations, such as divergent diagnostic protocols, that limit its generalizability to a certain extent, as noted by the authors in the paper. Patients who had MRIs and/or were evaluated for IIH may not have had CT scans, so many SSWAs were likely overlooked. Additionally, there is no mention of TSS as a cause, though other studies report this as being a very common finding in patients with venous PT or an independent cause of PT [4–6]. Additionally, it is unclear how an acoustic neuroma would cause PT. Aside from the fact that there is no route for transmission of vascular pulsations to the cochlea with that pathology, most patients with those tumors have sensorineural hearing loss that would make it less likely for them to hear a faint transmitted pulsation. Some of those patients may have simply had a non-vascular, rhythmic tinnitus that is not uncommonly seen with sensorineural hearing loss [7]. Furthermore, the authors don't specify what types of aneurysms, and in what locations, were thought to be causing PT. A cervical carotid aneurysm could certainly do that, but intracranial aneurysms are less likely to do so, if at all. The rate of successful diagnosis also differs between studies, ranging from 40–90%, in part depending on their methodology [3, 7–12].

A retrospective imaging-based study by Formeister *et al.*, as noted above, found a high incidence of TSS as the presumed cause of PT in their population [4]. Both this and the Lynch study found an incidence of migraine of about 20%. This is slightly higher than the incidence in the general population (15%), but not much higher than the incidence in females, who make up the bulk of patients with PT. Despite the high incidence of TSS in this study, there was a remarkably low incidence of SSWAs, despite the known strong association and co-incidence between these two entities. This suggests that many SSWAs co-existing with TSS may have been overlooked. This is supported by unpublished data from our institution showing SSWA as the most common cause of PT when no etiology was evident on clinical exam (e.g., paraganglioma or carotid bruit), and the fact that in our population most patients with SSWAs also have TSS [6].

Irrespective of the order of prevalence, it is helpful to categorize the various causes of PT as those caused by abnormal production of sound versus those due to abnormal perception of otherwise normal vascular sounds. Examples of the former include SSWA and TSS, arteriovenous malformations, carotid artery stenosis, and paragangliomas. Hyperdynamic states such as anemia and hyperthyroidism also fall in this category.

Examples of the latter include third mobile windows, such as superior semicircular canal dehiscence, and causes of conductive hearing loss, such as cerumen impaction, Eustachian tube dysfunction, middle ear effusion, and otosclerosis. Not uncommonly, particularly with diagnostically challenging patients, the final phenotype can be due to a combination of causes, such as TSS combined with a third mobile window or a vascular lesion eroding into the labyrinth. It is also possible that causes of increased sound perception only result in symptomatic PT when there is also some abnormal sound production. One large study found that most patients had more than one significant finding on imaging [8]. A more comprehensive, though still not exhaustive, list of etiologies based on this categorization can be found in Table 1.

Table 1. Classification of pulsatile tinnitus etiologies into different categories.

Abnormal Sound Production	Abnormal Sound Perception	Combinations	Unknown Mechanism
Acquired dural vascular lesions (malformations and fistula	Third mobile window syndromes	Superior petrosal sinus-superior semicircular canal fistula	Migraine
Temporal bone and cerebellopontine angle vascular neoplasms	Conductive hearing loss, including otosclerosis	Cochlear-carotid fistula	Meniere's disease
Fibromuscular dysplasia	Eustachian tube dysfunction	Dehiscent tegmen tympani with dura or cephalocele in contact with ossicles	Idiopathic intracranial hypertension
Carotid artery or aortic stenosis or dissection	Patulous Eustachian tube	Perigeniculate neoplasm with cochlear erosion	Intracranial hypotension
Aberrant carotid artery and other middle ear vascular anomalies			
Sigmoid sinus wall anomalies			
Transverse sinus stenosis			
Anemia			
Hyperthyroidism			

History

The history is an important component of the evaluation of any patient with PT. This includes both the history of the present illness, i.e., the PT itself, and the past medical history. Importantly, it must first be established that the patient is hearing what seems to be a vascular somatosound, i.e., their pulse or heartbeat. This will usually sound like a whoosh or a thump, but it can sometimes be described as grinding, humming, pulsing, or even screeching. The latter, however, can also be a description of an intense central tinnitus, which can present a challenge. A high-pitched ringing or cricket-like sound is not usually vascular, even if it is pulse synchronous. If this sound is associated with sensorineural hearing loss, then that is most likely the cause. A diagnostic evaluation for a vascular etiology will be an expensive and sometimes invasive exercise in futility, one that we see all too often in patients seeking additional opinions.

The sound must, of course, be synchronous with the patient's palpable pulse (see below, Physical Exam), and it will typically increase in intensity and/or rate with physical activity, as would be expected. It is important to ask if the patient can do anything to increase or decrease the intensity of the sound. The intensity of PT from venous etiologies can commonly be transiently decreased with jugular venous compression, such as pressing on or rotating the neck in a certain location or direction, and patients with PT typically figure this out quickly since it may be the only way they can fall asleep at night or concentrate on something during the day. Some patients with acquired dural vascular lesions can press over a portion of the periauricular calvarium or the occiput and decrease flow through the lesion, decreasing the intensity of the sound. Did the sound start gradually or suddenly? This should be asked, but for unclear reasons, patients with SSWA will often notice the abrupt onset of the sound having started at a specific time, even though the bony dehiscence and/or diverticulum must have taken some time to develop. This could be related to their having reached a certain threshold of stenosis or other factors leading to turbulent flow. Sudden onset PT associated with facial pain, headache, or cervical trauma should raise concern for an arterial dissection.

The patient should be asked about other otologic symptoms, such as aural fullness, vertigo, hearing loss, otalgia, otorrhea, and other types of tinnitus. Hearing loss is commonly a complaint. This can either be due to a masking effect from the PT itself and would therefore be improved by maneuvers that made the noise abate, or it could be a primary symptom indicative of a specific cause, such as otosclerosis, Eustachian tube dysfunction with or without middle ear effusion, Meniere's disease or a middle ear vascular neoplasm. The clinician should also inquire about other possible abnormal somatosounds the patient may hear, such as their breathing (autophony)- which suggests a patulous Eustachian tube- or their eye movements or footsteps- which suggests the possibility of a third mobile window syndrome. Generalized sound sensitivity may indicate primary hyperacusis or misophonia. Neither of these typically causes PT independently, but both can exacerbate it.

Other non-otologic symptoms that should specifically be sought include headaches and palpitations. Headaches can be associated with migraine, intracranial hypertension or hypotension, certain arteriovenous malformations, or a large jugular foramen vascular neoplasm with intracranial extension, amongst other things. They may also simply be due to tension related to stress from the PT itself. The pattern of the headache may help distinguish some of these etiologies. If headaches worsen with bending over that suggests IIH, if they are better with bending over or lying supine it suggests intracranial hypotension. Headaches associated with photophobia, phonophobia, or nausea suggest migraine. Palpitations may be associated with anemia or hyperthyroidism. The clinician should specifically inquire about rapid weight loss or weight gain, and other neurologic or cardiovascular diseases. It is important to recognize that not all SSWAs are associated with elevated BMI, particularly in females of far-East Asian ethnicity and males, in general, but even in Caucasian or African-American females. The past medical history may suggest risk factors for atherosclerotic carotid artery disease, such as diabetes mellitus, smoking, hypertension, or hyperlipidemia.

Physical Exam

The examination begins with a complete head & neck exam including, of course, otoscopy to look for evidence of Eustachian tube dysfunction, middle ear neoplasms, vascular lesions, aberrant vessels, or any other abnormalities that could cause or contribute to PT. A paraganglioma that causes PT should be visible on otoscopy. This conclusion is supported by the anecdotal data on the success of subtotal excision of the middle ear component of a jugular foramen paraganglioma in alleviating PT, a common current treatment approach. Tuning fork tests, including Rinne, Weber, and the lateral malleolar sign, can aid in the diagnosis of conductive hearing loss and third mobile window syndromes. The pulse should be palpated while the patient is asked to reproduce their perceived PT to confirm synchronicity. Gentle jugular venous compression and neck rotation should be done to gauge their effects on the PT. Periaural and intraural auscultation are helpful. Their absence doesn't exclude an objective vascular source, but their presence provides great reassurance to the patient since somebody else can hear what they are hearing and will narrow the focus of the evaluation to that region. We perform periaural auscultation with a digital stethoscope (Thinklabs One, Thinklabs LLC, https://www.thinklabs.com/) connected to noise-cancellation headphones. Alternatively, it can be done with a traditional stethoscope in an audiology booth for the exclusion of ambient noise. Intraaural auscultation is done with a simple Toynbee tube. Auscultation of the neck for a bruit should also be performed. If a vascular somatosound from aortic valve stenosis was causing PT, it would also be audible in the neck, so chest auscultation is not routinely performed. Periocular auscultation may also be considered and may reveal evidence of a cavernous-carotid fistula or other acquired vascular lesion in that region.

A brief review of the history and physical highlights is summarized in Table 2.

Table 2. Summary of key history questions and physical examination findings in patients with pulsatile tinnitus.

History	Physical Examination
What does it sound like?	Complete head & neck exam
Is it in sync with your heartbeat?	Otoscopy
Onset: gradual or sudden?	Tuning fork tests including Weber, Rinne, and lateral malleolar sign
Maneuvers that decrease or increase the intensity?	Palpation of the pulse to confirm synchronicity with the pulsatile tinnitus
Other otologic symptoms?	Jugular venous compression on both sides
Aural fullness, vertigo, hearing loss, otalgia, otorrhea, non-rhythmic tinnitus, autophony, audible respirations, audible footsteps or eye movements, hyperacusis?	

(*Continued*)

Table 2. (*Continued*)

History	Physical Examination
Other symptoms? Headache, palpitations, diaphoresis, visual changes, syncope, or pre-syncope	Neck rotation to both sides
Past medical history Diabetes, smoking, hypertension, hyperlipidemia	Periaural and intraural (Toynbee tube) auscultation
	Neck auscultation for carotid bruit
	Complete cranial nerve exam

Diagnostic Testing

After completion of the head and physical diagnostic testing is usually indicated. An audiogram should be obtained on all patients. In addition to showing evidence of otosclerosis or a third mobile window patients will often have a mild, low-frequency sensorineural hearing loss due to a masking effect on the side of the perceived PT [13]. If the loss is larger than typically seen from PT masking (mean 6 dB, almost always less than 20 dB) then other causes of asymmetric sensorineural hearing loss should be explored. Additionally, since these data suggest that the intensity of venous PT is almost always less than 20 dB, patients with a low-frequency air-conduction threshold greater than 20 dB are unlikely to be able to perceive that sound. If they are complaining of PT and their hearing is symmetrically reduced, it is more likely that they are perceiving a rhythmic central tinnitus rather than a vascular somatosound. Complete blood count, thyroid function tests, and serum lipid levels should be obtained in the proper setting, if not recently done elsewhere. Carotid, and possibly cardiac, ultrasound is indicated for patients with an audible bruit.

The most common scenario facing the clinician evaluating a patient with PT is that no definitive etiology has been identified after the above tests and a decision has to be made about the next best step. If a paraganglioma is observed on otoscopy then appropriate imaging, or no further imaging, must be obtained. If the lesion is visible circumferentially then no further imaging is routinely indicated. If the lesion extends below the inferior tympanic annulus then typically a non-contrast CT is obtained. If jugular foramen involvement is identified an MRI with contrast must be performed. Analogously, if the initial testing suggests a third-mobile window then appropriate windowed (W-=4000, L=900) non-contrast temporal bone CT (0.6 mm cuts or less) is the next test of choice, along with vestibular evoked myogenic potentials and possibly electrocochleography. None of this is controversial. The controversy over the best test for evaluation of PT surrounds the situation in which no cause is evident after initial testing. Aspects of this controversy have been previously reviewed [7,9,14]. Herein we present our institutional approach that is based on a protocol previously published [15]. The protocol was developed as a means of capturing most of the most common causes of PT with a single imaging study. It is a CT-based study that is windowed like a temporal bone CT, with contrast administered as for a CT angiogram but with image acquisition delayed ensuring both arterial and venous imaging. This technique permits the identification of subtle bony abnormalities, such as otosclerosis and third mobile windows, major venous sinus stenosis, and almost all of the soft signs of an acquired dural vascular lesion [16]. The most common causes of PT, regardless of which study is reviewed, are shown in Table 3, not in order of frequency, since that is a matter of controversy.

As shown in the table, the modified CTA/V protocol we describe is the only single-modality study that can reliably identify each of these common causes. Additionally, if the image acquisition is extended into the

Table 3. The most common causes of pulsatile tinnitus across various studies, listed without order of frequency due to controversy. MRI, magnetic resonance imaging; MRA, magnetic resonance angiography; CTA, computed tomography angiography; CTA/V, computed tomography angiography/venography.

	CT temporal bone	MRI	MRA	CTA	Modified CTA/V
Dural vascular lesion	–	+/–	+/–	+	+
Middles ear or jugular foramen paraganglioma	+/–	+	–	+	+
Sigmoid sinus wall anomalies	+	–	–	+/–	+
Transverse sinus stenosis	–	+	–	–	+
Otosclerosis	+	–	–	–	+
3rd mobile windows	+	–	–	+/–	+
Middle ear vascular anomalies	+	–	+/–	+	+
Signs of idiopathic intracranial hypertension	+/–	+	–	+/–	+

upper neck, then the typical "string of beads" sign associated with fibromuscular dysplasia can also be identified, as can jugular vein stenosis at the skull base, a still controversial entity that may be associated with PT. Inspection of Table 1 reveals that with this minor modification, all the listed causes that have typical imaging findings can be reliably identified. Yet unpublished data from our institution confirm that this protocol is extremely reliable in detecting almost any cause of PT that has imaging findings and that the rate of detecting a cause with further imaging when this protocol is negative is extremely small. A similar protocol published by a different group showed similar reliability [4].

Specific Causes of Pulsatile Tinnitus

Paragangliomas

Paragangliomas are a class of rare catecholamine-secreting neuroendocrine tumors derived from the embryonic neural crest cell layer [17]. As chromaffin cell tumors, paragangliomas are closely related to pheochromocytomas with a shared cellular morphology and clinical presentation and similarly arise from a sporadic etiology or hereditary syndromes such as multiple endocrine neoplasia types 2A and 2B, neurofibromatosis 1, and von Hippel Lindau [18]. However, paragangliomas can be differentiated from pheochromocytomas based on their classification as extra-adrenal paraganglia tumors [19].

Paragangliomas are characterized as either sympathetic or parasympathetic paraganglia, which are indistinguishable on a cellular level but differ widely in their anatomic distribution and clinical presentations [19]. Sympathetic paraganglia are primarily found along the sympathetic chain of the neck and surrounding pelvic organs, with clinical symptoms most commonly stemming from the hypersecretion of catecholamines [20]. In contrast, parasympathetic paragangliomas form within the head and neck region [21], and are often discovered incidentally as a result of mass effect rather than hormonal secretion [22].

The parasympathetic paraganglia can be divided into four distinct categories based on tumor location: carotid body paraganglia, vagal paraganglia, laryngeal paraganglia, and jugulotympanic paraganglia. Of the four subtypes of parasympathetic paragangliomas, jugulotympanic paraganglia are associated with PT and cranial nerve palsies due to their

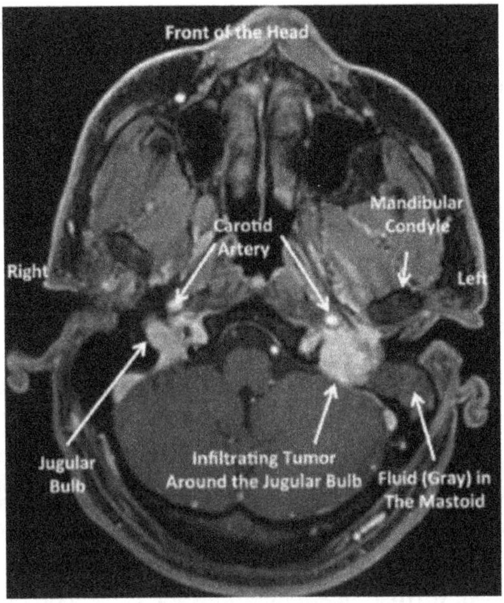

Figure 3. Axial T1 post gadolinium MRI of the internal auditory canals at the level of the jugular bulb showing the enhancement lateral to the left jugular foramen compatible with a left glomus.

location within the middle ear and/or jugular foramen (Figure 3) [21]. Jugulotympanic paragangliomas most commonly form either within the wall of the jugular bulb, called jugular glomus tumors (glomus jugulare), or on the middle ear promontory, called tympanic glomus tumors (glomus tympanicum) [19]. A retrospective chart review of 67 patients with jugular glomus tumors found that 89.6% initially presented with symptoms of PT, followed closely by conductive hearing loss at 80.6% [23].

Treatment of paragangliomas includes a variety of modalities ranging from surgical intervention, radiation therapy, and combined modality treatment with subtotal excision of the middle ear component and radiation for the jugular foramen. In a small retrospective review of 10 patients with PT who underwent single fraction gamma knife radiosurgery for resection of jugular glomus tumors, complete PT resolution was observed in 40% of patients, with improvement of symptoms noted in an additional 40% [24].

The high incidence of early onset PT in patients with jugulotympanic paraganglia, despite the overall low prevalence of paragangliomas, underscores the importance of considering parasympathetic paragangliomas as a possible etiology in patients presenting with PT.

Idiopathic Intracranial Hypertension, Sigmoid Sinus Wall Abnormalities, and Transverse *Sinus stenosis*

IIH, also known as pseudotumor cerebri, is defined as an elevation of intracranial pressure (ICP) in the absence of a mass lesion causing obstructive hydrocephalus. IIH primarily affects obese young women of childbearing age [25]. Though the pathophysiological mechanisms underlying IIH are speculative, the clinical presentation of this disease is well-characterized, commonly manifesting as papilledema and progressive visual deterioration, diffuse headache, and PT [25]. The presence of PT in IIH is thought to arise from increased vascular pulsation as a result of elevated intracranial pressure, though that is unclear [26]. Notably, PT is one of the most common symptoms initially reported, affecting an estimated 52% to 65% of patients with IIH and presenting bilaterally in up to two-thirds of cases [27, 28].

The current mainstay of medical treatment, which prioritizes symptom alleviation and vision preservation, includes weight loss and acetazolamide, a carbonic anhydrase inhibitor thought to decrease cerebrospinal fluid production and therefore reduce intracranial pressure [29]. In addition to pharmacologic treatments, procedural techniques such as lumbar puncture (LP) may transiently decrease intracranial pressure for short-term symptom management [30]. This is supported by one prospective study in patients with elevated intracranial pressure, which compared the efficacy of LP and acetazolamide treatments in reducing PT severity as defined by the Tinnitus Handicap Inventory. Investigators found that LP resulted in a transient yet statistically significant reduction in PT severity 30 minutes after the procedure, with symptom recurrence 3 days post-procedure [30]. However, acetazolamide offered long-term symptomatic relief, with significantly reduced PT severity 1 month after treatment initiation [30].

Discussions regarding the relationship between IIH and PT would be incomplete without considering two potentially interconnected conditions that likewise manifest with PT: SSWA (Figure 4) and TSS (Figure 5).

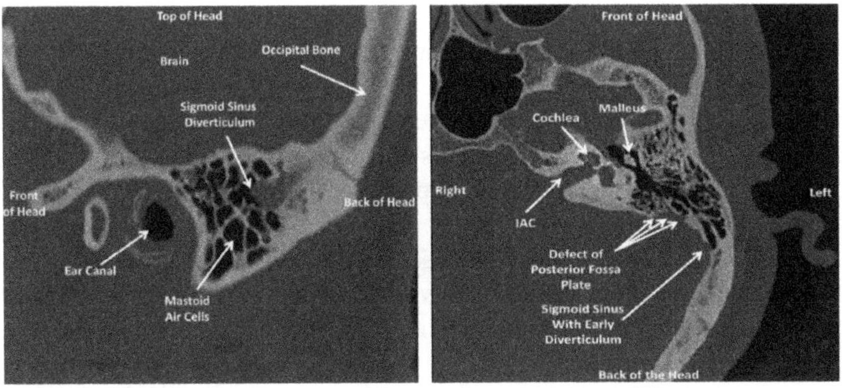

(a) (b)

Figure 4. (a) Sagittal CT showing sigmoid sinus diverticulum. (b) Axial CT of the left temporal bone showing a defect of the posterior cranial fossa and early sigmoid sinus diverticulum development.

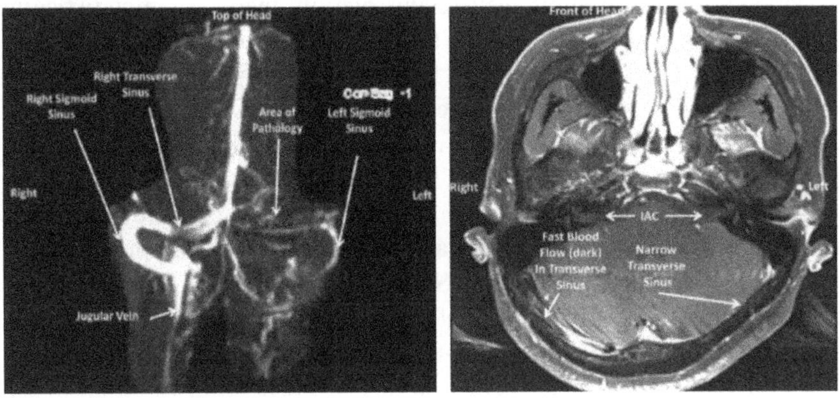

(a) (b)

Figure 5. (a) MRV venography showing areas of narrowing of the left transverse sinus. (b) Axial T1 post gadolinium MRI of the internal auditory canals showing the narrowing of the transverse sinus on the left. The right sinus demonstrates fast blood flow which creates a dark artifact.

SSWA encompasses two anatomic defects within the sigmoid sinus, sigmoid sinus diverticulum, and sigmoid sinus dehiscence. Much like IIH, SSWA predominantly occurs in females with an elevated BMI, and is likewise associated with an increased spinal opening pressure [2, 31, 32]. The potential pathophysiological connection between SSWA and IIH is further strengthened by their shared imaging characteristics [2, 33]. In a retrospective case-control study using T1-weighted volumetric MRI to explore the association between IIH, SSWA, and PT, patients with PT and SSWA were found to have a significantly higher prevalence of imaging characteristics associated with IIH, such as protrusion and vertical tortuosity of the optic nerve, a flattened posterior sclera, and transverse sinus stenosis [33].

Currently, surgical and endovascular modalities are utilized as the mainstay of treatment in patients with sigmoid sinus diverticulum and dehiscence [34]. Surgical correction of SSWA with both endoluminal and transtemporal approaches has a high degree of success in alleviating PT, however, confounding factors such as concomitant IIH and TSS have been found to significantly influence outcomes [2, 35]. In a recent retrospective review, surgical outcomes were analyzed in a cohort of 40 patients with SSWA who underwent transtemporal sinus wall reconstructions for correction of PT [35]. While the successful resolution of PT occurred in 90% of patients following sinus wall reconstruction, there was a statistically significant association between the absence of TSS and failure to alleviate PT following surgical intervention. Furthermore, the presence of an empty sella on imaging, which is a characteristic finding in IIH, was associated with a higher successful surgical response. These findings suggest that SSWA alone may not be sufficient to account for the presence of PT and that other factors, such as stenosis of the transverse sinuses, elevated intracranial pressure, or both, may play a critical role in the development of PT.

The increased prevalence of TSS in patients with IIH further complicates attempts to elucidate the pathophysiological relationship between these three conditions. Independently, TSS is a well-established cause of PT, with venous stent placement resulting in successful resolution of PT in the absence of increased intracranial pressure [36]. However, multiple studies have identified an increased prevalence of TSS in patients with IIH, with bilateral venous stenosis present in over 90% of individuals with IIH compared to a rate of 6.8% in the control cohort [37, 38]. Venous stenting has been described as a highly effective treatment for the resolution of PT in IIH, providing both immediate relief and in several cases, complete symptomatic resolution 2 years following the procedure [39]. Additionally, and importantly, many of the studies of stenting for PT associated with TSS have been based on MR imaging alone. It is likely that many of those patients also had sigmoid sinus wall dehiscence, or even diverticula, that were overlooked and might have benefited from transtemporal approaches just as well. In our institutional practice, all patients being considered for either transverse sinus stenting or sigmoid sinus wall reconstruction have imaging that will diagnose both. If a patient has an SSWA with TSS and has PT alone, transtemporal surgery may be preferred. If they also have symptoms associated with elevated intracranial pressure, then transverse sinus stenting is the recommended approach.

Despite intensive research efforts, the precise relationship between IIH, SSWA, and TSS remains a source of debate. Whether anatomic defects such as TSS result in increased CSF pressure, or if TSS and SSWA result directly or indirectly from IIH is unclear [32]. The highly interconnected nature of these conditions reinforces the possibility that the manifestation of PT may result from a combination of these three etiologies rather than a single pathology alone [37, 40]. However, previous attempts to elucidate a potential connection have proved unsuccessful. A retrospective review of 79 patients diagnosed with IIH found the incidence of TSS and SSWA did not significantly differ between patients with IIH PT and those without PT [32]. These findings suggest that the presence of TSS or SSWA, either independently or together, may not be sufficient to explain the cause of PT in patients with IIH. Given the interconnected nature of IIH, SSWA, TSS, and the manifestation of PT, identification of a single pathology alone should prompt further investigation for co-occurring conditions to guide medical and surgical management decisions.

Dural Vascular Lesions

Dural vascular lesions, also known as dural arteriovenous fistulas (dAVF), are an important cause of arterial PT, accounting for an estimated 10–15% of all intracranial arteriovenous malformations (Figure 6) [41, 42]. Fistula formation arises from an abnormal connection between dural arteries and a venous sinus and is often found at the vertebral venous plexus or transverse, sigmoid, cavernous, and marginal sinuses [43]. The etiology of dAVFs is predominantly idiopathic and manifests in the fifth to sixth decade of life, though a small number of cases are associated with dural sinus thrombosis, craniotomy, and prior trauma [44]. Pediatric malformations are also possible and are thought to develop as a result of birth trauma, infection, in-utero venous thrombosis, or congenital abnormalities [43, 45].

The clinical presentation of dAVFs varies, ranging from completely asymptomatic to severe manifestations of intracranial hemorrhage and

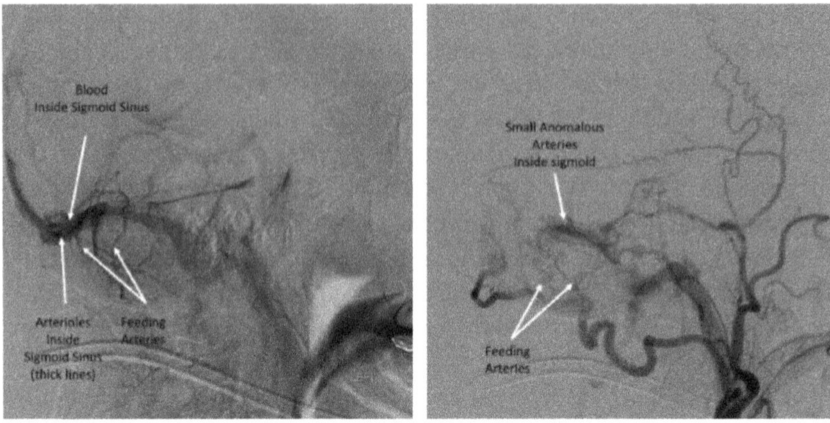

(a) (b)

Figure 6. (a) Lateral view angiography of the left external carotid demonstrating feeding arteries arising from the right middle meningeal, posterior auricular, and occipital arteries. (b) Lateral view angiography of the left external carotid during arterial and venous phase showing blood and small arterioles inside sigmoid sinus.

neurologic defects, such as cognitive decline and seizures [42, 43]. The heterogeneous symptom presentation is largely dependent on dural vascular lesion location and the type of lesion, as specified by the Borden and Cognard classification systems. These two classification systems serve to stratify the severity of lesions based on the direction of venous drainage, the presence or absence of cortical venous drainage, and venous outflow architecture. Type I lesions are considered the most benign and commonly present with symptoms such as PT and exophthalmos, while types II and III are associated with more severe findings [43].

Several studies have assessed the relationship between the location of dAVFs and the presentation of PT. In a study of 220 patients diagnosed with dAVFs, PT was identified as the only symptom in 13.6% of cases, with the transverse-sigmoid sinus being the most common location (70%), followed by the hypoglossal canal (10%) and middle cranial fossa (6.7%) [41]. The high proportion of PT in dural vascular lesions of the transverse and sigmoid sinus is thought to be attributed to their anatomic proximity to the inner ear. Notably, while dAVFs of the hypoglossal canal (HCDAVF) are a rare subtype comprising 3.6%–4.2% of all intracranial dAVFs, up to 75% of all patients with HCDAVF patients present with PT [48]. For HCDAVFs with solely anterograde venous drainage, PT was often the only symptom. The perception of PT in patients with HCDAVFs is theorized to arise from direct transmission of the venous bruit to the inner ear via the temporal bone [41].

Management of dAVFs includes conservative treatment, endovascular embolization, and in refractory disease, surgical intervention and radiosurgery [41]. Conservative treatment of dAVF is often indicated for low-grade type I fistulas, but the presence of debilitating PT may necessitate endovascular therapy [43]. The endovascular intervention of arterial causes of PT has a high degree of success in the resolution of PT, though symptom resolution is often a secondary priority to the prevention of intracranial hemorrhage and stroke [9]. Elimination of the arteriovenous shunt is achieved through transarterial and/or transvenous embolization, with a transarterial approach being preferred in the vast majority of dAVFs [43]. In a cohort of 26 patients who underwent endovascular embolization for PT, 80.8% achieved complete symptom resolution while an additional 11.5% reported significant improvement in symptom severity [41]. In conclusion, dAVFs comprise a relatively small but significant portion of patients with PT, with PT presenting as the only initial symptom in over 10% of cases. In patients with dAVFs, transarterial embolization is a highly effective approach for the resolution of debilitating PT when conservative management fails.

Third Mobile Windows

Third window syndrome is characterized by the formation of a pathological third mobile window- that is, a mobile area in addition to the oval and round windows- within the bony labyrinth of the inner ear, resulting in vestibular and auditory dysfunction [49]. While a third window syndrome can arise from several etiologies such as enlarged vestibular aqueduct, cochlea-facial nerve dehiscence, cochleo-carotid dehiscence, dehiscence of the posterior and lateral semicircular canals, and otosclerosis, the most common cause of third window syndrome is superior semicircular canal dehiscence (SSCD) (Figure 7) [50–52]. First described by Minor *et al.* in a series of articles in 1998, SSCD is caused by an absence or thinning of bone overlying the superior semicircular canal [52, 53]. The original case report of 8 patients identified symptoms such as vertigo and oscillopsia induced by changes in sound and pressure, though later studies have expanded upon the initial characterization to include hyperacusis, hearing loss, and PT as additional manifestations [49, 53].

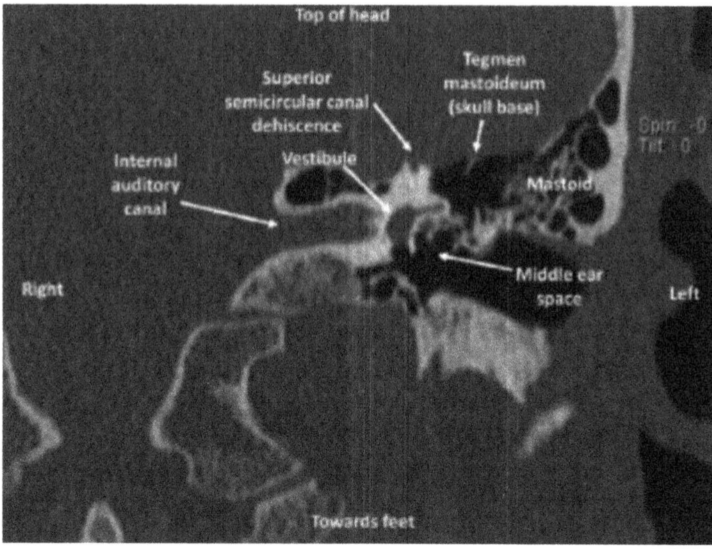

Figure 7. Coronal CT of the left temporal bone demonstrating dehiscence of the bony plate overlying the left superior semicircular canal. Note the very thin tegmen.

The diversity of clinical presentations may be partially attributed to differences in the degree of dehiscence, ranging from frank dehiscence to slight thinning of the bone (near dehiscence) [54]. In one retrospective case-control study, the prevalence of auditory and vestibular symptoms was compared between patients with frank superior canal dehiscence versus those with near dehiscence [54]. The prevalence of PT did not differ significantly between the two groups, with 94% of near dehiscence cases presenting with PT compared to 82% of frank dehiscence cases. In comparison, vestibular symptoms such as nystagmus were significantly more prevalent in those with frank dehiscence (94%) compared to those with only near dehiscence (19%). The high overall prevalence of PT in both groups suggests that the presence of anatomic dehiscence is more important than the degree of dehiscence in the manifestation of PT.

While the etiology of SSCD is unknown, its underlying biomechanisms relating to symptom presentation are relatively well-understood. The inner ear is normally characterized as a closed hydraulic system, with the transmission of acoustic waves into the scala vestibuli through the round window, followed by the release of mechanical energy from the scala tympani out of the inner ear. The formation of dehiscence in the superior semicircular canal results in redirection of the hydroacoustic pressure wave from the cochlea into the vestibular system, inappropriately stimulating the superior canal crista in response to sound [49, 52]. The presence of PT in SSCD is thought to arise from increased inner ear compliance and sensitivity to bone-conducted sounds [49, 55] as well as potentially having the third window transmitting dural pulsations to the cochlea.

The gold standard for diagnosis of SSCD is high-resolution temporal bone CT showing dehiscence of the arcuate eminence or dehiscence in the superior petrosal sinus [56]. In addition to imaging modalities, a diagnosis of SSCD should be supported by some physiologic evidence of a third mobile window such as lateralization of the Weber tuning fork to the affected side, supernormal bone thresholds, and low-frequency conductive hearing loss on audiometry, hearing a 256 Hz tuning fork placed on the ankle, as well as lower thresholds and increased amplitudes in ocular vestibular evoked myogenic potentials [57, 58].

Conservative management is often initially recommended for mild to moderate cases of SSCD, with common modalities such as vestibular rehabilitation, trigger avoidance, hearing aids, and medications including antiemetics, antihistamines, tricyclic depressants, benzodiazepines, and migraine prophylaxis medication [59]. In patients with SSCD refractory to conservative management and those with debilitating symptoms, surgical intervention may be required. Several surgical approaches can be used for the repair of SSCD, though techniques differ significantly in morbidity and success rates [49]. The middle cranial fossa (MCF) approach offers the potential advantage of direct visualization of the arcuate eminence defect without requiring labyrinthine bone removal and exposure of the surrounding cranial base. In contrast, a transmastoid route involves greater difficulty in dehiscence visualization but has lower morbidity and a shorter hospital stay compared to the MCF approach. In addition to the different approaches described above, a combination of plugging, capping, and resurfacing techniques can be used for sealing the defect, each with differences in long-term symptom control.

Given the variety of surgical approaches and techniques available for treating SSCD, it is unsurprising that success rates for surgical resolution of tinnitus also vary widely. In one study utilizing an MCF approach with a combination of plugging and resurfacing, only 12% of patients with complete dehiscence reported persistence or worsening of tinnitus symptoms, which encompassed both pulsatile and central tinnitus together [54]. However, in another study utilizing a middle fossa craniotomy approach, symptomatic resolution of tinnitus was reported in only 42% of cases, with one of the lowest percentages regarding symptom improvement compared to other symptoms [60].

Migraine

In patients with a normal imaging workup and normal neuro-ophthalmologic exam, otologic migraine should be considered as a potential etiology for PT. In a cohort of 1204 patients with migraine, 12% (n=145) were found to have been diagnosed with tinnitus. After other causes were ruled out, the prevalence of PT among all patients with migraine was 1.9% [61]. Of those with migraine and PT who underwent migraine treatment, 11 out of 16 reported resolution or improvement of their PT. Migraine is thought to lead to PT due to increased central sensitivity and amplification of internal sounds. Given the common nature of migraine, and a frequently negative imaging workup, some centers start migraine lifestyle/diet/medical management from the initial visit while performing the imaging studies which take time to get insurance approval, etc. This allows the patient to start treatment while getting imaging, neuro-ophthalmology exam, etc.

Conclusion

- PT is defined as the self-perception of one's blood flow, accounting for about 4% of all patients with a chief complaint of tinnitus.
- PT etiologies can be classified into different categories, including abnormal sound production such as sigmoid sinus wall and transverse sinus anomalies, abnormal sound perception like third mobile window syndromes and Eustachian tube dysfunction, a combination of both, or causes of unknown mechanisms such as migraine and Meniere's disease.
- It is important to explain to patients that PT is a symptom of an underlying disease.
- The history of PT and the patient's medical history are important components of the evaluation of any patient with PT.
- Patients should be asked about other associated otologic symptoms, such as aural fullness, vertigo, hearing loss, otalgia, otorrhea, and other types of tinnitus. Non-otologic symptoms such as headaches indicating

migraine, IIH, hypotension, AVM, and palpitations large jugular foramen vascular neoplasm and palpitations associated with anemia or hyperthyroidism should be considered.

- Otoscopy and hearing evaluation should be obtained on all patients.
- Based on these findings, appropriate imaging such as CT of the temporal bone, MRI, CTA, and MRA can be ordered to aid in the diagnosis.
- Parasympathetic paragangliomas, specifically jugulotympanic paraganglia, are associated with PT due to their location within the middle ear and/or jugular foramen.
- PT is one of the most common presenting symptoms in patients with IIH.
- Sigmoid sinus wall abnormalities and transverse sinus stenosis, frequently associated with IIH, can also cause PT.
- Dural vascular lesions, specifically type I lesions of the transverse sigmoid sinus, are an important cause of arterial PT.
- A dehiscence in the superior semicircular canal is another cause of PT by increasing inner ear compliance and sensitivity to bone-conducted sounds as well as potentially having the third window transmitting dural pulsations to the cochlea.
- Patients with a normal imaging workup and normal neuro-ophthalmologic exam, otologic migraine should be considered as a potential etiology for PT.
- For each of the aforementioned conditions, several treatment options are available. However, when a specific cause cannot be found, nonspecific treatments, such as tinnitus habituation therapies, masking, and cognitive therapies, like cognitive behavioral therapy, can be employed.

References

1. Stouffer JL, and Tyler RS. Characterization of tinnitus by tinnitus patients. *Journal of Speech and Hearing Disorders.* **55**:439–453 (1990).
2. Harvey RS, Hertzano R, Kelman SE, and Eisenman DJ. Pulse-synchronous tinnitus and sigmoid sinus wall anomalies: Descriptive epidemiology and the idiopathic intracranial hypertension patient population. *Otolaryngology–Head and Neck Surgery.* **35**:7–15 (2014).
3. Lynch P, Mitton T, Killeen DE, Kutz JW, and Newcomer M. Diagnosing Pulsatile Tinnitus: A Review of 251 Patients. *Otolaryngology–Head and Neck Surgery.* **43**:128–136 (2022).

4. Formeister EJ, *et al.* Combined arterial and venous phase computed tomographic imaging of the skull base in pulsatile tinnitus. *Otolaryngology–Head and Neck Surgery.* **43**:1049–1055 (2022).
5. Abdalkader M, *et al.* State of the Art: Venous causes of pulsatile tinnitus and diagnostic considerations guiding endovascular therapy. *Radiology.* **300**:2–16 (2021).
6. Hewes D, Morales R, Raghavan P, and Eisenman DJ. Pattern and severity of transverse sinus stenosis in patients with pulsatile tinnitus associated with sigmoid sinus wall anomalies. *Laryngoscope.* **130**:1028–1033 (2020).
7. Sismanis A. Pulsatile tinnitus: Contemporary assessment and management. *Current Opinion in Otolaryngology & Head and Neck Surgery.* **19**:348–357 (2011).
8. Dong C, Zhao P.-F, Yang J.-G, Liu Z.-H, and Wang, Z.-C. Incidence of vascular anomalies and variants associated with unilateral venous pulsatile tinnitus in 242 patients based on dual-phase contrast-enhanced computed tomography. *Chinese Medical Journal.* **128**:581–585 (2015).
9. Narsinh KH, *et al.* Diagnostic approach to pulsatile tinnitus: A narrative review. *JAMA Otolaryngology–Head and Neck Surgery.* **148**:476–483 (2022).
10. Sonmez G, Basekim CC, Ozturk E, Gungor A, and Kizilkaya E. Imaging of pulsatile tinnitus: A review of 74 patients. *Clinical Imaging.* **31**:102–108 (2007).
11. Vattoth S, Shah R, and Curé JK. A Compartment-based approach for the Imaging evaluation of tinnitus. *American Journal of Neuroradiology.* **31**:211–218 (2010).
12. Sismanis A. Pulsatile tinnitus. A 15-year experience. *American Journal of Otology.* **19**:472–477 (1998).
13. Bhatnagar K, Lataille AT, and Eisenman DJ. Patterns of audiometric threshold shifts from pulsatile tinnitus due to sigmoid sinus wall anomalies. *American Journal of Otolaryngology.* **41**:102647 (2020).
14. Ahsan SF, Seidman M, and Yaremchuk K. What is the best imaging modality in evaluating patients with unilateral pulsatile tinnitus? *Laryngoscope.* **125**:284–285 (2015).
15. Reardon MA, and Raghavan P. Venous abnormalities leading to tinnitus: Imaging Evaluation. *Neuroimaging Clinics of North America.* **26**:237–245 (2016).
16. Narvid J, Do HM, Blevins NH, and Fischbein NJ. CT angiography as a screening tool for dural arteriovenous fistula in patients with pulsatile tinnitus: Feasibility and test characteristics. *American Journal of Neuroradiology.* **32**:446–453 (2011).
17. Sandow L, Thawani R, Kim MS, and Heinrich, MC. Paraganglioma of the Head and Neck: A Review. *Endocrine Practice.* **29**:141–147 (2023).

18. Plouin P-F, and Gimenez-Roqueplo A-P. Pheochromocytomas and secreting paragangliomas. *Orphanet Journal of Rare Diseases.* **1**:49 (2006).
19. Barnes L, Weltgesundheitsorganisation, International Agency for Research on Cancer, eds. Pathology and Genetics of Head and Neck Tumours: ... Reflects the Views of a Working Group That Convened for an Editorial and Consensus Conference in Lyon, France, July 16–19, 2003. Reprinted. IARC Press; 2007.
20. Al-Harthy M, Al-Harthy S, Al-Otieschan A, Velagapudi S, and Alzahrani AS. Comparison of pheochromocytomas and abdominal and pelvic paragangliomas with head and neck paragangliomas. *Endocrine Practice.* **15**:194–202 (2009).
21. Erickson D. *et al.* Benign paragangliomas: Clinical presentation and treatment outcomes in 236 patients. *Journal of Clinical Endocrinology & Metabolism.* **86**:5210–5216 (2001).
22. Lee JA, and Duh Q-Y. Sporadic paraganglioma. *World Journal of Surgery.* **32**:683–687 (2008).
23. Fayad JN, Keles B, and Brackmann DE. Jugular foramen tumors: Clinical characteristics and treatment outcomes. *Otolaryngology–Head and Neck Surgery.* **31**:299–305 (2010).
24. Dobberpuhl MR, Maxwell S, Feddock J, St Clair W, and Bush ML. Treatment outcomes for single modality management of glomus jugulare tumors with stereotactic radiosurgery. *Otolaryngology–Head and Neck Surgery.* **37**:1406–1410 (2016).
25. Raoof N, and Hoffmann J. Diagnosis and treatment of idiopathic intracranial hypertension. *Cephalalgia.* **41**:472–478 (2021).
26. Jensen RH, Radojicic A, and Yri H. The diagnosis and management of idiopathic intracranial hypertension and the associated headache. *Therapeutic Advances in Neurological Disorders.* **9**:317–326 (2016).
27. Skau M, Sander B, Milea D, and Jensen R. Disease activity in idiopathic intracranial hypertension: A 3-month follow-up study. *Journal of Neurology.* **258**:277–283 (2011).
28. Wall M. *et al.* The idiopathic intracranial hypertension treatment trial: Clinical profile at baseline. *JAMA Neurology.* **71**:693–701 (2014).
29. Thurtell MJ, and Wall M. Idiopathic intracranial hypertension (pseudotumor cerebri): Recognition, treatment, and ongoing management. *Current Treatment Options in Neurology.* **15**:1–12 (2013).
30. Guo P, Sun W, Shi S, and Wang W. Patients with pulse-synchronous tinnitus should be suspected to have elevated cerebrospinal fluid pressure. *Journal of International Medical Research.* **47**:4104–4113 (2019).
31. Grewal AK. *et al.* Clinical presentation and imaging findings in patients with pulsatile tinnitus and sigmoid sinus diverticulum/dehiscence. *Otolaryngology–Head and Neck Surgery.* **35**:16–21 (2014).

32. Lansley JA, Tucker W, Eriksen MR, Riordan-Eva P, and Connor SE. J. Sigmoid sinus diverticulum, dehiscence, and venous sinus stenosis: Potential causes of pulsatile tinnitus in patients with idiopathic intracranial hypertension? *American Journal of Neuroradiology*. **38**:1783–1788 (2017).
33. Liu Z. *et al.* Association between idiopathic intracranial hypertension and sigmoid sinus dehiscence/diverticulum with pulsatile tinnitus: A retrospective imaging study. *Neuroradiology*. **57**:747–753 (2015).
34. Eisenman DJ. Sinus wall reconstruction for sigmoid sinus diverticulum and dehiscence: A standardized surgical procedure for a range of radiographic findings. *Otolaryngology–Head and Neck Surgery*. **32**:1116–1119 (2011).
35. Eisenman DJ, Raghavan P, Hertzano R, and Morales R. Imaging findings of sigmoid sinus diverticulum and dehiscence in patients with pulsatile tinnitus: A review of 56 patients. *Laryngoscope*. **128 Suppl 2**:S1–S13 (2018).
36. Baomin L, Yongbing S, and Xiangyu C. Angioplasty and stenting for intractable pulsatile tinnitus caused by dural venous sinus stenosis: A case series report. *Otolaryngology–Head and Neck Surgery*. **35**:366–370 (2014).
37. Farb RI. *et al.* Idiopathic intracranial hypertension: The prevalence and morphology of sinovenous stenosis. *Neurology*. **60**:1418–1424 (2003).
38. Riggeal BD. *et al.* Clinical course of idiopathic intracranial hypertension with transverse sinus stenosis. *Neurology*. **80**:289–295 (2013).
39. Boddu S. *et al.* Resolution of pulsatile tinnitus after venous sinus stenting in patients with idiopathic intracranial hypertension. *PLoS One*. **11**:e0164466 (2016).
40. Geng W, Liu Z, and Fan Z. CT characteristics of dehiscent sigmoid plates presenting as pulsatile tinnitus: A study of 23 patients. *Acta Radiologica*. **56**:1404–1408 (2015).
41. An Y-Hm, *et al.* Dural arteriovenous fistula masquerading as pulsatile tinnitus: Radiologic assessment and clinical implications. *Scientific Reports*. **6**:36601 (2016).
42. Kwon BJ, Han MH, Kang H-S, and Chang K-H. MR imaging findings of intracranial dural arteriovenous fistulas: Relations with venous drainage patterns. *American Journal of Neuroradiology*. **26**:2500–2507 (2005).
43. Gandhi D. *et al.* Intracranial dural arteriovenous fistulas: Classification, imaging findings, and treatment. *American Journal of Neuroradiology*. **33**:1007–1013 (2012).
44. Chung SJ. *et al.* Intracranial dural arteriovenous fistulas: Analysis of 60 patients. *Cerebrovascular Diseases*. **13**:79–88 (2002).
45. Morita A, Meyer FB, Nichols DA, and Patterson MC. Childhood dural arteriovenous fistulae of the posterior dural sinuses: Three case reports and literature review. *Neurosurgery*. **37**:1193–1199; discussion 1199–1200 (1995).

46. Cognard C. *et al.* Cerebral dural arteriovenous fistulas: Clinical and angiographic correlation with a revised classification of venous drainage. *Radiology*. **194**:671–680 (1995).
47. Borden JA, Wu JK, and Shucart WA. A proposed classification for spinal and cranial dural arteriovenous fistulous malformations and implications for treatment. *Journal of Neurosurgery*. **82**:166–179 (1995).
48. Spittau B, *et al.* Dural arteriovenous fistulas of the hypoglossal canal: Systematic review on imaging anatomy, clinical findings, and endovascular management. *Journal of Neurosurgery*. **122**:883–903 (2015).
49. Palma Diaz M, Cisneros Lesser JC, and Vega Alarcón A. Superior Semicircular canal dehiscence syndrome — Diagnosis and surgical management. *International Archives of Otorhinolaryngology*. **21**:195–198 (2017).
50. Wackym PA, Balaban CD, Zhang P, Siker DA, and Hundal JS. Third window syndrome: Surgical management of cochlea-facial nerve dehiscence. *Frontiers in Neurology*. **10**:1281 (2019).
51. Shim YJ. *et al.* Involvement of the internal auditory canal in subjects with cochlear otosclerosis: A less acknowledged third window that affects surgical outcome. *Otolaryngology–Head and Neck Surgery*. **40**:e186–e190 (2019).
52. Iversen MM and Rabbitt RD. Biomechanics of third window syndrome. *Frontiers in Neurology*. **11**:891 (2020).
53. Minor LB, Solomon D, Zinreich JS, and Zee DS. Sound- and/or pressure-induced vertigo due to bone dehiscence of the superior semicircular canal. *Archives of Otolaryngology–Head and Neck Surgery*. **124**:249–258 (1998).
54. Baxter M. *et al.* Clinical and physiologic predictors and postoperative outcomes of near dehiscence syndrome. *Otolaryngology–Head and Neck Surgery*. **40**:204–212 (2019).
55. Minor LB. Clinical manifestations of superior semicircular canal dehiscence. *Laryngoscope*. **115**:1717–1727 (2005).
56. Belden CJ, Weg N, Minor LB, and Zinreich SJ. CT evaluation of bone dehiscence of the superior semicircular canal as a cause of sound- and/or pressure-induced vertigo. *Radiology* **226**:337–343 (2003).
57. Castellucci A, Brandolini C, Piras G, and Modugno GC. Tympanometric findings in superior semicircular canal dehiscence syndrome. *Acta Otorhinolaryngologica Italica*. **33**:112–120 (2013).
58. Zuniga MG, Janky KL, Nguyen KD, Welgampola MS, and Carey JP. Ocular vs. Cervical VEMPs in the diagnosis of superior semicircular canal dehiscence syndrome. *Otolaryngology–Head and Neck Surgery*. **34**:121–126 (2013).

59. Cozart AC, Kennedy JT, and Seidman MD. A basis for standardizing superior semicircular canal dehiscence management. *Ear, Nose & Throat Journal*. **100**:NP444–NP453 (2021).
60. Mozaffari K. *et al.* Superior semicircular canal dehiscence revision surgery outcomes: A single institution's experience. *World Neurosurgery*. **156**:e408–e414 (2021).
61. Weinreich HM, and Carey JP. Prevalence of pulsatile tinnitus among patients with migraine. *Otolaryngology–Head and Neck Surgery*. **37**:244–247 (2016).

https://doi.org/10.1142/9789819807345_0014

Chapter 14

Childhood Tinnitus: Treatment Strategies

Randa Al Barazi[*,‡], **Yara Yammine**[*], **and Karen Tawk**[†]

[*]*Department of Otolaryngology-Head and Neck Surgery, American University of Beirut, Beirut, Lebanon*

[†]*Department of Otolaryngology-Head and Neck Surgery, University of California, Irvine, CA, USA*

[‡]*ra110@aub.edu.lb*

Abstract

Childhood tinnitus is characterized by 'phantom auditory perceptions' in the absence of external stimuli, which poses significant challenges for pediatric healthcare providers. This chapter offers an overview of the underlying factors and treatment modalities available for children with tinnitus. Despite the broad estimates for its prevalence, childhood tinnitus often goes underreported due to communication barriers by children. Various factors such as hearing loss, noise exposure, and psychological stressors contribute to its onset, underscoring the importance of comprehensive assessment and tailored interventions. Treatment options for childhood tinnitus include counseling, psychological therapies, sound therapy, and amplification devices. Counseling serves as a cornerstone, educating children and parents about tinnitus while addressing associated psychological distress. Cognitive Behavioral Therapy (CBT) aims to modify unhelpful thoughts and behaviors, although its efficacy in children necessitates further investigation. Additionally, biofeedback training, relaxation techniques, and mindfulness-based interventions offer alternative avenues for managing tinnitus-related stress. Sound therapy, including masking and Tinnitus Retraining Therapy (TRT), aims to alleviate tinnitus perception and distress. However, evidence supporting their effectiveness in children remains limited, highlighting the need for future research. Amplification devices like hearing aids and cochlear implants may provide relief for children with concurrent hearing loss, though their efficacy specifically for tinnitus management requires further exploration. Pharmacotherapy and migraine management strategies are emerging as potential avenues for tinnitus relief, emphasizing the multifaceted approach needed to address childhood tinnitus. Moreover, managing tinnitus-related challenges in the classroom is essential for supporting children's educational progress. In conclusion, childhood tinnitus presents a complex clinical scenario requiring tailored interventions addressing both auditory and psychological aspects. Further research is warranted to elucidate optimal treatment strategies and improve outcomes for affected children.

Keywords: Childhood, Tinnitus, Hearing Loss, Psychological, Training Therapy, Sound Therapy, Amplification.

Introduction

Tinnitus is an 'auditory phantom perception' of a sound in the absence of an external source or acoustic stimulation [1]. It is often described as a buzzing, ringing, or humming sound. This condition can affect children of any age with nearly 90% of childhood tinnitus being subjective. Tinnitus can be either idiopathic or secondary to an underlying problem. While it is usually not a serious condition, it can range from mild to severe and be bothersome. Due to its effect on sleep, concentration and attention, hearing, and emotional health, tinnitus has the potential to negatively affect children's quality of life and/or their functional health status [2, 3].

The prevalence of tinnitus varies widely among children, with estimates indicating that Prevalence estimates range from 4.7% to 46% of the general pediatric population may be affected and rises to from 23.5 to 62.2% among children with hearing loss [4]. Other studies reported a broader prevalence ranging from 3.6% to 66.8% [5], depending on the definitions and characteristics of the studied cohort [6]. It is important to note that statistics in children might be underestimated due to communication difficulties. Moreover, tinnitus is not spontaneously reported by children as they are easily distracted by their surroundings or they may consider it as normal, not realizing its clinical significance.

Tinnitus in children can be caused by several underlying factors, including sensorineural hearing loss, noise exposure, conductive hearing loss, and otitis media [7]. Stress, anxiety, depression, and hyperacusis are also found to be associated with tinnitus, although these factors are less common in children [8, 9].

Treatment Options

Although tinnitus in children has received little attention in the literature, a few therapeutic studies have addressed this condition. Currently, tinnitus treatment in children involves approaches similar to those used in adults but with interventions tailored to each child's developmental level, age, audiological status, and life circumstances. Early identification and treatment can help ameliorate its negative impact on the child's health in terms of their psychological well-being, educational progress, and quality of life and prevent it from becoming a chronic problem.

In 2014, the American Academy of Otolaryngology–Head and Neck Surgery (AAO-HNS) established the Clinical Practice Guidelines (CPG) on Tinnitus [11]. That same year, the British Society of Audiology (BSA) also published guidelines and recommendations specifically addressing tinnitus management during childhood [12, 13]. In March 2020, the UK National Institute for Health and Care Excellence (NICE) issued guidelines for the assessment and management of tinnitus in children, young people, and adults [14]. However, the evidence to support these guidelines, recommendations, and clinical practices is limited and the recommendations are often based on the experience and opinion of the guideline development committee members [3, 14].

Several treatment modalities for tinnitus in children have been proposed, including counseling and education, psychological therapies, sound or masking therapy, pharmacological treatments, and the use of hearing devices, among others. However, the evidence supporting these approaches in children is scarce, indicating a need for further research.

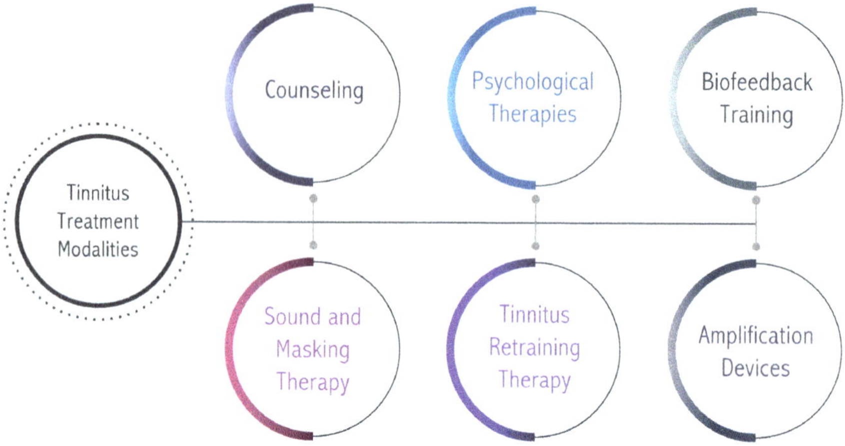

Counseling

Counseling is a fundamental part of therapy and should be recommended as the basis of treatment for chronic tinnitus. Clinicians need to engage the child or the child's parent in understanding tinnitus and the possible management options [10, 15]. Early detection and treatment of psychological problems through counseling are essential in preventing these aggravating factors of tinnitus.

Explaining tinnitus to children

The development of counseling protocols tailored to the degree of understanding of children helps provide appropriate support for managing tinnitus in this population. Very young children appreciate very simple explanations to which they can relate. For example, tinnitus can be explained as the sound their ears occasionally make when they are working, like tummy rumble or the sound of breathing. For instance, once the child notices the sound of their breathing, this observation can be used as a comparison to their experience of tinnitus [13]. Older children are more likely to have developed the linguistic and cognitive skills to understand the complex relationship between tinnitus symptoms and thoughts, emotions, physiological reactions, and life events.

A detailed example of this approach is the "Child-Friendly Tinnitus Model" [16]. In a recent study by Tegg-Quinn *et al.* the strategy stating "audiologists/clinicians are able to confidently discuss tinnitus and explain it to children and families" showed the most benefit in reducing the impact of tinnitus in children experiencing it [17].

Reassurance

It is important to ensure that clinicians and patients understand that management strategies for bothersome tinnitus do exist. Despite a lack of a definitive cure, many treatment approaches can improve tinnitus and relieve the associated distress. Therefore, it is crucial to explain to the patient that tinnitus represents a symptom rather than a dangerous disease, and a comprehensive audiological and medical assessment can exclude medical conditions requiring prompt treatment [10, 18].

Normalizing tinnitus

Children are often surprised and relieved to learn that many other children also hear noises in their ears and that they are not the only ones experiencing them [13].

Education

Counseling aims to provide a clarifying psychoeducational explanation, attentive listening to patients' experiences, answering their questions, and discussing lifestyle factors that can positively or negatively affect their tinnitus, such as noise exposure, dehydration, hunger, caffeine or preservative-containing foods, or sleep quality.

If the physician lacks the time or resources or comprehensive counseling, they should refer the patient to a psychotherapist specialized in tinnitus therapy. Alternatively, they can provide brochures, suggest self-help books, and discuss other medical and complementary/alternative therapies, while also noting the lack of proven benefits of questionable supplements or online "cures" [11]. Patients should be encouraged to return for follow-up and reassessment if their tinnitus persists and remains bothersome or if it worsens over time.

Bae *et al*. showed a very successful treatment outcome (30/32 children, 93.8%) using only counseling in the pediatric tinnitus population with normal hearing or minimal hearing loss [19]. In addition, in a recent review of the effects of counseling in children by Lee *et al*., there was an overall 83.3% subjective improvement in tinnitus by counseling, and the benefit seemed to be greater in younger patients and children without hearing loss [20].

Healthy Hearing Education and Noise Avoidance

Noise avoidance is an important component of managing tinnitus in children by reducing the severity of their symptoms and preventing further damage. Exposure to loud noises can worsen tinnitus and prolonged exposure can cause damage to the child's hearing; thus, providing education about the impact of noise (cumulative or sudden) on hearing is essential. Children should be instructed to avoid exposure to loud noises, such as fireworks, and to minimize exposure to loud music. When indicated, providing appropriate hearing protection (e.g., custom ear plugs) is recommended [13]. Parents and caregivers play an important role in teaching children about noise avoidance and providing them with the necessary tools to protect their hearing.

Psychological Therapies

Tinnitus can lead to psychological symptoms, although the causality of this relationship remains unclear in children. Psychoemotional factors have been previously associated with tinnitus, and their role was observed as early as 1948 [21]. Recent studies have reported that psychoemotional factors are more prevalent in tinnitus patients. Several psychiatric disorders, especially depression and anxiety, along with sleep disturbance, suicidal thoughts, somatoform disorders, and attention deficit hyperactivity disorder (ADHD), appear to affect the development of tinnitus [22, 23]. A significant association between tinnitus and anxiety has been observed [8], with the severity of tinnitus strongly correlating with the severity of the psychiatric disorder [22]. Anxiety and depression are important factors that should be considered in the management of tinnitus in the pediatric population.

Cognitive Behavioral Therapy

Cognitive behavioral therapy (CBT) is a psychological intervention designed to alleviate anxiety by assisting patients in modifying their unhelpful cognitions and safety-seeking behaviors [24, 25]. It focuses on "confronting, disputing, and restructuring maladaptive thought patterns to develop more adaptive patterns of thought, leading to more adaptive emotional and behavioral responses" [26]. Therefore, by reducing psychological distress, CBT aims to reduce the intrusiveness of tinnitus and the pervasiveness of its impact [27].

CBT has proven to be effective in the treatment of tinnitus in adults, hence improving the quality of life of those affected. According to the AAO-HNS clinical practice guidelines (CPG) for tinnitus, CBT should be "recommended" to those experiencing bothersome persistent tinnitus [11]. Several systematic reviews have suggested a beneficial effect of CBT for patients with tinnitus.

A recent Cochrane Review by Fuller *et al.* reported that CBT may reduce the impact of tinnitus on quality of life at the end of treatment, as evidenced by lower Tinnitus Handicap Inventory scores compared to the non-CBT group. Additionally, they found that CBT may result in a slight reduction in depression; however, it remained uncertain whether CBT reduces anxiety or biased interpretations of tinnitus [28]. Moreover, CBT proved to be superior to no intervention/waiting, auditory intervention, or tinnitus retraining therapy. The largest meta-analysis to date, including 15 RCTs, found that CBT has a positive impact on tinnitus-related distress with effect sizes ranging from 0.44 to 0.7. Furthermore, CBT exhibited a positive effect on mood, and these benefits were sustained for up to 18 months [29].

No studies to date have established the effectiveness of CBT as a treatment modality for childhood tinnitus. However, its efficacy in managing anxiety and depression in pediatric populations has been well-documented. One large systematic review by James *et al.* showed that CBT may increase post-treatment remission rates for primary anxiety disorders as well as for all anxiety disorders [30]. Furthermore, a more recent review by Kunas *et al.* emphasized that primary symptom severity acts as a negative predictor of CBT outcomes for both anxious and depressed youths [31].

Cognitive behavioral therapy can be delivered to individuals or a group. Most studies on CBT and tinnitus have investigated group therapy for people with persistent tinnitus. It is typically provided by mental health professionals; however, audiologists or other health professionals trained in cognitive behavioral intervention are also capable of providing this treatment. Additionally, CBT can be performed remotely using online resources.

It is important to note that the effectiveness of CBT can vary greatly between individuals, and it may not be suitable for all children with tinnitus. Little is known about the effectiveness of CBT with children under the age of 7. For younger children, there tends to be a greater focus on behavioral aspects, whereas older children are more capable of engaging with the cognitive aspects [13]. Also, various factors such as the child's age, the severity and type of tinnitus, along with their overall mental health and coping abilities can all impact the effectiveness of CBT.

Biofeedback Training

Introduced in the 1970s for tinnitus by William F. House, biofeedback training is a self-regulation technique that teaches individuals to control their involuntary body processes according to their body feedback [32]. Electrophysiological recordings (e.g., muscle tension, skin temperature, etc.) are converted to visual and auditory signals on a screen. Patients are then encouraged to modify their thoughts and emotions and observe the correlated changes on the display, creating a general state of relaxation. Numerous studies have revealed the benefits of biofeedback training for tinnitus treatment, particularly in reducing annoyance and distress, and improving feelings of controllability [33].

Neurofeedback, an emerging sub-field of biofeedback, is a neuromodulation technique whereby the subjects learn to voluntarily change their brain activity in the desired direction, typically through operant conditioning (i.e., rewarding wanted changes and inhibiting unwanted ones). Even though neurofeedback has already shown promising results for chronic tinnitus treatment, further research is needed to develop more sophisticated protocols that more precisely address the specific needs of tinnitus patients [34, 35]. The complexity of this therapy makes it inappropriate and inapplicable in the pediatric population.

Other Management Strategies

Relaxation

Currently, there is limited evidence supporting the use of relaxation in the management of tinnitus in children with little evidence of its effectiveness as a stand-alone approach. However, considering that stress can exacerbate tinnitus, incorporating relaxation techniques is widely suggested as part of a holistic approach to tinnitus management. Lamontagne *et al.* reported that "relaxation can be learned by children and may aid in coping with stress" [36]. Relaxation can be a useful self-management tool, reducing physiological arousal in response to stress and promoting a sense of calm. For example, simple breathing techniques could be carried out with ongoing guidance. It is important to monitor and support the progress to ensure that these techniques are implemented effectively and provide maximum benefit [13].

Mindfulness techniques

Mindfulness techniques propose an "acceptance-based" approach to tinnitus, focusing on "allowing" the presence of tinnitus rather than attempting to "fight" it [37].

While evidence of the effectiveness of mindfulness techniques in the management of tinnitus is still emerging, recent studies have shown promising results [37]. Mindfulness-based interventions (MBIs) are beginning to be offered and applied [14]. A pilot study in 2013 showed positive findings in treating chronic tinnitus and its comorbid symptoms after an 8-week mindfulness-based tinnitus stress reduction program [38]. A recent systematic review noted a decrease in tinnitus distress scores immediately after MBIs [39]. A more recent randomized clinical trial (RCT) found a decrease in tinnitus handicap inventory (THI) and Hospital Anxiety and Depression scores, as well as tinnitus intensity post-therapy, leading to an overall improvement in quality of life [40]. Mindfulness may have a place in tinnitus therapy, although the long-term effects require further investigation.

The evidence supporting the use of mindfulness-based approaches with young people is preliminary, yet mindfulness practice skills have been proven to be useful in children learning how to manage their stress and anxiety, which may in turn contribute to more effective tinnitus management [13]. In the UK, mindfulness has been introduced in some schools through 4 different course curricula tailored to each age category, ranging from 3 to 18 years old [41]. In one study, the Mindfulness in School Programme (MiSP) was offered for secondary school students, resulting in fewer depressive symptoms, lower stress levels, and greater well-being among participants [42]. In an evaluation of pediatric tinnitus services in the UK National Health Service audiology department, 28% offered mindfulness as a treatment modality. Nonetheless, providing psychological therapy for children with tinnitus-related psychological problems requires optimization [43].

Narrative therapy

Narrative therapy refers to a technique that involves listening to others' stories or narratives. It is based on the fact that individuals create stories to make sense of their lives and experiences. For children experiencing tinnitus, these stories are often filled with negative connotations and emotions such as "tinnitus can be scary and frightening" or "tinnitus can be a monster". By reshaping their narrative about tinnitus, we can help children perceive it in a more positive aspect. For instance, we can help them see tinnitus as a monster that can be befriended. This therapy emphasizes the importance of acknowledging that negative information on which a child focuses, can reinforce existing beliefs such as "tinnitus always keeps me awake". Narrative therapy helps children explore alternative narratives, like recalling a time when they were able to sleep despite their tinnitus, thereby constructing a new narrative that challenges their belief that "tinnitus stops me from sleeping" [44]. Narratives about tinnitus relate to the broader context of children's lives, including troublesome events at home or school. These stressful situations can become linked to their tinnitus narratives. Therefore, narrative therapy helps in untangling these associations, aiming to understand why tinnitus becomes a significant problem for some children and their families [45]. Although to date, there is no research demonstrating the effectiveness of narrative therapy for children with tinnitus, its suitability for young children is notable. The playful, imaginative, and fun nature and reliance on storytelling align with the way children interact with their environment.

Masking or Sound Therapy

Sound therapy is a form of non-invasive treatment where individuals are exposed to calming or soothing sounds to induce habituation. This process involves "masking" the tinnitus by reducing the contrast between the surrounding environment and the patient's perception of the tinnitus, making it less noticeable. It may also involve providing interesting sounds to distract the patient's attention away from the tinnitus (active attention diversion) [11]. It is thought to induce a sense of relief from the tension caused by tinnitus by reducing the emotional reactions toward it [46]. Some individuals may experience residual inhibition following total or partial masking (i.e., tinnitus suppression or temporary disappearance of the tinnitus sensation after exposure to an external sound) by promoting habituation to the tinnitus. Sound therapy can be applied using a variety of acoustic device s including, sound (noise) generators, hearing aids, a combination of hearing aids and noise generators, or environmental enrichment devices, such as a smartphone, TV, or fans [11].

Sound generators or smartphone applications can produce broadband noise, such as white or pink noise. A white noise is a constant low-level background sound encompassing all audible frequencies. Examples include the sound of nature or rainfall, adjusted at an intensity slightly below that of the perceived tinnitus. The volume of the external masking sound can be increased to the point where it does not interfere with the environmental sounds that a patient needs to hear [18]. However, evidence supporting the suppression of tinnitus only through acoustic stimulation is either lacking or of poor quality [47, 48]. In 2012 a systematic review by Hobson *et al.*, on tinnitus masking therapy in adults, concluded that sound therapy alone does not provide significant benefit. The authors noted that the lack of conclusive evidence should not be interpreted as evidence of ineffectiveness and suggested that optimal management might require a combination of strategies [49]. The lack of high-quality research in this area, coupled with the frequent use of combined approaches in the management of tinnitus, contributes to the lack of conclusive evidence. A more recent review by Liu *et al.* reiterated the reduction of tinnitus severity with sound therapy; however, combination treatments proved a more effective modality of therapy [50].

Bartnik *et al.* reported a significant improvement in tinnitus in 48 out of 59 children using sound generators as part of tinnitus retraining therapy (TRT). This report, however, provided limited clarification on the counseling input and other strategies used alongside the devices, or how its benefits compared to other treatment strategies for children with tinnitus [51]. Further research is required to determine if certain pediatric populations, such as those with special needs, autism spectrum disorder, or limited language and communication skills, might particularly benefit from wearable sound generators. Until we gain more knowledge on the development of bothersome tinnitus in the pediatric population, audiologists have to rely on reports based on adult studies [13].

Tinnitus Retraining Therapy

Tinnitus retraining therapy (TRT), introduced in the late 1980s and first outlined in 1990 by Jastreboff [52, 53], is a form of habituation therapy achieved by retraining the auditory, limbic and autonomic processing in the brain [54]. This habituation technique comprises two major components: (1) the use of maskers set at the mixing point (i.e., where the masking noise and tinnitus blend) or slightly below the patient's perceived tinnitus (i.e., partial masking or total masking); and (2) directive counseling, which primarily involves educational content [55]. TRT is mainly aimed at reclassifying tinnitus into a category of neutral percepts by weakening the intensity of the tinnitus signal and reducing the negative tinnitus-related reactions. In other words, it helps patients gradually habituate themselves to the sound of their tinnitus. Phillips and McFerran performed a systematic review to assess the efficacy of TRT and included trials that compared TRT with either no treatment or other forms of tinnitus therapy [56]. Most studies were excluded because they used modified versions of the TRT protocol. The single included study found benefits for TRT in the treatment of tinnitus based on validated outcome measures (THI, THQ, and TSI); however, it was noted to have methodological flaws [57]. Recent studies [58, 59] showed that TRT with noise device therapy was not superior to treatment with placebo noise generators. However, the specifically defined counseling sequence seemed to be of importance in differentiating the TRT protocol from other forms of therapy for tinnitus. Nonetheless, a 2021 review by Han *et al.* mentioned that TRT was an effective treatment for tinnitus, albeit based on low-quality evidence with a high risk of bias. Therefore, larger and high-quality randomized RCTs are needed [60].

TRT application in the pediatric population is still limited. A 2021 systematic review by Dullart *et al.* included 3 RCTs that applied simplified TRT, reporting improvements in tinnitus outcomes in 68 out of 82 children after 3–6 months of treatment [61]. The interventions in these studies varied, with one involving a single counseling session of 30 min and sound therapy for 3 months [19], in another, patients received a single directive counseling (i.e., counselor-centered) session and sound therapy using hearing aids, or sound generators for 6 months with follow-up counseling as per the individual's needs [51], and a third study provided a single counseling session followed by sound therapy with radio music for 6 months [62]. In conclusion, TRT can be considered a long-term therapeutic intervention for chronic tinnitus, although further research, particularly in the pediatric population, is warranted.

Amplification Devices

Hearing aids

Although tinnitus has been strongly associated with hearing loss, the degree of hearing loss cannot linearly predict tinnitus severity. Only 50% of patients with hearing loss experience tinnitus and 10% of tinnitus sufferers have normal hearing as measured by pure tone audiometry [15]. Hearing aids (HAs) are used primarily to improve hearing, but they also often reduce the perception of tinnitus and its impact. Hearing aids will primarily benefit patients with hearing loss and hearing aids are limited to improving tinnitus when some noise is present. In silence, a masking feature needs to be available on the hearing aid to provide benefit to patients with tinnitus.

According to the CPG and NICE guidelines, hearing amplification for patients with hearing loss and persistent, bothersome tinnitus [11, 14] is proposed as a weak recommendation. Given the improvement in communication functions and health-related QoL following the provision of amplification coupled with the potential benefits of tinnitus relief with minimal risks, HA use is a reasonable option for those with concurrent tinnitus and hearing loss.

Recent studies established the effectiveness of HAs in reducing tinnitus and related discomfort in patients with associated hearing loss [63–65]. A 2021 review reported that up to 50% of patients experienced improvement due to HA fitting, with the proportion of those showing some improvement ranging between 40% to 85% [66]. However, a previous Cochrane review on assessing the impact of HA fitting on tinnitus in adults with both hearing loss and tinnitus concluded that given only one clinical study met the quality standards, no definitive conclusion could be reached [67]. Therefore, the role of HAs remains contentious, the evidence base is limited, and currently no evidence to support or refute the provision of HAs as a primary intervention in the management of tinnitus in patients with co-existing hearing loss [67].

It is mentioned that the use of amplification devices in the absence of hearing loss is unlikely to improve tinnitus and may not be tolerated due to unnecessary sound amplification [14]. While this has never been documented and no studies have specifically evaluated the use of HAs for the management of tinnitus in the absence of hearing loss, HAs are sometimes included in tinnitus treatment plans.

Despite these considerations, HAs play an important role in the treatment of hearing loss, which significantly overlaps with tinnitus in children. It is thus it is important to recommend HAs to children with tinnitus when hearing loss is present [13]. For instance, children with severe to profound hearing loss and tinnitus may experience more noticeable tinnitus when not using their HAs, especially in quiet environments such as bedtime. Currently, no studies have specifically evaluated the effectiveness of HAs in children with tinnitus, highlighting a gap in the research [3].

Cochlear implants

Cochlear implantation (CI) is recommended only for patients meeting the hearing loss criteria. However, a few studies reported tinnitus reduction following the electrical activation in the auditory pathway after CI. So far, several systematic reviews reported the positive, but not conclusive, effect of intracochlear electrical stimulation on tinnitus, in patients with a primary indication for CI due to bilateral hearing loss [68, 69] or single-sided deafness [70–72] to restore hearing outcome. However, studies investigating intracochlear stimulation for tinnitus relief in patients with tinnitus as the primary indication, rather than hearing loss, are still lacking. Assouly *et al.* conducted a review focused on CI outcomes in patients with tinnitus as a primary complaint. They found a statistically significant reduction in tinnitus distress and loudness with a large difference in pre- and post-CI THI scores. In addition, they reported cases of complete suppression and tinnitus reduction at 3 months follow-up and beyond [73].

A more recent study investigating the prevalence of tinnitus pre- and post-CI showed a decrease from 55.8% to 44.3% with a median TRQ score decreasing from 12.0 to 3.5. Moreover, the proportion of patients experiencing moderate to catastrophic tinnitus distress decreased from 14.6% before the implant to 6.3% post-implantation [74].

Indications for CI have been established in the pediatric population for hearing loss, but none have studied tinnitus as a primary indication for

CI. One study by Chadha *et al.* reported a 38% prevalence of tinnitus in children post-CI. It occurred most commonly in the implanted ear when the implants were not in use, such as at night. The incidence of tinnitus was highest among children aged 6–8 years (47%) and in those who received bilateral implants with a delay of at least 2 years between surgeries (60%). The lowest reports of tinnitus were in children who received simultaneous bilateral implants (17%) and those aged 5 years or younger (27%). No obvious relationship was identified between the prevalence of tinnitus and the etiology of deafness, age of implantation, or time elapsed since implantation [75]. To further characterize the onset, variability, and impact of tinnitus in children with CI, future large-scale prospective cohort studies are needed.

Pharmacotherapy and Dietary Supplements

Many classes of drugs have been used or trialed for the treatment of tinnitus, including antiarrhythmics, anticonvulsants, anxiolytics, glutamate receptor antagonists, antidepressants, muscle relaxants, and others [76], with little evidence of benefit over harm [11]. Generally, no drug was recommended for the treatment of tinnitus. Nonetheless, medications may be beneficial in managing psychiatric comorbidities associated with tinnitus. As for the dietary supplements studied for the management of tinnitus, the aggregate data were mostly focused on Ginkgo biloba, melatonin, and zinc. However, these studies exhibited methodological flaws and conflicting results regarding their benefit, with significant and well-documented potential side effects. Consequently, there is insufficient evidence to recommend dietary supplements for the treatment of tinnitus [77].

In the pediatric population, these medications have not been studied for the primary treatment of tinnitus and are used with caution for the treatment of underlying diseases. It is important to note that the use of pharmacological treatments in children should be carefully monitored and managed by a healthcare professional to ensure safety and efficacy. Similarly, the use of dietary supplements in the pediatric population has not been investigated.

Migraine Treatment

With the new understanding of the role of migraine in the exacerbation of tinnitus [78] some centers use supplements found to be of benefit in migraine to treat patients with tinnitus as part of a comprehensive medical management of tinnitus which includes lifestyle and dietary changes. The migraine approach to treating tinnitus includes avoiding dehydration and hunger, a set sleep schedule (no change between weekend and weekdays), treatment of sleep apnea if present, managing stress, and avoiding foods that trigger migraine (Table 1). In the pediatric population, sleep schedule

Table 1. Recommended migraine diet for patients with loud bothersome tinnitus or those with fluctuating levels of tinnitus.

Category	Avoid, Reduce, or Limit	Acceptable Foods
Caffeine*	Coffee, tea Caffeinated soda	Decaffeinated tea, Herbal tea Caffeine-free soda Fruit juice (non-citrus)
Snacks*	Chocolate Nuts Dry fruit (raisins, etc.)	Seeds Sherbet, ice cream Cakes
Alcohol*	Fermented alcohol (e.g., wine, beer, etc.)	Non-alcoholic beverages Highly distilled alcohol (e.g., vodka)
Dairy*	*Most cheeses*: Brie, Boursault, Blue, Cheddar, Brick, Camembert, Mozzarella, Romano, Emmental, Gouda, Parmesan, Provolone, Swiss, Roquefort, Stilton Sour cream	*Other cheeses*: American, Cream cheese, Velveeta, Cottage, Farmer, Ricotta) Milk, rice milk Oatmeal Egg substitutes, egg yolk
Cereals and Grains*	Fresh bread, Bagels, Donuts, Yeast products, Yeast extracts, Sourdough, Brewer's yeast	Commercial breads (White, Wheat, Multicereal, Rye, Italian) English muffins Crackers, Toast Potatoes Spaghetti, Noodles, Rice
Meats*	Aged, canned, cured, pickled, salted, and dried meats Processed meats (Pepperoni, Hot dogs, Bologna, Salami, Pre-packaged deli meats, Jerky, Sausages) Any processed protein that contains Tyramine	Fresh/unprocessed meats, Fish, Poultry, Veal, Lamb, Tuna

Table 1. (*Continued*)

Category	Avoid, Reduce, or Limit	Acceptable Foods
MSG* (Monosodium Glutamate)	Soy sauce, Bouillon cubes, Canned soups, Tenderizers, Seasoned salts Foods containing "hydrolyzed protein products" or "autolyzed yeast", "autolyzed protein", "hydrolyzed yeast" Pickled, preserved, or marinated foods Asian food sauces Off-the-shelf salad dressings	Salt and other spices Butter, Margarin White vinegar
Sweeteners*	Aspartame (Equal, Nutrasweet)	Sucrose (sugar), High-fructose corn syrup, Sucralose, Saccharin
Vegetables*	*Beans*: Pole or broad, Italian, Lima, Navy, Fava, Pinto, Garbanzo, Snow peas, Lentils, Pea pods Sauerkraut, Pickles, Kimchi (Yyramine) Onions Olives	Asparagus, Broccoli, Garlic, Beets, Carrots, Pumpkins, Squash, Lettuce, Spinach, String beans, Tomatoes, others
Fruits*	Avocados, Papaya, Figs, Passion fruit Bananas and Citrus fruit (Orange, Grapefruit, Lime, Lemon) Overly ripened fruits (high in Tyramine	Apples, Berries, Hard peaches, Hard pears

Note: *Level of Evidence 1.

changes between weekends and weekdays, dehydration, and dietary triggers tend to be the most common triggers of migraine/tinnitus in our experience and that of others [79].

Tinnitus Management in the Classroom

Tinnitus can significantly affect children's concentration skills in the classroom, potentially impairing their educational progress. Teachers and support staff may notice an unexplained change in the level of attention, reduced focus, and increased student frustration. Therefore, assessing tinnitus in the classroom, applying classroom-based strategies, and providing professional support is crucial. However, discussing these measures falls beyond the scope of this chapter.

Conclusion

- Tinnitus affects children of varying ages and can range from mild to severe, impacting various aspects of their lives.
- The prevalence of childhood tinnitus varies widely, with estimates ranging from 4.7% to 66.8%, often underestimated due to communication challenges in children.
- Tinnitus in children can stem from different factors including hearing loss, noise exposure, and psychological factors such as stress and anxiety.
- Therapeutic approaches include counseling, psychological therapies, sound therapy, pharmacological treatments, and amplification devices, among others.
- Counseling plays a vital role in educating children and parents about tinnitus, addressing psychological factors, and providing coping strategies tailored to the child's developmental level.
- Cognitive Behavioral Therapy (CBT) aims to modify unhelpful thoughts and behaviors associated with tinnitus, though its efficacy in children requires further research.
- Biofeedback training and relaxation techniques are proposed as methods to reduce tinnitus distress, while mindfulness techniques and narrative therapy offer alternative approaches for managing tinnitus-related stress.
- Sound therapy aims to mask or distract from tinnitus, while Tinnitus Retraining Therapy (TRT) focuses on habituation to the tinnitus sound, though its efficacy in children needs more investigation.
- Hearing aids and cochlear implants may alleviate tinnitus in children with hearing loss, although the evidence for their effectiveness in tinnitus management specifically remains limited.
- Pharmacological treatments and migraine management strategies are explored as potential avenues for tinnitus relief, while addressing tinnitus-related challenges in the classroom is essential for supporting children's educational progress.

References

1. Tyler, RS. In *Tinnitus Handbook* (1st Ed.). San Diego: Singular. Thomson Learning, (2000).
2. Hall DA, *et al.* A narrative synthesis of research evidence for tinnitus-related complaints as reported by patients and their significant others. *Health and Quality of Life Outcomes.* **16**:61 (2018).
3. Smith H, *et al.* A scoping review to catalogue tinnitus problems in children. *International Journal of Pediatric Otorhinolaryngology.* **122**:141–151 (2019).
4. Rosing SN, *et al.* Prevalence of tinnitus and hyperacusis in children and adolescents: A systematic review. *BMJ Open.* **6**(6):e010596 (2016).
5. Tegg-Quinn S, *et al.* The use of patient-report measures and intervention strategies for children and adolescents with chronic tinnitus: A scoping review. *International Journal of Audiology.* **12**:1–9 (2023).
6. Baguley D, *et al.* In *Tinnitus: A Multidisciplinary Approach* (Second Edition), New York, NY: Wiley-Blackwell, p. 324 (2013).
7. Coelho CB, *et al.* Tinnitus in children and associated risk factors. *Progress in Brain Research.* **166**:179–191 (2007).
8. Kim YH, *et al.* Tinnitus in children: association with stress and trait anxiety. *Laryngoscope.* **122**(10):2279–2284 (2012).
9. Humphriss R, *et al.* Prevalence and characteristics of spontaneous tinnitus in 11-year-old children. *International Journal of Audiology.* **55**(3):142–148 (2016).
10. Baguley DM, and McFerran DJ. Tinnitus in childhood. *International Journal of Pediatric Otorhinolaryngology.* **49**(2):99–105 (1999).
11. Tunkel DE, *et al.* Clinical practice guideline: Tinnitus. *Otolaryngology-Head and Neck Surgery.* **151**:S1–S40 (2014).
12. Kentish R, *et al.* In *Tinnitus in Children and Teenagers. Practice Guidance.* London: British Tinnitus Association, (2014).
13. Benton C, *et al.* In *British Society of Audiology Good Practice Guidance for the Management of Children with Tinnitus*, (2015).
14. Lewis SE. Assessment and management of tinnitus: Summary of NICE guidance. *British Medical Journal.* **368**:m976 (2020).
15. Cima RFF, *et al.* A multidisciplinary European guideline for tinnitus: Diagnostics, assessment, and treatment. *HNO.* **67**(Suppl 1):10–42 (2019).
16. Emond A, and Kentish R. Tinnitus counselling with children. *Audacity.* **27**(2):26–29 (2013).
17. Tegg-Quinn S, *et al.* Reducing the impact of tinnitus on children and adolescents' lives: A mixed-methods concept mapping study. *International Journal of Pediatrics.* **2021**:5534192 (2021).

18. Han BI, *et al.* Tinnitus update. *Journal of Clinical Neurology.* **17**(1):1–10 (2021).
19. Bae SC, *et al.* Childhood tinnitus: Clinical characteristics and treatment. *American Journal of Otolaryngology.* **35**(2):207–210 (2014).
20. Lee DY, *et al.* Management of tinnitus in children: Review of literature and effect of counseling. *Auris Nasus Larynx.* **45**(4):667–672 (2018).
21. Fowler EP. The emotional factor in tinnitus aurium. *Laryngoscope.* **58**:145–154 (1948).
22. Rhee J, *et al.* Prevalence, associated factors, and comorbidities of tinnitus in adolescents. *PLoS One.* **15**(7):e0236723 (2020).
23. Kim SY, *et al.* Characteristics of tinnitus in adolescents and association with psychoemotional factors. *Laryngoscope.* **127**(9):2113–2119 (2017).
24. Beck AT. In *Cognitive Therapy and the Emotional Disorders*. New York: International Universities Press, (1976).
25. Clark DA, *et al. Scientific Foundations of Cognitive Theory and Therapy for Depression.* John Wiley & Sons Inc., pp. 100–106 (2000).
26. Field T, *et al.* The new ABCs: A practitioner's guide to neuroscience-informed cognitive-behavior therapy. *Journal of Mental Health Counseling.* **37**(3):206–20 (2015).
27. McKenna L, *et al.* Current validated medical treatments for tinnitus: Cognitive behavioral therapy. *Otolaryngologic Clinics of North America.* **53**(4):605–615 (2020).
28. Fuller T, *et al.* Cognitive behavioural therapy for tinnitus. *Cochrane Database of Systematic Reviews.* **1**(1):CD012614 (2020).
29. Hesser H, *et al.* A systematic review and meta-analysis of randomized controlled trials of cognitive-behavioral therapy for tinnitus distress. *Clinical Psychology Review.* **31**(4):545–53 (2011).
30. James AC, *et al.* Cognitive behavioural therapy for anxiety disorders in children and adolescents. *Cochrane Database of Systematic Reviews.* **2015**(2):CD004690 (2015).
31. Kunas SL, *et al.* Psychological predictors of cognitive-behavioral therapy outcomes for anxiety and depressive disorders in children and adolescents: A systematic review and meta-analysis. *Journal of Affective Disorders.* **278**:614–626 (2021).
32. House JW. Treatment of severe tinnitus with biofeedback training. *Laryngoscope.* **88**(3):406–412 (1978).
33. Tang D, *et al.* Advances in understanding, diagnosis, and treatment of tinnitus. *Advances in Experimental Medicine and Biology.* **1130**:109–128 (2019).
34. Güntensperger D, *et al.* Neurofeedback for tinnitus treatment — Review and current concepts. *Frontiers in Aging Neuroscience.* **9**:386 (2017).
35. Barrenechea FV. Efficacy of neurofeedback as a treatment for people with subjective tinnitus in reducing the symptom and related consequences:

A systematic review from 2010 to 2020. *Acta Otorrinolaringológica Española.* **17**:S2173-5735(22)00111-9 (2022).

36. Lamontagne L, *et al.* Effects of relaxation on anxiety in children: Implications for coping with stress. *Nursing Research.* **34**(5):289–292 (1985).
37. Marks E, *et al.* I wasn't at war with the noise: How mindfulness-based cognitive therapy changes patients' experiences of tinnitus. *Frontiers in Psychology.* **11**:483 (2020).
38. Gans JJ, *et al.* Mindfulness-based tinnitus stress reduction pilot study: A symptom perception-shift programme. *Mindfulness.* **12**(Suppl 1):P76 (2013).
39. Rademaker MM, *et al.* The effect of mindfulness-based interventions on tinnitus distress: A systematic review. *Frontiers in Neurology.* **10**:1135 (2019).
40. Jariengprasert C, *et al.* The effectiveness of mindfulness-based cognitive therapy in the management of tinnitus. *International Tinnitus Journal.* **26**(1):1–10 (2022).
41. http://mindfulnessinschools.org/research/research=evidence-mindfulness-schools-project/.
42. Kuyken W, *et al.* Effectiveness of the mindfulness in schools programme: Non-randomised controlled feasibility study. *British Journal of Psychiatry.* **203**(2):126–131 (2013).
43. Smith H, *et al.* An evaluation of paediatric tinnitus services in UK National Health service audiology departments. *BMC Health Services Research.* **20**(1):214 (2020).
44. Kentish R and Crocker S. Scary monsters and waterfalls: Tinnitus narrative therapy with children. In Tyler R, (Ed.) *Tinnitus Treatment Clinical Protocols*, Thieme Medical Publishers (2006).
45. Kentish R. Drawing pictures and telling stories: Treating tinnitus in childhood. *ENT and Audiology News.* **22**(6):95–96 (2014).
46. Henry JA, *et al.* Using therapeutic sound with progressive audiologic tinnitus management. *Trends in Amplification.* **12**(3):188–209 (2008).
47. http://www.evidence.nhs.uk/topic/tinnitus. Published 2011.
48. Hoare DJ, *et al.* Systematic review and meta-analyses of randomized controlled trials examining tinnitus management. *Laryngoscope.* **121**:1555–1564 (2011).
49. Hobson J, *et al.* Sound therapy (masking) in the management of tinnitus in adults. *Cochrane Database of Systematic Reviews.* **11**:CD006371 (2012).
50. Liu H, *et al.* Efficacy of sound therapy interventions for tinnitus management: A protocol for systematic review and network meta-analysis. *Medicine (Baltimore).* **100**(41):e27509 (2021).
51. Bartnik G, *et al.* Troublesome tinnitus in children: Epidemiology, audiological profile, and preliminary results of treatment. *International Journal of Pediatrics.* **2012**:945356 (2012).

52. Jastreboff PJ, *et al.* In *Tinnitus Retraining Therapy: Implementing the Neurophysiological Model*. New York, NY: Cambridge University Press (2004).
53. Jastreboff PJ, *et al.* Tinnitus retraining therapy. In Snow JB, (Ed.) *Tinnitus: Theory and Management*, Hamilton, London: BC Decker, pp. 310–313 (2004).
54. Jastreboff PJ. 25 Years of tinnitus retraining therapy. *HNO.* **63**(4):307–311 (2015).
55. Tyler RS, *et al.* Tinnitus retraining therapy: Mixing point and total masking are equally effective. *Ear and Hearing.* **33**:588–594 (2012).
56. Phillips JS, and McFerran D. Tinnitus retraining therapy (TRT) for tinnitus. *Cochrane Database of Systematic Reviews.* **2010**(3):CD007330 (2010).
57. Henry JA, *et al.* Outcomes of clinical trial: Tinnitus masking versus tinnitus retraining therapy. *Journal of the American Academy of Audiology.* **17**:104–132 (2006).
58. Piccirillo JF, *et al.* Tinnitus. *JAMA.* **323**(15):1497–1498 (2020).
59. Grewal R, *et al.* Clinical efficacy of tinnitus retraining therapy and cognitive behavioural therapy in the treatment of subjective tinnitus: A systematic review. *Journal of Laryngology and Otology.* **128**:1028–1033 (2014).
60. Han M, *et al.* Efficacy of tinnitus retraining therapy in the treatment of tinnitus: A meta-analysis and systematic review. *American Journal of Otolaryngology.* **42**(6):103151 (2021).
61. Dullaart MJ, *et al.* Treatment of tinnitus in children — A systematic review. *Frontiers in Neurology.* **12**:726803 (2021).
62. Lee K, *et al.* Evaluation of tinnitus retraining therapy for patients with normal audiograms versus patients with hearing loss. *Auris Nasus Larynx.* **45**(2):215–221 (2018).
63. Simonetti P, *et al.* Hearing aid effectiveness on patients with chronic tinnitus and associated hearing loss. *Brazilian Journal of Otorhinolaryngology.* **88**(Suppl 3): S164–S170 (2022).
64. Lee HJ, *et al.* Hearing aid effects and satisfaction in patients with tinnitus. *Journal of Clinical Medicine.* **11**(4):1096 (2022).
65. Yokota Y, *et al.* Retrospective evaluation of secondary effects of hearing aids for tinnitus therapy in patients with hearing loss. *Auris Nasus Larynx.* **47**(5):763–768 (2020).
66. Kikidis D, *et al.* Hearing aid fitting in tinnitus: A scoping review of methodological aspects and effect on tinnitus distress and perception. *Journal of Clinical Medicine.* **10**(13):2896 (2021).
67. Hoare DJ, *et al.* Amplification with hearing aids for patients with tinnitus and co-existing hearing loss. *Cochrane Database of Systematic Reviews.* **2014**(1):CD010151 (2014).

68. Ramakers GG, *et al.* The effect of cochlear implantation on tinnitus in patients with bilateral hearing loss: A systematic review. *Laryngoscope.* **125**(11):2584–2592 (2015).
69. Ramakers GGJ, *et al.* Tinnitus after simultaneous and sequential bilateral cochlear implantation. *Frontiers in Surgery.* **4**:65 (2017).
70. Levy DA, *et al.* Cochlear implantation for treatment of tinnitus in single-sided deafness: A systematic review and meta-analysis. *Otolaryngology-Neurotology.* **41**(8):e1004–e1012 (2020).
71. Peter N, *et al.* The influence of cochlear implantation on tinnitus in patients with single-sided deafness: A systematic review. *Otolaryngology-Head and Neck Surgery.* **161**(4):576–588 (2019).
72. Idriss SA, *et al.* Short- and long-term effect of cochlear implantation on disabling tinnitus in single-sided deafness patients: A systematic review. *Journal of Clinical Medicine.* **11**(19):5664 (2022).
73. Assouly KKS, *et al.* Cochlear implantation for patients with tinnitus: A systematic review. *Progress in Brain Research.* **260**:27–50 (2021).
74. Assouly KKS, *et al.* Analysis of a cochlear implant database: changes in tinnitus prevalence and distress after cochlear implantation. *Trends in Hearing.* **26**:23312165221128431 (2022).
75. Chadha NK, *et al.* Tinnitus is prevalent in children with cochlear implants. *International Journal of Pediatric Otorhinolaryngology.* **73**:671–675 (2009).
76. Langguth B, *et al.* Current pharmacological treatments for tinnitus. *Expert Opinion on Pharmacotherapy.* **13**(17):2495–2509 (2012).
77. Coelho C, *et al.* Survey on the effectiveness of dietary supplements to treat tinnitus. *American Journal of Audiology.* **25**(3):184–205 (2016).
78. Lee A, Abouzari M, Akbarpour M, Risbud A, Lin HW, and Djalilian HR. A proposed association between subjective nonpulsatile tinnitus and migraine. *World Journal of Otorhinolaryngology-Head and Neck Surgery.* **9**(2):107–114 (2022).
79. Umemoto KK, Tawk K, Mazhari N, Abouzari M, and Djalilian HR. Management of migraine-associated vestibulocochlear disorders. *Audiology Research.* **13**(4):528–545 (2023).

https://doi.org/10.1142/9789819807345_0015

Chapter 15

Stem Cell Therapy for Tinnitus

Ghaidaa S. Khlaifat*, **Karen Tawk***, **and Mehdi Abouzari**†

Department of Otolaryngology-Head and Neck Surgery, University of California, Irvine, CA, USA

†*mabouzar@hs.uci.edu*

Abstract

Hearing loss is frequently linked to tinnitus. It is believed that damage to the auditory system, such as hair cell loss in the inner ear, can lead to hearing loss and tinnitus. Despite advances in noninvasive and invasive neuromodulation techniques for tinnitus management, a permanent solution remains elusive. During the early stages of life and growth, stem cells possess an exceptional ability to differentiate into a wide spectrum of cell types within the body. They can also divide and differentiate to replenish other cells, serving as the body's built-in repair system. One of the key advantages of stem cell treatment is its capability to provide long-lasting and possibly permanent cures for diseases and injuries that were formerly considered untreatable. Stem cell therapy has become a promising strategy for managing tinnitus, providing the potential to recover impaired hair cells and auditory neurons. with promising results in preclinical studies showing the ability of stem cells to repair damaged auditory pathways and reduce tinnitus. Ongoing research in stem cell biology, coupled with precise delivery techniques and thorough clinical

---

*These authors contributed equally to this manuscript.

trials, continues to advance the potential role of stem cell therapy as a groundbreaking approach for alleviating tinnitus. To further improve their therapeutic potential, stem cells may be used in conjunction with growth factors and gene editing methods. This emerging therapeutic approach offers hope for those who suffer from chronic and debilitating tinnitus, potentially enhancing the quality of life for countless patients. Nevertheless, many challenges and ethical concerns associated with stem cell therapy still need to be addressed, such as the risk of immune rejection, tumor formation, and the ethical implications of using embryonic stem cells.

Introduction

Hearing loss is widely recognized as a major risk factor for acquiring tinnitus [1]. The search for a permanent solution for tinnitus continues despite the presence of various treatment options, such as sound therapy, cognitive behavioral therapy, and pharmaceutical interventions. Due to the complexity of the auditory system, hearing loss may arise from any anatomical or physiological abnormality along the hearing pathway [2]. The majority of individuals with irreversible hearing loss have malfunctioning cochlear cells (HCs) [3, 4].

The Science Behind Stem Cells

In humans, cochlear HCs fail to regenerate after damage or dysfunction [5]. In the mature organ of Corti, once an HC dies, it is not replaced. Instead, the loss of a hair cell leads to rapid alterations in the morphology of the surrounding cells to rapidly expand and seal the reticular lamina, the barrier separating endolymph from perilymph, which results from the loss of an HC [3, 4]. Therefore, injury to these cell types results in irreversible loss of hearing. In contrast, birds and other vertebrates have demonstrated the ability to continuously produce and/or replace HCs, maintaining auditory function throughout their lifespans [6–8]. In the early 2000s, the successful isolation of stem cells (SCs) from the organ of Corti and the sensory vestibular apparatus in infant mice pointed to the potential use of these endogenous stem cells for restoring damaged HCs [9].

While HCs cannot regenerate in mammals, other mammalian organs, such as the liver, skin, and blood cells are regularly regenerated.

Tissue-specific SCs have been identified in these organs [10–12]. These findings highlighted the significant potential for using SCs as therapeutic agents [13, 14]. However, the regenerative ability of each of these SC populations varies widely. The application of SCs to regenerate injured tissues represents one of the most promising areas of regenerative medicine. Bone marrow hosts a population of hematopoietic stem cells (HSCs), derived from early embryonic stem cells, which differentiate into various types of blood cells throughout life [15].

Two key characteristics of SCs are self-renewal, allowing them to undergo endless cycles of cell division and differentiation, enabling them to transform their phenotype and differentiate into a more specialized cell type [16]. This differentiation capability has been extensively studied in efforts to regenerate auditory neurons and HCs. Primarily, three types of stem cells were used in these regeneration studies: (i) embryonic stem cells (ESCs), (ii) adult stem cells (ASCs), and (iii) induced pluripotent stem cells (iPSCs) (Figure 1) [17–19]. ESCs are isolated from the inner cell mass of the mammalian blastocyst. These cells have the remarkable ability to proliferate indefinitely both *in vitro* and *in vivo*, and they can differentiate into various types of tissue originating from the three primary layers of cells in the embryo, making them pluripotent [17]. However, human ESCs are associated with numerous challenges and ethical concerns, such as potential immune rejection by the recipient and the ethical implications of embryo destruction s for their isolation [20].

Regeneration of Hair Cells

The importance of HC regeneration research has been highlighted by the recent discovery of cochlear progenitor cells in the neonatal mammalian ear [21]. When provided with the appropriate promoting factors, cochlear progenitor cells exhibited the ability to differentiate into new HCs [21]. Studies of inner ear development have identified numerous genes, such as growth factors, notch activation molecules, transcription regulatory factors, and cell division factor-dependent kinase inhibitors, whose expression products can regulate HC differentiation [22]. Among these progenitor cells, Lgr5+ cells are currently under investigation as potential precursors for HC regeneration [23, 24]. Lgr5, a recognized marker of ASC, has been used to confirm that supporting cells serve as progenitors of HCs. Isolated Lgr5+ supporting cells from neonatal mice have shown the ability to generate self-renewing neurospheres and differentiate into myo7a+HCs [25].

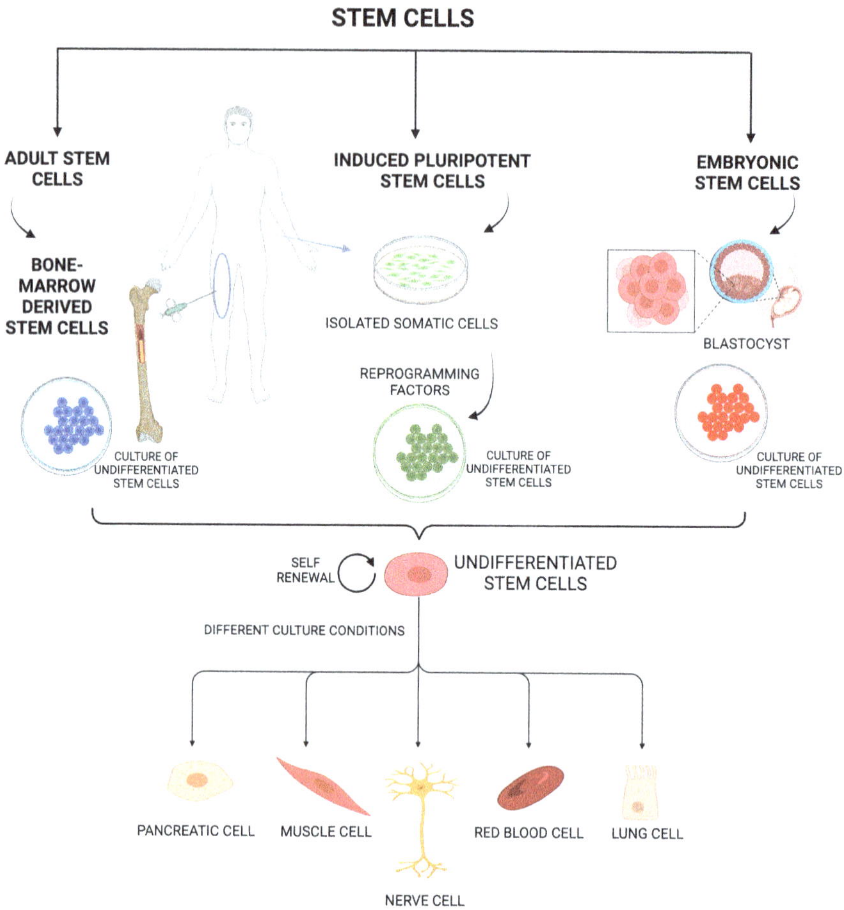

Figure 1. The three primary types of stem cells used in regeneration studies: adult stem cells (ASCs), induced pluripotent stem cells (iPSCs), and embryonic stem cells (ESCs). ASCs, commonly recognized as somatic stem cells, are found in different tissues, such as bone marrow, and are solely responsible for preserving and restoring the tissue in which they reside. iPSCs are generated by reprogramming adult somatic cells to revert to a pluripotent state, thus contracting the ability to develop into any cell type, like ESCs, but without the accompanying ethical concerns. ESCs originate from the inner cell mass of the blastocyst and possess pluripotent capacity, allowing them to differentiate into any cell type from the three germ layers. These stem cell types each offer distinct potential and challenges in the realm of therapeutic applications and regenerative medicine.

Research indicated that the regulation of the Notch and Wnt signaling pathways, which control cell proliferation, enabled Lgr5+ cells to differentiate into HCs [23, 24].

In recent years, SC therapy has emerged as a promising new frontier in the pursuit of effective tinnitus treatments [26]. Having the ability to reduce neuroinflammation and oxidative stress, both of which are related to the pathophysiology of tinnitus [27, 28]. Although the development of SC differentiation and transplantation is still in its early stages, positive future outcomes could potentially aid in the restoration of hearing function [26]. Consequently, this may reduce the risk of tinnitus in patients with permanent hearing loss. According to a review of recent research, SC transplantation and *in situ* growth factor injection were two viable strategies for the regeneration and differentiation of cochlear tissues. These approaches appeared to stimulate the development of resident cochlear cells through various pathways, rather than directly repairing dead cells [29].

Stem-Cell Transplantation

In the past decade, the utilization of exogenous stem cell transplantation to replace damaged or regenerate new spiral ganglion neurons (SGNs) has achieved significant traction. This growing field focuses on transplanting SCs to recover injured inner ear cells and regenerate hearing function, addressing the inability of mammalian HCs and SGNs to regenerate after cell death caused by trauma, disease, or genetic mutations [30, 31]. To a certain extent, the therapeutic outcome of cochlear implants relies on the quality and quantity of residual SGNs [32].

Several techniques for SGN regeneration have been explored. One promising study demonstrated that ESCs could differentiate into glutamatergic neurons *in vitro* following treatment with glial cell line-derived neurotrophic factor (GDNF) and brain-derived neurotrophic factor (BDNF) [33]. Remarkably, these findings were also confirmed *in vivo*, with 50% to 70% of the transplanted ESCs expressing Tuj1 and glutamatergic neural markers [34]. Other research has investigated the potential use of mesenchymal stem cells, dorsal root ganglion (DRG) cells, and neural progenitor cells as candidates for transplantation [35–37]. Nevertheless, these studies have been carried out *in vitro*, and their capabilities *in vivo* have yet to be established.

Challenges and Considerations

The procedure of transplanting SCs into the cochlea is a complex and challenging process, requiring careful consideration of numerous factors, including the types of SCs to be used, their differentiation status, transplantation methods and sites, and different surgical approaches [22]. Furthermore, the cochlear environment must be taken into account, as its ionic conditions significantly impact the likelihood that transplanted cells will survive [38]. For instance, high concentrations of endolymphatic potassium have been shown to activate necrosis and apoptosis in transplanted SCs [39]. Additionally, the cochlea's complex environment renders it difficult to precisely control the proliferation of certain SCs, posing risks of lesion and cancer cell formation [40]. Research has shown that the CRISPR/Cas9 gene-editing technique can improve the accuracy of SC regulation *in vivo*. In animal models, this technology has been utilized to treat autosomal dominant hearing loss [40].

The differentiation of SCs into HCs and auditory neurons is highly successful, achieving rates of up to 82% and 100%, respectively. These differentiated cells can form new synapses and interact with the HCs and auditory neurons of cochlear explants [41]. Combining SC therapies with precise genetic editing procedures represents the future of auditory HC regeneration research [26]. Zheng *et al.* conducted a promising work wherein they extracted cochlear SCs from newborn rats and introduced them into the scala tympani of gentamicin-deafened rats [42]. The transplanted HC-like cells migrated from the injection site to the basal membrane and organ of Corti, resulting in hearing recovery [42]. This breakthrough paves the way for potential clinical applications in treating hearing loss in humans.

While preclinical experiments have demonstrated promising outcomes, translating these results into clinical applications presents substantial challenges, including ensuring the precise delivery of SCs to target tissues, limiting the risk of immune rejection, and ensuring long-term efficacy and safety.

Conclusion

- Hearing loss is a primary risk factor for acquiring tinnitus, which can originate from any dysfunction along the auditory pathway.
- The most recent advances in tinnitus management modalities are promising, but there is a lack of strong evidence, partly due to the heterogeneity of tinnitus.

- Even though noninvasive strategies are simpler, less expensive, and pose fewer risks than invasive treatments, invasive procedures can achieve long-term therapeutic effects with low participant/patient engagement.
- Hearing loss in humans is irreversible due to the inability of cochlear HCs to recover following damage or dysfunction. Conversely, birds and other vertebrates can endlessly generate and/or replace HCs and preserve the auditory function.
- The recent discovery of cochlear progenitor cells in the neonatal mammalian ear has emphasized the importance of HC regeneration research. In the right conditions, cochlear progenitor cells can differentiate into new HCs.
- SCs have two key features: self-renewal and differentiation, which hold promise for limiting neuroinflammation and oxidative stress, both of which are linked to the pathophysiology of tinnitus.
- Although the application of SC therapy in hearing restoration is still in its early stages. this treatment approach offers great potential in restoring hearing loss when combined with genetic editing techniques.
- With the advancement of inner ear SC therapy, it is becoming more likely that SC therapy will replace surgery and hearing aids as the main treatment option for sensorineural hearing loss.
- SCs can be targeted to the cochlea, where they may stimulate the regeneration of HCs, auditory neurons, and synaptic connections disrupted by noise-induced or age-related hearing loss, which are common causes of tinnitus.
- The process of transplanting SCs into the cochlea is complicated, demanding consideration of SC types, differentiation status, and transplantation methods. Additionally, the cochlear environment makes it challenging to precisely control stem cell proliferation, carrying risks of cancer cell formation.
- Promising research has demonstrated that transplanted cochlear SCs can migrate and aid in hearing recovery. Positive results from SC differentiation and transplantation could hopefully regenerate hearing function and reduce tinnitus risk.

References

1. Baguley D, McFerran D, and Hall D. Tinnitus. *Lancet Lond Engl.* **382**:1600–1607 (2013).
2. Mittal R. *et al.* Indispensable role of ion channels and transporters in the auditory system. *Journal of Cellular Physiology.* **232**:743–758 (2017).

3. Raphael Y. Cochlear pathology, sensory cell death and regeneration. *British Medical Bulletin.* **63**:25–38 (2002).
4. Lenoir M, *et al.* Morphological and molecular changes in the inner hair cell region of the rat cochlea after amikacin treatment. *Journal of Neurocytology.* **28**:925–937 (1999).
5. Jacob S, Johansson C, and Fridberger, A. Noise-induced alterations in cochlear mechanics, electromotility, and cochlear amplification. *Pflugers Archiv: European Journal of Physiology.* **465**:907–917 (2013).
6. Corwin JT. Postembryonic production and aging in inner ear hair cells in sharks. *Journal of Comparative Neurology.* **201**:541–553 (1981).
7. Corwin JT. Perpetual production of hair cells and maturational changes in hair cell ultrastructure accompany postembryonic growth in an amphibian ear. *Proceedings of the National Academy of Sciences USA.* **82**:3911–3915 (1985).
8. Corwin JT, and Cotanche, D. A. Regeneration of sensory hair cells after acoustic trauma. *Science.* **240**:1772–1774 (1988).
9. Oshima K, *et al.* Differential distribution of stem cells in the auditory and vestibular organs of the inner ear. *Journal of the Association for Research in Otolaryngology (JARO).* **8**:18–31 (2007).
10. Kim CFB, *et al.* Identification of bronchioalveolar stem cells in normal lung and lung cancer. *Cell.* **121**:823–835 (2005).
11. Landgren H, and Curtis MA. Locating and labeling neural stem cells in the brain. *Journal of Cellular Physiology.* **226**:1–7 (2011).
12. Schwartz RE, and Verfaillie C. Hepatic stem cells. *Methods in Molecular Biology.* **640**:167–179 (2010).
13. Hanna JH, Saha K, and Jaenisch R. Pluripotency and cellular reprogramming: Facts, hypotheses, unresolved issues. *Cell.* **143**:508–525 (2010).
14. Saha K, and Jaenisch R. Technical challenges in using human induced pluripotent stem cells to model disease. *Cell Stem Cell.* **5**:584–595 (2009).
15. Huang X, Cho S, and Spangrude GJ. Hematopoietic stem cells: Generation and self-renewal. *Cell Death and Differentiation.* **14**:1851–1859 (2007).
16. Biehl JK, and Russell B. Introduction to stem cell therapy. *Journal of Cardiovascular Nursing.* **24**:98–105 (2009).
17. Hu Z, and Ulfendahl M. The potential of stem cells for the restoration of auditory function in humans. *Regenerative Medicine.* **8**:309–318 (2013).
18. Peng T, Dong Y, Zhu G, and Xie D. Induced pluripotent stem cells: Landscape for studying and treating hereditary hearing loss. *Journal of Otology.* **10**:295 (2015).
19. Clarke DL, *et al.* Generalized potential of adult neural stem cells. *Science.* **288**:1660–1663 (2000).
20. Volarevic V, *et al.* Ethical and safety issues of stem cell-based therapy. *International Journal of Medical Sciences.* **15**:36–45 (2018).

21. Chen J and Streit A. Induction of the inner ear: Stepwise specification of otic fate from multipotent progenitors. *Hearing Research.* **297**:3–12 (2013).
22. Mittal R, *et al.* Recent advancements in the regeneration of auditory hair cells and hearing restoration. *Frontiers in Molecular Neuroscience.* **10**:236 (2017).
23. Bramhall NF, Shi F, Arnold K, Hochedlinger K, and Edge ASB. Lgr5-positive supporting cells generate new hair cells in the postnatal cochlea. *Stem Cell Reports.* **2**:311–322 (2014).
24. Chai R, *et al.* Wnt signaling induces proliferation of sensory precursors in the postnatal mouse cochlea. *Proceedings of the National Academy of Sciences USA.* **109**:8167–8172 (2012).
25. Shi F, Kempfle JS, and Edge ASB. Wnt-responsive Lgr5-expressing stem cells are hair cell progenitors in the cochlea. *Journal of Neuroscience.* **32**:9639–9648 (2012).
26. White PM. Perspectives on human hearing loss, cochlear regeneration, and the potential for hearing restoration therapies. *Brain Sciences.* **10**:756 (2020).
27. Haider HF, *et al.* Tinnitus, hearing loss, and inflammatory processes in an older Portuguese population. *International Journal of Audiology.* **59**:323–332 (2020).
28. Angeloni C, Gatti M, Prata C, Hrelia S, and Maraldi T. Role of mesenchymal stem cells in counteracting oxidative stress-related neurodegeneration. *International Journal of Molecular Sciences.* **21**:3299 (2020).
29. Simoni E, *et al.* Regenerative medicine in hearing recovery. *Cytotherapy.* **19**:909–915 (2017).
30. Ciorba A, Astolfi L, and Martini A. Otoprotection and inner ear regeneration. *Audiology and Medicine.* **6**:170–175 (2009).
31. Tan BTG, Lee MMG, and Ruan R. Bone-marrow-derived cells that home to acoustic-deafened cochlea preserve their hematopoietic identity. *Journal of Comparative Neurology.* **509**:167–179 (2008).
32. Kujawa SG, and Liberman MC. Adding insult to injury: Cochlear nerve degeneration after "temporary" noise-induced hearing loss. *Journal of Neuroscience.* **29**:14077–14085 (2009).
33. Reyes JH, *et al.* Glutamatergic neuronal differentiation of mouse embryonic stem cells after transient expression of neurogenin 1 and treatment with BDNF and GDNF: *In vitro* and *in vivo* studies. *Journal of Neuroscience.* **28**:12622–12631 (2008).
34. Coleman B, *et al.* Fate of embryonic stem cells transplanted into the deafened mammalian cochlea. *Cell Transplantation.* **15**:369–380 (2006).
35. Bas E, *et al.* Adult human nasal mesenchymal-like stem cells restore cochlear spiral ganglion neurons after experimental lesion. *Stem Cells and Development.* **23**:502–514 (2014).

36. Hu Z, Ulfendahl M, Prieskorn DM, Olivius P, and Miller JM. Functional evaluation of a cell replacement therapy in the inner ear. *Otolaryngology and Neurotology*. **30**:551–558 (2009).
37. Chen J, *et al.* Transplantation of mouse-induced pluripotent stem cells into the cochlea for the treatment of sensorineural hearing loss. *Acta Otolaryngologica*. **137**:1136–1142 (2017).
38. Matsuoka AJ, Kondo T, Miyamoto RT, and Hashino E. Enhanced survival of bone-marrow-derived pluripotent stem cells in an animal model of auditory neuropathy. *The Laryngoscope*. **117**:1629–1635 (2007).
39. Wang M, Qiu J, Mi W, Wang F, and Qu J. *In vitro* effect of altering potassium concentration in artificial endolymph on apoptosis and ultrastructure features of olfactory bulb neural precursor cells. *Neuroscience Letters*. **487**:383–388 (2011).
40. Gao X, *et al.* Treatment of autosomal dominant hearing loss by *in vivo* delivery of genome editing agents. *Nature*. **553**:217–221 (2018).
41. Dufner-Almeida LG, *et al.* Stem-cell therapy for hearing loss: Are we there yet? *Brazilian Journal of Otorhinolaryngology*. **85**:520–529 (2019).
42. Zheng H, Tai H, and Zeng S. Cochlear stem cell transplantation for hearing recovery in a rat model of sensorineural hearing loss. *Neural Regeneration Research*. **5**:833–837 (2010).

https://doi.org/10.1142/9789819807345_bmatter

Index

S

www.ingramcontent.com/pod-product-compliance
Lightning Source LLC
Chambersburg PA
CBHW070832230725
29833CB00014B/19
* 9 7 8 9 8 1 9 8 0 7 3 3 8 *